Technical Progress on Sustainable Hydropower Development and Roller Compacted Concrete Dams

Jia Jinsheng, Zhou Jianping, Jose POLIMON,
Wu Gaojian, Xiang Jian, Chen Mao

黄河水利出版社
·Zhengzhou China·

图书在版编目(CIP)数据

水电可持续发展与碾压混凝土坝建设的技术进展 = Technical Progress on Sustainable Hydropower Development and Roller Compacted Concrete(RCC) Dams Co:英文/贾金生等主编.—郑州:黄河水利出版社,2015.8
 ISBN 978-7-5509-1220-5

Ⅰ.①水… Ⅱ.①贾… Ⅲ.①水利水电工程-可持续发展-中国-学术会议-文集-英文②碾压土坝-混凝土坝-水利工程-学术会议-文集-英文 Ⅳ.①TV-53

中国版本图书馆 CIP 数据核字(2015)第 204487 号

出 版 社:黄河水利出版社
　　　　地址:河南省郑州市顺河路黄委会综合楼 14 层　　邮政编码:450003
发行单位:黄河水利出版社
　　　　发行部电话:0371-66026940、66020550、66028024、66022620(传真)
　　　　E-mail:hhslcbs@126.com
承印单位:河南省瑞光印务股份有限公司
开本:787 mm×1 092 mm　1/16
印张:43.75
字数:1092 千字　　　　　　　　　　　　　印数:1—1 000
版次:2015 年 9 月第 1 版　　　　　　　　印次:2015 年 9 月第 1 次印刷

Price:180.00 US $

Organizations

Sponsors

Chinese National Committee on Large Dams (CHINCOLD)

Spanish National Committee on Large Dams (SPANCOLD)

Organizers

Dadu River Hydropower Development Co., LTD.

Chengdu Engineering Corporation Limited

Sinohydro Bureau 5 Co., LTD.

Sinohydro Bureau 7 Co., LTD.

Sinohydro Bureau 10 Co., LTD.

China Institute of Water Resources and Hydropower Research(IWHR) Sichuan University

Co - Sponsors

International Commission on Large Dams (ICOLD)

State Grid Corporation of China

China Three Gorges Corporation

China Huaneng Group

China Datang Corporation

China Huadian Corporation

Yalong River Hydropower Development Company, LTD.

Kunming Engineering Corporation Limited

Yellow River Engineering Consulting Co., LTD.

China Huadian Corporation Sichuan Company

Organizing Committee

Chairmen

WANG Shucheng	Former Minister, Ministry of Water Resources P. R. China
J. Polimón	Vice President of ICOLD and President of SPANCOLD

Vice – Chairmen

KUANG Shangfu	President of IWHR and Vice President of CHINCOLD
JIA Jinsheng	Honorary President of ICOLD
	Vice President and Secretary General of CHINCOLD
	Vice President of IWHR
ZHANG Qiping	Chief Engineer, State Grid Corporation of China
	Vice President of CHINCOLD
ZHOU Hougui	Vice President, China Energy Engineering Co., LTD.
	Vice President of CHINCOLD
QU Bo	Chief Engineer, China Datang Corporation
	Vice President of CHINCOLD

Members (Name list in alphabet)

CHAI Fangfu	DeputyDirector, Hydropower and New Energy Industry of China Huadian Corporation
CHEN Jiankang	Party Secretary, College of Water Resources&Hydropower, Sichuan University
	Executive Director of CHINCOLD
CHEN Yong	Party Secretary, Sinohydro Bureau 10 Co., LTD.
J. C. de Cea	Secretary General of SPANCOLD
LI Wenxue	President, Yellow River Engineering Consulting Co., LTD.
	Executive Director of CHINCOLD

LUO Xiaoqian	General Manager, China Huadian Corporation Sichuan Company
REN Junyou	General Manager Assistant, Sinohydro Bureau 5 Co., LTD.
WANG Renkun	Vice President and Chief Engineer, Chengdu Engineering Corporation Limited
	Director of CHINCOLD
WANG Yongxiang	Chairman, Huaneng Lancang River Hydropower Co., LTD.
WU Shiyong	Deputy General Manager, Yalong River Hydropower Development Company, LTD.
	Executive Director of CHINCOLD
XIANG Jian	Chief Engineer, Sinohydro Bureau 7 Co., LTD.
	Director of CHINCOLD
YAN Jun	Vice President, Dadu River Hydropower Development Co., LTD.
YAN Xinchun	DeputyDirector, Department of Engineering Construction, China Huaneng Group
	Director of CHINCOLD
YANG Jun	Director of News Center, China Three Gorges Corporation
	Vice Secretary General and Director of CHINCOLD
ZHANG Guoxin	Director of Department of Structures and Materials, IWHR
	Vice Secretary General and Director of CHINCOLD
ZHANG Shujun	Deputy Director, Department of Engineering Construction, China Guodian Corporation
	Director of CHINCOLD
ZHANG Zhiqiang	General Manager, Guizhou Wujiang Hydropower Development Co, LTD
ZOU Lichun	Vice President, Kunming Engineering Corporation Limited
	Director of CHINCOLD

Advisory Committee

Chairmen

LU Youmei	Academician, Chinese Academy of Engineering
	Honorary President of CHINCOLD
Adama Nombre	Honorary President of ICOLD

Vice – Chairmen

QIAO Yong	Vice Minister, Ministry of Water Resources P. R. China
	Vice President of CHINCOLD
ZHANG Ye	Vice Director General, South – to – North Water Diversion Project Construction Commission of the State Council Office
	Vice President of CHINCOLD
ZHOU Dabing	Honorary President, China Society for Hydropower Engineering
	Vice President of CHINCOLD
YUE Xi	Vice President of CHINCOLD
YAN Zhiyong	Chairman and General Manager, Power ConstructionCorporation of China
	Vice President of CHINCOLD
LIN Chuxue	Vice President, China Three Gorges Corporation
	Vice President of CHINCOLD
KOU Wei	Deputy General Manager, China Huaneng Group
	Vice President of CHINCOLD
CHENG Niangao	General Manager, China Huadian Corporation
	Vice President of CHINCOLD

ZHANG Zongfu	Chief Engineer, China Guodian Corporation Vice President of CHINCOLD
XIA Zhong	Deputy General Manager, State Power Investment Corporation Vice President of CHINCOLD
XIE Heping	President, Sichuan University Academician, Chinese Academy of Engineering

Members (Name list in alphabet)

CHEN Yunhua	General Manager, Yalong River Hydropower Development Company, LTD.
FENG Junlin	General Manager, Kunming Engineering Corporation Limited.
HE Pengcheng	General Manager, Sinohydro Bureau 5 Co., LTD.
HE Qigang	General Manager, Sinohydro Bureau 10 Co., LTD.
LI Wenpu	Deputy Director, Department of Project Management, ChinaDatang Corporation
SHEN Maoxia	General Manager, Sinohydro Bureau 7 Co., LTD.
TU Yangju	General Manager, Dadu River Hydropower Development Co., LTD. Director of CHINCOLD
ZHANG Jianyue	General Manager, Chengdu Engineering Corporation Limited

Technical Committee

Chairmen

CHEN Houqun	Academician, Chinese Academy of Engineering
	Executive Director of CHINCOLD
Luis Berga	Honorary President of ICOLD

Vice Chairmen

ZHANG Jianyun	President, Nanjing Hydraulic Research Institute
	Academician, Chinese Academy of Engineering
	Vice President of CHINCOLD
NIU Xinqiang	President, Changjiang Institute of Survey, Design and Research
	Academician, Chinese Academy of Engineering
	Vice President of CHINCOLD
ZHONG Denghua	Vice President, Tianjin University
	Academician, Chinese Academy of Engineering
	Executive Director of CHINCOLD
WEI Shanzhong	Deputy Director, Changjiang Water Resources Commission
	Vice President of CHINCOLD
SU Maolin	Deputy Director, Yellow River Conservancy Commission
	Vice President of CHINCOLD
LIU Zhiming	Vice President, Water Resources and Hydropower Planning and Design General Institute
	Vice President of CHINCOLD
LI Sheng	Vice President, China Renewable Energy Engineering Institute
	Vice President of CHINCOLD
ZHOU Jianping	Chief Engineer, Power Construction Corporation of China
	Vice Secretary General and Executive Director of CHINCOLD

Members (Name list in alphabet)

Brasil P. Machado	President, Brazilian Committee on Large Dams
Brian Forbes	Project Manager, Australia GHD Pty Ltd (formerly known as Gutteridge Haskins & Davey)
CHEN Mao	Deputy General Manager and Chief Engineer, Sinohydro Bureau 10 Co., LTD.

DING Guangxin	Director, Department of Engineering Construction, State Grid Corporation of China
Ersan Yildiz Temelsu	Project Manager, International Engineering Consulting Company
F. Ortega	President, Spanish FOSCE Company
GUO Xuyuan	Director, Department of Planning and Development, Yalong River Hydropower Development Company, LTD.
JIANG Changfei	Vice Hydropower Section Chief, Department of Project Management in ChinaDatang Corporation
JING Laihong	Chief Engineer, Yellow River Engineering Consulting Co., Ltd. Director of ICOLD
Joji YANAGAWA	Director, Japan Dam Center Vice President, Japan Commission on Dams
LI Jia	Dean of College of Water Resources & Hydropower, Sichuan University
LIN Peng	Vice Section Chief, Department of Engineering Construction, China Huaneng Group Director of CHINCOLD
Malcolm. R. H. D.	President, Britain Management's Discussion and Analysis Company
Michael ROGERS	Vice President of ICOLD
Michel Lino	Vice President, ICOLD Committee on Cemented Material Dams
R. IbáñezD. A. D.	Director, Technology Department of American Company
WEN Xuyu	Vice Chief Engineer, General Institute of Water Resources and Hydropower Planning & Design Vice Secretary General and Director of CHINCOLD
WU Gaojian	Deputy General Manager and Chief Engineer, Sinohydro Bureau 5 Co., LTD. Director of CHINCOLD
WU Xu	Vice Chief Engineer, Sinohydro Bureau 7 Co., LTD.
WU Yuandong	Hydropower Section Chief, Hydropower and New Energy Industry of ChinaHuadian Corporation
XU Zeping	Professor of IWHR Vice Secretary General and Director of CHINCOLD
YAO Fuhai	Vice Chief Engineer, Dadu River Hydropower Development Co., LTD.
YU Ting	Chief Engineer, Chengdu Engineering Corporation Limited
ZHAI Endi	Professional Chief Engineer, China Three Gorges Corporation
ZHANG Zongliang	Deputy General Manager and Chief Engineer, Kunming Engineering Corporation Limited.
ZHONG Guodong	Deputy General Manager, China Huadian Corporation Sichuan Company

Preface

In order to achieve a better solution worldwide for water, energy and food security, and cope with global climate change in a proactive manner, many countries have stepped up their efforts on construction and rein for cement of reservoirs and dams in recent years. The call for less carbon emission and more clean energy expedites hydropower development, as many countries accelerate domestic hydropower development and draft new planning and objectives so as to promote the coordinated development of the local economy, society and hydropower.

The new Chinese government has launched the new idea on water management and energy strategies, demanding that water saving be top prioritized, the balance between population, economy, resource and environment be maintained, integrated water management be adopted, and both the roles of government and the market be paid importance to safeguard national water security. Energy production and consumption revolution will be promoted, with more focus on non-coal energy, and anenergy strategy that emphasizes "thrift, cleanness and safety" being aggressive enforced, so as to safeguard national energy safety and stability. China's vigorous pursuit of a sustainable hydropower development and energy strategy will make a big difference for its long-term economic growth as well as the security of water, energy and food.

The hydropower has profound economic and social benefits in terms of its flood-prevention, electricity generation, navigation, irrigation functions as well as its ecological functions. As a complicated system engineering, it is hence essential to promote the application of new technologies and materials in order to ensure its high standard performance in construction and management of reservoirs and dams.

Aiming at showcasing the new development in dam construction techniques worldwide, providing a platform for water and hydropower industry to communicate in a wider scope, broader field and higher level, and promoting sustainable development of reservoirs and dams, Chinese National Committee on Large Dams(CHINCOLD) holds successively academic conferences and international symposium on reservoirs and damsin recent years, which are fruitful and highly appreciated by officials and member organizations.

By the virtues of quick construction, low cost and high quality, Roller Compacted Concrete (RCC) dam has been widely developed in many countries. After 30 years of practice and research, the design and construction technology of RCC dams have been considerably improved all over the world. CHINCOLD and Spanish National Committee on Large Dams (SPANCOLD) have jointly organized international symposiums for many years to summarize experience and further promote the development of RCC dams. Following those successful and fruitful symposiums held in Beijing (China) in 1991, Santander (Spain) in 1995, Chengdu (China) in 1999,

Madrid (Spain) in 2003 and Guiyang (China) in 2007, Zaragoza (Spain) in 2012, the 7th symposium will continue to contribute significantly to the knowledge and application of RCC technology with a wide range of contents and international scopes.

It is my honor and privilege to welcome you to the 2015 CHINCOLD Annual Academic Congress/7th International Symposium on RCC Dams holding in Sept. 2015 in Chengdu, China. Theexperts and scholars all over the world will gather to learn from each other.

68 papers have been published and other papers are collected in the official CD Proceeding. Topics of this symposium include:

(1) The key technologies and management for high dam, involving optimal design, construction management, quality control, monitoring, repair and reinforcement.

(2) Technical progress on construction of RCC dams, focusing on materials, mix proportion, structure design, operation, monitoring and new technology application of RCC dams.

I sincerely hope that the publication of this symposium proceeding will lay a solid foundation for success of this conference and provide a valuable reference for decision makers, investors, designers, researchers and engineers engaged in water and hydropower industry.

This congress co-hosted by CHINCOLD and SPANCOLD is jointly organized by Dadu River Hydropower Development Co., LTD., Chengdu Engineering Corporation Limited, Sinohydro Bureau 5 Co., LTD, Sinohydro Bureau 7 Co., LTD, Sinohydro Bureau 10 Co., LTD, China Institute of Water Resources and Hydropower Research (IWHR) and Sichuan University, and supported by International Commission on Large Dam (ICOLD), State Grid Corporation of China, China Three Gorges Corporation, China Huaneng Group, China Datang Corporation, China Huadian Corporation, Yalong River Hydropower Development Company, LTD., Kunming Engineering Corporation Limited, Yellow River Engineering Consulting Co., LTD, and China Huadian Corporation Sichuan Company, etc. My thanks goes to all of you.

President of LOC
Sept. 2015
Beijing

Table of Contents

Preface *Wang Shucheng*

Part I: Technology Progress on Roller Compacted Concrete Dams

RCC Dams in Spain: Development, Innovation and International Experience
.... *De Cea, J. C., Ibáñez, R., Polimón, J., Yagüe. J., etc.* (3)
The First 30 Years of RCC Dams .... *Malcolm Dunstan* (19)
A New Test Method for Studying Hydraulic Fracturing Influence on High Concrete
    Gravity Dam Design ...... *Jia JinSheng, Wang Yang, FengWei, Liu ZhongWei, etc.* (32)
Review on RCC Dams in Turkey .... *Dr. Ersan YILDIZ, Dinçer AYDOĞAN* (50)
Immersion Vibrated RCC .... *Ortega F* (58)
The Major Use of Lagoon Ash in RCC Dams
.... *Malcolm Dunstan, Nguyen Hong Ha, David Morris* (69)
Construction of Enciso Dam .... *Allende M, Ortega F, Cañas JM, Navarrete J* (79)
Not all Flyashes are the Same .... *Mark Turpin, Todd Smith, Malcolm Dunstan* (91)
Design and Construction of Beyhan – 1 RCC Dam in Turkey
.... *Ersan YILDIZ, A. Fikret GÜRDİL, Nejat DEMİRÖRS* (102)
The Influence of Size and Layer Face Effects on The Deformation Performance of
    Fully – graded RCC .... *Zhang Sijia, Feng Wei, Ji Guojin* (111)
Update of ICOLD Bulletin 126 on RCC Dams .... *Shaw QHW* (121)
Features of the Russian Technology of Dam Construction with the Use of Roller Compacted
    Concrete .... *Vadim Sudakov, Alfons Pak* (131)
Design and Quality Control of the IV – RCC Mix USED at Enciso Dam in Spain
.... *Allende M, Ortega F, Blas C* (136)
3D Geomechanical Model Test on Global Stability of Lizhou RCC Arch Dam on Muli River
.... *Chen Yuan, Zhang Lin, Yang Gengxin, Yang Baoquan* (147)
How Fast should an RCC Dam be Constructed? .... *Malcolm Dunstan* (158)
Earthquake Pipe – Burst Protection of Dam Penstock by Using a VAG Pipe Burst Safety
    Valve with Integrated Earthquake Sensor .... *Anton Rienmüller* (169)
Raw Materials Selection and Temperature Control Design for Ludila RCC Dam
.... *Ji Peimin, Huang Tianrun, Wang Tianguang, Yang Xinping, etc.* (175)
Geomembranes on RCC Dams: A Case History After 13 Years of Service

·················· *Scuero A，Vaschetti G，Jimenez M J，Cowland J*(189)
Inga Project in Drc – design Considerations to Facilitate the Construction in Two Stages
of the Bundi Dam ·················· *Arnaud ROUSSELIN*(199)
The First two RCC Arch/Gravity Dams in Turkey ·················· *Shaw QHW*(209)
Practical Application of Super High – Volume Flyash Roller – Compacted Concrete
·················· *Liu Xianjiang*(220)
Design and Construction of Cerro Del Águila Gravity Dam in Peru
·················· *SayahS. M. , BiancoV. , Ravelli M. , Bonanni S.*(230)
Two – dimensional (2D) and Three – dimensional Dynamic(3D) Analysis of Bushehr Baghan
Rolled Concrete Dam by Considering Interaction Effects of Dam, Reservoir and Lake
·················· *Majid Gholhaki，Ali Mohammad Mahan Far，SeyedTaherEsmaili，*
Fereidoon Karampoor, etc.(240)
Thermal Stress Analysis for the Lai Chau RCC Gravity Dam
·················· *Marco Conrad，Marko Brusin，David Morris，Pham Van Trong*(257)
Simulation Analysis on the Thermal Field and the Stress Field for Gongguoqiao RCC Dam
·················· *Chen Hongjie，Wang Zhaoying*(268)
Thermal Study and Implemenation of Short Joints at Upper Paunglaung RCC Dam
·················· *U Aye Sann，U Zaw Min San，Christof Rohrer，Marko Brusin*(275)
RCC Construction Aspects and Quality Control of Spring Grove Dam (South Africa)
·················· *Nyakale J，Badenhorst D，Mohale N，Trümpelmann M，etc.*(284)
Kahir RCC Dam Thermal Analysis
·················· *Araghian H. R. , Hajialikhani M. R. ,Jafarbegloo M.*(294)
Design Feature of the Nam Ngiep 1 Hydropower Project
·················· *Makoto ASAKAWA，Mareki HANSAMOTO*(304)
Construction Challenges of the Portugues Arch – Gravity RCC Dam in Puerto Rico
·················· *Rafael Ibáñez – de – Aldecoa，David Hernández，Eskil Carlsson*(317)
The Key Technology of Cement Grouting for Concrete Interlayer Crack Seepage or Leakage
of RCC Dam ·················· *Li Yan，Chen Weilie，Li Geng，Yue Mingtao*(330)
Concrete Production in the Required Quality,Temperature and Output Capacity,Considering
the Different Climatic and Geographic Conditions ·················· *Reinhold Kletsch*(337)
Speed of Construction of Evn's Son La and Lai Chau RCC Dams
·················· *Nguyen Hong Ha，Marco Conrad，David Morris，Malcolm R. H. Dunstan*(348)
Lessons Learnt from Operations of Some RCC Dams ·················· *François Delorme*(357)
Research and Application of Cinder and Ash from Coal – fired Power Plant Used as Admixture
in RCC of Guanyinyan Hydropower Station
·················· *Xu Xu，Yi Junxin，Li Xiaoqun，LiuYingqiang*(369)
The Relationships between the In – situ Tensile Strength Across Joints of an RCC Dam and the
Maturity Factor and Age of Test ·················· *Malcolm Dunstan，Marco Conrad*(376)

Investigation on Pulse Velocity Changes in RCC with Different Cement Content and
Different Types of Admixtures ················· *Shabani, N. , Araghian H. R.* (386)
Conclusion of and Suggestions for Engineering Materials and Test, Application and Practice
for Longtan Roller Compacted Concrete Dam ···················· *Ning Zhong* (394)
The In-situ Properties of the RCC at Lai Chau
············ *Nguyen Hong Ha, Nguyen Pham Hung, David Morris, Malcolm Dunstan* (403)
Stress Meter and Strain Gauge Measurements at Upper Paunglaung RCC Dam
················ *U Maw Thar Htwe, U San Wai, Christof Rohrer, Stuart L. L. Cowie* (415)
The Revival of Dam Building in Afghanistan: Ministry of Energy and Water Islamic Republic
of Afghanistan ··· *Sayed Karim Qarloq* (424)
Analysis of the Running State of Guizhou Guangzhao Roller Compacted Concrete Gravity Dam
··· *Yang Ning'an* (431)
Innovative Technologies for Construction of RCC Arch Dams
·· *Chongjiang Du, Bernhard Stabel* (444)
The Soil Treated Dam (STD) a New Concept of Cemented Dam
·· *Michel Lino, François Delorme, Daniel Puiatti* (461)
A Study of Lift Joint Shearing Strength of RCC Dams
··· *Zhang Jianguo, Zhou Yuefei, Xiao Feng* (472)
Kahir Dam: First FSHD Experience in Iran
····························· *Ebrahim Ghorbani, M. Lackpour, A. Mohammadian* (480)
The Paradox of RCC Research in Developing Countries (Libyan Example)
·· *S. Y. Barony* (493)
The Quality Control of Lai Chau RCC Dam
················· *Nguyen Hong Ha, Marco Conrad, David Morris, Pham Thanh Hoai* (497)
Laboratory Study of Shear and Compressive Strength of Construction Joints in two Ways
Categorization Maturity Factor and Time of Setting in RCC Dams
································ *Sayed Bashir Mokhtarpouryani, Shahram Mahzarnia* (507)

Part II: The Key Technology and Management of High Dam Construction

Rapid Construction Method of High Gravel-soil Core Wall Rockfill Dam of Changheba
Power Station ·· *Wu Gaojian* (519)
Reviewing and Enlightenments of more than one Million Resettlers System Engineering
Management in Three Gorges Project
································ *Liang Fuqing, Sun Yongping, Zhou Hengyong* (534)
Real-time Intelligent Tracking Method of Fresh Concrete Vibration Status
·· *Tian Zhenghong, Bian Ce, Xiang Jian* (543)
Study on a Distressed High CFRD with Face Slab Rupture

.... Xu Yao, Jia Jinsheng, Hao Jutao, Li Rong(552)

Structural Characteristic Analysis on Hardfill Dam Based on Model Test Method
.... Yang Baoquan, Zhang Lin, Chen Yuan, Dong Jianhua, etc. (563)

Technical Features of Lower Reservoir for Panlong Pumped Storage Power Station and
its Design .... Liu Chun, Shi Hanxin, Xia Yueyi, Xie Liang (573)

Durability Test on Concrete Using Fiber-reinforced Air-entraining Fly Ash
.... Zhang Jinshui, Chen Meng, Yang Jingliang(582)

Risk Analysis of Dams during Construction Based on Catastrophe Evaluation Method
.... Ge Wei, Li Zongkun, Li Wei, Guan Hongyan(588)

The Application of Underwater Construction Techniques in the Project of Adding the New
Maintenance Gate Slots in Yaotian Hydropower Station ...... Shan Yuzhu, Chen Ye(595)

Study on Adsorption and Release Characteristics of Phosphorus in the Bottom Sediment of Yellow
River Xiaolangdi Reservoir
.... Deng Congxiang, Zhao Qing, Wu Guangqing, Li Xiaobing(606)

Study on the Relationship between the Mass Concrete Temperature and Cooling Water Parameters
.... Tan Kaiyan, Duan Shaohui, Yu Yi, Hu Shuhong, etc. (612)

Analysis of the No. 6 main Transformer's Temperature Rise in XLD Hydropower Station
.... Cui Peilei, Lu Feng, Zhang Yang, Chen Meng, etc. (620)

Practical Application of Rapid Dam Construction Techniques in Construction of Tingzikou
Hydroproject Dam .... Kong Xikang, Liu Hui(625)

Discussion on Technological Development Direction of Grout Enriched Vibrated RCC
.... Zhang Zhenyu(640)

Development and Application of Digital Huangdeng Dam's Construction Management
Information System ...... Xiang Hong, Yang Mei, Zheng Aiwu, Gong Yongsheng, etc. (645)

I Index and Application in Risk Monitor of Water Conservancy and Hydropower Engineering
.... Zhang Xiangyu(656)

Quality Management by the Owner in Hydropower Project
.... Zhang Jinshui, Chen Meng, Wu Guangqing(663)

The Impact of Primary Frequency Regulation to Henan Grid by Xiaolangdi Power Station
.... Li Yindang, Zheng Wei, Li Yiding, Cheng Chao, etc. (672)

On-line Monitoring of High-speed Flow Sediment Concentration in the Sediment Tunnels
of the Xiaolangdi Project on the Yellow River by Vibration Method
.... Song Shuke, Zhang Jinshui, Ma Zhihua, Xin Xingzhao(678)

Part I: Technology Progress on Roller Compacted Concrete Dams

RCC Dams in Spain: Development, Innovation and International Experience

De Cea, J. C.[1], Ibáñez, R.[2], Polimón, J.[3], Yagüe. J.[4], Berga, L.[5]

(1. SPANCOLD, Spain; 2. DRAGADOS USA, USA;
3. President of SPANCOLD & Vice-president of ICOLD. Spain;
4. Ministry of Agriculture, Food, and Environment, Spain;
5. Honorary President of ICOLD, Spain)

Abstract: Spain was one of the pioneering countries in RCC dams, with Castillblanco de los Arroyos dam built in 1985. Since then Spain has been the European country with the largest RCC dam experience. There are currently 27 RCC dams in operation and one under construction. The extensive RCC experience accumulated along the past 30 years by the dam owners and by Spanish engineering and construction companies has created a vast knowledge that is worthwhile preserving and exporting to other countries.

This paper reviews the main features of the Spanish RCC dams and its long-term performance, summarizing the design philosophies, construction methodologies and their essential characteristics: cross section, main structural aspects and mixtures. Based on this experience, the technology has evolved considerably bringing innovation and improvements in the efficiency and quality of modern RCC dams. *The Technical Guidelines for Dam Safety* published by the Spanish National Committee on Large Dams (SPANCOLD) have been updated in 2012. The main objective of this updating has been the incorporation of latest innovations and recommendations for RCC dam design and construction. Its new content is summarised in this paper. Finally, the paper shows the application of the Spanish experience in other countries

Keywords: Spanish RCC know-how, innovation, guidelines, international experience

1 Introduction

Theconstruction of dams in Spain began in Roman times. Proserpina and Cornalvo dams (2nd Century A. C.), still remain in operation. At the present time there are 1 087 large dams in Spain, 16 under construction, with a total reservoir capacity of about 61 000 hm^3. With this number of large dams Spain occupies the first place among the European countries, and the seventh in the World ranking. A great part of the Spanish dams are concrete dams (72%), mainly due to the good quality of the foundations. When at the beginning of the eighties of the last century the construction of RCC started, Spain had a great activity on dam construction, so this new technology was rapidly implemented in the country, and in 1984 the left abutment of Erizana dam was the first RCC structure completed. Between 1990 and 1994, a total of 14 RCC dams were built. Currently, 27 RCC dams have been completed in the country and one is under construction, Enciso dam. Fig.1 shows the location of RCC dams, and Table 1, their main characteristics.

Compared with other types of dams, RCC dams had several advantages, among which its economy and speed of construction. For those reasons RCC technology had become very popular in Spain compared to traditional concrete dams. The aspect that mostly contributed to this economy was the reduction of the construction time. Construction of Cenza (1993) and more recently La Breña II (2009) have been good examples of what can beachieved as regards the reduction of the construction time of a concrete dam.

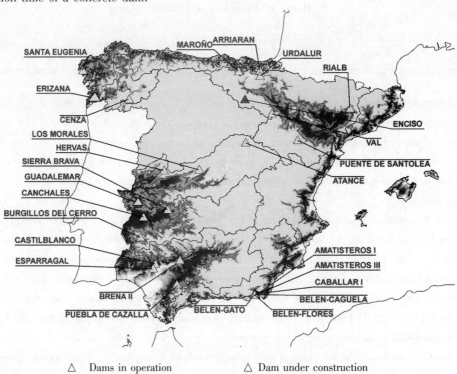

△ Dams in operation △ Dam under construction

Fig.1 Location of RCC dams in Spain

2 Typical cross section

All dams constructed up to now in Spain are gravity dams, with standard cross section designed to facilitate the new technology of continuous placing of the concrete; only in some occasions the triangular profile has been approached to a trapezoidal one. *Hard fill and arch RCC dams* have been constructed by Spanish companies abroad (Moncion dam (Dominican Republic) and El Portugués dam (Puerto Rico), respectively), but none in Spain.

Table 1　Main characteristics of the Spanish RCC dams

Year of Completion	DamName	Height (m)	Crest Length (m)	Reservoir Capacity ($\times 10^6$ m^3)	Slopes ($H:V$)			Concrete Volume ($\times 10^3$ m^3)		Rate RCC/Total (%)
					Upstream	Downstream	Total	CC	RCC	
1984	Erizana (Bayona)	12.0	115.0	0.48	0.10	0.60	0.70	2.0	9.7	82.9
1985	Castillblanco	25.0	124.0	0.87	0.00	0.75	0.75	6.0	14.0	70.0
1988	Los Morales	28.0	200.0	2.84	0.00	0.75	0.75	3.5	22.0	86.3
1988	Sta. Eugenia	83.0	280.0	16.60	0.05	0.75	0.80	29.0	225.0	88.6
1990	Maroño	53.0	182.0	2.23	0.05	0.75	0.80	11.0	80.0	87.9
1990	Hervas	33.0	210.0	0.22	0.15	0.70	0.85	19.0	24.0	55.8
1991	Los Canchales	32.0	240.0	15.00	0.00	0.50~0.80	0.50~0.80	29.0	25.0	46.3
1991	Burguillo del Cerro	24.0	167.0	2.50	0.00	0.60	0.60	8.0	25.0	75.8
1991	Belen Gato	34.0	158.0	0.25	0.25	0.75	1.00	5.0	38.0	88.4
1992	Puebla de Cazalla	71.0	220.0	7.40	0.00~0.20	0.80	0.80~1.00	15.0	205.0	93.2
1992	Belen – Cagüela	31.0	167.0	0.20	0.05	0.75	0.80	5.0	24.0	82.8
1992	Belen Flores	27.0	87.0	0.30	0.05	0.75	0.80	2.0	10.0	83.3
1992	Caballars	16.0	98.0	0.03	0.05	0.75	0.80	1.0	6.0	85.7
1992	Amatisteros I	11.0	91.0	0.03	0.05	0.75	0.80	0.5	3.0	85.7
1992	Amatisteros III	15.0	78.0	0.01	0.05	0.75	0.80	1.0	5.0	83.3
1993	Urdalur	58.0	396.0	5.40	0.00	0.75	0.75	90.0	150.0	62.5
1993	Arriaran	58.0	206.0	3.20	0.05	0.70	0.75	13.0	110.0	89.4
1993	Cenza	49.0	608.0	4.30	0.00	0.75	0.75	8.5	215.0	96.2
1994	Sierra Brava	53.0	800.0	232.00	0.05	0.75	0.80	63.0	277.0	81.5
1994	Guadalemar	13.0	400.0	4.00	1.00	1.00	2.00	5.0	50.0	90.9
1997	Boquerón	58.0	290.0	15.00	0.05	0.73	0.78	8.0	150.0	94.9
1998	Val	94.0	379.0	25.30	0.00~0.02	0.80	0.80~1.00	120.0	630.0	84.0
1998	Atance	45.0	185.0	35.30	0.00	0.80	0.80	6.5	63.0	90.6
2000	Rialb	101.0	604.0	402.00	0.15~0.35	0.40~0.65	0.55~1.00	150.0	1050.0	87.5
2003	Esparragal	20.7	390.5	4.00	0.30	0.90	1.20	5.0	125.0	96.2
2009	La Breña II	119.0	685.0	823.00	0.05~0.30	0.75	0.80~1.05	200.0	1438.0	87.8
2011	Puente de Santolea	35.0	203.0	17.70	0.20	0.60	0.80	5.0	65.0	92.9
Under Construction	Enciso	103.1	375.6	46.50	0.00	0.80	0.80	77.0	641.0	89.3

The most representative dam's cross sections and the evolution of the distribution of the different types of the RCC of some of them are shown in Fig. 2. The experience on constructing RCC dams in Spain during the last 30 years has shown that the design of the dam body should be as simple as reasonably possible (RCC – friendly dam design) taking into account the following points:

(1) In terms of speed of construction, it is preferable to make the whole section with only one RCC very rich in paste (cement plus fly ashes and water); that is to say a true all RCC dam (El Atance, La Breña II, Puente de Santolea and Enciso dams).

(2) The downstream face is designed steeped with a step height equal to a multiple of the layer thickness, which is more consistent with the construction process and facilitates the task of the formwork.

(3) Intakes, outlets and river diversion are now concentrated in one block.

(4) The number of galleries should be limited to a minimum and their construction should not interfere with the process of execution of the concrete. It is important to ensure the construction of at least one perimeter gallery in the dam body if corrective foundation works are needed or if it is necessary to strengthen the impervious or drainage curtain.

(5) The crest's width has been increased to allow the transit of vehicles and machinery without any interference. Overall width of more than 8 m (in large dams usual width of 10 m), will facilitate the construction.

3 Imperviousness

The imperviousness of the Spanish RCC dams has been generally entrusted to the dam body and very specifically to its upstream zone. Until 2 000, two trends were used:

(1) In the first RCC dams it is used a conventional high quality concrete in the upstream face with a minimum width of 1.5 m to wrap up the "water – stop" bands.

(2) Use of two types of different RCC, one of them near the upstream face (usually in a strip of a minimum width of 3 m but increasing with the head of water), of a great quality, water tightness and durability, with a higher paste content and a smaller maximum size of aggregate (M. S. A.).

Also in some cases (Puebla de Cazalla, Cenza and Atance dams) use of a strip of bedding mortar, 80 cm wide, placed between layers near the upstream face. After this methodologies it was usual (in 16 RCC dams) the use of only one RCC type in the whole dam body. This is possible using a high paste content concrete and limiting the M. S. A. This procedure presents some advantages: fast construction, reduced cost and no problems from the point of view of the union between the layers.

During the last decade the GEVR was used. In Esparragal dam Grout – Enriched Vibratable RCC (GEVR) was used for the first time in Spain. It consists in the addition on the top of the previous layer, near the face of the formworks on a strip of 50 cm wide, of a grout with a dosage of about 6 L/m, that allowed to vibrate the RCC. The GEVR improved the impermeability and the final aspect of the face. The excess of paste comes up to the surface easily if the RCC has a good workability. In la Breña II dam instead GEVR Mortar – Enriched Vibratable RCC (MEVR) was

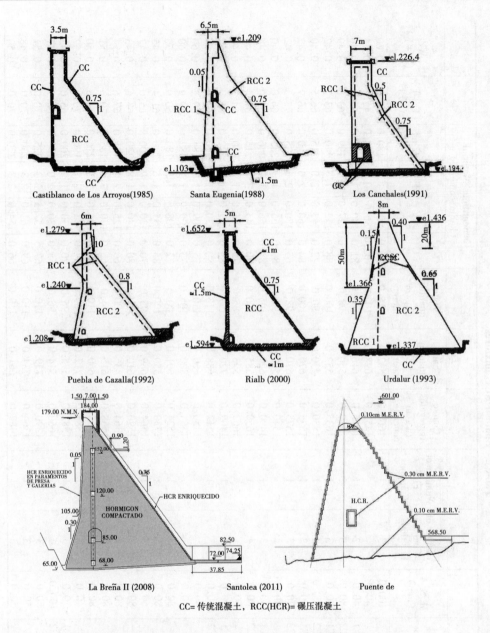

CC = Conventional Concrete, RCC (HCR) = Roller Compacted Concrete

Fig. 2   Representative cross sections of some Spanish RCC dams

Name (Year of construction)

used also with very good results. A new method is being used in Enciso dam construction with a very workable mixture, with a consistency of 8 to 12 seconds Vebe time. In this case RCC could be directly vibrated without any previous enrichment. The preliminary tests carried out at the beginning show the success of this new technology.

4   Materials and mix

Table 2 show the characteristics of aggregates andmixes used in RCC dams in Spain.

Table 2 Spanish RCC mixtures

Year of Completion	DamName	Type	M.S.A (mm)	Number of Coarse Fractions	Number of Fine Fractions	Coarse Aggregates (kg/m³ concrete)	Sand quantity (kg/m³ concrete)	Water quantity (kg/m³ concrete)	Cementitious Material (kg/m³ concrete) C	F	F/C	C+F	F/(C+F) (%)	W/(C+F)
1985	Erizana (Bayona)		100.0	3	1	1668.0	532.0	115.0	90.0	90.0	1.0	180.0	50.0	0.60
1985	Castilblanco		40.0			1452.0	628.0	102.0	102.0	86.0	0.8	188.0	45.7	0.54
1988	Los Morales	RCC1	40.0	2	1	1415.0	616.0	108.0	81.0	140.0	1.7	221.0	63.3	0.46
1988		RCC2	80.0	3	1	1519.0	560.0	98.0	72.0	127.0	1.7	199.0	63.8	0.49
1988	Sta. Eugenia	RCC1	70.0	3	1	1635.0	552.0	100.0	88.0	152.0	1.7	240.0	63.3	0.42
		RCC2	100.0	4		1830.0	430.0	90.0	72.0	143.0	2.0	215.0	66.5	0.40
1990	Maroño	RCC1	70.0	3	1	1575.0	670.0	100.0	80.0	160.0	2.0	240.0	66.7	0.42
		RCC2	70.0	3	1	1575.0	670.0	98.0	65.0	170.0	2.6	235.0	72.3	0.42
1990	Hervas		80.0			1540.0	540.0	95.0	80.0	155.0	1.9	235.0	66.0	0.40
1991	Los Canchales	RCC1	40.0	3	1	1490.0	620.0	105.0	84.0	156.0	1.9	240.0	65.0	0.44
		RCC2	80.0	4	1	1650.0	585.0	100.0	70.0	145.0	2.1	215.0	67.4	0.46
1991	Burguillo del Cerro		60.0			1662.0	593.0	85.0	75.0	135.0	1.8	210.0	64.3	0.40
1992	Puebla de Cazalla	RCC1	40.0	2	1	1409.0	720.0	127.0	85.0	137.0	1.6	222.0	61.7	0.57
		RCC2	80.0	3	1	1512.0	688.0	113.0	80.0	130.0	1.6	210.0	61.9	0.51
1992	Belen – Cagüela		40.0			1450.0	660.0	110.0	75.0	109.0	1.5	184.0	59.2	0.60
1992	Amatisteros I		40.0			1364.0	800.0	105.0	73.0	109.0	1.5	182.0	59.9	0.60
1993	Urdalur		80.0	3	1	1524.0	691.0	90.0	72.0	108.0	1.5	180.0	60.0	0.50
1993	Arriaran		80.0	3	1	1730.0	550.0	100.0	85.0	135.0	1.6	220.0	61.4	0.45
1993	Cenza		60.0	3	1	1564.0	689.0	95.0	70.0	130.0	1.9	200.0	65.0	0.47
1994	Sierra Brava		80.0	3	1	1590.0	610.0	95.0	80.0	140.0	1.8	220.0	63.6	0.43
1994	Guadalemar		80.0	2	1	1364.0	836.0	100.0	60.0	125.0	2.1	185.0	67.6	0.54
1997	Boquerón		80.0	3	2	1568.0	615.0	94.0	55.0	130.0	2.4	185.0	70.3	0.51
1998	Val		80.0	4	2	1552.0	660.0	110.0	80.0	146.0	1.8	226.0	64.6	0.50
1998	Atance		40.0	3	1	1384.0	712.0	109.0	57.0	133.0	2.3	190.0	70.0	0.57
2000	Rialb	RCC1	70.0	3	1	1582.0	575.0	95.0	70.0	130.0	1.9	200.0	65.0	0.43
		RCC2	100.0	4	1	1660.0	514.0	90.0	65.0	130.0	2.0	195.0	66.7	0.47
2003	Esparragal		50.0	3	1	1390.0	739.0	110.0	67.5	157.5	2.3	225.0	70.0	0.51
2009	La Breña II		50.0	3	1	1364.0	734.0	110.0	69.0	161.0	2.3	230.0	70.0	0.48
2011	Puente de Santolea		50.0	3	1	1482.0	661.0	110.0	66.0	154.0	2.3	220.0	70.0	0.50

4.1 Aggregates

In general the M. S. A. has been as maximum 80 mm, only being greater in Erizana, Sta. Eugenia and Rialb dams, 100 mm. But since 2000, in accordance with the current trends, and contrary to the common practice in previous years, the M. S. A. has been reduced to 50 ~ 60 mm for crushed aggregate and about 40 ~ 50 mm for natural gravel, for diminishing segregation, for increasing the workability and for a good finishing of the faces of the dam body.

4.2 Cementitious content and admixtures

In RCC Spanish dams theRCC mixes used have a high cementitious content (high paste content), with an average value of 210 kg/m^3, and a minimum of 180 kg/m^3. The binding material has been generally a mixture of Portland cement and flyash, class F (type silica – aluminous), with a flyash content higher to that of the cement (average, 1.8 times higher). Rock powder filler, as one component more of the RCC, for improving the filling of the voids, has been added in some occasions: in the construction of La Breña II 46 kg (20% of the total) of high quality limestone filler with cementitious properties. Other additions, such as ground granulated blast – furnace slag, have been used in Urdalur dam. In Puebla de Cazalla and La Breña II dams setting retarders were successfully used. Nowadays setting retarders, plasticizers, and water reducers are of a more general use.

4.3 Mix

In Spain, the concrete technology has been used to designthe RCC mixes. The result is a grading for the coarse aggregates with a minimum volume of voids, which subsequently will be filled with the mortar, and also to determine the paste content needed to exceed the volume of voids of the sand. It is preferable very well grading of the aggregates and to measure the consistency of the concrete with the Modified Vebe method. In general, it is recommended that the ratio paste/mortar should be a 10% ~15% higher than the void content of the compacted sand (typically between 0.25 and 0.30), so it is common to design mixes with ratios paste/mortar between 0.35 and 0.45. The excess of paste flows up to the surface once compacted the concrete contributing to improve the bond between successive layers.

In the majority of dams the strength design age has been 90 days but in recent cases (La Breña II dam), it has been used a longer time, 180 days. Additionally, also in recent cases (Esparragal and La Breña II dams), the critical design criterion has been the direct tensile strength of jointed cores across lift joints.

5 Construction

5.1 Mixing, transport and placing

The mixing of the concrete has been carried in batching plants. In some cases with a continuous batching plant, as was the case of Rialb dam.

Conventional trucks with pneumatic tires and high speed belt conveyors as well as their combinations, have been the usual means of transport of the concrete. High speed belt conveyors and interior distribution by trucks was the method used in Sierra Brava, Maroño, Cenza, Val,

Boquerón, Atance andLa Breña II dams, among others. Only by trucks in Los Canchales, Puebla de Cazalla, and Santa Eugenia dams. In Rialb dam an "all – conveyor" system was implemented. In a recent work (Puente de Santolea dam) has been used successfully for the first time in Spain a vacuum chute, which is suitable if the mixture is rich in paste, cohesive and not prone to segregation. The transport of the concrete is an important decision factor to take into account, as it affects to the quality and final cost of the works. The decision should be taken depending on the topography of the site and the concrete volume and production rates.

The compaction of the concrete has been almost always carried out in Spain with self – propelled smooth vibrating tandem rollers from 10 to 16 t. In some dams single drum vibrating rollers have been used (i. e. La Breña II dam). Other light units of some 3 t or pneumatic plates and tampers are used close to exterior faces and contacts with galleries and conduits. For usual, layer thicknesses (about 30 cm when compacted) a number of 4 to 6 passes, back and forth, is normal, with the first and last without vibration and the rest with vibration.

5.2 Faces – formworks

Have been usual in Spain RCC dams with upstream flat faces and downstream stepped. Esparragal dam (2003) was the first dam in the country with a stepped upstream face. After that, Santolea dam has been constructed also with a steeped upstream face.

The first Spanish RCC dams were constructed using conventional climbing formwork, with design details depending on the number and type of the joints, on the rhythm of the construction and on the exposure time. The height of the forms is a multiple of the thickness of the layer (2 to 2.40 m) generating at its upper edge a cold joint. This type of formwork has been especially indicated when the dam had been divided into shuttered blocks, as when the formwork goes up in one of them, the concrete is placed into other. In Cenza Dam a special climbing formwork was used, which allows greater speed of placing and in this manner, the layer could be continuous from side to side avoiding the cold joint. In La Breña II dam this same type of climbing formwork has been used. Also the slip formed curb of concrete has been employed at Sierra Brava and Burguillo dams. Faces without formwork have been carried out in Guadalemar Dam (trapezoidal cross section). In Los Canchales Dam no formwork has been employed on the lower part of the downstream face since it was then covered by an embankment.

5.3 Joints

5.3.1 Transversal joints

In the first RCC dams in Spain all the joints were formworked and they were separated between 40 m to 60 m. The blocks created allowed to place the formworks on the face of one of them, whilst other was being under RCC placement. Later on, the blocks were made longer, and they were carried out in a continuous way from side to side depending on the concrete production and the maximum temperature. In the case of El Esparragal, the dam was constructed through three hyperblocks (large blocks 100 m, 130 m and 150 m wide) executed from the foundation to the crest on a continuous way. The left abutment of La Breña II dam was constructed also in a similar manner (Fig. 3). But in all cases, the blocks have to be divided into other intermediate sub-blocks in order to avoid cracking due to the hydraulic and thermal shrinkage. This division of the blocks is now made inducing a crack on the layer driving in a plate or sheet trough equipment which inserts by vibration a synthetic film or a galvanized sheet.

Whatever the type of transverse joint employed it has to be waterproofed next to the upstream face with one or two "water-stop". Up to the end of the 1990s, in between the water stops one or two conduits are left moulded in the joint, one of them connected to the inspection and drainage gallery. In Atance Dam, an external water-stop was installed on the upstream face, which undoubtedly allows the placing of the RCC with greater speed and quality. Currently these types of joints don't differ from those used in conventional concrete dams (15 m to 20 m width).

Fig. 3 Hyperblocks at Esparragal dam andLa Breña II dam

5.3.2 Horizontal joints between layers

These joints are the most controversial and weak zones on RCC dams. Usually the joints are spaced 0.30 m (thickness of layer), so their number is very high. These joints are classified as hot joints (no treatment needed), cold joints (treatment needed), and warm joints (any treatment if desired). There are several criteria for determining if a joint could be considered as hot, warm or cold. One of them is the Maturity Factor (M. F.), defined as the product of the mean hourly temperature, measured on the surface of the layers in Centigrade degrees, by the time in hours lapsed between the placing of two successive layers: $T(h) \times t$ (℃) = M. F. In the beginning, the M. F. was fixed between (150~250) ℃ × h. With these values it was necessary in many cases to divide the dam in several blocks trough form-worked joints, depending on the size of the facilities and the climate. The experiences in Spain show that this approach should be applied with flexibility. Every dam is a prototype that has a M. F variable with time according to the environmental conditions of temperature and relative humidity. So, the experimental data, obtained on the trial slab, will provide an M. F. to be applied as a practical control of the constructive process. Depending on the cases the M. F. varies between 80 and 300, showing the lack of uniformity. It supposes an exposure time (ET) from 6 to 9 hours (8 hours if $T > 30°$ in the case of Esparragal dam), although, for instance, in Puebla de Cazalla and La Breña II dams, a time of 16 hours to 20 hours was reached due to the use of a setting retarder.

One of the usual questions in RCC dams is the necessity or not of use of a bonding mortar between concrete layers in RCC rich in paste. The Spanish experience in this important question shows that the M. F. approach should be considered as an orientation. It will be the daily practice which will determine the time of exposure; samples obtained by drilling in the trial slab and during the work will provide useful information about the quality of the join.

6 Test sections

It seems mandatory that before the construction of RCC dama trial slab should be carried out, and it has been done in all the Spanish RCC dams. Before starting the placing of the concrete in the dam, a full scale trial should be constructed, on which the data obtained from the laboratory tests are corrected and optimized as well as those others that are imposed by the technical specifications of the project. On the test section the conditions of placing will be tested: faces, thickness of layers, segregation, treatment of joints between layers, etc. It will serve for the personnel of the job site to acquire experience also. In La Breña II dam one test section was constructed, with 12 layers and an approximate volume of 3 000 m^3. From the test sections, cores will be obtained by drilling boreholes to measure concrete densities, bond between layers and "in situ" permeability tests, by filling boreholes with water.

7 Behaviour of Spanish RCC dams

Through more than 30 years of design and construction of RCC in Spain, Spanish engineers have learned many valuable lessons about the behaviour of this type of dams. Some pathology observed could be attributed to the concrete placing and to the numerous joints existing in the dam body. The main defect observed has been the appearance of small cracks, basically, due to sudden temperature changes. But also shrinkage, cooling and gravity may be involved in the genesis of all of these cracks. Table 3 shows the behaviour of some RCC dams with data of measured seepage and cracks. It should be noted that, in general these cracks are facing cracks produced mainly by concrete cooling during long stops. They are not very deep (a few tens of centimetres) and usually the subsequent placement of concrete stops their evolution. In general, we could consider that the behaviour of these dams has been very satisfactory as consequence of the careful and deep studies and thermal analysis carried out. Roller compacted concrete, on the whole, could be considered as highly impermeable, but locally very permeable due to a bad bonding between layers, to segregation, to curing defects, or to long surface exposure to environmental agents. However, despite this, the leakage associated to all of these defects, has been usually very small. Furthermore, the outflow rates are larger during the first filling of the reservoir, decreasing dramatically when water circulates due to a natural sealing of the percolation channels.

Table 3 Basic behaviour of some Spanish RCC Dams

Year of Completion	Name	Heigth(m)	Seepage(l/sg)	Remarks
1988	Sta. Eugenia	83.0	1.5	Some cracks
1990	Maroño	53.0	<1	Some cracks. Repaired
1991	Belen Gato (*)	34.0	—	
1992	Puebla de Cazalla	71.0	20	Repaired. Dam body grouted
1992	Belen – Cagüela (*)	31.0	—	
1992	Belen Flores (*)	27.0	—	
1992	Caballars (*)	16.0	—	
1993	Urdalur	58.0	17	Some cracks. Repaired. Value corresponding to a individual drain hole
1993	Arriaran	58.0	<1	Some cracks
1993	Cenza	49.0	> 45	Repaired. Upstream face waterproofed with epoxy resins
1994	Sierra Brava	53.0	1.7	Without any crack
1994	Guadalemar	13.0	1.5	Cracks every 30 ~ 40 m
1997	Boquerón (*)	58.0	—	
1998	Val	94.0	4.5	Some cracks.
1998	Atance	45.0	1.7 ~ 0.56	Repaired. Injections of the dam body
2000	Rialb	101.0	7.5	One crack. Repaired
2003	Esparragal	20.7	0.1	Without any crack
2009	La Breña II	119.0	>50	Some cracks. Repaired.
2011	Puente de Santolea	35.0	0.25	Without any crack

(*) Flood mitigation dams (empty dams)

8 Technical innovations in recent RCC Spanish dams

The development of high – paste RCC mix extensively used in Spain in the past towards even more workable and super retarded mixes has brought undoubted benefits in the quality of recently completed RCC dams (El Esparragal, Puente de Santolea). In Puente de Santolea dam the initial set criteria was used for the first time to design the degree of retardation of the layers that were placed in 'hot' condition up to an 'exposure' time of 16 hours. These criteria replaced the traditional specification of the maturity factor that was traditionally used. Cores extracted from this dam showed an excellent RCC quality with no visible joints. The dam has proven to be impermeable and the finishes of the faces are excellent (Fig. 4).

At the time of construction of Puente de Santolea dam, the design of mixes for the construction of Enciso dam (103.1 m high) was being developed. This is the second highest RCC dam in Spain (and in Europe) after La Breña II. At the end of June 2015, progress was up to 50% of

Fig. 4 General view of the completed Puente de Santolea RCC dam (2011)

the total RCC volume placed in the dam (nearly 400 000 m³), that were placed in less than 6 months (Fig. 5).

The mix design development carried out prior to the start of construction of Enciso dam was presented in the last International Symposium on RCC dams, held in Zaragoza in 2012. Very workable RCC mixes with VeBe times between 7 and 12 seconds have been used for the construction of this dam. In addition the setting time has been retarded to achieve initial set up to 20 hours. Improvements in the consistency and limitation of the amount of the coarser aggregate sizes have been extremely important measures to minimise segregation. The optimization of the mix design has been achieved following a material specification which considers the quality of fine aggregate a key factor to reduce water demand. Thanks to this optimization the desired consistency has been achieved with a relatively low water content of 93 litres/m³. Core samples from the trial section and from the dam have been tested under compressive and tensile strength at different ages matching the design values. Most of these samples included horizontal joints between layers with 15 to 20 hours exposure time and no treatment, and they show that the joints were absolutely invisible and that a perfect interpenetration of aggregate and paste from both the upper and the lower layers was evident.

Fig. 5 Enciso RCC dam under construction (2015)

Another innovative experience at Enciso dam has been the successful application of the Immersion Vibrated RCC (IV – RCC) asfacing concrete, instead of grout – enriched or mortar – enriched RCC (GEVR and MEVR) methods. This technology has been investigated, tested and its use has been implemented in this project with successful results, being possible to work with the range of relatively low VeBe times, despite the low water content of the RCC mix. (Fig. 6). At Enciso dam, the only use of a different mix than the RCC is the mortar applied as bedding mix at cold joints.

Fig. 6 Immersion vibration and roller compaction of the same RCC mix at Enciso dam

The introduction of IV – RCC implies a great simplification as only one type of mix is fabricated, transported and placed. This allows achieving the objectives of a rapid placement at a very good quality. Due to the high level of retardation of the concrete, depth of immersion vibration is extended beyond the thickness of the layer until at least half of the previous one. This procedure assures that the monolithic vertical construction is guaranteed not only in areas that have been roller compacted but also in those that are consolidated by immersion vibration.

9 Update of the technical guideline of SPANCOLD

An update of the RCC section of the Spanish Guidelines for Dam Safety published by SPANCOLD was carried out in 2012, with the aim to incorporate the latest innovations and recommendations for RCC dam design and construction. The more noteworthy concepts updated have been the following ones:

(1) In very narrow valleys, the possibility to use self – raising systems of high speed belt conveyors. When the RCC mix is a cohesive high paste content mix, vacuum chutes can be a cheap option.

(2) It is advisable to locate the outlet works (bottom outlets, mid – level outlets and intakes) outside the RCC areas, in order to avoid interferences with the possibility of continuous RCC placement.

(3) It is desirable that the width of the crest is at least 8 m, and in large dams even 10 m, to facilitate the execution of the upper part of the non – spillway sections of the dam.

(4) In low cementitious content RCC dams, in which is not assured a good bonding between RCC layers, the upstream lining with geomembranes can solve the problem of the needed impermeability. But in seismic areas the good bonding between layers is imperative.

(5) The use of an upstream lining of CVC creates functional and constructive drawbacks. It

is advisable to substitute this technique with the new of the GEVR or GERCC.

(6) The more advisable design for an RCC dam corresponds to the concept of the "All RCC Dam", using for the faces the GEVR (or MEVR) or GERCC techniques.

(7) When high impermeability and strength at the horizontal joints between layers is required, the volumetric ratio paste/mortar of the RCC mix should be above 5% in excess of the compacted sand void ratio and, better, above 10%.

(8) The maturity factor for the hot joint limit could be extended up to a 500 ℃ × h and more with the use of retarding admixtures.

(9) The slip - formed facing elements can be a good solution in dams in wide U - shaped valleys.

(10) It is desirable that the distance of the galleries to the upstream face of the dam is at least 6 m or 8 m, to facilitate the execution of this critical part of the dam.

(11) The inspection galleries should not be constructed with methods that hide the RCC, as for example precast panels, metallic pipes left in place, etc.

(12) 20 $m^3/(s \cdot m)$ is a reasonable limit for the unit flow value for the implementation of a stepped spillway.

(13) The mix designer must always bear in mind that there are the in - situ properties, including most especially those of the horizontal joints between lifts, which are important, and not the ones that can be reached in the laboratory.

(14) The recommended Vebe consistency is 10 to 15 seconds at the moment of compaction.

(15) The sand should have a high non - plastic fines content; 5% to 18 % are the recommended limits.

(16) The recommended MSA is 50 ~ 60 mm for crushed aggregate and 40 ~ 50 mm for natural rounded.

(17) High proportion of mineral admixtures in the total cementitious content, in the range of 60% ~ 70% is advisable.

(18) With high proportion of mineral admixtures in the RCC mix, the design age should be at least 180 days or, preferable, 365 days.

(19) Especially in seismic areas and/or where high stresses of thermal origin are anticipated, the critical design criterion for an RCC dam should be the in - situ vertical direct tensile strength across horizontal joints.

Additionally, two new chapters have been included: one referring to RCC arch dams and another dealing with other uses of RCC in dam construction.

10 Synthesis of RCC dams constructed by Spanish companies abroad

Many Spanish contractors have been collaborating abroad, usually with their own technologies, in the execution of a great number of dams. Since the 1970s of last century, more than 60 dams have been built by Spanish companies abroad. Some of them have been RCC dams, and they have served to disseminate the know - how and state of the art on this matter of the Spanish companies. Table 4 collect the main characteristics of the RCC dams built by Spanish companies

Table 4 Experience of Spanish contractors in RCC Dams abroad

Dam	Year completion	Country	Purpose	Spanish contractor	Dimensions					Facings				Cementitious material (kg/m³)	
					Height (m)	Length (m)	Volume (×10³ m³)			Upstream		Downstream			
							RCC	Total		Slope	Type	Slope	Type	Cement	Pozzolan
San Rafael	1994	México	F/H	Acciona	48	168	85	110		V	7	0.66/0.8	(3) *	90	18
Pangue	1996	Chile	H	Dragados	115	410	670	740		V	1	0.8	(3) *	80	100
Beni Haroun	2000	Algeria	W/I	Dragados	118	714	1 690	1 900		V	1	0.8	(1) *	82	143
Contraembalse Monción	2000	Dominican Republic	I/W/H	Ferrovial Agroman	28	273	130	175		0.7	14/12	0.7	(14) *	80	0
Porce II	2001	Colombia	H	Dragados	123	425	1 305	1 445		0.1	14	0.75	(14) *	132	88
Villarpando	2007	Dominican republic	I	Ferrovial Agroman	6.5	563	20	31		V	14	0.75	(14) *	90	0
Portugués	2013	Puerto Rico (USA)	F	Dragados	67	375	280	300		V	3 **	0.35	(3 **) *	114	51
Zapotillo	UC	México	W	FCC	134	395	1 456	1 542		V	1	0.8	(3) *	65 / 85	45 / 65
Bajo Frío	UC	Panamá	H	FCC	56	238	86	223		0.2	3	0.85	(1) *		

* = Stepped face
** = GEVR

Method of forming faces of dam

1	Traditional concreting. against formwork	12	Reinforced conventional concrete cast after RCC placement
3	RCC against formwork	13	Reinforced concrete cast against precast units or slip - formed facing elements
7	Traditional. concreting against panels	14	Slip - formed / extruded facing elements

PURPOSES   H: Hydropower   I: Irrigation   W: Water Supply   F: Flood Mitigation abroad, some of them having been very important projects from the point of view of their technical features.

References

[1] Alonso-Franco, M. Compacted Concrete dams in Spain. Evolution and constructive details. Proceedings of the 2nd RCC International Symposium. Santander (Spain),1995.

[2] Alonso-Franco, M., Yagüe, J. The Spanish approach to RCC dam engineering[J]. The International Journal on Hydropower and Dams. 1995.5.

[3] Alonso-Franco, M., Yagüe, J., Berga, L. RCC Dams in Spain. Proceedings of the 3rd RCC International Symposium Chengdu (China),1999.

[4] Alonso-Franco, M., Jofré, C. RCC dams in Spain. Present and future. Proceedings of the 4th RCC International Symposium. Madrid. Spain,2003.

[5] De Cea, J.C., Berga, L., Yagüe, J., et al. RCC Dams in Spain. Proceedings of the 5th RCC International Symposium. Guiyang. China,2007.

[6] De Cea, J.C., Ibañez de Aldecoa, R., Polimón, J.,30 years constructing RCC Dams in Spain. Proceedings of the 6th RCC International Symposium. Zaragoza. Spain,2012.

[7] Ortega, F. Lessons learned and innovations for efficient RCC dams. Proceedings of the 6th RCC International Symposium. Zaragoza. Spain,2012.

The First 30 Years of RCC Dams

Malcolm Dunstan

(Malcolm Dunstan and Associates, U. K.)

Abstract: In RCC1991 (in Beijing), the first of the RCC Dam Conferences, a paper was presented in association with the greatly – missed late Dr Shen Chonggang called "The Development of RCC Dams throughout the World". In RCC1995 (in Santander), a General Report was given on the "Expansion of RCC dams throughout the World and the Construction Methodologies being used". In RCC1999 (in Chengdu) a General paper was given titled "Latest Developments in RCC Dams". In RCC2003 (in Madrid), a Lecture was given called "The State – of – the – Art of RCC dams in 2003——an update of ICOLD Bulletin No. 125". In RCC2007 (in Guiyang) another Lecture was given titled "Overview of RCC at the end of 2006" and in RCC2012 (in Zaragoza) a further Lecture was given on "New developments in RCC dams". This Paper is an extension and update of the above series of papers and traces the development of RCC dams over the past 30 years. During this period approaching 700 RCC dams have been completed, or are under construction, in more than 60 countries. These dams have been constructed in all types of conditions for all sorts of purposes and have essentially superseded traditional concrete dams. The paper describes how the types of RCC being used in dams has changed and developed over the years and the various new techniques that have been introduced.

Key words: Key words: RCC, dams, history, design, construction

1 Introduction

This paper is based up on the data held in MD&A's RCC dam database that holds details of most of the RCC dams in the World. A summary of the database is published each year in the World Atlas of *Hydropower & Dams*[1]. A searchable copy of the summary is held on MD&A's website, www. rcccdams. com. The data is updated each year in July/August in time for the publication of the World Atlas. Unfortunately at the time of drafting of this paper, the 2014 update had not been completed and thus the paper is based on data at the end of 2013. Nevertheless by the time of the RCC2015 Conference, the 2014 update will have been completed and the presentation to the Conference will be based on the latest data.

2 Number of RCC dams

At the end of 2013 there were at least 554 completed RCC dams (average height 63 m and average volume 506 000 m^3 of which 356 000 m^3 was RCC) and there were a further 86 or more under construction (average height 86 m with an average volume of 1 166 000 m^3 of which

988 000 m³ was RCC). The numbers of RCC dams completed each year are shown in Fig. 1, which also shows a five-point moving average of those numbers.

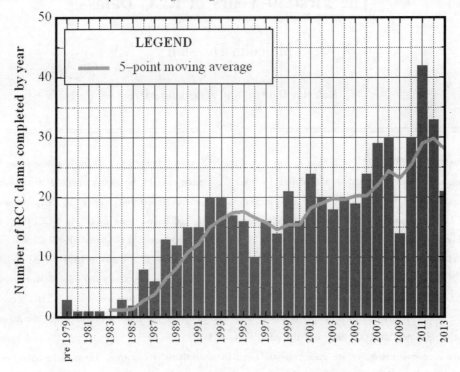

Fig. 1 Total number of RCC dams completed each year

It can be seen from the Figure that there was an "experimental" stage in the development of RCC dams up to about 1983, at which point there was a decade of rapidly increasing numbers of RCC dams being constructed. From circa 1993 to 2003 there was a consolidation period during which confidence was being built for the construction methodology. Since 2003 there has been a further rapid increase in the number of RCC dams being constructed and now an average of circa 30 RCC dams are beingcompleted each year——some 30 years after the real start of the method of construction. Given that there were at least 86 RCC dams under construction at the end of 2013, it is probable that this rapid increase in the number of RCC dams being constructed each year will continue for at least the next few years.

It is estimated that the total volume of RCC dams that have been completed, or are under construction, is circa 304 Mm³ of which 223 Mm³ is RCC. The average mixture proportions of all these dams is a Portland cement content of 82 kg/m³ and a pozzolan content of 65 kg/m³, i.e. a total cementitious content of 147 kg/m³.

3 Distribution of RCC dams around the world

China has by far the greatest number of completed RCC dams at 174. It can therefore be considered to be the sole member of the Premier Division of RCC dam countries. There are three countries, Japan, Brazil and the USA each of which has 45 ~ 55 completed RCC dams. They might be considered to be in the First Division of RCC dam countries. There are a further seven countries, Spain, Turkey, Morocco, South Africa, Vietnam, Australia and Mexico each of which have 15 ~ 25 RCC dams. These are in the Second Division. Finally there are four countries in the Third Division with 5 ~ 10 completed RCC dams, Greece, France, Iran and Peru. The number of dams completed each year in the Premier, First and Second Divisions have been plotted in Fig. 2. It is clear that China has completed far more dams than any other country; indeed they have completed more RCC dams than the three countries in the First Division combined. It is of interest that China has determined that the best solution for all their RCC dams is high – paste content RCC (see Section 4).

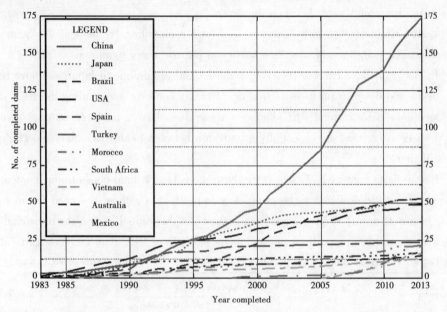

Fig. 2 Number of RCC dams completed each year by country

In the First Division, the USA initially constructed more RCC dams but Japan overtook in the mid – 1990s and has been leading the Division for the rest of the time. Brazil started later than the other two countries but caught up in the mid – 2000 s and is now likely to move ahead into overall second place behind China.

There is a similar situation in the Second Division; initially Spain was ahead but was caught by Morocco and then there is a group of three countries, South Africa, Australia and Mexico. However the two countries of most interest are Turkey and Vietnam, they both started late and have now had a very rapid growth in the number of RCC dams and both, in particular Turkey, are likely to make the transition from the Second Division to the First Division.

4 Design philosophy of RCC dams

4.1 Design philosophies

All the RCC dams that were either complete or under construction at the end of 2013 have been classified as either:

(1) Hard – Fill dams (also called CSG (cemented sand and gravel) dams in Japan and CMD (cemented materials dam) in China)——these dams have symmetrical (or near symmetrical) upstream and downstream slopes, usually between 0.60:1 and 1.20:1 ($H:V$). These dams require a rather lower Quality Control than other forms of RCC dam and do not rely upon the performance at the horizontal joints between the layers.

(2) Lean RCC dams——these dams have a total cementitious content of 99 kg/m^3 or less and generally have some form of upstream barrier to provide water tightness.

(3) RCDs (Rolled – Concrete Dams)——these dams are unique to Japan and have relatively thick upstream and downstream faces of traditional facing concrete surrounding the RCC itself.

(4) Medium – paste content RCC dams——these dams have a total cementitious content between 100 ~ 149 kg/m^3 and are essentially a transition between Lean RCC dams and High – paste content RCC dams.

(5) High – paste content RCC dams——these dams have a total cementitious content of 150 kg/m^3 or more and are designed to be completely impermeable without the need for an upstream membrane of any form and to have a good performance at the joints between the horizontal layers.

The average mixture proportions of each of these forms of RCC dam are shown in Table 1.

Table 1 Average mixture proportions of the paste of the various forms of RCC dams

Items	Mixture proportions of the paste (kg/m^3)				Water/cementi-tious ratio
	Cement	Pozzolan	Cementi-tious	Water	
Hard – fill	66	9	75	132	1.76
Lean RCC	72	9	81	122	1.51
RCD	86	36	123	94	0.76
Medium – paste content RCC	80	37	117	114	0.98
High – paste content RCC	87	109	196	111	0.57

It can be seen from the Table 1 that the average PortlandCement content of the different design philosophies does not vary significantly, generally being between 65 ~ 85 kg/m^3. The different total cementitious contents are obtained by the use of widely different contents of pozzolan. What is also of interest is that, in spite of high – paste content RCC being very much more workable, the average water content is significantly lower than with Hard – Fill and Lean RCC dams. This is a function of the use of high proportions of pozzolan, usually low – lime flyash, which act as ball bearings to produce the more workable RCC. In addition the use of better – controlled aggregates in high – paste content RCC also reduces the water demand. Thus the higher cementitious contents and lower water contents lead to widely different water/cementitious ratios, the average for Lean RCC dams being over two and a half times that of the average for high – paste content RCC, and this leads to widely different strengths for each of the different forms of RCC.

4.2 Use of the different forms of RCC

Fig. 3 shows the proportion of the number of RCC dams that were completed or under construction at the end of 2013 that used the various design philosophies described above.

Approximately half of all the RCC dams contain a high – paste content RCC, approximately 15% use medium – paste RCC, approximately 10% lean RCC and just 2.7% are hard – fill dams and the design philosophy is not known for circa 15%.

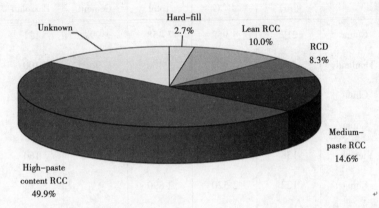

Fig. 3 Proportion of the number of RCC dams with various design philosophies

5 Highest, largest and fastest RCC dams

5.1 Highest RCC dams completed at the end of 2013

The ten highest completed RCC dams as shown in Table 2. All are gravity dams except Wanjiakouzi, which is the highest RCC arch dam. Of these ten highest dams, five are in China and eight of the ten are in Asia. To be in the top ten highest RCC dams, the dam has to have a height in excess of 140 m.

Table 2 Ten highest RCC dams in the World

Dam	Country	Height (m)	Volume ($\times 10^3$ m^3)		Cementitious content (kg/m^3)		Type of pozzolan
			RCC	Total	Cement	Pozzolan	
Longtan	China	217	4 952	7 458	99	121	(F)
Guangzhao	China	201	2 420	2 870	71	87	(F)
Miel I	Colombia	188	1 669	1 730	85 to 160	0	(–)
Guandi	China	168	2 970	4 710			(F)
Wanjiakouzi	China	161					
Jin'anqiao	China	160	2 400	3 920	96	117	(F)
Miyagase	Japan	156	1 537	2 060	91	39	(F)
Urayama	Japan	156	1 294	1 750	91	39	(F)
Ralco	Chile	155	1 596	1 640	133	57	(N)
Murum	Malaysia	141					

5.2 Largest RCC dams completed at the end of 2013

The ten dams with the greatest volume of RCC are shown in Table 3. Six of these dams are in China and nine of the ten are in Asia, the exception beingTaum Sauk, a hard - fill dam in the USA. To get into the top ten dams with the largest volumes of RCC, the volume has to be in excess of 2.40 Mm^3.

Table 3 Ten largest (in terms of RCC volume) RCC dams in the World

Dam	Country	Height (m)	Volume ($\times 10^3$ m^3) RCC	Volume ($\times 10^3$ m^3) Total	Cementitious content (kg/m^3) Cement	Cementitious content (kg/m^3) Pozzolan	Type of pozzolan
Longtan	China	217	4 952	7 458	99	121	(F)
Tha Dan	Thailand	95	4 900	5 400	90	100	(F)
Guandi	China	168	2 970	4 710			(F)
Longkaikou	China	116	2 840	3 853	83	101	(F)
Son La	Vietnam	138	2 677	4 800	60	160	(F)
Kalashuke	China	122	2 520	2 890			(F)
Yeywa	Myanmar	135	2 473	2 843	75	145	(N)
Taum Sauk	USA	49	2 448	2 500	59	59	(F)
Guangzhao	China	201	2 420	2 870	71	87	(F)
Jin'anqiao	China	160	2 400	3 920	96	117	(F)

5.3 Fastest RCC dams completed at the end of 2013

The ten RCC gravity dams with the highest average rates of placement are shown in Table 4 together with two hard - fill dams, Beydag from Turkey andTaum Sauk from the USA, thus making 12 dams in the Table. The hard - fill dams have been highlighted because the rates of placement cannot really be compared to those of an RCC gravity dam because the Quality Control required for a gravity dam is far in excess of that required for a hard - fill dam, as the latter does not usually require any properties at the horizontal joints. To get into the top ten fastest RCC dams, the average rate replacement of the RCC needs to be in excess of 76 500 m^3/month.

Upper Stillwater and Olivenhain dams, both in the USA, have extremely fast average rates of placement compared to other RCC dams of a similar volume. The only two other dams with similar (or faster) rates of placement, i.e. Longtan and Tha Dan, both have volumes well in excess of four times that of Upper Stillwater and Olivenhain.

Table 4　Ten (12) RCC dams with the fastest average rates of placement

Dam	Country	Height (m)	RCC vol. (×10³ m³)	Placement time (months)	Monthly placement (m³)		Max. day (m³)
					Average	Peak	
Longtan	China	217	4 623	32.4	1 427 58	400 755	18 475
U. Stillwater	USA	91	1 125	9.0	125 324	204 430	8 415
Tha Dan	Thailand	95	4 900	40.1	1 222 66	201 490	13 280
Olivenhain	USA	97	1 070	8.8	121 895	224 675	12 250
Beydag	Turkey	96	2 350	20.9	112 566	165 000	
Beni Haroun	Algeria	121	1 690	16.4	102 860	175 000	9 100
Taum Sauk	USA	49	2 448	24.9	98 492	189 470	11 330
Guangzhao	China	201	2 420	27.9	86 598		
Son La	Vietnam	138	2 677	31.5	84 995	200 075	9 980
Guandi	China	168	2 970	36.0	82 500		
M. Vaitarna	India	103	1 202	15.5	77 548	134 125	7 536
Ralco	Chile	155	1 596	20.9	76 449	147 600	6 860

6　Cementitious materials used in RCC dams

6.1　Portland cement

The various types of Portland cement that have been used in the RCC dams that had been completed or were under construction at the end of 2013 are shown in Fig. 4.

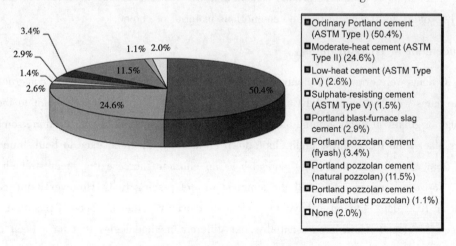

Fig. 4　Portland cements used in RCC dams

It can be seen from the Figure that Ordinary Portland cement (ASTM C150[2] Type I) has been used in approximately 50% of all RCC dams, Moderate – heat cement (ASTM C150 Type II) in approximately 25% and various other cements in the remaining 25%. Approximately half of the latter has been Portland pozzolan cement containing a natural pozzolan.

6.2 Pozzolans

The various pozzolans, manufactured and natural, that have been used in the RCC dams that had been completed or were under construction at the end of 2013 are shown in Fig. 5.

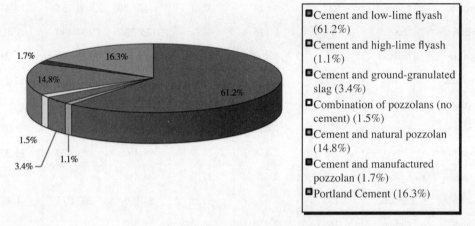

Fig. 5 Pozzolans used in RCC dams

It can be seen in the Fig. 5 t that by far the most popular pozzolan is low – lime flyash (ASTM C618[3] Type F) having been used in just over 60% of all RCC dams. Approximately 15% of all RCC dams have used a natural pozzolan in the cementitious content, approximately 15% have used just Portland cement without a pozzolan and the remaining 10% covers all the other forms of pozzolan (including a combination of pozzolans without any Portland cement——albeit only 1.5%). A recent innovation has been the use of "conditioned" ash (i. e. ash retrieved from lagoons at Thermal Power Stations) that is suitable treated and dried[4].

A combination of Ordinary Portland cement and low – lime flyash has been used in the majority of RCC dams and seems to be the cementitious material of choice.

7 Placement of RCC in dams

RCC dams were originally conceived to provide a simple and rapid method of construction for concrete dams with the layers of RCC being placed horizontally from one abutment to the other (although originally at the beginning of the development of the method of construction, some practitioners placed the RCC with a slight slope down towards the upstream face to both drain excess water (from rainfall and/or curing) and also as an effort to increase the shear strength at the joints). Horizontal placement is still the simplest way of placing RCC. However if the concrete plant at an RCC dam is found to have an inadequate capacity, other methods of placing the RCC have been developed. There are essentially four different methodologies that have been used to place RCC in dams:

(1) Horizontal placement (Fig. 6)——the simplest and still the best methodology as long as the layers can be placed sufficiently rapidly.

(2) Slope – layer placement (Fig. 7)——this method was developed at Jiangya dam in Chi-

Fig. 6 Horizontal placement at Upper Stillwater dam in the USA

na when it was found that the concrete plant had insufficient capacity to place a layer of RCC from one abutment to the other within a reasonable time period. The RCC is placed in 300 - mm layers on a slope of approximately 1∶10 ($H:V$) in a lift of 1.2 ~ 3.0 m. This has the advantage of decreasing the size of the "working face" of the RCC and thus allowing the RCC to be placed "fresh on fresh". Unless the steps on the downstream face are the same height as the lifts, there is a major complication with the interaction of the sloped placement and the horizontal steps.

Fig. 7 Slope - layer placement at Al Wehdah dam on the Jordan/Syria border

(3) Split - level placement (Fig. 8)——this method was introduced in a large RCC dam at Beni Haroun in Algeria where, because of the shape of the valley, for a short period the volume of the horizontal layers was very high and in excess of the capacity of the concrete plant. The dam was thus split into two halves and 14.4 - m lifts placed alternatively on one or other half of the dam. This method has the great advantage that the formwork for the gallery can be erected on one

half of the dam (Fig. 8) while the RCC is placed in the other half.

Fig. 8 Split - level placement at Ghatghar Lower dam in India

(N. B. the formwork being erected on the right half of the dam while the placement is on the left half)

(4) Block placement (Fig. 9)——on very large dams (generally with volumes in excess of 2 Mm^3), the placement can be split into blocks and each individual block placed horizontally (NB The placements of RCC are not necessarily in adjacent blocks).

Fig. 9 Placement in blocks at Longtan dam in China

8 Forming the faces of RCC dams

Until the early 1990s by far the most popular method of forming the faces of RCC dams was by using a traditional facing concrete. This had the disadvantage that a different concrete plant (or if the same plant was used, the production of RCC had to be stopped) had to be erected and a separate concrete had to be transported to the placement area on the dam. There was also the problem of the two concretes having rather different properties and thus the potential of thermal cracking due to the facing concrete requiring contraction joints at a closer spacing than the RCC in the dam body.

In the early 1990s two different methods of forming the faces were developed in Europe (GEVR (Grout – Enriched Vibratable RCC)) and in China (GE – RCC (Grout – Enriched RCC)) independent of each other and without the knowledge one of the other. Both had the same concept, to add grout to the RCC on the placement to make it suitable for immersion vibration. However there was a fundamental difference between the two concepts, with GEVR (see Section 8.1) the grout was added at the bottom of the layer and with GE – RCC (see Section 8.2) the grout was added on top of the layer. Since then a further development has been undertaken and that is to design the RCC to be even more workable such that it can be consolidated with immersion vibrators without the addition of grout, this has been called IVRCC (Immersion – Vibrated RCC) (see Section 8.3).

8.1 GEVR

Grout is lighter than RCC and thus it is relatively easy to vibrate the grout upwards through the RCC and this is the philosophy of GEVR. The grout, which has a water/cementitious ratio similar to that of the RCC, is first spread on top of the previous layer ahead of the RCC of the new layer. The RCC is then spread over the grout and the latter is vibrated with immersion vibrators. A small vibratory roller is then used to consolidate the interface between the RCC and the GEVR (that is usually 400 ~ 600 mm wide). Finally a vibratory plate is used to finish the surface of the GEVR.

8.2 GE – RCC

The concept of GE – RCC is to spread the grout on the surface of the RCC after it has been spread, to let it soak into the uncompacted RCC and then to vibrate the grout through the thickness of the layer. As such the grout has to have a relatively high water/cementitious ratio and unless a super – plasticiser is used, this can lead to the GE – RCC having a lower in – situ strength than the RCC itself.

8.3 IVRCC

With the increased workability of most modern RCCs (the loaded VeBe time has dropped from circa 30 ~ 40 seconds in the 1980s to eight to 12 seconds in modern RCCs), it was only one step further before it was possible to design an RCC that not only could be consolidated with a large vibrator roller, but could also be consolidated at the face with immersion vibrators[5].

8.4 Forming the upstream faces of RCC dams

The various methods of forming the upstream faces of the RCC dams that were complete or under construction at the end of 2013 are shown in Fig. 10.

It can be seen that 40% of all RCC dams still use traditional facing concrete for the upstream face. However this proportion has dropped from 55% only ten years ago. Nearly 30% now use RCC. directly against formwork (practically all using GEVR or GE – RCC), an increase from just 13% only ten years ago. There have been few changes to the other methods of forming the face that make up some 30% of RCC dams. Approximately 10% of RCC dams (usually lean RCC

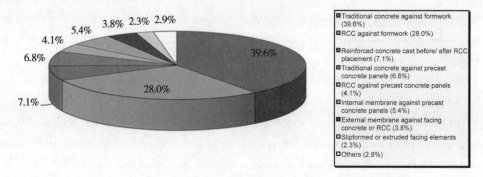

Fig. 10 Methods of forming the upstream faces of RCC dams

dams) use an upstream (either internal behind precast concrete panels or external) membrane for water – tightness.

8.5 Forming the downstream faces of RCC dams

The various methods of forming the downstream faces of the RCC damsthat were complete or under construction at the end of 2013 are shown in Fig. 11. As with the upstream face, in the last decade there has been a decline in the use of traditional facing concrete (from circa 48% to 40%) and an increase in the use of RCC (again usually using GEVR or GE – RCC) directly against the formwork (from circa 28% to 38%). The difference between the two figures is made up by a reduction in the use of RCC against precast concrete blocks (reduced by 2%) and a reduction in the number of RCC dams using unformed faces (by 4%). The latter figure probably means that this method it is hardly being used in modern RCC dams given the increase in the number of dams completed in the last ten years.

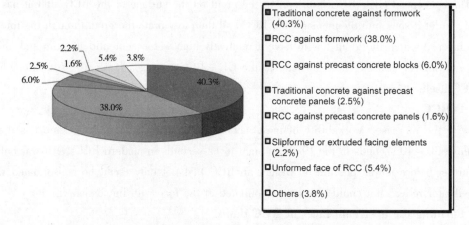

Fig. 11 Methods of forming the downstream faces of RCC dams

8.6 Forming the spillways of RCC dams

The various methods of forming the spillways of the RCC damsthat were complete or under construction at the end of 2013 are shown in Fig. 12.

As with both the upstream and downstream faces, in the last decade there has been a decline in the use of traditional concrete to form the spillway (from circa 60% to just over 50%) and an increase in the use of RCC (again using GEVR or GE – RCC) directly against formwork (from 8.

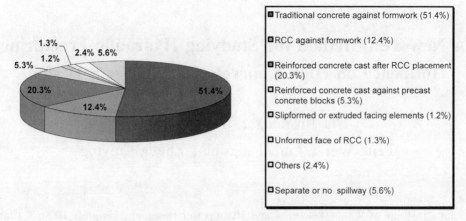

Fig. 12　**Methods of forming the spillways of RCC dams**

5% to 12.5%). However the largest increase has been in the use of reinforced concrete being placed after the placement of the RCC dam body; this has increased from just under 17% ten years ago to 25% in 2013. The reason for this is simply the increasing size of RCC dams and thus the increasing capacity of the spillways that are required for those dams. The only other significant change is the halving of dams having a separate or no spillway; again this is probably a function of the increasing size of RCC dams.

Acknowledgements

Acknowledgement must be accorded to the many engineers who kindly update the RCC dam database each year. Without their inputs, some more regular than others, the tremendous amount of data that is now contained in the database would not exist.

References

[1] M. R. H. Dunstan. The World's RCC dams, World Atlas & Industry Guide, Hydropower & Dams, London, August 2014.
[2] American Society for Testing and Materials, Portland cement, Standard Specification C150, ASTM, Philadelphia.
[3] American Society for Testing and Materials, Fly ash and raw or calcined natural pozzolan for use as a mineral admixture in Portland cement concrete, Standard Specification C618, ASTM, Philadelphia.
[4] M. R. H. Dunstan, N. H. Ha and D. Morris, The major use of lagoon ash in RCC dams, Proceedings of 7th International Symposium on Roller – Compacted Concrete (RCC) Dams, Chengdu, China, September 2015.
[5] F. Ortega, Key design and construction aspects of immersion vibrated RCC, Hydropower and Dams, London, 2014.21(3).

A New Test Method for Studying Hydraulic Fracturing Influence on High Concrete Gravity Dam Design

Jia Jinsheng[1], Wang Yang[2,1],
Feng Wei[1], Liu Zhongwei[1], Zheng Cuiying[1]

(1. China Institute of Water Resources and Hydropower Research, Beijing, 100038, China;
2. Depart. of Hydraulic Engineering, Tsinghua Univ., Beijing, 100083, China)

Abstract: To evaluate the safety of high concrete gravity dams when considering hydraulic fracturing, a new test method of concrete hydraulic fracturing is presented. Simulation tests have been conducted using a large fully - graded concrete specimen (ϕ450 mm × 1 280 mm) with embedded joint (crack) inside it under uni - axial compressive and tensile stresses. High water pressure can be applied to the embedded joint to test the hydraulic fracturing of concrete. A series of tests at various ages and under different uni - axial stress conditions were performed with the proposed test method. A formula for evaluating the safety of concrete structure against hydraulic fracturing has been achieved based on tests.

Key words: Hydraulic fracturing, gravity dam, concrete joint (crack), high water pressure, physical test

1 Introduction

Some failures or distresses of concrete dams are due to hydraulic fracturing. It was the case for the 200 m high Kolnbrein arch dam in Austria. When the reservoir impounded toits normal water lever at altitude 1 860 m EL. In 1978, the leakage in its gallery suddenly increased to 200 L/s, and foundation seepage uplift of several dam monoliths located in the riverbed reached full water head. Analysis demonstrated that small initial cracks caused by thermal stress may become severe under high water pressure and the dam may be fractured at its heel (Feng and Pekau et al. 1996). Rehabilitation cost of Kolnbrein dam was very high and it was near the fee for its construction. Another case of hydraulic fracturing was Zhexi concretebuttress dam in China. The vertical cracksinitially occurred at upstream surface were propagated deep inside the dam body by water pressure. It was endangered for the dam safety and had to be rehabilitated (Liang and Qiao et al. 1982; Liu and Zhou et al. 1989). There are many papers of investigation on problems related to hydraulic fracturingbut mainly focused on calculation analysis (e. g., Chen and Feng et al. 2006; Utili and Yin et al. 2008; Jiang and Ren et al. 2011; Secchi and Schrefler 2012; Guan and Li et al. 2014).

No tensile stress at upstream surfaceas one of the stress principle has been required for designing gravity damin China (Chinese standard No. SL 319—2005) and there are no difference in the design principle for gravity dams with different height. The design principle is similar for most of countries but the stress principle is different in USA and Switzerland compared with other countries. It is required that the upstream vertical compressive stress must be greater than the water pressure subtracting concrete strength when hydraulic uplift pressure is disregarded (USBR 1976) in USA. For Grand Dixence damin Switzerland, it is required that the compressive stress on the upstream face must be at all levels larger than 85% of the water pressure disregarding the uplift pressure (Swiss COD 1985). In general, the highest Longtan RCC gravity dam ($H = 192$ m, to be heightened to 216.5 m, RCC gravity dam) designed based on the principle in China, compared with Dworshak (USA, $H = 219$ m, conventional concrete gravity dam) or Grand Dixence (Switzerland, $H = 285$ m, conventional concrete gravity dam) is quite differentin the cross section of gravity dam. Its dam volume is about 6% smaller. Comprehensive test on its safety takes a very long time because it is delayed for heightening to the final stage. Mass concrete damsmay be cracked due to the low tensile resistance of concrete or with weak layer interfaces especially constructed by RCC technique (e. g. , Bhattacharjee and Léger 1994; Zhu 2006). Although cracksorweak layers on dams are very popular, damhistory has proven that most of the structure can work safely with cracks for dams lower than 200 m. However, there are two questions to be answered:

(1) May upstream surface cracks (or weak layers) be fractured by high water pressure for gravity dam higher than 200 m?

(2) Which design principle is safe and economic Principle in China, USA or the Switzerland?

Analysis on hydraulic fracturing at upstream surface of the gravity dam has been discussed in this paper in order to answer the first question and to develop a more reasonable criterion for dams higher than 200 m for next step.

Slowik and Saouma (2000) conducted numerous tests on wedge splitting specimens with hydrostatic pressure along the crack. The uplift pressure in a concrete gravity dam during and after an earthquake has been investigated. By artificially controlling the opening and closing speed of the crack, they studied the distribution of hydraulic pressure in concrete crack. It was observed that, for the tested specimens, the crack opening rate has important consequences on the internal water pressure distribution. Several techniques have been developed to study the dynamic fracture behavior of concrete and concrete structures, like modified Charpy impact test, Split Hopkinson pressure bar test, drop weight impact test and explosive test (ACI 1989; Zhang et al. 2010). The investigation on concrete hydraulic fracturing at gravity dam heel is so far mainly based on the numerical method. Silva and Einstein (2014) numerically analyzed the effect of the ratio between a vertical load, or stress, and the hydraulic pressure applied in existing flaws on the stress field in the vicinity of the flaw tips. The study showed that the ratio between the water pressure applied in the flaws and the vertical load/stress (WP/VL) plays a crucial role in the magnitude and shape of the stress field around a flaw tip, and therefore in the location of tensile and shear fracture initiation.

The concrete gravity dam is usually considered as a plane stress structure. For the concrete at the upstream surface(when designed as vertical) of the gravity dam, there are only water pressure perpendicular to the dam surface and vertical force perpendicular to the groundas the main loads. The upstream surface crackis suffered uni – axial load in vertical direction, when the crack is horizontal. The horizontal crack may be hydraulic fractured under compressive, tensile stress or the state without stress.

In order to investigate in detail, a new test method of concrete hydraulic fracturing is proposed in this paper. Test techniques related have been developed to evaluate the hydraulic fracturing behavior of concrete with cracks (joints) under different uni – axial stress states and different concrete strengths. First, the main components of the testing apparatus are introduced. Then the step – by – step testing procedures are described. Finally, results of a series of tests on a fully – graded concrete under differents tress states and different concrete strengths are presented.

2 Testing apparatus

For the concrete gravity dam, the fracture process has been investigated for developing under a combination of flexure – and shear – type forces (Bhattacharjee and Léger 1994). But it is refer to the deep crack inside the gravity dam. This paper focuses on surface crack at the dam heel subjected to uni – axial stress in vertical direction, and there is very little shear – type force because the gravity dam section can be considered as a plane stress structure based on no – tensile stress design principle and gravity dam block design principle.

In order to simulate the concrete surface crack of gravity dam heel, the general layout of the testing model is designed as Fig. 1. And a testing apparatus was developed. It is composed of a set of components for making crack, a uni – axial system, a high water pressure supply system. The detailed description of each component is presented in the following three sections.

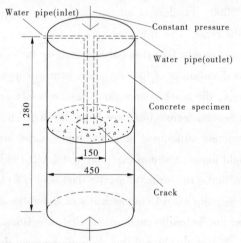

Fig. 1 Specimen of hydraulic fracturing test under uni – axialstress(unit: mm)

2.1 A set of components for making crack

A special set of components for making crack in the concrete specimen was created, as shown in Fig. 2. The members include one cylindrical steel mould (High 1 350 mm, Inner diameter 450 mm), onerim which plays a supporting role (Inner diameter 180 mm, Outer diameter 230 mm, High 45 mm), one circular steel plate (Thickness 25 mm, Diameter 450 mm, and with a ϕ80 mm hole in the center), anchor rods (Length 290 mm, they are needed in the uni-axial tension specimen, and not necessary in the uni-axial compression specimen. This will be explained below.), some metalpipes and two thin steel sheets to make up a crack and water pipelines in the concrete specimen.

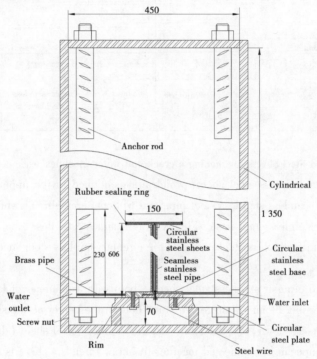

Fig. 2　Components for making crack

The details of how metal pipes and steel sheetsto form a crack and water pipelines in the concrete specimen is presented as the following,

(1) Prepare a circular stainless steel basewitha diameter of 150 mm and a thickness of 15 mm, as shown in Fig. 3. Firstly, inside this base, channel 1 and channel 2 both with a diameter of 2 mm were drilled. Secondly, channel 3 with a diameter of 4 mm was drilled in this base, and channel 3 is a threaded hole. Thirdly, four threaded holes were drilled around the circular stainless steel base, they were used to fix the base on the circular steel plate, as shown in Fig. 2.

(2) Prepare a circular stainless steel sheet with a diameter of 11 mm and a thickness of 3 mm. Two holes were drilled on the sheet, one is 2 mmin the diameter and another is 4 mm. A brass pipe which is 587 mm in length and 4 mm in outer diameter was welded together with the 2 mm diameter hole. The other end of the brass pipe was welded together with channel 1 inside the circular stainless steel base, as shown in Fig. 3.

(3) Prepare a seamless stainless steel pipe with an outer diameter of 17 mm and a length of 590 mm, and lathe screw threadson one end of the pipe. Then the pipe was sleeved over the brass

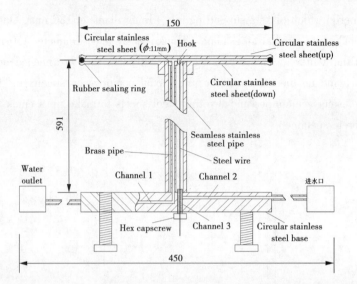

Fig. 3 Pipes and Steel Sheets for making a crack and water pipelines(when pouring concrete)

pipe and the circular stainless steel sheet, which are mentioned in the upper section. It is worth mentioning that the seamless stainless steel pipe end without screw threads should be close together with the circular stainless steel base. Finally, the seamless stainless steel pipewas welded together with the circular stainless steel base and the circular stainless steel sheet ($\phi = 11$ mm), as shown in Fig. 3.

(4) Prepare two circular stainless steel sheets, both with a diameter of 150 mm and a thickness of 2 mm. There is a 17 mm diameter threaded hole in the center of one sheet, namely: circular stainless steel sheet (down). So that, this circular stainless steel sheet (down) and the seamless stainless steel pipecan be connected together by screw threads. There is a hook at the center of another sheet, namely: circular stainless steel sheet (up). Edges of these two sheets were forged into arc-shaped grooves, which were 2 mm in radius and 1 mm in depth. In the grooves, a 4mm diameter rubber sealing ring was set, as shown in Fig. 3. After these two sheets stacked close together, the gap between them is the crack of the concrete specimen. In this case, the crack is disc-shaped, its diameter and thicknessis 150 mm and about 1 mm respectively. Due to the convenience of the threaded connection between circular stainless steel sheet (down) and seamless stainless steel pipe, these twocircular stainless steel sheets can be replaced by other shaped sheets in order to form crack with other shape.

(5) Prepare a steel wire with a diameter of 0.61 mm, a no cap screw, a hex cap screw with a hole in the middle. And the steel wireis allowed to pass through the hole. Before pouring concrete, the steel wire should be hung on the hook of the circular stainless steel sheet (up), then the steel wire pass through the 590 mm long seamless stainless steel pipe, channel 3and the hex cap screw, finally it was tensed by the hex cap screw, as shown in Fig. 3. This leaded to the two circular stainless steel sheetstightly stacked close together. One effect of this is to prevent dislocation of thecircular stainless steel sheet (up) in the process of pouring concrete, another effect is toincrease the sealing contact of the rubber sealing ring. After seven days of pouring concrete, it is

the time to dismantle the hex cap screw and the steel wire, then channel 3 should be blocked by the no cap screw as shown in Fig. 4.

After going through the steps above, a crack and two water pipelines were made inside the concrete specimen. One water pipeline is for water entering, another is for draining. The diagram of the crack and water pipelines which are fully filled with water is shown in Fig. 4.

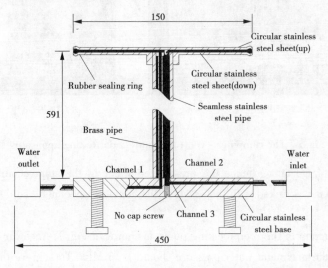

Fig. 4 Pipes and Steel Sheets for making a crack and water pipelines (During testing procedures)

2.2 Uni – axial system

A computer – controlled uni – axial testing apparatus was used to allow the independent control of stress state of the concrete specimen for investigating the initiation and development of concrete hydraulic fracturing. This apparatus can exert compression and tension load. The compression and tension load range 0 ~ 15 000 kN and 0 ~ 2 000 kN, respectively. The precisions of the vertical load and effective stress are 100 N and 629 Pa, respectively. The range of uniform time – dependent load is 0.25 ~ 25 kN/s. Resolution in strain measurement is 0.2 μm, its error is ± 1%. The range of uniform time – dependent deformation is 0.000 25 ~ 0.025 mm/s, its error is ±1%. Fig. 5 is photographs of the uni – axial testing apparatus.

The final stress state of the concrete specimen can be approached by setting a stress path through the control system automatically. During the hydraulic fracturing process, the uni – axial testing apparatus can maintain constant pressure or tension load. A linear variable differential transformer (LVDT) with a precision of 0.000 2 mm is employed to measure the total verticaldisplacement of the specimen.

2.3 High water pressure supply system

To supply high pressure water in concrete specimen, a computer – controlled high water pressure supply system which contains one transparent acrylic water pool 200 mm in diameter, 400 mm in height, one stainless steel high pressure storage tank 150 mm in diameter, 450 mm in length, and one reciprocating pump, as shown in Fig. 6, is used to provide sufficient high pressure water. The pressure value is computer controlled. A pressure gauge and a pressure regulator are

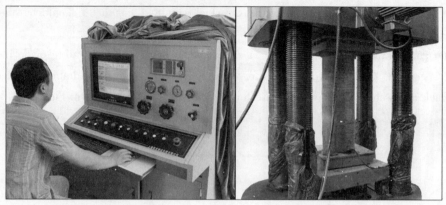

(a)Computer controller (b)Loading machine

Fig. 5 The computer – controlled uni – axial testing apparatus

connected to the storage tank; hence the pressurized water in the storage tank can be regulated first. The storage tank is linked with the crack in the concrete specimen, so that water pressure in them is equal.

Schematic diagram of the reciprocating pump is shown in Fig. 7. According to the experimental requirements, the maximum water pressure should be 6 MPa. To achieve this goal, two non – return valves are employed upstream and downstream of the pump, respectively. Details of non – return valves is shown in Fig. 8. Each non – return valve is composed by a PTFE gasket 15 mm in diameter and a stainless steel ball 5 mm in diameter. Each cycle, the reciprocating pump provide 0. 47 ml water to the storage tank, and this meets the precision of the pressure gauge.

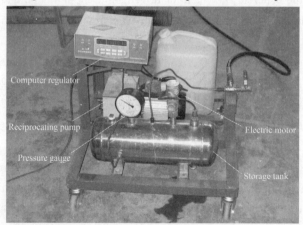

Fig. 6 The computer – controlled highwater pressure supply system

The precision of the pressure gauge is 0. 1 MPa; thus the hydraulic gradient can be changed at a precision of 0. 1 for a concrete specimen. During the test, the inlet water pressure can be either increased gradually or kept constant.

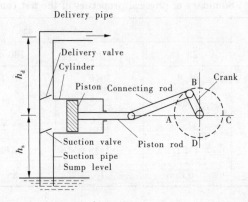

Fig. 7 Schematic diagram of the reciprocating pump

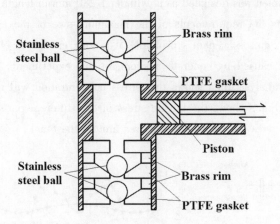

Fig. 8 Details of non - return valves

3 Concrete specimen preparation

3.1 Concrete specimen type

The concrete for this study was obtained by mixing three different grain - sizes of aggregates, namely fully - graded concrete. The aggregates was extracted from a construction site of LongTan RCC gravity dam with designed height of 216.5 m. This means that there is more than 2 MPa water pressure at the damheel. The grain - size distribution (GSD) of aggregates and the proportion of other materials are shown in Table 1. The physical properties of the test concrete are summarized in Table 2.

Table 1 Proportion of materials for mixing concrete

Bill of material (BOM) for one cubic meter of concrete (kg/m^3)								
Water	Cement	Flyash	Artificial sand	Stone			Admixture (JM - II, liquor)	Admixture (ZB - 1G, liquor)
				Small	Middle	Large		
79	86	109	743	437	583	437	5.85	3.900

Table 2 Summary of physical properties of the test concrete

Items	Concrete age			
	28 d	90 d	180 d	365 d
Compressive strength (MPa)	25.4	35.6	40.7	42.6
Splitting tensile strength (MPa)	1.54	2.58	2.71	2.88
Elastic modulus (GPa)	37.7	43.2	45.5	
Tensile strength (MPa)	1.87	2.25	2.79	
Ultimate tensile strain (10^{-6})	63	76	85	

The concrete specimen was designed as a cylinder 1 280 mm in length, 450 mm in diameter. On one hand, this meets the requirements of the maximum size of the aggregate, on the other hand, according to the Saint – venant's principle, it is long enough to eliminate the stress concentration around crack caused by concrete anchors. In this experiment, when we apply tensile force through anchors, analysis data shows that stress concentration will not influence the stress state around the crack. Fig. 9 shows the tensile stress of two different parts of the specimen, and the stress concentration does not transfer to the area around the crack.

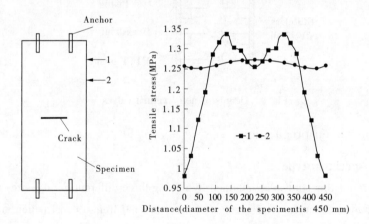

Fig. 9 The stress concentration caused by anchor

3.2 Specimen preparation

Each cylindrical specimen, 450 mm in diameter and 1 280 mm in height should be prepared by 0.21 cubic meters of concrete. Laboratory HZ – 2 rotating concrete mixer was employed to mix the concrete, it can produce 0.05 cubic meters of concrete each time. Making two concrete specimens will take three workers almost 8 hours. When pouring concrete, the circular stainless steel sheet should be protected by a metal pail 160 mm in diameter, as shown in Fig. 10. So that the impact of concrete vibrator on the circular stainless steel sheet can be further reduced.

After concrete poured, the specimen with concrete mould should be kept still for one week. Afterwards, the mould was dismantled and the hex cap screw and the steel wire were removed,

Fig. 10 **Vibrating concrete**

then channel 3 was blocked by the no cap screw, as shown in Fig. 4. Moreover, testing whether or not the water pipeline in the specimen can be used should be made immediately. If it can be used, the inlet and outlet of the pipeline should be blocked for protection, and the screw threads on concrete anchors should also be wrapped up for metal rust prevention.

Thereafter, the specimen was carefully transported to a concrete curing room – a humidity and temperature controlled room. According to Chinese Standard (Chinese standard NO. SL 352—2006), standard curing conditions are as follows: temperature of 20 ℃ ±5 ℃, relative humidity above 95% RH. In this case, according to the design age of concrete of gravity dam, the specimen should be maintained for 90 d(Chinese standard No. SL 319—2005).

4 Testing procedures

To investigate the effect of stress state on the hydraulic fracturing in concrete specimen and to study the general regularity ofhydraulic fracturing subjected to strength of concrete, aseries of tests was conducted in four steps: Specimen installation, Debugging experiment, Hydraulic fracturing testing, and post – test analysis. The details of each step are presented in the followingsections.

4.1 Specimen installation

The concrete specimen was took out from the concrete curing room on the 90th day. After removing all protective measures, the specimen was installed on the uni – axial testing apparatus. To ensure the air in the water pipeline can be completely excluded and to avoid water remaining in the pipeline going into crack, after hydraulic fracturing occurs, the concrete specimen bottom which has water inlet and outlet should be downward. This can also avoid unnecessary interference from water remaining in the pipeline, if staining method fornoting the path of crack is adopted.

4.2 Debugging experiment

After the concrete specimen was installed on the uni – axial system, low confining stress (i.e., 2 kN) was applied to fasten the specimenin order to prevent dislocation of the specimen, which could cause injury during debugging experiment. Then the water inlet of the specimen was connected with the high pressure water supply system, and the water outlet was connected with a high – pressure hydraulic hose 1 m in length, which was connected with a pressure gauge and a valve. In order to exclude the air in the water pipeline, four steps were executed: open the valve near the water outlet; launch the high water pressure supply system; keep the pressurized water supply system working until water flows out from the valve; meanwhile, close the valve and stop the high pressure water supply system. Before hydraulic fracturing testing, the surface of the concrete specimen should be dried by airing in order to making it easy to observe water leaking on the specimen surface.

4.3 Hydraulic fracturing testing

First of all, a constant load was applied on the specimen by the uni – axial system, in this case, as shown in Fig. 11, the tensile load was 113 kN, and this is equivalent to the tensile pressure of 0.7 MPa. After the proposed stress state had been applied to the specimen, the HP water supply system was launched to supply high water pressure into the crack of the concrete specimen.

Fig. 11 Constant load was applied on the specimen

The hydraulic pressure difference value, j, was increased in stages to the final value (i.e., 0.5 MPa per hour for $j < 1.0$ MPa, 0.2 MPa per hour for 1.0 MPa $< j <$ 2.0 MPa, and 0.1 MPa per hour for $j > 2.0$ MPa). A typical increasing process of the water pressure added step is shown in Fig. 12.

The selection of the difference value was based on the following considerations:

(1) Some researches shows that water head under 100 m generally do not cause hydraulic fracturing on gravity dam, the hydraulic pressure value that initiates the hydraulic fracturing of a concrete specimen is usually higher than 1 MPa (Jia et al. 2013), thus the hydraulic pressure increases to 1 MPa in two steps at the beginning;

(2) Reservoir impoundment is a very long process, often lasts several months or several years, thus the hydraulic pressure is increased at a low rate (i.e., 0.1 per hour) when the water pressure is larger than 2.0 MPa in order to capture the initiation value more accurately;

(3) In this case, the predicted water pressure that initiates the hydraulic fracturing is higher than 2 MPa, so the hydraulic pressure is increased at a higher rate (i.e., 0.2 per hour) when the water pressure is between 1.0 MPa and 2.0 MPa to shorten the testing period.

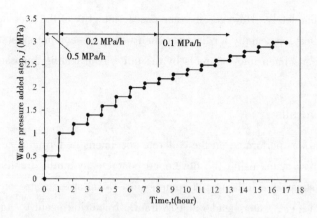

Fig. 12 Illustration of water pressure added step

It must be ensured that water pressure be kept constant for a long period, because the water pressure changes very little and it is always a long-term loading in a real project, such as in a dam; another important reason is the effects of long-term loading, saturation and water penetration into the micro cracks may reduce the characteristic values obtained under short-term conditions (Brooks 2015). Two major reasons were considered during determining the procedure of applying water pressure. On one hand, for concrete specimen which is 150 mm high, it is considered that water can penetrate through the specimen in 8 hours according to test code of concrete permeability test of Chinese standard (Chinese standard NO. SL 352—2006); on the other hand, in this hydraulic fracturing test, fracture process zone of concrete is the key zone significantly affected by penetration, and fracture process zone of concrete is usually 20 cm in length (Galouei and Fakhimi 2015). So that, it is considered reasonable each water pressure stage should be kept for at least one hour.

For each stage, the hydraulic pressure was increased successively within the first 1min, and then it was kept constant during the following 59 min. During the constant pressure period, the computer will regulate hydraulic pressure to the correct value when the sensor detects a deviation of 0.05 MPa. Sometimes, leakage of the specimen will cause serious pressure loss, however, the computer will always regulate hydraulic pressure to the correct value.

Once the HP water supply system can't maintain steady hydraulic pressure, at that moment leakage of the specimen has reached more than 1 mL/s and hydraulic pressure declines sharply, or when the specimen is ruptured, it means that hydraulic fracturing occurs. The maximum hydraulic pressure that the HP water supply system can reach is hydraulic fracturing pressure of the concrete specimen.

4.4 Post-test analysis

After the hydraulic fracturing test, the specimen was disassembled from the uni-axial system, and then it was divided into two parts along the fracture plane. Details of the fracture plane should be recorded clearly. If staining method for noting the path of cracks had been adopted, the area of the fracture plane could be acquired. In general, hydraulic fracturing can split most part of the specimen into two parts, thus the turbulence while disassembling the specimen is able to divide the specimen completely. However, in some cases, integrity of the specimen is still good after hydraulic fracturing test, especially for the specimen with pre - compressive stress, and it is necessary to stretch the specimen into two parts by the uni - axial testing apparatusas the last step of the test.

5 Experimental results

A series of tests was performed on the concrete specimens undermulti - concrete age and different uni - axial stress states using the developed concrete hydraulic fracturing apparatus. The purposes are to investigate the influence of complexuni - axial stress states on the development of hydraulic fracturing (i. e. , water head when hydraulic fracturing occur), and to study the relationship between concrete strength and hydraulic fracturing.

The details of the testing program are summarized in Table 3. All the test specimens are made by the same concrete materials with concrete grade of C25W10F100. According to the test report of concrete of LongTan gravity dam (216.5 m high), the concrete impermeability is good, and its freeze - thaw cycles is as high as 100. The concrete is located at a base layer from dam heel to dam toe, inevitably, there are cracks on its upstream surface because of influence of thermal stressand dry shrinkage (Zhu,2006).

Table 3   Summary of hydraulic fracturing for the unstressed, uni - axial tensile and compressive specimens

Specimen type	Specimens with different stress states	Age of concrete (d)	Uni - axial tensile strength(MPa)	Pressure on both ends of the specimen(MPa)	Test value of hydraulic fracturing pressure(MPa)
I	Unstressed	7	2.15	0	1.5
I - R	Unstressed	7	2.15	0	1.6
II	Unstressed	28	3.4	0	2.4
III	Unstressed	90	4.1	0	2.7
IV	Uni - axial compressive	28	3.1	1.0	3.2
V	Uni - axial tensile	180	4.4	-0.71	2.3

5.1 General feature of hydraulic fracturing of concrete

Take specimen IV (uni - axial compressive stress $q = 1.0$ MPa) as an example. A compressive load of 159 kN was applied at both ends of the specimen as shown in Fig. 13. When the hydraulic pressure increases to 3.2 MPa, the water started to percolate from the specimen surface as shown inFig. 14. Several minutes later, large amount of water suddenly flowed from cracks at the specimen surface, and the hydraulic pressure declined sharply at that moment. It meant that hydraulic fracturing occurred, and test value of water pressure of hydraulic fracturing was 3.2 MPa.

After the hydraulic fracturing test, pipes connected with water inlet and outlet of the speci-

Fig. 13 Hydraulic fracturing test of concrete

Fig. 14 Leakage at specimen surface

men should be removed firstly, and then water remained in pipes was drained out immediately. It is important to make sure that the water was drained completely, because the remained water can contaminate surface of the specimen. Afterwards, the specimen was disassembled from the uni-axial system, and then it was divided into two parts along the fracture plane. The area of hydraulic fracturing noted by the staining method was 785 cm^2, 56% of the total area.

5.2 Hydraulic fracturing under different concrete strengthsand different stress states

The crack propagation is dependent on the loading conditions and material strength (Ohlsson et al. 1998). Specimens Ⅰ, Ⅱ, Ⅲ, were tested under the same uni-axial stress, but different concrete strengths. Specimens Ⅳ, Ⅴ were tested under different uni-axial stress. Results showed that the better the concrete strength, the higher the water pressure against hydraulic fracturing, and with the increase of uni-axial compressive pressure, water pressure of hydraulic fracturing increase linearly (Ito and Hayashi 1993).

The test value of water pressure of hydraulic fracturing of specimens Ⅰ, Ⅱ, and Ⅲ are 1.5 MPa, 2.4 MPa, and 2.7 MPa, respectively. And the test value of specimens Ⅳ and Ⅴ are 3.2 MPa and 2.3 MPa, respectively, meanwhile pressure on both ends of these two specimens are 1 MPa and −0.71 MPa, respectively. Here the negative values mean the tensile stress. According to some research (e.g., Moseley and Ojdrovic et al. 1987; Ojdrovic and Stojimirovic 1987; Li and Joan 1990), the concrete uni-axial tensile strength positively correlated with fracture toughness, so that a higher tensile strength is associated with higherwater pressure against hydraulic fracturing, as responded by specimens I, II, and III. The load on both ends of specimens enhanceor reduce the stress concentrationat the edge of concrete crack. Therefore, the fracturing resistance of the concrete structure becomes higher or lower, as responded by specimens IV and V.

Fig. 15 shows relationships of uni – axial tensile strength and fracturing pressure in the five tests. Triangle markers in Fig. 15 are results of I, II, III, and round markers are results of IV, V. The value of Y – axis is $p_w - p_0$, where p_w refers to water pressure against hydraulic fracturing, p_0 refers to compressive pressure on both ends of the specimen. Least – squares fitting procedures were commonly used in data analysis, and a formula was fitted, as follow.

$$p_w - p_0 = 0.6f_t + a \tag{1}$$

where, p_w—refer to water pressure of hydraulic fracturing (MPa); p_0—refer to compressive pressure on both ends of the specimen (MPa); f_t—refer to uni – axial tensile strength of the concrete (MPa); a—constant coefficient, more data in the future may be able to fit a more accurate coefficient.

Taking into account the is 0.32 fitted by the five specimens in this paper, the rule revealed by this formula is coincidental with the test and numerical simulation results reported by Jia and Wang etc. (Jia and Wang et al. 2013; Wang and Zou et al. 2014).

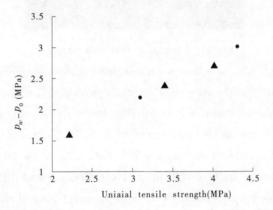

Fig. 15　Relationships of uni – axial tensile strength and fracturing pressure

The observation suggests that the concrete structure's ability to resist hydraulic fracturing, such as concrete dam, is low. The test condition was designed to conform to the dam heel concrete of the gravity dam designed by current principle. And results show that concrete under the state without stress is easy suffering hydraulic fracturing. Therefore, gravity dam higher than 200 m designed by Chinese standard may be not safe. Dams designed by principles in USA and Switzerland are safer compared with dams designed by Chinese principle. But it may have problems when considering hydraulic fracturing for some cases.

Moreover, under different stress state, hydraulic fracture surface appearance is different, as shown in Fig. 16. As the compressive stress on two ends of the specimen increases, the fractures typically penetrate through the gravels completely rather than propagating around the gravels, and the water pressure of hydraulic fracturing decreases with increasing tensile stress on two ends of the specimen.

(a)Specimen with uni-axial tension   (b)Specimen wion uni-axial compression

Fig. 16 Appearance of fracture surface

5.3 Test Repeatability

To investigate the repeatability of the test results and to validate the testing apparatus, specimen I - R was tested under the same concrete strength and stress state as those of specimen I. Specimen I - R was prepared and tested following the same testing procedures for specimen I. The water pressure of hydraulic fracturing was 1.6 MPa, which was very close to the result of specimen I. The difference is approximately 6.7%. That error is acceptable in a concrete experiment. If the precision of the HP water supply system is higher, such as 0.01 MPa, that error may be smaller.

6 Conclusions

A new test method for studying hydraulic fracturing of gravity dam was developed to evaluate the influence of hydraulic fracturing for concrete structure with different concrete strength and uni - stresses.

Using concrete with mix design of Longtan RCC dam and simulating the stress condition of dam heel, a series of tests was carried out under multi - concrete ages and different uni - axial stress states using the developed apparatus. The testing observation suggested that, with the increase of high water pressure, there exist a hydraulic pressure at which hydraulic fracturing of the specimen occurs. When the applied high water pressure reaches the critical value that causes the hydraulic fracturing, there is a roughly split of the specimen parallel to the preset crack.

According to the results of five tests, a formula $p_w - p_0 = 0.6 f_t + a$, was recommended. It approximately revealed rules of hydraulic fracturing of concrete specimen under uni - tensile or compressive stress which is similar to the state of surface concrete at gravity dam heel. The formula shows that only about 0.6 times concrete tensilestrength is contributed to resisting the hydraulic fracturing. Generally, tensile strength of concreteis low. For RCC dam, the weaklayer will decrease the tensile strength. Therefore, hydraulic fracturing is a key issue for high concrete dam.

It is clear that stress principle for gravity dam should be investigated in detail based on results achieved. The design code in USA and the Switzerland is safer compared with in China for dams higher than 200 m. But the hydraulic uplift pressure was not considered when design Grand Dixence dam in Switzerland, and the principle may be not safe for dams considering hydraulic fracturing if the uplift pressure was included.

Acknowledgements

The research was substantially supported by the National Basic Research Program of China (973 Program), 2013CB035903.

References

[1] ACI Committee 544, Measurement of properties of fiber reinforced concrete (Reapproved 2009), ACI Committee 544 report 544.2R – 89. Detroit: American Concrete Institute; 1989. 6-8.

[2] Bhattacharjee, S. S., Léger, P.. Application of NLFM Models to Predict Cracking in Concrete Gravity Dams [J]. Journal of Structural Engineering, 1994,120(4):1255-1271.

[3] Brooks, J. J.. Concrete and Masonry Movements, Butterworth Heinemann, 281-348.

[4] Chen, Z. H., Feng, J. J., Li, L., Xie, H. P.. Fracture analysis on the interface crack of concrete gravity dam Key Engineering Materials, 2006(324 ~ 325 I):267-270.

[5] Chinese standard, 2005, "Design specification for concrete gravity dams," SL319—2005.

[6] Chinese standard, 2006, "Test code for hydraulic concrete," SL 352—2006.

[7] Feng, L. M, Pekau, O. A., Zhang, C.. Cracking analysis of arch dams by 3D boundary element method [J]. Journal of Structural Engineering, 1996,122(6). 691-699.

[8] Galouei, M., Fakhimi, A.. Size effect, material ductility and shape of fracture process zone in quasi – brittle materials[J]. Computers and Geo – technics, 2015(65):126-135.

[9] Guan, J. F., Li, Q. B., Wu, Z. M. Determination of fully – graded hydraulic concrete fracture parameters by peak – load method[J]. Gongcheng Lixue/Engineering Mechanics, 2014,31(8):8-13 (in Chinese).

[10] Ito, T., Hayashi, K.. Analysis of crack reopening behavior for hydro – frac stress measurement," Haimson B. Rock mechanics in 1990 s: Pre – print proceedings of the 34th U. S. Symposium on Rock Mechanics [C]. Madison: University of Wisconsin Madison, 1993. 335-338.

[11] Janssen, D. J.,Snyder, M. B.. SHRP C – 391: Resistance of Concrete to Freezing and Thawing, (Washington, DC: Transportation Research Board)

[12] Jia J. Sh., Wang, Y. etc.. Simulation method of hydraulic fracturing and discussions on design criteria for super high gravity dams[J]. Shuili Xuebao/Journal of Hydraulic Engineering, 2013. 44(2):127-133 (in Chinese).

[13] Jiang, Y. Z., Ren, Q. W., Xu, W., Liu, S.. Definition of the general initial water penetration fracture criterion for concrete and its engineering application[J]. Science China Technological Sciences, 2011. 54(6):1575-1580.

[14] Li, V. C., Joan, H.. Relation of concrete fracture toughness to its internal structure[J]., Engineering Fracture Mechanics, 1990,35(1 ~ 3):39-46.

[15] Liang, W. H., Qiao, C. X., Tu, C. L.. Summarized Study on Cracks in Zhexi Concrete Buttress Dam [J]. Shuili Xuebao/Journal of Hydraulic Engineering, 1982(6): 1-10 (in Chinese).

[16] Liu, Z. G., Zhou, Z. B., Yuan, L. W.. Calculation of vertical crack propagation on the upstream face of the Zhexi single buttress dam[J]. Shuili Xuebao/Journal of Hydraulic Engineering, 1989(1):29-36 (in

Chinese).
[17] Moseley, M. D. , Ojdrovic, R. P. , Petroski, H. J. . Influence of aggregate size on fracture toughness of concrete[J]. Theoretical and Applied Fracture Mechanics, 1987,7(3):207-210.
[18] Ohlsson, U. , Nystr m, M. , Olofsson, T. , Waadaard K. . Influence of hydraulic pressure in fracture mechanics modelling of crack propagation in concrete[J]. Materials and Structures,1998(31):203-208.
[19] Ojdrovic, R. P. , Stojimirovic, A. L. , Petroski, H. J. , 1987, "Effect of age on splitting tensile strength and fracture resistance of concrete," Cement and Concrete Research, Vol. 17, Issue 1, 70-76.
[20] Secchi, S. , Schrefler, B. A. . "A method for 3 – D hydraulic fracturing simulation," International Journal of Fracture, 2012,178(1~2):245-258.
[21] Silva, B. G. , Einstein, H. H. , 2014, "Finite Element study of fracture initiation in flaws subject to internal fluid pressure and vertical stress," International Journal of Solids and Structures, 51 (2014), 4122-4136.
[22] Slowik, V. , Saouma, V. E. . Water Pressure in Propagating Concrete Cracks[J]. Journal of Structural Engineering, 2000,126(2):235-242.
[23] SwissCOD, 1985, Swiss dams – Monitoring and Maintenance, Swiss national committee on dams, 1985.
[24] USBR, 1976, "DESIGN OF GRAVITY DAMS – DESIGN MANUAL FOR CONCRETE GRAVITY DAM", A Water Resources Technical Publication, Denver, Colorado, 1976.
[25] Utili, S. , Yin, Z. Y. , Jiang, M. J. . Influences of hydraulic uplift pressures on stability of gravity dam [J]. Yanshilixue Yu Gongcheng Xuebao/Chinese Journal of Rock Mechanics and Engineering, 2008,27 (8):1554-1568 (in Chinese).
[26] Wang, X. Z. , Zou, H. F. , etc. . Finite element simulation and comparison of hydraulic splitting fracturing test of concrete[J]. Applied Mechanics and Materials, 2014,(678):551-555.
[27] Zhang, X. X. , Ruiz, G. , Yu, R. C. . A new drop – weight impact machine for studying fracture processes in structural concrete[J]. Strain 2010, 2010,46(3):252-257.
[28] Zhu, B. F. . On Some Important Problems about Concrete Dams[J]. Engineering Science, 2006,8(7): 21-29 (in Chinese).

Review on RCC Dams in Turkey

Dr. Ersan YILDIZ[1], Dinçer AYDOĞAN[2]

(1. Temelsu Int. Eng. Services Inc., Turkey;
2. DSİ 26. Bölge Müdürlüğü(26. Reg. Dir. of State Hydraulic Works), Turkey)

Abstract: The aim of this paper is to present a summary about the design and construction methodologies and key characteristics of Roller Compacted Concrete (RCC) dams in Turkey. In parallel to the development of hydropower projects, there has been a significant increase in the number of RCC dams in Turkey in the recent years. At the moment, there are 22 RCC dams in operation and 24 under construction. It is intended to give brief information about the key properties of these dams including height, RCC volume, cross-section and geometry, RCC material, foundation characteristics, upstream face for watertightness, joints, lift joints, placement procedures, foundation-RCC contact and instrumentation. Moreover, some detailed information will be given about the design and construction of two important dams (which are in operation) in Turkey which are : Cindere Dam ($H = 107$ m) as the highest hardfill dam in the world and the first hardfill dam in Turkey; and Ayvalı Dam ($H = 177$ m) as the highest RCC dam in Europe.

Key words: RCC, Hardfill, Turkey, Cindere, Ayvalı

1 Introduction

In Turkey, the use of Roller Compacted Concrete (RCC) technology in the dam body construction started in 1998 for Çine Dam Project ($H = 137$ m) followed by the use of RCC with low cementitious content (also referred as hardfill) in 2000 for Cindere Dam ($H = 107$ m). In the last 15 years, increasing water demand and the development of hydropower projects have led to a significant increase in the number of dams, where RCC method has also been widely adopted due to its several advantages including rapid, practical and economical construction. At the moment, there are 22 RCC dams in operation and 24 under construction, which makes Turkey one of the leading countries in RCC technology (Table 1).

In this paper, it is aimed to present a general summary about the characteristics of RCC dams in Turkey. Moreover, the highest hardfill dam in the world, Cindere Dam and the highest RCC dam in Europe, Ayval₁ Dam are presented in more detail.

2 General Characteristics

The height of the RCC dams in operation varies between 36 ~ 177 m, and 4 of these dams arehigher than 100 m. Among the dams in construction, 4 dams are over 100 m and the height is

Table 1 Summary on RCC Dams in Turkey (Not all included)

No.	Name	Status	Purpose	River	Height	Length	RCC Vol.	Total Conc. Vol.	Upstream Slope	Downstream Slope	Cement kg/m³	Pozzolan kg/m³	Pozzolan Type
1	Ayvalı	In Operation	H	Oltu	177	405	1,650	1,900	V	0.75	50	115	F
2	Çine	In Operation	HWI	Çine	137	300	1,560	1,650	0.1	0.85	85	105	F
3	Cindere	In Operation	HWI	B. Menderes	107	280	1,500	1,685	0.7	0.7	50	20	F
4	Köprü	In Operation	H	Göksu	103	413	880	1,050	0.6 - V	0.8 - 1.0	85	45	F
5	Beyhan-I	In Operation	H	Murat	97	361	1,480	1,661	0.7	0.7	130	0	M
7	Beydağ	In Operation	HWI	K. Menderes	96	800	2,350	2,650	0.35	0.8	60	30	F
9	Menge	In Operation	H	Göksu	73	304	321	384	0.25	0.80	80	40	F
10	Güllübağ	In Operation	H	Çoruh	72	97	160	175	V	0.5 - 0.7	70	70	F
11	Feke II	In Operation	H	Göksu	71	256	194	227	0.07	0.8	60	60	F
13	Gökkaya	In Operation	H	Göksu	69	118	96	122	0.07	0.8	55	55	F
15	Şırnak	In Operation	H	Ortasu	67	198	245	265	0.1	0.8 - 1.0	95	0	-
17	Akköy II (Aladereçam)	In Operation	H	Karaovaçk	60	260	101	196	V	0.8	85	85	N
18	Akköy I	In Operation	H	Harşit	53	146	46	101	0.1	0.8	100	100	N
19	Çamlıca 3 HES	In Operation	H	Zamantı	52	186	160	182	0.7	0.7	88	37	F
22	Su Çatı	In Operation	H	Güredin	36	192	55	60	V	0.8	50	100	S
43	Yukarı Kaleköy	Under Construction	H	Murat	150	404			0.25	0.8			
40	Göktas	Under Construction	H	Zamantı	135	200			0.3	0.6			
24	Melen	Under Construction	HWI	Melen	124	944			0.7	0.7			
35	Ergenli	Under Construction	I	Ilıca	107	540			0.2	0.7			
44	Kargı	Under Construction	H	Sakarya	95	270			0.1	0.8			
32	Naras	Under Construction	HWI	Manavgat	78	448			0.15	0.7			
28	Kavşaktepe	Under Construction	H	Robozik	71	268			0.1	0.8			
30	Musatepe	Under Construction	H	Robozik	66	165			0.1	0.8			
34	Akçakoca	Under Construction	W	Sarma	66	148			0.1	0.8			
38	Kelebek	Under Construction	I	Kelebek	60	193			0.1	0.8			
33	Ardıl	Under Construction	HWI	Ardıl	54	246			V	0.8			
39	Ergani	Under Construction	I	Gölkum	54	229			0.1	0.8			
31	Gölgeliyamaç	Under Construction	H	Güzeldere	52	150			V	0.8			
27	Ballı	Under Construction	H	Robozik	51	188			0.1	0.8			
36	Karareis	Under Construction	HWI	Camıboğazı	50	298			0.05	0.8			
29	Çetintepe	Under Construction	H	Robozik	39	173			0.1	0.8			
37	Gümüşören	Under Construction	HWI	Zamantı	37	842			0.7	0.75			
41	Devecikonağı	Under Construction	H	Emet	29	313			0.1	0.75			

Nots: Purpose:H(Hydropower), W(Water Supply),I(Irrigation)
Pozzolan Type:F(Class F Fly Ash), M(Milled Sand), N(Natural Pozzolan), S(Ground Granulated Blast Furnace Slag)

between 29 ~ 150 m. A table including the summary of general information about the RCC dams in operation or in construction is given at the end of the paper.

2.1 Dam geometry and RCC volume

The cross – section of the RCC dams and the slopes of the upstream and downstream faces show a significant variation, there is not a common cross – section. Dependingon the highly variable seismicity, geotechnical characteristics of the foundation rock and the design strength, a wide range of slopes have been designed for the faces of RCC dams. The upstream face slope varies between vertical and $0.7(H):1(V)$ whereas the downstream slope has a narrow range and is generally between $0.7(H):1(V)$ and $0.8(H):1(V)$. Especially for high seismic regions, a faced symmetrical dam geometry (which is common for hardfill dams) has been adopted in some RCC dams.

The maximum volume of total concrete and RCC in a dam project are 2 650 000 m^3 and 2 350 000 m^3 respectively for Beydağ Dam with a height of 96 m, that is in operation since 2008. For six RCC dams in Turkey, the volume of the RCC is higher than 1 500 000 m^3. The percentage of RCC volume over the total concrete generally varies between 80% ~ 95% for large projects.

2.2 RCC material

The design strength of RCC material is generally chosen between 10 MPa and 16 MPa. Use of pozzolanic material in RCC mixture is adopted in order to increase workability, reduce heat of hydration of concrete, increase the strength and durability in the long term. The cementitious material is generally a mixture of Portland cement and fly ash (Class F). Fly ash content compared to Portland cement shows variations but is generally less. Althogh the use of pozzolan is a common approach in RCC mixtures, in a few projects the cementitious content included only Portland cement with no use of pozzolan. The selection and process for aggregates used in RCC are similar to those employed in conventional concrete. Natural deposits or crushed stone are either used depending on the availability and cost.

2.3 Foundation characteristics

As RCC can be regarded as mass concrete as a material, the foundations which are suitable for mass concrete are considered as appropriate for RCC dams. In this regard, the general approach is to build the RCC dam on the rock foundation with sufficient strength and stiffness properties. For this purpose, the final elevation and layout of the foundation are determined based on the findings of detailed geological and geotechnical investigations.

The shear strength of the rock material is generally greater required for the stability of the dam and the critical failure surface is either the dam – foundation contact or the lift joints within the dam body. For this reason, special attention is paid to dam – foundation contact and a careful cleaning of the exposed rock surface is carefully cleaned and bedding mortar is applied to the whole contact area. Stiffness of the foundation is another important criterion and the foundation rock properties should satisfy deformations small enough which will not lead to cracks in the dam body at the end of construction or in operation with full reservoir level.

It can be said that all RCC dams in Turkey are founded on rock foundations which are well sufficient in terms of strength and stiffness as a foundation for a mass concrete structure. Significant improvements by grouting or dental excavation and concrete backfilling of local poor zones have also been employed in some projects. None of the RCC dams (including hardfill dams with relatively low strength and stiffness) are founded on very weak rock or soil foundations partially or completely.

2.4 Watertightness

Thewatertightness of the dam is mainly provided by the upstream face of the dam. The most common technique to provide imperviousness is the use of a conventionally vibrated concrete (CVC) strip at the upstream face (generally around 1 m) with the use of formworks. The second most common method is the application of grout – enriched RCC against formworks. In a few prjects including Cindere and Beydağ dams, use of precast PVC panels have also been adopted at the upstream face to provide impermeability.

Behind the CVC or grout – enriched RCC, a strip zone (of around 10 m) where bedding mortar is applied between lifts is also provided in many projects both to reduce the possible leakage and to create a zone with higher tensile and shear strength in the critical area in terms stresses. The typical cross – section of Menge Dam, where a bedding mortar zone is applied behind CVC facing is shown in Fig. 1.

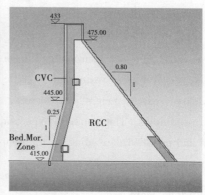

Fig. 1 Typical Cross Section of Menge Dam

2.5 Joints, Lift Joints and Placement Procedures

All RCC dams bodies in Turkey are divided into blocks by placing vertical construction joints, in order to avoid cracking due to thermal shrinkage. The interval of the verticaljoints are determined from a thermal analysis considering the material thermal properties and placement procedures. The common procedure for the division of the blocks is driving a plate or sheet to the continuously placed and compacted RCC layer. One or two layers of waterstops are installed at the joints near the upstream face to prevent seepage through these joints.

The locations of the vertical joints are also evaluated by taking into account the variations in the foundation rock. Additional joints are installed or the joint intervals are adjusted to place thevertical joints at the locations of abrupt changes in foundation rock stiffness, in order to prevent potential cracks due to differential settlement in the foundation.

The horizontal joints (lift joints) between the successive RCC layers are critical to the seepage performance or stability of the dam body. The treatment for the lift surfaces is a matter of the condition of the joints, characterized as either "hot" or "cold". This characterization is mainly based on the maturity factor based on the time in hours lapsed between two successive layers. In hot joints, no treatment is necessary as long as the surface is kept clean and moist.

For cold joints, the surface treatment includes the application of pressurised air or water, cleaning with vacuum equipment and application of a layer of bedding mortar. In some projects, bedding mortar application is held mandatory at a limited width of the upstream face (about 10 m) for seepage and strength performance.

The mixing of RCC is carried out in batching plants with a capacity which will allow continious placing depending on the speed of placement. The transfer of RCC from the plant to the site is performed by using conveyor systems, conventional trucks or a combination of both. The common method in recent projects is to use a conveyor belt for transfer from the plant and load the trucks at dam site forsurface distribution.

Bulldozers are used to spread theRCC mix poured at site. The compaction of RCC is carried out with vibrating rollers of typically 10 ~ 15 tons. Other light rollers (3 tons) or compaction equipment are used for areas close to faces, galleries or structural elements. The thickness of the layer is typically 30 cm.

2.6 Instrumentation

All RCC dams are instrumented according to common pratice of DSİ (State Hydraulic Works of Turkey). A typical instrumentation for a RCC dam includes : pressure cells, piezometers, strain gages, extensometers, joint meters, temperature sensors, accelerometers and normal or inverted pendulums. The number, type and locations of these instrumentation devices are determined and installed depending on the dam properties like the size, purpose, seismicity, foundation conditions etc.

3 Cindere Dam

Cindere Dam is the first hardfill dam in Turkey, located in Denizli Province. With its maximum height of 107 m from the foundation, it is the highest hardfill dam in the world. The crest length and width are 280 m and 10 m respectively. The total hardfill volume is 1 500 000 m^3 whereas the total concrete volume is 1 685 000 m^3.

Cindere Dam has been designed as Faced Symmetrical Hardfill Dam with the slope ratio of $0.7(H):1(V)$. Hardfill can be defined as a lean RCC with relatively low cementitious content. This type of dam geometry has better characteristics with regard to the stability, low sensitivity to differential settlements and high resistance against earthquake excitations. Low modulus ofdeformation provides better adaptation to settlements. Flatter upstream slope increases the stability by making use of water weight and wider foundation base reduces the base pressures and settlements enhancing the sliding stability which is more important in high seismic zones.

The foundation rock consists of schist masses. The schist masses include mica schist, calc schist, sericite schist and graphite schist. The geotechnical evaluation based on rock mass classifications has resulted in adeformation modulus of 5 000 MPa for the foundation rock which is sufficient for a hardfill dam.

The design strength of the RCC is 6 MPa (180 days compressive strength). After trials with several mixes with different cement ratios, this strength has been achieved with a cementitious material of 50 kg/m^3 Portland cement + 20 kg/m^3 fly ash. Three different well graded aggregate sizes have been used for maximum density where 55% of the aggregates were between 0 ~ 10 mm, 25% between 10 ~ 25 mm and 20% between 25 ~ 75 mm. The RCC mix has been transported by trucks, spread by bulldozers and placed in layers of 30 cm thick and compacted by vibrating rollers.

The upstream face is formed by precast panels. A PVC membrane stuck on the inner side of precast concrete panels provides the imperviousness on the upstream face. Precast panels have been used as a formwork for hardfill placement and also a protection for the impervious membrane.

The dam has been constructed in the years between 2002 and 2009. The maximum daily and monthly production have been recorded as 6 620 m^3 and 140 000 m^3 respectively. Fig. 2 shows a downstream view of Cindere Dam.

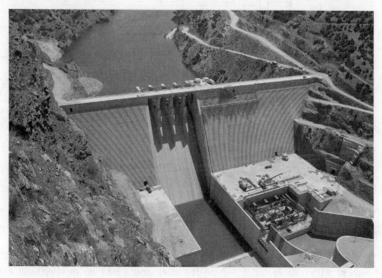

Fig. 2 Cindere Dam

4 Ayvalı Dam

Ayvalı Dam is a RCC dam located in Erzurum Province. The maximum height of the dam is 177 m from the foundation, that makes it the highest RCC dam in Europe. The crest length and width are 390 m and 8 m respectively. The total RCC volume is 1 650 000 m^3 and the total concrete volume is 1 900 000 m^3.

The upstream face of the dam is vertical and downstream face has a slope of 0.75 $(H):1(V)$.

The cross section of the dam is shown in Fig. 3.

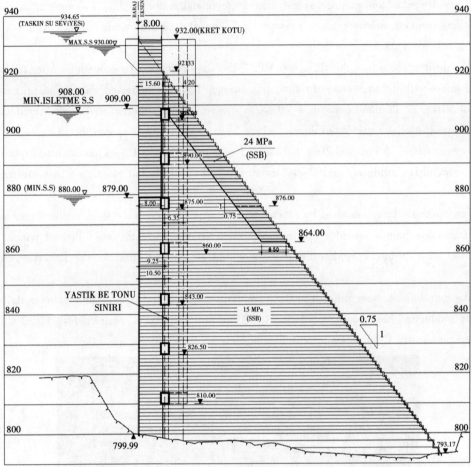

Fig. 3   Cross – Section of Ayvalı Dam

The dam is founded on a volcanic rock mass which is sufficient in terms of strength and stiffness as a foundation rock for a RCC dam.

Two different RCC design strengths are used in Ayvalı Dam. A compressive strength of 15 MPa is designed for the lower portion of the dam body whereas a higher strength of 24 MPa is used at the upper portion of the body, as indicated in Figure 3. The cementitious material of (50 kg/m^3 Portland cement with 110 kg/m^3 F – type fly ash) and (100 kg/m^3 Portland cement with 120 kg/m^3 F – type fly ash) are used for the RCC mixes of 15 MPa and 24 MPa strengths respectively.

The upstream face is grout – enriched RCC against formwork to providewatertightness. Two layers of PVC water stops have been installed at the vertical construction joints within the grout – enriched zone.

The concrete aggregate for dam body RCC has been obtained from the upstream river bed by washing and crushing. The aggregates have been stocked with preventive measures against segregation and transported to RCC plant by a 900 m long conveyor band. RCC batching plant included 2 units and 4 mixers. The produced RCC mix has been transferred to dam site by a 350 m long conveyor system and loaded to trucks for surface distribution. Bulldozers and vibrating rollers (18 tons and 22 tons) have been used for spreading and compaction of RCC layers of 30 cm in thickness.

During continuous placement of RCC, no bedding mortar application has been performed. Only for cold joints, the lift surfaces have been treated by cleaning and bedding mortar application. At the rock – RCC contact, the rock surface has been carefully cleaned and bedding mortar has been applied at the whole interface.

RCC placement has been stopped in the daytime and nights for hot and cold weather conditions respectively. RCC mix temperature has always been kept over 10 ℃ in cold weather and below 28 ℃ in hot weather. RCC lift surfaces have always been kept moist in hot weather and covered with insulators in cold weather.

The RCC construction started in Dec. 01. 2012 and ended on Dec. 31. 2014. The maximum RCC production rates have been recorded as 6 000 m^3/day and 130 000 m^3/month.

Fig. 4 shows the downstream view of Ayvalı Dam.

Fig. 4　Ayvalı Dam.

Acknowledgements

The authors would like to express their sincere thanks to Özkar Construction Company for the information about Ayvalı Dam.

The authors are also grateful to Dr. A. Fikret Gürdil and Mr. Nejat Demirörs from Temelsu for the information about Cindere Dam.

The suggestions and help of Mr. Ferruh Anık on the paper are also acknowledged.

References
[1] Batmaz S.. Cindere Dam – 107 m high Roller Compacted Hardfill Dam(RCHD)[A] // Proceedings of the Fourth International Symposium on Roller Compacted Concrete (RCC) Dams[C]. Madrid:2003 ;121-126.

Immersion Vibrated RCC

Ortega F

(FOSCE Consulting Engineers)

Abstract: Roller – Compacted Concrete (RCC) technology has replaced in many cases the traditional way of building concrete dams, which was based on the use of Conventional Vibrated Concrete (CVC). This development is mainly motivated by lower construction costs of RCC compared with CVC dams. However, from beginning, in – situ quality has been a point of controversy. The concrete mix (or combination of mixes) used in some RCC dams, has not achieved the level of performance which is typically required in concrete dams, namely: density, strength, impermeability and durability.

A new mass concrete mix which is in the limit range of consistency between CVC and RCC has been developed recently. This concrete is called Immersion Vibrated – Roller Compacted Concrete (IV – RCC) and, when designed and placed in the dam following certain specifications and construction procedures, the same mix is suitable for consolidation either by traditional immersion vibration or by roller compaction. This finding has brought major simplifications in the design and construction of modern concrete dams. Both, quality and economy are derived from this simplicity.

Details of IV – RCC materials, typical mixture proportions, construction, and dam design benefits are discussed in this article. In addition, application of IV – RCC in several dam projects is presented. These experiences show that the end product has similar quality, if not better, than traditional mass concrete that was used in dam construction before RCCs broke into the dam world.

Key words: Immersion vibration, mix design, placement procedures, IV – RCC

1 Introduction

The design and construction of concrete dams worldwide follows two primary approaches: the conventional concrete (CVC) dam and the roller – compacted concrete (RCC) dam. Advantages and suitability of either option are analyzed for each particular dam project. Geometrical constrains, high quality and well – proven long – term performance seems to have been the main criteria for the selection of the CVC option in the past, especially in countries with a large tradition in concrete dam construction. However RCC dam construction has now a history of more than 35 years. The economy of RCC dams is the key of the success against CVC. The potential for cost reduction of RCC dams is based on speed of construction and materials optimization[1]. Improvements introduced in the last 15 years in the design and construction of RCC dams have shown that the in – situ quality and concrete properties of RCC can be as good (if not better) than those of dams that have been built with traditional immersion vibrated mixes (CVC).

Experience in construction of CVC and RCC dams has led recently to a combination of both technologies in the same structure, summing up their respective advantages. This is a new concept of RCC dam which is called IV – RCC (Immersion Vibrated RCC) dam and there are already some experiences around the world[1,2].

Fig. 1 De Hoop dam (h = 90 m, vol. = 1.0 Mm^3)
IV – RCC first experience (South Africa, 2013)

This interesting development has been achieved through optimization of RCC mixes towards a much more workable concrete than the typical dry mixes used in the past. These improvements have been required in RCC dams to guarantee bond between layers, the elimination of the potential for segregation and the quality of the dam faces.

2 Definition of IV – RCC

The term Immersion Vibrated Roller – Compacted Concrete (IV – RCC) is a variation of a workable RCC mix which has a consistency similar to a stiff traditional mass concrete of zero slump. IV – RCC mixes are produced, transported and spread as RCC, but the consolidation can be accomplished either by external roller compaction or by internal immersion vibration.

The decision on where to apply in the dam one or the other method of vibration is made on site based on both, design specification and construction practicality. Factors such as the proximity to formed faces or rock abutments, narrow areas of difficult access to large equipment, or especially critical dam zones like embedded instruments, drains or waterstops, are typical places where IV – RCC is immersion vibrated. The rest of the concrete mass is consolidated by roller – compaction.

It must be noted that the main difference from previous RCC dams is that the same concrete mix, without any additional enrichment of grout or mortar, is placed across the entire block. When designed and placed according to certain specifications, IV – RCC will have the same composition and long-term performance in all points of the dam regardless of the method of vibration. As such, the IV – RCC dam will exhibit at the long term many similarities with the traditional concrete dam.

3 Mixture proportions of IV – RCC

A typical range of mix proportions of IV – RCC mixes is included in Table 1. All IV – RCC mixes that have been used or tested at dam projects to date include pozzolanic mineral admixtures like fly – ash, natural pozzolan or slag. Up to now, the design of IV – RCC mixes excluding such components has not been successful. Further investigations are still ongoing in this regard.

Table 1 Mix proportions of IV – RCC

Material	Code	Kg/m^3
Coarse aggregate (>5 mm)	[CA]	1 350 ~ 1 550
Fine aggregate (<5 mm)	[FA]	700 ~ 850
Cement (type I)	[C]	50 ~ 70
Fly – ash / Natural pozzolan	[F/N]	70 ~ 150
Free water	[W]	85 ~ 130
Admixture	[A]	0.5% ~ 2%

Mixes containing fly – ash will generally require lower content of water and cementitious materials than those incorporating other mineral admixtures. Fine aggregate for IV – RCC mixes shall include up to 10% ~ 18% of non – plastic fines passing #200 mm (0.075 mm) sieve. The optimum proportion is a balance of cost and fines quality. A relatively high dose of a retarder admixture is required to extend the setting time of the mix. This chemical agent usually also has an important effect as water reducer. Table 2 shows a list of key – specifications of the IV – RCC materials and mixes.

Table 2 Materials and mix parameters of IV – RCC

Material / Mix parameter	Typical range	Remarks
Maximum size of aggregate (100% crushed)	38 ~ 50 mm	To avoid segregation
Proportion of fine aggregate [FA/(FA + CA)]	32% ~ 38%	Optimized gradation[1]
Voids in compacted fine aggregate [VC]	0.26 ~ 0.30	ASTM C29
Per cent of fines (<#200 – 0.075mm) in [FA] = [FI]	10% ~ 18%	Non – plastic fines
Water/Cementitious ratio [W/(C + F/N)]	0.45 ~ 0.70	by weight
Modified VeBe time (12.5 kg total surcharge)	6 ~ 12 s	At time of vibration
Volume of cementitious paste [C + F/N + W + A] = [P]	16% ~ 22%	160 ~ 220 litres/m^3
Volume of total paste [C + F/N + W + A + FI]	18% ~ 25%	180 ~ 250 litres/m^3
Cementitious paste/mortar ratio [P/(P + FA)]	0.36 ~ 0.42	by volume (> [VC] + 0.1)
Initial set (mortar sieved from the IV – RCC mix)	20 ~ 24 h	ASTM C403
In – situ density (% of theoretical – air – free density)	98% ~ 99.5%	Nuclear densiometer

These parameters refer to typical mixes that comply with most of the design criteria for medium and large size concrete dams in which direct tensile strength values up to 1.5 MPa or higher might be required at the long term. This level of in-situ concrete performance will, in addition to the required strength, guarantee the impermeability, density and durability of the mass concrete structure.

4 IV – RCC dam concept

4.1 General

While conventional mass concrete (CVC) dams are constructed in separate vertical approximately 15 m wide monoliths, RCC dams are constructed in principle in much larger blocks, ideally placing concrete in horizontal 300 mm thick layers from one abutment to the other. The availability of large construction areas allows that full advantage can be taken of the high capacity production, delivery and placement equipment and plant. Forward planning of the construction methodology and logistics is a key element of RCC dams. In addition, the degree of simplicity of the dam design will also play a major role in the efficiency of RCC construction.

Similarly to traditional CVC dams, IV – RCC dams follow the "overall" approach, i.e. the concrete is designed to be impervious and no additional upstream water – tight barrier or membrane is required to assure impermeability. The point of concern of early RCC dams has been the bond capacity between layers. Mix design optimization and construction techniques have since been developed that can improve significantly the quality of the lift joints.

4.2 Inter – layer bonding

Among other types of RCC dams, IV – RCC concept offers full guarantee of inter – layer bonding. IV – RCC is a super – retarded mix with long setting times, up to 20 or 24 hours of initial set. Placing successive layers within this time will assure interpenetration between the aggregate and paste of both layers in a similar way as it is done in the sub – layers placed in the lifts of a traditional CVC dam block. Cores taken from CVC and IV – RCC dams are very similar in this regard, as no planes of discontinuity can be appreciated. In CVC cores, larger size of aggregate, less mortar and a bit more of air can be seen than in IV – RCC, but the lack of visible joints is a common important factor in both cases.

The interpenetration between layers is achieved in IV – RCC dams regardless of the method of compaction that has been used. Both the effect at deeper levels of a large vibratory roller (RCC) and the length of the immersion vibration (CVC) are sufficient to achieve such bonding and to eliminate the potential horizontal planes of discontinuity. In fact the only discernible joints in cores are in both cases those surfaces that have required treatment. In terms of permeability and strength, these treated joints could be the weakest points of both types of dams (CVC and IV – RCC), if treatments are not fully accomplished as specified. The number of treated joints should be much less in IV – RCC dams than in CVC dams. In CVC dams the surface of the lifts are cleaned up to an exposed aggregate finish in every dam block and a mortar is spread as bedding mix ahead of the next lift. This happens systematically every 2.5 or 3.0 meters depending on the height of the forms. However, treatment of lift joints in CVC dams is normally a well controlled operation as the size of the block and the time available for such activities are usually not critical in the construction programme.

When a joint need to be treated in an RCC dam, the size of the block is much larger, and if not correctly planned, the activity may become critical. This is another important positive aspect of IV – RCC dams in which the continuous placement and the level of mix retardation reduces to a minimum the creation of such joints.

4.3 Facing and interface concrete

Many different ways of forming the faces of an RCC dam have been put into practice throughout the past years. In most cases, a CVC concrete mix, or an RCC with additional grout (GEVR) or mortar, has been used against the forms or as interface with the rock abutment, in addition to the RCC mix that was being used in the core of the dam. The use of different mixes and methodologies is against the principle of simplicity that should govern the construction of any RCC dam. However, this was necessary as previous RCC mixes directly placed in these areas would not achieve a durable and pleasant surface finish of the dam faces, or a good contact with the rock or other embedded structures.

The introduction of IV – RCC, a concrete mix that is suitable for immersion vibration in a similar way as stiff traditional mass concrete, and that in addition, can be easily roller compacted, is a significant further step in the development of RCC dams. The same mix fits either as CVC interface concrete or as RCC in the rest of the dam.

The advantages derived from this new approach of RCC dams are evident to dam designers, as in this way a great homogeneity is achieved in the concrete mass across the dam. In this scenario, the IV – RCC dam can essentially be assumed to behave under operating conditions in a similar manner to traditional CVC dams.

Fig. 2 Core testing of IV – RCC vs. CVC/GEVR as facing concrete at Enciso dam trial section (Spain, 2011)

5 Specification for construction

There are several methods and procedures that are specific of IV – RCC construction, which need to be well understood and trained prior to the start of concrete placement in the dam. The concrete production plants and delivery systems are similar to those commonly used in RCC dams. The RCC plants shall be well – controlled and regularly inspected and calibrated in order to assure a perfect control on the water content and the consistency of the mix within the specified limits. IV – RCC is very sensitive to moisture content, especially in mixes which are in the lower values of the water/cementitious ratio.

Aggregate plant may require incorporating a vertical – shaft impact (VSI) type – crusher to meet the usual specifications of the fine aggregate in terms of shape and gradation (see Table 2 above).

5.1 Spreading and compaction

Spreading of IV – RCC is made with a type Cat D4 or D5 laser – guided dozer in 320 ~ 350 mm thick layers to achieve the standard 300 ± 30 mm thickness after compaction. Crawler dozers provided with low – ground pressure tracks are recommended. Workers with shovels will always need to assist spreading at narrow areas and around embedded elements. This same team shall remove any spot of segregation that may have occurred during spreading, and before compaction, especially against forms or at the interfaces.

Roller compaction should follow immersion vibration of the edges of the layer and not the other way round. This procedure will guarantee a better finish of the faces and a good uniformity. Smooth single – drum vibratory rollers with a static weight between 8 and 10 t are required. Larger rollers are not recommended. A first static pass is followed by 4 to 6 passes in the dynamic (high – frequency) modus. Compaction is concluded when most of the surface is covered with paste that has been driven up from the concrete mass. Density will never be a critical factor in IV – RCC dams.

All areas that will not be reached by the large roller shall have been previously consolidated by immersion vibration. Small rollers and vibratory plates are just surface finishers and will neither improve density nor lift bonding. Smooth surface finish is advisable to facilitate cleaning but it is not necessary to improve the joint strength of IV – RCC.

5.2 Immersion vibration of IV – RCC

In addition to the good bonding of RCC lifts that have been roller – compacted, similar interpenetration can be achieved also between successive layers at areas which are consolidated by immersion vibration. To that end, equipment with enough vibrating capacity and well – trained procedures need to be implemented.

Up to now the vibration of the faces and interfaces of RCC dams has been made in general by hand – held pokers with quite variable outcomes. Backhoes mounted with a gang of vibrators as those required in traditional mass concrete dams, have seldom been used in RCC dams to date. This should be a standard unit in the list of equipment of any IV – RCC dam.

According to the maximum size of aggregate the diameter of the poker for IV – RCC mixes can be less than in traditional CVC dams. Manual pokers are 3 in (76 mm) diameter and those mounted on equipment might be up to 4 inches (100 mm) or more. The latter should be installed in line or staggered to improve efficiency. In all cases, a vibration depth of 500 mm is required to assure penetration of the poker in the upper half of the second lower layer that has not reached initial set. This procedure eliminates the physical joint and creates a similar effect as the one achieved in adjacent areas that are roller compacted, or as in CVC dam blocks.

The figure below shows a typical sequence of the placement of IV – RCC:

(1) a conveyor system is used to deliver the concrete mix from the plant to the dam,

(2) trucks in the dam bringing the concrete to the point of placement,

(3) the dozer spreads the mix in lanes parallel to the dam axis,

(4) a hydraulic backhoe mounted with a gang of vibrators is used to consolidate the facing and interface concrete by immersion vibration,

(5) roller compaction of the core, and

(6) insertion of crack inducers at transverse joints following roller compaction.

Fig. 3   **Enciso dam** (h = 103 m, vol. = 0.8 Mm3)
View of the IV – RCC placement (Spain, 2014)

The dotted line in the figure indicates the approximate limits between the respective areas where the same mix is consolidated by the two different methods. After consolidation no discernible interface can be appreciated between both areas.

5.3 Formation and sealing of transverse joints

RCC dams benefit from large placement areas. Therefore formed transverse joints should be reduced to a minimum. Various methods have been used to date to avoid traditional shutters at the joints of RCC dams, from very simple ones to quite sophisticated. When working with very workable mixes, like is the case in IV – RCC, the most popular way of forming the transverse joints is by inserting a crack inducer (either plastic or galvanized steel) in the fresh fully – compacted concrete (Fig. 4, left). Using this method, the induced crack might open during the service life of the dam in a more irregular shape than the traditional formed joints of CVC dams. This might induce additional shear forces between adjacent blocks. The intersections of these vertical joints with the dam faces or galleries should be centered with the waterstops, which are installed at the theoretical position of the joints. Therefore it is important to provide enough stability to the de-bonding sheets that are installed to that end in these areas close to the external faces.

Fig. 4 **Formation of transverse joints (left) and structure fixing the upstream sealing elements (right)**

The fixing system guiding the upstream sealing system (including waterstops, drain and crack inducers) to the correct position should not move while concrete is being placed around it. Ideally a temporary structure like those shown in Figure 4 (right) should be fixed to the supporting bracing or panels of the upstream formwork.

The aim of this element is to avoid movement during vibration which could cause potential weak planes in the lower fresh IV – RCC, and the development of cracking by – passing the sealing system. Therefore none of these temporary fixing elements should remain embedded in the lower layers, and the structure should be designed in a cantilever configuration without support on the fresh concrete.

5.4 Finishing the horizontal steps at the downstream face

The typical design arrangement of the downstream face of RCC dams is the stepped face. While IV – RCC remains very workable during a long period after compaction, finishing the top surface of the steps should be made as soon as possible following vibration to assure a high – quality finish and an even geometry. At the upper level, the immersion vibration should be extended slightly beyond the width of the step. The horizontal surface might be finished manually or aided by a vibrating screed (Figure 5, left).

If the right procedures are followed on time while the surface of the concrete is still fresh, an excellent finish can be achieved, similar to those which are typical of traditional concrete (Figure 5, right). Due to the high degree of retardation, IV – RCC mixes start developing some strength after 48 hours. The design of the formwork, length and size of the anchors, should be checked and tested in advance to avoid undesirable movements and loss of paste.

6 Advantages of IV – RCC

IV – RCC dams have in principle the same advantages as RCC dams. Some additional benefits have been identified as follows:

Fig. 5 Excellent finishing of the downstream stepped face of IV – RCC dams

● Greater simplicity of construction, associated with a reduced number of activities,

● More rapid construction, derived from this simplicity and from the availability of large construction areas,

● Bonding between successive layers of the immersion vibrated concrete used as facing concrete up to the same level as is achieved with roller – compaction. Extension of the initial set up to 20 hours or more has led to improved in – situ quality of super – retarded RCC[2], which includes IV – RCC,

● Cost saving in the equipment involved in the production, delivery and placement of facing and interface concrete,

● Cost saving in materials and mixes (CVC, grout or mortar) which have been traditionally used in RCC dams to improve the quality of the external concrete skin,

● Same concrete class can be used across the entire block, which leads to a greater uniformity in the development with the time of the mechanical, elastic and thermal properties of the concrete mass,

● The simplification of the dam construction material to just one type of concrete mix has many similarities with the CVC dam concept. In this scenario, IV – RCC dams might exhibit a steadily uniform behavior under operating conditions,

● According to the two points above, numerical models of IV – RCC dams will be simplified in comparison with any other RCC alternative, especially against those combining RCC of different mix compositions, bedding mixes and including additional interface and/or facing CVC concrete mixes,

● IV – RCC brings significant improvement in the consolidation of the mass concrete embedding the waterstops at the upstream face, eliminating one of the traditional critical areas as regards the dam permeability, and

• The fact that IV – RCC mixes can be consolidated by immersion vibration opens new possibilities to using the same mix in reinforced concrete which might be specified by the designers as skin concrete of embedded access shafts or galleries in RCC dams.

7 Conclusions

A new understanding of concrete dams has been developed recently creating a technology which closes the loop between traditional CVC dams and the more modern RCC dams. This new approach is based on the use of a very workable and super – retarded RCC mix that, due to its specific consistency, can be vibrated either by internal immersion or by external roller compaction.

Fig. 6 IV – RCC full – scale trial at Spring Grove dam (South Africa, 2012)

Some design and construction specifications have been summarized in this paper, which need to be taken into account when further developing this concrete dam concept. IV – RCC material is extremely friendly and easy to handle. Its application will bring additional benefits to the dam engineering, as has been demonstrated already in several projects.

References

[1] F. Ortega. Key design and construction aspects of immersion vibrated RCC[J]. International Journal of Hydropower & Dams, Volume 21, Issue 3, UK (2014).

[2] F. Ortega. Lessons learned and innovations for efficient RCC dams[C]. Proceedings, 6th International Symposium on RCC Dams, Zaragoza, Spain (2012).

[3] J. Nyakale, D. B. Badenhorst and F. Ortega. The Optimisation of the RCC Mix Design of Spring Grove Dam in South Africa[C]. Proceedings, 6th International Symposium on RCC Dams, Zaragoza, Spain (2012).

[4] Spanish National Committee on Large Dams (SPANCOLD). Technical Guidelines for Dam Safety, Nr. 2, Volume I (up-date on RCC), Spain (2012).
[5] F. Ortega. Key design and construction aspects of immersion vibrated RCC[J]. International Journal of Hydropower & Dams, Volume 21, Issue 3, UK (2014).
[6] F. Ortega. Lessons learned and innovations for efficient RCC dams[C]. Proceedings, 6th International Symposium on RCC Dams, Zaragoza, Spain (2012).

The Major Use of Lagoon Ash in RCC Dams

Malcolm Dunstan[1], Nguyen Hong Ha[2], David Morris[3]

(1. Malcolm Dunstan and Associates, U. K. ;
2. Son La Management Board (EVN), Vietnam;
3. AF – Consult Switzerland Ltd. Switzerland)

Abstract: Coal – fired Thermal Power Stations produce large quantities of ash that is frequently stored in lagoons and which potentially can have a hazardous impact on the environment. Dry flyash from these Power Stations has, for a long time, been used as a part of the cementitious content in concrete but large quantities of ash are still stored in lagoons around the World. Experimental studies have shown that mixes with flyash forming up to 65% of the cementitious content produces an acceptable strength that could be suitable for use in the base course of concrete pavements[1]. In the USA untreated lagoon ash has been used in the RCC of two dams with up to 70 kg/m^3 of ash in the mix[2].

For the RCC dams of Son La and Lai Chau in Viet Nam a search for a suitable pozzolan resulted in the choice of a low – lime (ASTM C618[3] Class F) lagoon ash from a Thermal Power Station that has a very variable and high carbon content. Tests on treated ash from this source showed that an RCC with ash representing 73% of the total cementitious content could produce an RCC of adequate strength for the dams which were up to 140 m high. The paper describes the treatment process, the tests carried out and the results leading to the use of this ash in the some four and a half million cubic metres of RCC in the two dams. The successful use indicates that such ash, when suitably understood and treated, can be an excellent source of pozzolan and its use helps reduce a potential environmental risk from its storage.

Key words: RCC, dams, flyash, lagoon ash, strength development

1 Introduction

Flyash from Thermal Power Stations is a well known and proven mineral admixture for use in concrete and such Power Stations have in the past produced more ash than could be used, the surplus being stored in lagoons. Unless carefully stored such ash can pollute ground water and there is a need to maintain a layer of water on the surface to prevent wind – blown particles affecting the environment. In addition, in an effort to reduce the effects of climate change caused by burning fossil fuels, there is a movement away from this source of energy reducing the quantity of ash available. Yet there are large quantities of ash still stored in lagoons all over the World that could be used but to date little conditioned ash, as it is known, has been used in concrete. Son La and

Lai Chau RCC dams are two of the first, if not the first, large-scale use of such ash in concrete.

For the construction of the Son La and Lai Chau RCC dams a suitable source of pozzolan was required and a search was made for any natural pozzolan sources that might exist within a reasonable distance from the site. None were found. A proposal was made that the basalt at the Son La site could be milled and used as it had similar chemical characteristics as some natural pozzolans. A series of trial mixes proved that, although the chemical composition was similar, the milled basalt had no reactive properties and thus would be no more than a filler. This would have led to an increase in the cement content and radical changes to the design of the dam leading to a longer construction time.

Fortunately a source of flyash from Pha Lai Thermal Power Station (one of only three in Viet Nam at that time of a suitable size to produce ash which could be considered for large-scale use in a structure as large as Son La) was located some distance from the site but at still what was an economic haul distance. However, not only was the quantity of dry flyash produced at the Power Station insufficient for the needs of Son La, the carbon content was very variable and generally ranged from 15% to 25%. Consequently, since the beginning of the operation of the Power Station, the ash had been discharged into lagoons (Fig. 1) and had never been used in concrete. In addition, to use such a lagoon ash, which contained both flyash and furnace-bottom ash, it would have to be treated and there was no precedent for the use of such a material in as important a structure as Son La. Some small plants existed around the lagoons that were screening and using a flotation process to remove some of the carbon to produce a product that was sold as a cement substitute for blended cement but the capacity was only 700 tonne per month. Nevertheless, in all other respects other than the carbon content, the ash conformed to the requirements of ASTM C618[3].

Fig. 1　The ash lagoon at Pha Lai Thermal Power Station

While conditioned ash had been used previously in fill applications, its use as a cementitious material was very unusual and an extensive Trial Mix Programme was undertaken to study in detail the performance of such an ash in RCC.

The effects of the carbon content, more strictly the Loss of Ignition (LoI), on concrete properties are recognised in International Standards and limits are placed on this value mainly because of the effects on absorption of admixtures, increased water demand and on air entrainment to protect the concrete from freeze/thaw damage. The latter is not really a significant problem in a tropical, or sub – tropical environment, such as Viet Nam. ASTM C618 allows a LoI of up to 12% if sufficient tests are carried out. However the Vietnamese Authorities decided to restrict the LoI to less than 6% in accordance with the basic requirements of ASTM C618.

It was not known what effect the storage of ash under water for several years where constituents could be leached out would have on the quality of the ash or on the long – term properties of the RCC. A series of laboratory trial mix tests were undertaken which showed that the ash when treated for the high residual carbon content was an excellent pozzolan and, in fact, contributed to the strength more than the Portland cement.

In all, nearly 750 000 tonnes of treated ash has been produced from the ash lagoons selected for use at Pha Lai for use in some 4.5 Million m^3 of RCC.

2 Treatment

The lagoon ash had a LoI of more than 15%, contained furnace – bottom ash and the moisture content was very high. The treatment comprised of:
- screening to remove the larger particles;
- a floatation process (with suitable admixtures) to remove the carbon;
- drying to reduce the moisture content. The recovered carbon was used as a source of heat in the drying process thus reducing costs.

At the start of the Project, two small treatment facilities were available that would not have been able to meet the peak requirements of Son La and would have required large – scale stockpiles at site. However a third and larger facility (Fig. 2) was commissioned shortly after the start of Son La and this allowed the supply requirements to be met with modest stockpiling at site to cover any delivery problems.

3 Trial Mix Programmes and Results

Three years prior to the start of the RCC placement at Son La, the first of two Stages of Trial Mix Programmes was started studying a range of mixture proportions and various forms of Pha Lai flyash. The main objective of the Stage – 1 Programme was to assess the performance of the cementitious materials, in particular at what LoI would there be an increase in water demand (there was in fact a negligible increase as the LoI increased) and at what LoI did the early – age and/or long – term strength drop.

During the Stage – 1 Programme two different PC40 Portland cements were considered (from But Son and Hoang Mai) and three different forms of ash from Pha Lai. The first of these was a screened ash from the Pha Lai 1 lagoons with an LoI of approximately 7.5% (the excess carbon was burnt off); this was classified as PL1. The second ash was a run – of – station flyash from Pha Lai 2 (the second Stage of the Power Station had just been commissioned) with an LoI of ap-

Fig. 2 Part of the main treatment facility for lagoon ash at Pha Lai Power Station

proximately 20%; this was classified as PL2R. The third ash was a reduced-carbon flyash from Pha Lai 2 with an LoI of approximately 5%; this was classified as PL2C. Also considered as mineral admixtures were a limestone dust, which could be sourced not far from the Son La site, and the milled basalt (Section 4). A range of Portland cement and ash/flyash contents was considered with a total cementitious content of 230 kg/m^3. The relationships between the But Son Portland cement and PL1 ash (i.e. lagoon ash) contents and the cube compressive strength are shown in Fig. 3 (in which the design cube compressive strength of 24 MPa is also shown at the design age of 365 days). Similar relationships for the PL2R (i.e. the run-of-station flyash) and for the PL2C (i.e. reduced-carbon flyash) are shown in Fig. 4 and Fig. 5.

With mixture proportions of 60 kg/m3 of But Son Portland cement and ash contents of 170 kg/m^3, the three ashes have 365-day cube compressive strengths of:

(1) PL1 (i.e. lagoon ash——LoI 7.5%), 24 MPa;
(2) PL2R (i.e. run-of-station flyash——LoI 20%), 27 MPa;
(3) PL2C (i.e. reduced-carbon flyash——LoI 5%), 27.5 MPa.

There was therefore negligible difference between the two flyashes from Pha Lai 2 with LoIs ranging from 5% to 20% (Fig. 4 and Fig. 5), and although the treated lagoon ash had a slightly lower strength, it was still more than sufficient for a 140 m high RCC dam such as Son La.

Further refinement of the mixture proportions was undertaken during the Stage-2 Trial Mix Programme, in which different total cementitious contents were tried, but the overall conclusions of the Stage-1 Programme did not change. If the impact on the strength of the LoI found at Pha Lai is general, it has to be asked why, when freezing and thawing is not a problem, such stringent limits have to be placed on the Loss on ignition of flyash. It has also been said that increased carbon contents can result in colour changes but at Son La there was no discernible variation in colour with ashes having LoIs ranging from 5% to 20%.

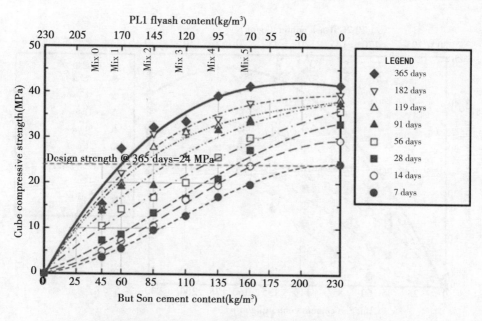

Fig. 3 Relationships between the But Son cement and PL1 ash contents and the cube compressive strengths at various ages up to 365 days

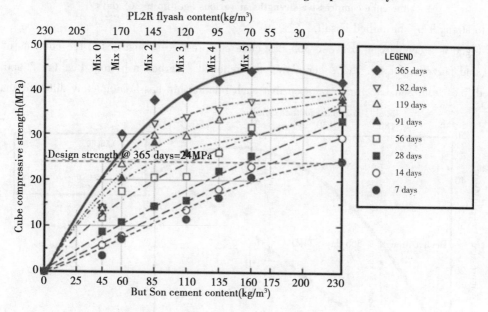

Fig. 4 Relationships between the But Son cement and PL2R flyash contents and the cube compressive strengths at various ages up to 365 days

4 Other mineral admixtures

As stated above, milling basalt from the site was considered as a possible alternative source for a mineral admixture because the chemical characteristics were similar to some natural pozzolans. However during the Stage – 1 Programme it was found that there was little, or no, contribu-

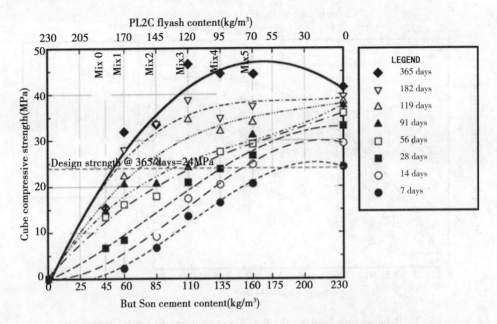

Fig. 5 Relationships between the But Son cement and PL2C flyash contents and the cube compressive strengths at various ages up to 365 days

tion to strength by the milled basalt.

Fig. 6 is a comparison between 365 day cube compressive strengths and the contents of cement and pozzolan for the RCCs containing the reduced-carbon ash from Pha Lai 2 and the milled fines from Son La. It can be seen that each relationship has a completely different shape.

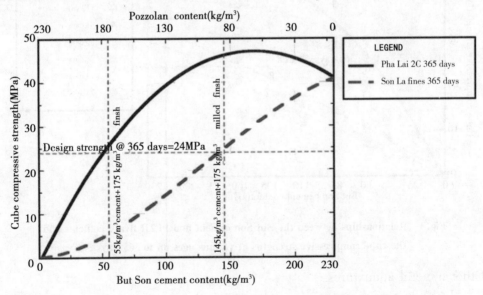

Fig. 6 Comparison between the 365-day compressive strengths of RCC containing the reduced-carbon Pha Lai 2 flyash and the milled basalt fines from Son La

The optimum mixture proportionsof the RCC with a design cube compressive strength of 24 MPa at a design age of 365 days was found to be a PC40 Portland cement (from But Son) content of 55 kg/m³ and 175 kg/m³ of the reduced - carbon Pha Lai flyash. This is a very economic RCC, not only because of the cost of the cementitious materials but more importantly because of the relatively low early - age heat generation of such a mix that would lead to considerably lower pre - cooling costs than would otherwise be needed with a less - efficient set of mixture proportions.

Fig. 6 clearly shows the separation of the contribution of the Portland cement from that of the reduced - carbon flyash. Below the Son La milled fines line is the contribution of the cement and between that line and the Pha Lai 2C line is the contribution of the reduced - carbon flyash. For example, an RCC with 100 kg/m³ of But Son PC40 cement and 130 kg/m³ of Pha Lai 2 reduced - carbon flyash has a strength at the age of a year of approximately 39.5 MPa. Of this, approximately 14.5 MPa comes from the 100 kg/m³ of cement (i.e. 0.145 MPa/kg/m³) and approximately 25.0 MPa comes from the 130 kg/m³ of flyash (i.e. 0.192 MPa/kg/m³). Thus the flyash is over 30% more efficient than the Portland cement. Similarly with an RCC having 150 kg/m³ of But Son PC40 cement and 80 kg/m³ of Pha Lai 2 reduced - carbon flyash, the strength at the age of a year is approximately 47.0 MPa, of which approximately 25.5 MPa comes from the 150 kg/m³ of cement (i.e. 0.172 MPa/kg/m³) and approximately 21.5 MPa comes from the 80 kg/m³ of flyash (i.e. 0.269 MPa/kg/m³). In this case the flyash is over 50% more efficient than the Portland cement. This is in line with work undertaken in the late 1970s[4].

5 Choice of final mixture proportions for Son La

The Trial Mix Programmes allowed relationships between the cube compressive strength and the cement and Pha Lai flyash contents to be developed for ages up to 365 days. Up to an age of 182 days there was little difference between the strengths of the run - of - station and the reduced - carbon flyash mixes. However between the ages of 182 and 365 days, there was an indication that the reduced - carbon flyash starts to provide a little more strength than the run - of - station flyash.

The final mix selected, based on all of the test results contained 60 kg/m³ of Portland cement and 160 kg/m³ of flyash i.e., a total cementitious content of 220 kg/m³. A very similar set of mixture proportions was also chosen for Lai Chau.

6 Performance during construction of Son La and Lai Chau

Although both reduced - carbon flyash from Pha Lai 2 and treated lagoon ash were used in the RCC in both Son La and Lai Chau (at the latter dam practically all the ash was treated lagoon ash), it was not possible to differentiate between the RCCs containing oneor the other ash. The developments with age of the equivalent cylinder compressive strength (modified from the cube compressive strength) from Son La and the cylinder compressive strength from Lai Chau are shown in Fig. 7. It can be seen that both are similar and both have average 365 day strengths (i.e. at the design age) somewhat in excess of 20 MPa. The required characteristic cylinder compressive strength of the RCC was 16.5 MPa. Taking into account the Coefficient of Variation of the RCC, this value was exceeded by nearly 20% by the actual results.

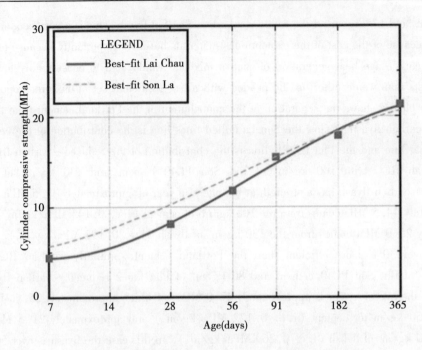

Fig. 7　Development of the cylinder of compressive strength
with age of the RCCs from Son La and Lai Chau

The average in-situ properties found from extensive coring of both Son La[5] and Lai Chau[6] are summarised in Table 1.

Table 1　Average in-situ properties of the RCCs from Son La and Lai Chau

		Son La	Lai Chau
Core compressive strength	(MPa)	19.2	17.9
Vertical direct tensile strength of unjointed RCC	(MPa)	1.33	1.24
Vertical direct tensile strength across joints			
Hot joints	(MPa)	1.27	1.15
Warm joints	(MPa)	1.14	—
Cold joints	(MPa)	1.52	—
Super-cold joints	(MPa)	1.31	1.32

As the average age at which the in-situ properties was measured at both Son La and Lai Chau was somewhat less than the design age, the true values at that design age will be somewhat higher than the values shown in Table 1. For example at Lai Chau, at the limiting Modified Maturity Factor for a hot joint, the vertical in-situ direct tensile strength across the hot joints at an age of 365 days was 1.23 MPa[6]. All the values in Table 1 (even not allowing for the age of test) exceed the design requirements by some margin, ranging from 15% to 30%. It should also be noted that the average vertical in-situ direct tensile strength across the joints are all close to the strengths of the unjointed RCC, indeed some of the joints have higher average strengths than the parent material——this is obviously not possible and is just a statistical anomaly but shows that practically all the RCC, whether containing joints or not, is effectively monolithic.

7 Conclusions

The extensive Trial Mix Programmes for the Son La RCC dam showed that ash reclaimed from lagoons, which had been treated to remove the carbon content higher than that permitted in ASTM C618 and dried to a moisture content that allowed it to be handled and stored without problems, was an excellent pozzolan that contributed significantly to the strength of the RCC. It permitted the cement content to be reduced to 60 kg/m^3 and thus produced a very economical mix.

Although the paper has concentrated on the test programmes for the Son La RCC dam, the same sources of cement and flyash were used for the Lai Chau RCC dam that followed immediately after the completion of Son La. The test programmes for Lai Chau showed that the same cementitious content as at Son La provided a mix of adequate strength for the dam. Part way through the construction of Lai Chau the source of Portland cement was changed but the strength obtained with the alternative cement was little changed.

For a number of reasons, parts of the World are running short of suitable flyash from Thermal Power Stations. The use of treated lagoon ash could be the breakthrough that will allow the next generation of very large RCC dams to be constructed rapidly and economically. However it will only be by undertaking studies well ahead of the start of construction of the dam, and by having an understanding of the potential problems involved, that treated lagoon ash can be provided that-could have the performance of those used at Son La and Lai Chau.

Acknowledgements

The authors would like to thank Viet Nam Electricity, the Son La Management Board for Lai Chau, and PECC1 for permission to publish this paper. Regrettably one of the authors, Mr Nguyen Hong Ha, Vice Director of EVN, passed away during the preparation of this paper and it is thanks to his untiring leadership and drive that both Son La and Lai Chau RCC dams were completed ahead of schedule and to a high standard.

References

[1] T. R. Naik, B. W. Ramme and R. N. Kraus. Performance of High - Volume fly ash concrete pavement construction since 1984. Indian Concrete Journal, Vol. 78, No3., March 2004.

[2] A. B. Aceves. The use of non - commercial fly ash in RCC structures. Proceedings of World of Coal Ash (WOCA) Conference, Denver, USA, May 2011.

[3] American Society for Testing and Materials. Fly ash and raw or calcined natural pozzolan for use as a mineral admixture in Portland cement concrete. Standard Specification C618, ASTM, Philadelphia.

[4] M. R. H. Dunstan. Relationship between properties generated by cement and by flyash. Proceedings of International Conference on Rolled concrete for dams´, CIRIA, London, June 1981.

[5] P. V. An, M. R. H. Dunstan, D. Morris and M. Conrad. The Son La RCC dam - testing of in - situ properties. Proceedings of 3rd International Symposium 'Water Resources and Renewable Energy Development in

Southeast Asia', Kuching, Malaysia, April 2010.

[6] N. H. Ha, N. P. Hung, D. Morris and M. R. H. Dunstan. The in-situ properties of the RCC at Lai Chau. Proceedings of 7th International Symposium on Roller-Compacted Concrete (RCC) Dams, Chengdu, China, September 2015.

Construction of Enciso Dam

Allende M[1], Ortega F[2], Cañas JM[3], Navarrete J[3]

(1. Ebro River Basin Authority (CHE); 2. FOSCE Consulting Engineers; 3. Dragados – FCC JV)

Abstract: Enciso RCC dam is at present under construction in the Ebro river basin in northern Spain. The main purpose of the project is to guarantee the irrigation and water supply to the local community. The dam is 103 m high and the total volume of RCC is about 800 000 m^3. Although the project has suffered several interruptions along its preliminary construction phases, the RCC placement could be planned as a continuous activity with a total duration of 10 months split in four (4) stages due to the winter and summer breaks. In the first stage some 170 000 m3 were placed in 2.5 months in the autumn of 2014 and the second stage has been completed recently (June 2015) reaching 50% of the total RCC volume.

This article describes the RCC construction aspects of the project. The mix design development and quality control issues are analysed in a separate work published in this same Symposium. In this second paper special attention will be paid to the description of the plant capacity and the logistics required for a continuous placement. The concept of the RCC mix that is being used at Enciso benefits from this continuity at high construction rates, and subsequently improves both quality and economy.

The aspects related with the RCC placement is another main topic in this article. Innovation through the use of an immersion vibrated RCC (IV – RCC) mix has simplified the placement activities to a great extent. Aspects describing this positive experience will be presented including specific construction procedures for good IV – RCC practice.

Key words: Construction, plants, logistics, IV – RCC placement, innovation

1 Project description and dam lay – out

Enciso dam is the main component of an irrigation and water supply project located in the region of La Rioja in the North of Spain. The Owner is the Ministry of Agriculture and Environment, who leads the works through a basin organization, the Ebro River Basin Authority. The dam body consists of a 103 m high RCC gravity dam and has a crest length of about 375m. Crest width is 8m. The upstream face is vertical and the downstream slope is 0.8V/1H. The dam crest is at 878.50 m ASL.

Various CVC mixes (conventional mass or reinforced concrete) and RCC (roller compacted concrete) are used in the different project structures. CVC is applied on dam foundations as leveling concrete and in the hydraulic structures (spillway, bottom outlet, intakes, etc.). The dam is built 100% with RCC.

Nominal discharge spillway capacity is about 600 m³/s, and nominal discharge of bottom outlets is about 125 m3/s. The hydraulic outlets are located in a CVC block on the right abutment at a distance of about 40m from the central section. The surface spillway is located in the middle of the dam.

Downstream, the chosen solution for the reinstatement of the channel flow is a stilling basin 25 m long, 43 m wide and 13 m deep. The channel discharge is stepped, dissipating water head, so stilling basin dimensions have been optimized. The bottom outlets have also a stilling basin downstream dissipating most of its energy, and preventing erosion of the downstream channel.

In the final dam construction design the former CVC concrete has been replaced by Immersion Vibrated RCC (IV – RCC). Figure 1 shows the typical cross section of the dam. The concrete zoning and main mix modifications are also commented in the Figure.

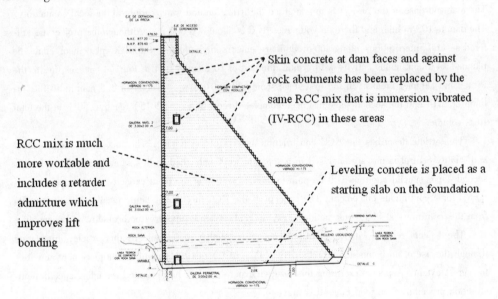

Fig. 1 Typical dam cross section and actual concrete zoning applied in the construction design

Two horizontal galleries and a perimetral one connecting with transverse galleries located at either abutment provide access at different levels in the dam. CVC forming the gallery faces has been also replaced by IV – RCC. Pre-cast reinforced concrete slabs are used as gallery roof.

In the longitudinal direction the dam is divided in blocks with a maximum width of 20 m. In the original design (1994) it was considered that one every three of these joints would be formed creating 60 m long blocks in which RCC would be placed. This was necessary to reduce the volume of the layers as the former 'hot' joint was specified with a maximum time of 6 hours between them. Initial set has been significantly extended in the actual mix design . 'Hot' joints are now specified up to 15 hours and in some cases up to 20 hours. With the actual capacity of the plants this new specification allows placement in much larger areas and the dam is either built in just one block in horizontal layers from one abutment to the other, or split with a formed joint intwo blocks at middle levels with the larger layer volumes.

The availability of large placement areas increases the rate of placement and provide enough space to safely and effectively develop the various activities on the lift.

2 Site installations

The finally selected source of aggregate is a limestone quarry located at about 18 km away from the dam site. A three – stage aggregate processing plant has been installed at the quarry area. Material is directly transported to the aggregate stockpiles near the hoppers of the batching plants or diverted to a main aggregate stockpile area that has been conditioned between the quarry and the dam. The main site installations are completed with the concrete production plants where both CVC and RCC can be manufactured and a 36 – in. (91 mm) wide high – speed conveyor system.

2.1 **Aggregate production**

The limestone quarry face is structured in 4 banks between 15 and 20 m high each. A large proportion of overburden and intercalations of loamy limestone need to be removed in order to feed the primary crusher with sound material adequate to produce concrete aggregate.

The plant runs in a dry operation process and is divided in two main units: (1) primary crushing and intermediate stockpile and (2) secondary/tertiary crushing, screening and classification. Monthly average production is ca. 100 000 t.

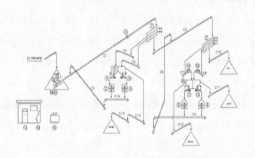

Fig. 2 Schematic lay – out and general view of the aggregate processing plant (2nd unit)

All crushers installed in this plant, including the primary, are impact – type crushers to ensure a better aggregate shape. A full – recirculation system has been installed at each crushing stage to have the possibility of re – crushing and adjusting the split proportions as required, depending on the variability of the raw material and the monthly demand. After the conclusions of the initial mix design stage the tertiary position has been modified with an additional screen and another impact – crusher. Without this modification an external natural sand would have been required to meet the gradation and shape of the fine aggregate specification. Meeting this specification was important to reduce the water demand of the mix and, subsequently the amount of cementitious materials in as much as 40 kg/m^3. A total of 40% of the RCC aggregate have been stockpiled prior to the start of the RCC placement. Several areas at an average distance of 7 km from the dam have been conditioned to stockpile the different aggregate sizes summing up a total surface of 35 ha.

2.2 Concrete production

The concrete plant consists of four (4) twin-shaft batch type horizontal mixers with a batch capacity of 4m3 – each of consolidated concrete. The estimated nominal overall production is 500 m3/hour.

Such a high capacity is required to ensure continuous production 24 hours/day, 7 days/week. The plant is split in two completely independent batching units with interchangeable materials feeding systems. This lay-out allows regular maintenance operations and any major repairs in part of the plant while the production continues on the rest of the installation.

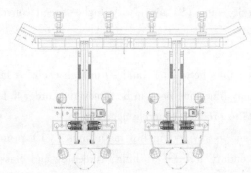 

Fig. 3 Schematic lay-out and general view of the concrete plant

The plant is located at the upstream side of the dam on the left bank at approximate middle elevation between the lower foundation level and the dam crest. This is the optimised arrangement in terms of access for the materials and lay-out of the staged conveyor system leading to the dam.

Aggregate hoppers are orientated in the same alignment for both batching units. A covered area has been additionally arranged at the back of the hoppers to keep aggregate protected from direct sun radiation and rainfall during several hours before entering the twelve (12) receiving hoppers of the batching plants.

Silos for cement and fly-ash with a total capacity of 3 000 tare located on the opposite side of the mixers at a lower level with independent access from the one reaching the aggregate bins. Silo trucks are feeding these silos continuously driving from the supplier's factory situated at 130 km (cement) and 275 km (fly-ash) distance from the site. The quality and safety of the road network between both points is quite adequate and capacity of silos installed at the site cover the demand of just four (4) days at peak production rate.

Water is pumped from the cofferdam reservoir up to a tank situated at a higher level on this abutment. This same tank supplies water to the plant (mixing water and cleaning) and also to the construction activities on the dam (curing and cleaning).

2.3 Conveyor system

Individual hoppers and reversible conveyors are located under each of the four (4) mixers. Both trucks or the main conveyor system can be loaded from either end of these conveyors. A collector conveyor located under the reversible conveyors runs in parallel with the plant axis feeding the high-speed conveyor.

The total length of the 36" (91 mm) – wide high – speed conveyor line leading to the dam is ca. 230 meters long split in three (3) conveyors and one (1) swinger at the end. A schematic view of the system including the stages of raising as the dam is built is shown in Figure 4.

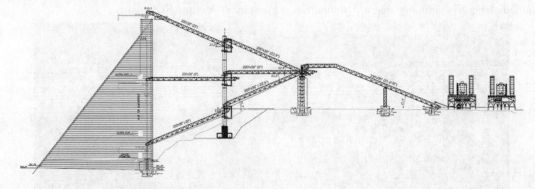

Fig. 4 Schematic lay – out of the high – speed conveyor system between the plant and the dam

The first conveyor is 90 m long and remains fixed along the whole construction period. The other two are 60 m long each and are raised with the dam as shown in the schematic lay – out, aided by a jacking system which also supports the transfer hopper between both conveyors. In this arrangement the maximum slopes of the conveyors at the beginning (downwards) and at the end (upwards) is 22°. With such inclination the capacity of the conveyor has some limitation but also the rate of placement in those sections of the dam is far from the required peak rate.

The system ends in a swinger conveyor installed on top of a self – climbing round post embedded in the dam. This post is raised every 4 layers of RCC (1.2 m) leaving a hole that will be filled up with self – compacting concrete when the conveyor is removed at the end of the construction.

The whole length of conveyor except the swinger is covered to protect the mix during transportation to the dam although the time of conveyance between the plant and the dam is just 2 minutes in normal operating conditions.

3 Dam construction planning

Enciso RCC dam construction has been planned as a continuous activity during the periods when concrete can be placed. RCC placement is limited by both environmental conditions and by other financial and contractual terms. In such scenario RCC can only effectively be placed in two periods of the year, between April and June, and between September and November.

First two stages have been completed (1) September – November 2014 and (2) April – June 2015 with a 4.5 months shutdown in between (Figure 4). Along this period 50% of the RCC volume has already being placed in five (5) months. Placement is planned to resume in September 2015 and completion is expected for the first half of 2016 after a second winter break.

In a first construction phase, prior to the start of RCC placement in the dam, several conventional concrete blocks have been placed as levelling and foundation blocks. Consolidation grouting has been done from the top of these blocks. The river diversion channel and culvert were also built during this preliminary phase in structural reinforced concrete. This culvert will be closed and used to install the bottom outlets at a later stage.

Therefore RCC placement in the dam remains a fairly clean operation, free of obstacles and impediments to continuity that are sometimes a hindrance in other RCC dams.

Fig. 5   General view of the dam and site installations, getting started after the first winter break

The volume of the maximum horizontal 300 mm thick layers is about 3 000 m^3. Although the capacity of the plants are enough to place them in less than 15 hours, it has been decided to split the dam in two blocks during the levels with maximum volume as an additional factor of safety to avoid unplanned joint treatments. In these elevations of maximum layer volume the dam is raised alternatively is two blocks split in a formed transverse joint situated in the centre of the dam. A ramp built in concrete is located at the downstream end of that joint and connects both sides to facilitate the movement of equipment and the distribution of concrete in the dam.

4   Concrete placement in the dam

4.1   RCC Placement

RCC transportation in the dam is made with dump trucks. Mainly articulated 25 t trucks with wide tires are used but in some cases in narrow areas a smaller unit has been necessary. Figure 6 (left) shows the position of the trucks at the loading point below the swinger conveyor. The concrete in this area is typically affected by the continuous traffic of the trucks. With a well trained movement of the trucks driving in and out of this position it is possible to reduce the depth of the tracks. In case of high alteration the conveyor swings to modify the position of this area. Removal of any segregated spots or additional passes of the roller can be given to flatten the surface and spray water curing. Cores extracted from this area show no differences in quality from the RCC placed in areas of less traffic[2]. An example of the maximum depth of the tire tracks that is usually tolerated in this kind of very workable and super-retarded mixes is shown in Figure 6 (right).

Layers are placed horizontally in lanes parallel to the dam axis. The starting point in each layer is usually located at the downstream side against one the abutments (opposite to the conveyor) and progressing towards the position of the swinger. Under the conveyor the trucks are removed and the RCC is directly dropped onto the surface. Segregation must be specially controlled at this time.

Fig. 6 Details of articulated 25 t dump trucks used for RCC transportation in the dam

A laser guided dozer is used to spread the RCC in 330 ~ 350 mm thick layers that will be later compacted to the specified depth of 300 ± 30 mm (Figure 7, left). Additional marks on the rock and forms also assistin spreading the right thickness at the edges of the layers (Fig. 7, right).

Fig. 7 Laser - guided dozer aided by marks at the edges closely followed by consolidation

4.2 IV - RCC

RCC is consolidated by immersion vibration in all areas that are not reached by the large 10 t vibratory rollers. From a construction point of view one of the main advantages of using this very workable mix is that no grout or mortar is additionally required to internally vibrate this concrete.

The procedures to achieve an adequate vibration on time have been tested previously at full - scale trials and can be summarized as follows (see Fig. 8):

• Spreading of fresh RCC at the edges and around embedded elements aided by hand - held shovels if required is a first priority in the placement activities,

• Dozer should then complete spreading the mix in the adjacent areas to leave space for the vibrating equipment as soon as possible,

- Both hand – held pokers and equipment mounted with a gang of vibrators are required to fully consolidate the RCC by immersion vibration. The latter shall be used in most areas to be vibrated. Manual vibration is applied in narrow spaces and around water stops (a),
- Most of the time vibration is progressing in two lines parallel to the forms or abutment. First the outside alignment is vibrated and then the inside one (b). Distance between pokers installed at the equipment may vary depending on their capacity, diameter and mix design. This distance might be adjusted within 350 ~ 500mm for 3 or 4 – in (75/100mm) diameter pokers (c). In any case it is important that the head of the vibrators can penetrate into the concrete a depth of 500 mm to ensure interpenetration and bonding with the lower layer,
- At least one person should attend the operation from the ground to observe the paste flowing up to the surface and indicating to the operator of the equipment the rate of raising the pokers. At times this person might need to stand close to any spots where coarse aggregate has not been covered by paste and assist (even just with the feet) to submerge it down inside the concrete. Same procedure is applied by hand vibration limited to narrow areas where the same operator stands where is being vibrated and makes sure that no aggregate but just paste covers the surface at the end of the vibration,
- Ideally IV – RCC should be fully vibrated within 30 ~ 45 minutes since mixing,
- Roller compaction of adjacent areas with the large roller follows overlapping a minimum width of 200 mm with the areas that have been immersion vibrated (d),
- If necessary a small 3 t twin smooth vibratory roller can be used to flatten the overlap, and
- Continuous curing of the surface until next layer is placed on top.

5 Other activities

5.1 Formation and sealing of transverse joints

Contraction joints are formed by insertion of a folded plastic sheet into the fully compacted RCC aided by a hydraulic vibrating device mounted on the arm of a small backhoe (Figure 8, left). Two water stops and a drain are located at the upstream end of the joint. A specially – designed frame supported on the formwork has been used to maintain alignment and stability of the various elements (Fig. 9, right).

5.2 Embedding of instruments

Instruments and cables are embedded in RCC at required location after compaction has been completed. A small trench ca. 150 ~ 200 mm depth and 100 ~ 150 mm wide is made from the surface. To that end several types of equipment have been tested. Figure 10 (left) shows one of them. After the elements have been installed the trench is filled again with fresh concrete and immersion vibration is applied in two alignments at either side of the instruments line. The area is then marked to avoid direct heavy traffic on top of it (Fig. 10, right). This protection measures are extended to upper layers as necessary depending on the type of instrument.

Fig. 8 Immersion vibrated RCC (IV – RCC) procedures

Fig. 9 Formation and sealing of transverse joints

5.3 Lift joint treatments

Most of the joints between layers don't need any special treatment apart from water curing with a thin spray to keep the surface of the layers in a saturated condition. Cleaning of water ponds and detritus is an activity concentrated just ahead of the placement of the next layer.

In case that the joints are older than 15 hours (within initial set of the lower layer) the surface is brushed with a soft plastic broom to remove a thin upper part of the paste. In some cases this limit has been extended up to 20 hours. This material is cleaned and removed to disposal. A more uniform treatment is achieved if the surface of the layer is well levelled and no major marks of tires are left after compaction. With the actual placement in two split blocks, this type of '

Fig. 10 Installation, embedding in IV – RCC and protection of instruments

warm' or 'intermediate' joints are no longer required.

Cold joints are traditional 'aggregate – exposed' finished joints. These are washed with high – pressure water jets. The pressure varies between 50 and 400 bars depending on the state of maturity of the joint. Most of the cold joints at Enciso (4% of the total surface up to now) are ' planned' cold joints and in such cases treatments become a well – controlled operation. Un – planned cold joints are time and cost consuming and should be avoided with an adequate mix and construction planning.

5.4 Formwork and finish of the dam faces

Specially designed formwork for the dam faces of continuous RCC placement has been developed in the past. This experience has been used in the design of formwork for Enciso dam. Additional measures have been taken to adapt thestepped system of the downstream face to the specific properties of this mix. For example longer tie rods anchoring the forms and some stronger bracing was required. However finishing of the steps is much easier than with drier mixes used in the past.

Fig. 11 Finish of the horizontal downstream steps just after consolidation is completed

5.5 Galleries

Horizontal galleries are built following well – known procedures with IV – RCC against formed faces and pre – cast roof slabs. The inclined abutment gallery is formed in a similar way (Figure

Fig. 12 General view of the upstream face after completion of the first dam section

13, left) but using variable lengths of shutters and slabs that are placed in steps every 300 mm following the slope. This gallery should be located at a distance of about 8 ~ 10 m from the rock abutment to get an efficient placement and consolidation around this area (Figure 13, right).

Fig. 13 Finish and location of the inclined gallerywith enough room between it and the rock

A well designed gallery with enough width to facilitate movement of equipment and personnel is a key issue in the future inspection and operation of any concrete dam.

6 Conclusions

There are many areas in the layer of an RCC dam that requires immersion vibration: dam face, rock interface, waterstops and drains, embedded instruments, galleries, and so on. At Enciso RCC dam the originally designed conventional mass concrete has been replaced by Immersion Vibration of the same RCC mix (IV – RCC) achieving a very good quality. Mix retardation also helps to increase the volume of layers and to reduce lift joint treatment. In sum it can be concluded that this is a very 'contractor's friendly' mix as it simplifies the production and placement activities. Well – known construction methods of RCC and CVC dams that were built in the past are both put together into practice at Enciso dam. In addition some specific procedures have been successfully incorporated.

References

[1] M. Allende, F. Ortega and C. Blas. Design and Quality Control of the IV – RCC Mix used at Enciso Dam in Spain[C]. 7th International Symposium on RCC Dams, Chengdu, China (2015).

[2] F. Ortega. Lessons Learned and Innovations for Efficient RCC Dams[C]. 6th International Symposium on RCC Dams, Zaragoza, Spain (2012).

[3] M. Sanz. Construction of Beni Haroun Dam (Algeria)[C]. 4th International Symposium on RCC Dams, Madrid, Spain (2003).

Not All Flyashes are the Same

Mark Turpin[1], Todd Smith[2], Malcolm Dunstan[3]

(1. Project Controls Specialists Inc, Canada; 2. SNC – Lavalin, Canada;
3. Malcolm Dunstan and Associates, U. K.)

Abstract: The paper describes a Trial Mix Programme in which two Portland cements and two low – lime flyashes were studied. The two Portland cements had similar chemical and physical properties and the two flyashes also had similar chemical and physical properties and similar Pozzolanic Activities.

During the Trial Mix Programme, it was found that the performance of the two Portland cements was quite similar – there were minor differences but these were not significant – but the performances of the two flyashes were completely different. The retarders investigated during the Programme required similar dosages for all four combinations of cementitious materials – the RCC was being retarded so that the Initial Setting Time was circa 21 hours – but the water demand of the two flyashes was different and the strength developments were completely different. Accelerated curing of all the specimens was used during the Programme to try to simulate the long – term strengths of the RCCs. The total cementitious content of all the mixes was 230 kg/m^3. With one flyash and Portland cement contents of 35 to 50 kg/m^3, the long – term strengths were between 30 and 35 MPa; this is probably amongst the best performance for a low – lime flyash that has ever been reported. However with the second flyash, for the same Portland cement contents, the long – term strength was between 10 and 13 MPa. At an age of 91 days, the first flyash with the same Portland cement contents, had cylinder compressive strengths between 25 and 28 MPa while with the second flyash the strengths were between 6 and 9 MPa; a very considerable difference.

Further research is needed to try to ascertain why there should be such a significant difference, but it seems the study of the chemical and physical properties and indeed the Pozzolanic Activity, does not allow the relative performance of flyashes in concrete to be predicted.

Key words: Trial Mix Programme, Portland cement, flyash, retarders, accelerated curing, compressive strength

1 Introduction

Two Portland Cements (classified as Cement A and Cement B) and two low – lime flyashes (classified as Flyash Y and Flyash X) were studied in some detail during a Trial Mix Programme undertaken prior to the award of a contract for the construction of an RCC dam. It was expected that a range of Portland Cement and flyashes contents could be determined so that the Contractors could price their Tenders on the basis of real data and thus in an equitable and fair way. This procedure has become the norm for RCC dams over the past many years due to the long – term design ages being used (182 to 730 days) and due the widely differing long – term strength developments (e. g. at Upper Stillwater in the USA where the average 365 – day strength was more than three times the 91 – day strength and at New Victoria where the five – year strength was over four times the 91 – day strength).

Accelerated tests were used for the mixes so that some understanding of the long – term strength might be obtained at a reasonably early age.

2 Objectives of the Stage – 1 Trial Mix Programme

The Trial Mix Programme was planned to be undertaken in two Stages. The objectives of the first Stage were:

(1) To train the laboratory staff (as they had no experience of RCC);

(2) To assess the performance of the various cementitious materials;

(3) To choose three (or four) retarders for the RCC (the Initial Setting Time was to be delayed until (21 ±3) hours);

(4) To initiate the calibration of the accelerated – curing procedures.

The second – Stage Programme will be used:

(1) For the refinement of the RCC mixture proportions;

(2) To develop the air entrainment of the RCC and GEVR;

(3) To develop the procedure for the air – entrained GEVR;

(4) To refine the mixture proportions of the levelling concrete (that is designed as the interface between the foundation and RCC).

3 Materials used in the Stage – 1 trial mix programme

3.1 Aggregates

A crushed basalt had been produced for a previous Contract. Although this did not conform to the Specification for the RCC, it was considered that it would not impact too greatly on the various assessments to be studied during the Stage – 1 programme.

3.2 Portland Cement

The two Portland Cements both conformed to the local Standard and there seemed to be no material difference between the two sets of physical and chemical properties. Both cements were thus expected to have a similar performance.

3.3 Flyash

Both low – lime flyashes also conformed to the local standard and were also expected to have a similar performance, based on the Pozzolanic Activities and the chemical and physical properties – the only difference was that Flyash X had a rather higher alkali content that Flyash Y. Both flyashes had rather high Specific Gravities.

4 Fresh properties of the RCC

4.1 Water demand

Having optimised the gradation of the aggregate, a series of mixes were undertaken to determine the optimum workability and thus optimum water content for the particular materials. Flyash X was found to have a significantly higher water demand than Flyash Y, although there were no indications from the properties of the flyash that this would be the case.

4.2 Retardation

There were very strict criteria that a retarder had to meet before it could be approved for testing in the final comparison in the Full – Scale Trial. These were:

(1) The Initial Setting Time (IST) should be (21 ±3) hours;

(2) At the optimum IST, the Final Setting Time (FST) should be between 30 and 48 hours;

(3) When the retarder is double dosed (which will happen at least once in every RCC dam), the IST must not exceed 48 hours;

(4) When the retarder is double dosed, the FST must not exceed 96 hours;

(5) The difference between the ISTs with dosages of 75% of the optimum and 125% of the optimum, must be less than 15 hours;

(6) The RCC that contains the retarder should not bleed. However this can only be determined during the Full – Scale Trial and thus there should be at least three, and ideally four, retarders available to be tested during the Full – Scale Trial.

Of the five admixtures tested during the Stage – 1 Programme only two conformed to the above requirements and even then in both cases each retarder just failed one of the requirements, but only by a very small amount. This showed the importance of testing sufficient retarders at an early stage so that there were at least three retarders that could be tested in the Full – Scale Trial.

5 Relationship between the cylinder compressive strength and the cement and flyash contents

5.1 Introduction

After a suitable dosage had been determined for the retarders and the water content optimized for each combination of cementitious materials, a series of mixes were designed assuming the Portland Cements and flyashes (which had no previous history in an RCC) had an "average" performance using MD&A's MacMix program[1]. Five sets of mixture proportions were chosen with a total cementitious content of 230 kg/m^3 with Portland Cement contents ranging from 50 to 80 kg/m^3. All five mixes were tried with the primary sets of cementitious materials (i.e. those that were considered the most likely to be the chosen by the Contractors) and three mixes (Mixes 1, 3 and 5) for the secondary cementitious materials. The specified characteristic cylinder compressive strength was defined as 17 MPa at the design age of 182 days. From this strength, the design strength (i.e. the strength that has to be exceeded by the average of all the strengths) was calculated to be 19.2 MPa.

5.2 Cement A and Flyash Y

The relationships between the normally-cured cylinder compressive strengths of the RCCs containing Cement A and Flyash Y (the primary materials) are shown in Fig. 1 for ages up to 182 days (the design age).

It was immediately apparent, after the first cylinders of the RCCs containing Cement A and Flyash Y had been tested, that the strengths being obtained were exceptional, so much so that two additional sets of mixture proportions were tried with Portland Cement contents of 35 and 42.5 kg/m^3. These are very low Portland Cement contents even for a well-designed RCC with a well-shaped aggregate and with very efficient cementitious materials.

In Fig. 1 it can be seen that at the design age of 182 days with Portland Cement contents ranging from 35 to 80 kg/m^3 the cylinder strengths range from 30 to 40 MPa. In these mixes the flyash is contributing very much more to the strength than the Portland Cement.

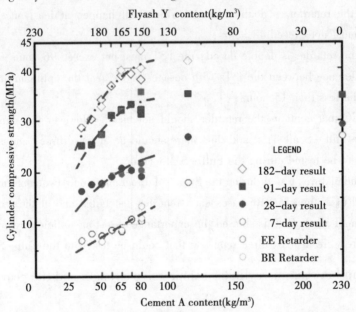

Fig. 1 Relationships between the Cement A and Flyash Y contents and the cylinder compressive strengths at various ages up to 182 days

Using the data from Fig. 1. the development of cylinder compressive strength with age is shown in Fig. 2. Also shown is the design strength of 19.2 MPa at the design age of 182 days. In addition typical developments of strength with age of two RCCs having mixture proportions of 30 + 190 and 60 + 160 (Portland Cement + flyash) are shown for comparison purposes. It is clear that the strengths of the Cement A/Flyash Y RCCs are very much higher than the typical RCCs. It is possible that the design strength could be achieved with a Portland Cement content of 20 to 25 kg/m^3 - an extraordinarily efficient mix not only in terms of cost (assuming typical cost for the Portland Cement of \$160/tonne - the average of well over 100 RCC dams - and flyash of \$60/tonne), in terms of the cooling (as the early-age heat generation will be very low) but also because an RCC with such a high content of flyash would be extremely cohesive and thus very easy to handle with negligible potential for segregation.

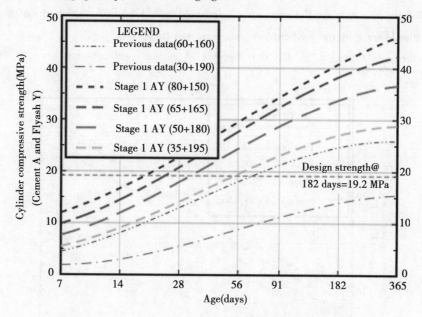

Fig. 2 Development of cylinder compressive strength with age of the RCCs containing Cement A and Flyash Y

5.3 Cement B and Flyash Y

The relationships between the normally-cured cylinder compressive strengths of the RCCs containing Cement B and Flyash Y are shown in Figure 3.

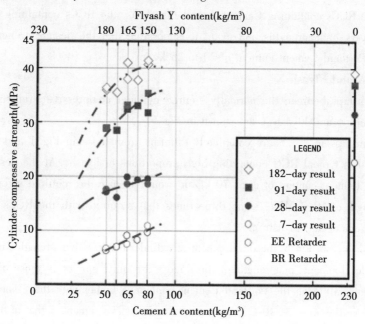

Fig. 3 Relationships between the Cement B and Flyash Y contents and the cylinder compressive strengths at various ages up to 182 days

It can be seen that there is little difference between Fig. 1 and Fig. 3; it seems the Portland Cements perform in a very similar way when combined with Flyash Y, although in this case the RCCs with the lower Portland Cement contents were not tried.

Using the data from Fig. 3, the development of cylinder compressive strength with age is shown in Fig. 4.

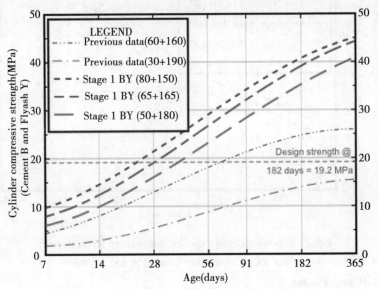

Fig. 4 Development of cylinder compressive strength with age of the RCCs containing Cement B and Flyash Y

As with the RCCs containing Cement A and Flyash Y, the RCCs containing Cement B and Flyash Y have very high strengths and again it is possible that the design strength could be achieved with a Portland Cement content of 20 to 25 kg/m^3.

5.4 Cement A and Flyash X

The relationships between the normally-cured cylinder compressive strengths of the RCCs containing Cement A and Flyash X are shown in Fig. 5.

The relationships in Fig. 5 are completely different from those in Fig. 1 and Fig. 3 and are more in line with a typical RCC containing high proportions of flyash. At the design age of 182 days a Portland Cement content of circa 70 kg/m^3 would seem to be required in order to achieve the design strength of 19.2 MPa, some three times that required with the RCCs containing the same Portland Cement and Flyash Y.

Using the data from Fig. 5, the development of cylinder compressive strength with age is shown in Fig. 6. The developments of strength in this Figure are quite similar to those of the 'typical' RCCs in the Figure although the Portland Cement content is higher with the Cement A/Flyash X RCCs, for example the 50 + 180 RCC has a similar strength development to that of the 'typical' 30 + 190 RCC and the 80 + 150 has similar developments of strength to the typical 60 + 160.

On the basis of the relationships in Fig. 6, it seems probable that a Portland Cement content of 70 kg/m^3 (a fairly typical figure) will be required in order to meet the design strength requirement.

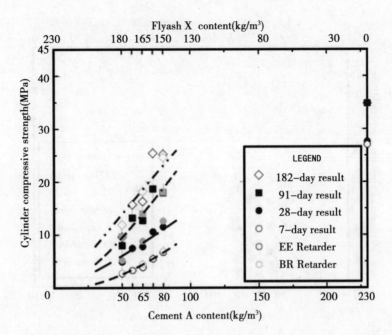

Fig. 5 Relationships between the Cement A and Flyash X contents and the cylinder compressive strengths at various ages up to 182 days

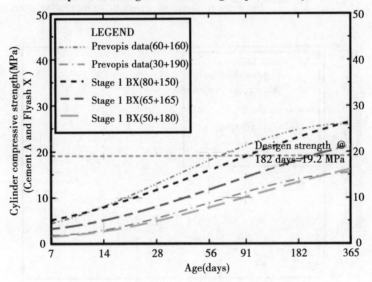

Fig. 6 Development of cylinder compressive strength with age of the RCCs containing Cement A and Flyash X

5.5 Cement B and Flyash X

The relationships between the normally – cured cylinder compressive strengths of the RCCs containing Cement B and Flyash X are shown in Fig. 7.

The relationships in Fig. 7 are very similar to those in Fig. 5. It is clear that the performance of the flyash is the major factor with these mixes.

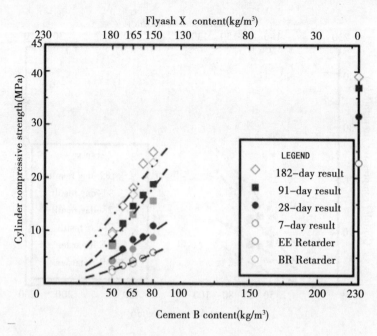

Fig. 7 Relationships between the Cement B and Flyash X contents and the cylinder compressive strengths at various ages up to 182 days

Using the data from Fig. 7, the development of cylinder compressive strength with age is shown in Fig. 8.

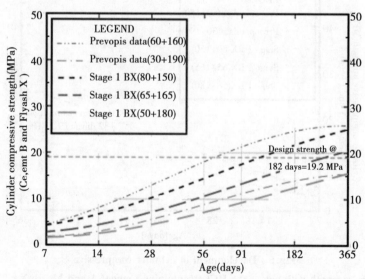

Fig. 8 Development of cylinder compressive strength with age of the RCCs containing Cement B and Flyash X

The developments of strength in Fig. 8 are fractionally lower than those in Fig. 6 and it is possible that a Portland Cement content of circa 72 kg/m^3 would be required to achieve the design strength.

6 Accelerated curing

6.1 Methodology

Two sets of specimens for each mix were subjected to accelerated curing. The 7 – day accelerated strengths are the result of curing the cylinders normally for six days followed by seven days in a hot water tank at 90 C and finally a day of normal curing. The 14 – day accelerated strengths are the same except the specimens spend 14 days in the hot water tank. It has been found that the 7 – day accelerated strengths are frequently equivalent to the 182 – day strengths of normally – cured specimens but the actual age can range from 56 to 365 days. Similarly the 14 – day accelerated strengths are frequently similar to the 365 – day strengths but it can be any age from 91 to 1 000 days.

6.2 Flyash Y

The strengths of all the 182 – day normally – cured RCCs containing Flyash Y have been compared to the equivalent 7 – day accelerated strengths in Fig. 9. It is clear that there seems to be little correlation. This is very surprising as it is one of the very few cases where there has not been some form of relationship. Nevertheless when the mixture proportions have been defined it might be possible to determine a particular 7 – day accelerated strength that has the equivalent 182 – day strength.

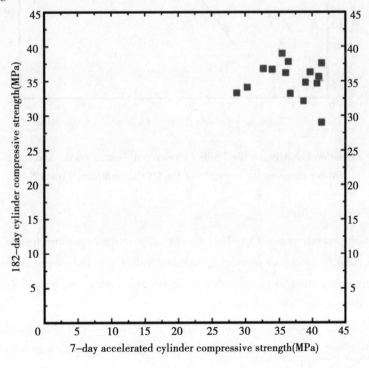

Fig. 9 **Relationship between the 7 – day accelerated – cured results and the 182 – day cylinder compressive strengths of the RCCs containing Flyash Y**

6.3 Flyash X

The strengths of all the 182 – day normally – cured RCCs containing Flyash X have been compared to the equivalent 7 – day accelerated strengths in Fig. 10. In this case there is a reasonable relationship – the Correlation Coefficient is 0.953 – and it should be relatively easy to predict the 182 – day compressive strength (i.e. at the design age) from the 7 – day accelerated strengths, that is 14 days after manufacture.

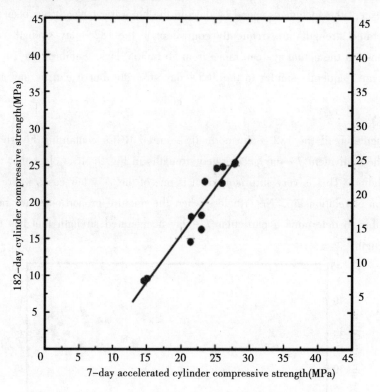

Fig. 10　Relationship between the 7 – day accelerated – cured results and the 182 – day cylinder compressive strengths of the RCCs containing Flyash X

7　Conclusions

Over the past several years, it has become normal practice to perform testing of various combinations of Portland Cement and pozzolans, together with a selection of admixtures, both retarders and air – entraining admixtures, in advance of issuing requests for proposals (RFP) for the construction of RCC dams. The results of such a programme, while often providing expected results, can, and on occasions do, provide unexpected results that can impact significantly on the design of the RCC and on the construction costs. Having this information in advance of the RFP can improve the quality of the proposals received, and potentially reduce the possibility of changes to the design and project costs compared to the scenario where this information is obtained following award of the construction of a Project. The results of this Trial Mix Programme demonstrate beyond a doubt why this is necessary.

References

[1] M. R. H. Dunstan. The optimization of the mixture proportions of RCC for dams[J]. Invited lecture in Proceedings of RCC Dam Workshop, IRCOLD, Tehran, Iran, February 2003.

Design and Constrution of Beyhan – 1 RCC Dam in Turkery

Ersan YILDIZ, A. Fikret GÜRDİL, Nejat DEMİRÖRS

(Temelsu Int. Eng. Services Inc., Turkey)

Abstract: This paper presents information about the design and construction of Beyhan – 1 Dam – a Roller Compacted Concrete (RCC) dam with a maximum height of 97 m from the foundation. The dam with a crest length of 365 m is located on Murat River, the largest tributary of the Euphrates River, at the Palu district of Elazğ province in Turkey. The dam location takes place in the east anatolian fault zone and very close to major active faults. In this regard, special attention has been paid to the seismic hazard evaluation performing detailed studies regarding the precise locations and characteristics of the active faults and surface rupture hazard. Concerning the high seismicity of the site supported by detailed studies, the dam geometry has been selected as faced symmetrical with a slope of 1 (V) : 0.7 (H). This geometry, which is commonly used for hardfill dams with relatively low strength characteristics, is adopted for a high design compressive strength of RCC as 15 MPa, in order to increase the stability by making use of water weight, and reduce the dynamic tensile stresses which are of major importance in high seismic zones. For the dynamic performance of the dam, advanced time – history calculations have been utilized in corporation with the conventional pseudo – static procedures. Linear and non – linear transient analyses have been conducted by the finite difference code, FLAC in order to determine the stress levels during the earthquake and estimate the post – earthquake damage. The RCC dam body construction started in May 2012 and ended in July 2014. The total RCC volume is 1 480 000 m^3. The average RCC placement was 2 250 m^3 per day, where as the maximum placement has been reported as 6 100 m^3 per day. The dam started operation in March 2015. The observations and monitoring data obtained from instrumentation devices have shown that the performance of the dam is satisfactory and the assumptions adopted in the design are in accordance with the real behaviour.

Key words: RCC, earthquake, faced symmetrical

1 Introduction

This paper presents brief information about the design and construction of Beyhan – 1 RCC Dam located in Elazğ Province of Turkey. The project characteristics are summarized in Table 1.

Table 1 Project infromation

Dam type	RCC
Purpose	Energy
Maximum height from foundation	97 m
Dam crest width	10 m
Dam crest length	360 m
Spillway gate No. / width	6 × 11.5 m
Installed Capacity	550 MW
Average annual generation	1 253 GW · h

2 Seismicity of dam location

The damsite is very close to the main active faults in the East Anatolian Fault Zone (EAFZ). The active faults around the dam location are shown in Fig. 1. In this regard, special attention has been paid to the seismic hazard evaluation of the site. Detailed studies and site investigations, including site mapping and examination of excavated trenches have been performed by the relevant specialists in order to define the active fault locations and characteristics, as well as the potential surface rupture hazard for the dam site. Based on these studies, it is concluded that no active fault crosses the dam axis, so there is no surface rupture potential for the dam site.

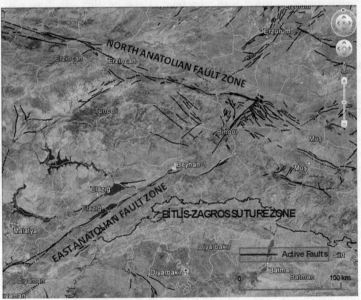

Fig. 1 Beyhan − 1 **Dam location and active faults**

The controlling seismic source for the dam location has been determinedas a fault segment at a minimum distance of 1.7 km to the dam site. The maximum earthquake magnitude that this strike − slip fault segment can generate has been estimated as M_w = 6.8. The other fault segments around the dam site with their characteristics and estimated maximum magnitudes have also been defined with this detailed study.

Based on the above mentioned seismic source data, the seismic hazard analysis has been carried out and the peak ground acceleration (PGA) levels have been calculated as 0.40 g and 0.88 g for Operating Basis Earthquake (OBE) and Safety Evaluation Earthquake (SEE) respectively. Both of these acceleration levels are significantly higher than usual values for dams, which makes the dam body design more important in terms of earthquake safety.

3 Dam body design

3.1 Foundation conditions

Detailed geological investigations including site observations, boreholes, in – situ and laboratory tests have been performed in order to reveal the foundation rock conditions. According to obtained data, the foundation rock is composed of andesite, mudstone and shale. Depending on the high tectonic movements at the region, different quality rock masses were encountered below the dam body, which were classfied into three as: PQR (tectonically disintegrated mudstone and shale mixtures, unsuitable as RCC foundation), MQR (tectonically disturbed andesite masses, can be suitable with improvements by grouting) and HQR (relatively undisturbed andesite masses, suitable).

The dam layout and base elevation have been determined so that a large portion of the foundation rock was composed of HQR. At areas where PQR material existed locally, it has been removed by excavating and backfilled with concrete. Consolidation grouting has been applied in MQR rock masses to increase the strength and stiffness of the rock. With these measures, a suitable foundation rock for RCC dam body in terms of strength and stiffness has been provided.

3.2 Design parameters and dam geometry

For the RCC dam body, the design compressive strength has been predicted as $\sigma_c = 15$ MPa. As bedding mortar application has been held as mandatory for the cold joints between lifts, the tensile strength in the vertical direction has been assumed as 5% of the compressive strength according to suggestions of USACE (2000). Similarly, the tensile strength of the parent RCC has been taken as 9% of compressive strength. The tensile strength has been increased by 35% for earthquake loading considering the dynamic response of the material based on common practice in literature.

Based on the geological and geotechnical evaluations on the foundation, the deformation modulus of the foundation rock has been estimated as 5 GPa. The shear strength of the rock has been determined as greater than that of rock – RCC contact or lift joints within RCC. Therefore, the stress and stability checks have been performed for the dam body only.

Considering the high seismicity of the dam site, a faced symmetrical dam geometry (Fig. 2) with a slope ratio of $0.7(H):1(V)$ has been adopted for the dam body. This type of geometry, which has commonly been used for hardfill dams constructed with relatively lower strength material, is adopted with a high strength material in order to cope with high shear and tensile stresses during dynamic loading.

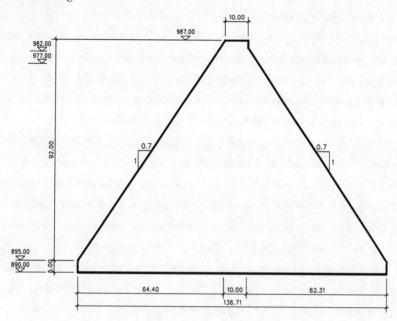

Fig. 2　Typical cross section of Beyhan – 1 Dam

The advantage of using a flatter upstream slope is to increase the initial compressive stresses and improve the stability by making use of reservoir water weight, and to reduce the level of dynamic tensile stresses in high seismic zones.

3.3　Stability analyses

The preliminary design of the dam body has been performed by conventional stability analyses. Three different loading conditions namely as "usual", "unusual" and "extreme" have been considered with the safety factors adopted as 3.0, 2.0 and 1.0 respectively. The operation condition has been considered as usual, earthquake loading for OBE as unusual, flood and eartquake loading for SEE as extreme. The stresses with the RCC and more realistic dynamic behaviour of the dam body have been evaluated by more rigorous numerical analyses explained as fllowing.

3.4　Dynamic analyses

ICOLD (Bulletin – 72, 1989) and USACE (2007) proposes two different earthquake levels to be considered in the earthquake design of dams, named as the operating basis earthquake, OBE and the maximum design earthquake, MDE (or safety evaluation earthquake, SEE as suggested by ICOLD Bulletin – 72, 2010). After an earthquake at OBE hazard level, the dam, appurtenant structures and equipment need to remain functional and the damage need to be easily repairable. USACE (1995) finds it necessary for the stress levels in RCC dam body to remain within the elastic region in case of OBE. After an earthquake of SEE level, reaching of the elastic limits, and the related damage are considered as acceptable (USACE, 1995), provided that a catastrophic failure which may lead to an uncontrolled release of reservoir water or life loss is prevented.

Time – history analyses are defined as the most effective method for the dynamic behaviour of structures (USACE, 2007) where the earthquake behaviour of the dam may well be examined for the whole earthquake duration, accordingly the time variations of the stresses and deformations may be inspected. A more realistic assessment of the seismic behaviour is possible with the interactions between the modelled foundation, reservoir water and dam. Besides the method gives accurate and reliable results for cases where concrete stresses are below the elastic limits; it is suggested for all dam designs to examine the behaviour beyond linear limits and estimate the potential damage by international standards (FEMA, 2005; USACE, 2007).

In a high intensity ground motion, cracks due to tension in the horizontal lift joints and concrete – rock contact or corners with stress concentrations are likely to occur (USACE, 2007). In this regard, most of the deformations will take place on these cracks and it is expected to have a few major crack surfaces in case of a large magnitude earthquake (Wieland, 2008). For such a case, there are two main approaches in order to investigate the nonlinear seismic behaviour of the dam body, referred as the smeared crack and discrete crack approaches.

In the discrete approach, joints to simulate the sliding and gap behaviour are defined at the lift joints or regions where cracking is expected. The nonlinear behaviour is taken into account by only the basic parameters for the joints. During an earthquake, the displacement of the blocks along these crack surfaces should be evaluated, and the post – earthquake stability should be checked by using residual strength parameters of the crack surface and unfavourable distribution of the uplift pressures. It has been found reasonable to adopt this approach for the nonlinear behaviour of Beyhan – 1 Dam.

The dynamic calculations have been made for OBE using linear elastic material models, and the maximum tensile stresses in the vertical and principal directions occurred during the earthquake have been determined. These resultsshown in Fig. 3 indicate that the vertical tensile stresses have occurred in a relatively small portion of the dam and the maximum vertical tensile stresses are below the tensile strength of lift joints. The tensile stresses in the principal direction exceed the strength in a small local area at the upstream base corner of the dam body. This has been considered as acceptable, as the local high stresses are mainly due to stress concentrations in the numerical model of maximum cross – section, and the subject area is within the improvement zone for concrete strength. Based on these data, the dynamic performance of the dam has been found to be satisfactory for OBE scenario.

The dynamic response for SEE has been first evaluated by using linear elastic material models and the obtained maximum vertical and principal tensile stresses are given in Fig. 4. It can be seen that the tensile strength of the material has been exceeded considerably, especially in the upstream and downstream base corners of the dam. This shows that cracks will likely develop at the subject area during this high level of seismic shaking, and a potential crack zone may be considered to be a horizontal joint along the base of the dam. Similarly, the overstressed regions at the upper elevations also indicate the potential forming of cracks and joints at these parts of the body. To examine the dynamic behaviour of the dam when cracks are formed at the overstressed zones, a jointed model is constructed by assuming three cracking horizontal joints as at the base of the

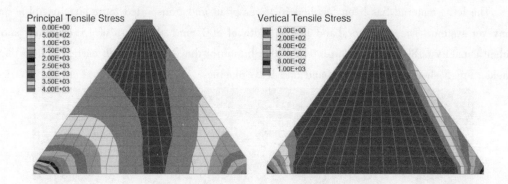

Fig. 3 Maximum tensile stress distributions for OBE

dam, at a height of 25 m from the base, and at a height of 50 m from the base. With this jointed model, the permanent displacements have been examined during and after the earthquake.

Obtainedresults show that, there is a permanent relative displacement between the dam body and the bedrock, and as well as at the joints defined in the upper elevations of 25 m and 50 m from the base. The permanent horizontal relative displacements are to the downstream side as 30 cm, 10 cm and 12 cm at the base joint, joint at $H = 25$ m from the base and joint at $H = 50$ m from the base respectively, resulting in a total relative displacement of about 50 cm between the bedrock and crest of the dam. The analysis has shown that the dam retains its stability during and after the earthquake, although considerable relative displacements have been obtained on the assumed cracking joints. To examine the post – earthquake safety of the RCC dam body, the stability of the block over the assumed crack joints is evaluated by considering the full uplift pressure and zero cohesion along the crack. The lowest safety factors has been found as $F_s = 2.81$ indicating that the safety of the dam is satisfied even for full uplift case. These results indicate that the dynamic response of the dam is satisfactory for SEE condition.

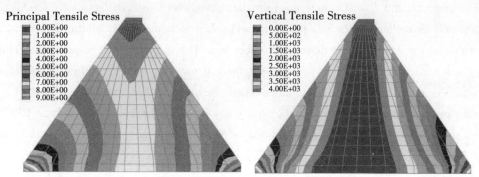

Fig. 4 Maximum tensile stress distributions for SEE

4 Dam construction

The construction of RCC dam body started in May 2012 and finished in April 2014. The total RCC volume is 1 480 000 m^3. The recorded average daily RCC production rate is 2 250 m^3. The maximum RCC production rate has been recorded as 6 100 m^3/day.

The RCC material has been batched in tower plant and transported to the site by a 4 - span conveyor system. The conveyor band had a width of 800 mm, thickness of 15 ~ 20 mm and a weight capacity of 80 kg/cm. The internal distribution of the RCC has been carried out by dump trucks. Fig. 5 shows the conveyor band and RCC placing.

Fig. 5 **RCC transport by conveyor band**

The spreading of RCC has been carried out by 2 D5N bulldozers. For the compaction of RCC, 2 vibrating rollers of 15 tons have been used. The compaction of RCC at limited areas have been performed by 3 rollers of 3 tons. Fig. 6 shows the spreading and compaction of RCC.

The upstream and downstream faces of the dam have been constructed as CVC. The CVC application with an inclined surface has been accomplished with the use of sliding formwork. The CVC widths at the upstream and downstream faces were 1.5 m and 1.0 m respectively. Vertical joints (with around 20 m intervals) to prevent thermal cracking have been formed in the RCC body and CVC, and two layers of PVC waterstops have been used in the joint within CVC. Between the primary vertical joints, secondary vertical joints have been installed only at CVC face where a single layer of PVC water stop has been used. CVC application against formwork is shown in Fig. 7.

5 Instrumentation and performance

A detailed instrumentation program has been adopted at RCC dam body for detailed monitoring and review on dam performance. The installed instrumentation included:

(1) Piezometers (40 in total) at dam foundation rock to observe the spatial variation of pore pressures along the dam base.

(2) Total pressure cells (10 in total) within the RCC at the upstream and downstream bases

Fig. 6 Spreading and compaction of RCC

Fig. 7 CVC application at face with formwork

of the dam along the valley.

(3) Strain gages (10 in total) within the RCC at upstream and downstream faces close to the foundation along the valley.

(4) Extensometers (4 in total) within the foundation rock in order to observe the deformations in the foundation (installed as inclined from the bottom gallery to both upstream and downstream directions).

(5) Jointmeters (14 in total) at the vertical joints between blocks in order to observe relative movement of blocks to each other.

(6) A normal pendulum in the middle of RCC body in order to observe movement and rotation.

(7) Temperature sensors to observe temperature changes along with all piezometers, total pressure cells and strain gages.

(8) Long – term temperature cells (53 in total) at the base and 2 higher levels in the middle of RCC body.

During the reservoir filling, all instrumentation devices have performed well. Obtained monitoring data was as expected, indicating that the design assumptions are in accordance with the real behaviour.

Some smallseepage has occurred at the right bank of the dam during the first filling, but has disappeared with time probably due to calcification and blocking of fine material within seeping water. No other problems have been observed related to cracking, seepage etc. and the dam performance is considered as satisfactory.

Acknowledgements

The suggestions and help of Mr. Ferruh Anık on the paper are acknowledged.

References

[1] FEMA. Federal Guidelines for Dam Safety, Earthquake Analysis and Design of Dams,2005.
[2] FEMA. NEHRP Recommended Seismic Provisions for New Buildings and Other Structures, 2009:750.
[3] ICOLD Bulletin 72. Selecting Seismic Parameters for Large Dams,1989.
[4] ICOLD Revision of Bulletin 72. Selecting Seismic Parameters for Large Dams,2010.
[5] Şaroğlu F., Perinçek D. Surface Rupture Hazard Investigations for Beyhan – 1 and Beyhan – 2 Dams and HEPP (Palu – Elazığ), Eastern Turkey, FN Ç Petrol Madencilik San. ve Tic. A. Ş., 2010.
[6] USACE. Gravity Dam Design, EM 1110 – 2 – 2200,1995.
[7] USACE. Roller Compacted Concrete, EM – 1110 – 2 – 2006, 2000.
[8] USACE. Earthquake Design and Evaluation of Concrete Hydraulic Structures, EM – 1110 – 2 – 6053, 2007.
[9] Wieland, M. Analysis Aspects of Dams Subjected to Strong Ground Shaking. International Water Power and Dam Construction, 2008.

The Influence of Size and Layer Face Effects on the Deformation Performance of Fully – graded RCC

Zhang Sijia, Feng Wei, Ji Guojin

(China Institute of Water Recourses and Hydropower Research, Beijing, 100038, China)

Abstract: This study conducts comparative tests of compressive elastic modulus, ultimate tension, autogenous volume deformation and creep, etc. on fully – graded RCC with and without layer face and wet – screened two – graded RCC. The results indicate that the ultimate tension, autogenous volume deformation and creep of fully – graded RCC are smaller than those of wet – screened two – graded RCC, which is affected by wet – screened size. The influence of layer face on ultimate tension of fully – graded RCC is larger than that on compressive elastic modulus.

Key words: Fully – graded RCC, Size effect, Layer face effect, Wet – screened concrete, Deformation performance

1 Introduction

The deformation performance of RCC is the key index for study and analysis of crack resistance of concrete[1]. In hydraulic mass RCC, three – graded RCC accounted for a significant proportion. At present, in the design technical requirements and construction quality monitoring of dam concrete, the performance indices of the two – graded RCC after wet screening (screened out the particles with aggregate size larger than 40 mm in fully – graded RCC) are usually adopted as the references. Because the eliminated aggregate by wet screening accounts for about 40% in the total original aggregate, and the mortar content is increased significantly, the change of the mix proportion will directly affect the deformation performances of the concrete. Accordingly, the results of each performance test of wet – screened two – graded RCC can not exactly reflect the real properties of fully – graded dam concrete[2-3].

The domestic and foreign research and engineering practice shows that the level is the weak link of RCC dam andthe interlayer combination properties directly affect the durability of concrete and the stability against sliding of dam. So the experimental studies on the deformation performance of fully – graded RCC with and without layer, to analyze the relation between the performance of wet – screened two – graded RCC and fully – graded RCC with and without layer. It is aimed at accurately indicating the actual properties of dam concrete, and providing references for deformation tests of fully – graded RCC and engineering application.

2 The size of fully – graded and wet – screened RCC sample and their forming method

Fully – graded RCC is different from normal concrete, and belongs to super rigid concrete. The new design of high – power flat – panel vibration molding machine is used for vibrated forming. Its frequency is 2 800 times/min, the power of 1.1 kW, 5.2 mm of amplitude and 5 000 N of exciting force, and two flat – panel size of ϕ290 mm and 290 mm × 290 mm. Wet – screened concrete sample are all vibrated with small flat – panel vibration forming machine. Its vibration frequency is (50 ± 3) Hz, and the amplitude is (3 ± 0.2) mm. The specimen size of fully – graded RCC and wet – screened two – graded RCC for elastic modulus test is ϕ300 mm × 600 mm and ϕ150 mm × 300 mm respectively. The specimen size of fully – graded RCC and wet – screened two – graded RCC for ultimate tension test is ϕ300 mm × 900 mm and 100 mm × 100 mm × 550 mm respectively. The specimen size of fully – graded RCC for autogenous volume deformation and creep test is ϕ300 mm × 900 mm. The specimen size of wet – screened two – graded RCC for autogenous volume deformation and creep test is ϕ200 mm × 600 mm and ϕ150 mm × 450 mm respectively. The strain gauges are buried into specimens for autogenous volume deformation and creep test.

3 Comparative tests of deformation performance of fully – graded and wet – screened RCC

3.1 Raw material and mix proportion

The raw material and mixproportion of concrete are shown as following table 1.

Table 1 Mixproportion of fully – graded RCC

Type of flyash	Grading	Mixproportion parameter				material amount (kg/m³)							Admixture and amount	
		$\frac{W}{C+F}$	W	F (%)	S (%)	Water	cement	Flyash	Sand	Stone (mm)			Water reducing agent (%)	Air entraining agent (%)
										5~20	20~40	40~80		
Grade I	Three	0.41	79	56	34	79	86	109	743	437	583	437	0.2	0.05
Grade II	Three	0.41	79	56	34	79	86	109	743	437	583	437	0.8	0.05

3.2 Molding condition of tests

Themolding conditions of compressive elastic modulus and ultimate tension of fully graded RCC include RCC with layer and without layer, and both conditions of the autogenous volume deformation and creep are RCC without layer. The condition of RCC without layer is the testing sample vibrated via one – time – concreting. The condition of RCC with layer is concreting on half of the testing sample first, and compacting, and then concreting and compacting on another half of the testing sample in certain intervals, in order to study the influence of molding condition on the deformation performance of fully – graded RCC, and provide reliable construction interval time of dam RCC concreting and scientific test data. The interval time for RCC with layer is determined as 6 h according to the setting time (table 2) and the construction requirements of RCC with different types of flyash, and all the wet – screened RCC has no layer.

Table 2 Setting time of RCC with differenttypes of flyash

Type of flyash	Grading	material amount (kg/m³)							Admixture and amount		setting time (h)	
		Water	Cement	Flyash	Sand	Stone (mm)			Water reducing agent (%)	Air entraining agent (%)	Initial set	Final set
						5~20	20~40	40~80				
Grade I	Three	79	86	109	743	437	0.2	0.05	0.2	0.05	8	13
Grade II	Three	79	86	109	743	437	0.8	0.05	0.8	0.05	26	32

3.3 Test results

3.3.1 Compressive elastic modulus

The results of the compressiveelastic modulus test are shown in table 3. The direction of application force should be vertical to the layer face of the sample, so the layer face effect is difficult to reflect in the test, therefore the effect coefficient of layer face is close to 1.0 (0.98 ~ 0.98).

Table 3 The results of compressive elastic modulus

Type of flyash	Type of sample	Condition and grading	Elastic modulus (GPa)		
			28 d	90 d	180 d
Grade I	D_1	Fully graded RCC without layer	37.7	43.2	45.5
	D_2	Fully graded RCC with layer	37.1	42.5	44.9
		Layer faceeffect coefficient: D_2/D_1	0.98	0.98	0.99
	S	Wet – screened two – graded RCC without layer	36.5	42.7	45.2
		Wet – screen size effect coefficient: D_1/S	1.03	1.01	1.01
Grade II	D_1	Fully graded RCC without layer	35.4	42.1	43.2
	D_2	Fully graded RCC with layer	35.2	41.9	43.0
		Layer faceeffect coefficient: D_2/D_1	0.99	1.00	1.00
	S	Wet – screened two – graded RCC without layer	34.8	41.3	42.8
		Wet – screen size effect coefficient: D_1/S	1.02	1.02	1.01

The difference of compressiveelastic modulus between fully – graded RCC and wet – screened RCC is mainly due to the coarse aggregate content. It is higher in fully – graded RCC. Compared with wet – screened RCC sample, in early age, the framework role of coarse aggregate in fully – graded RCC is much obvious, so the compressive modulus is much larger. As the growing of the age, the mortar strength is gradually increased, and the framework role of coarse aggregate of the fully – graded RCC is decreased gradually, and the difference of compressive elastic modulus between fully – graded RCC and wet – screened RCC is much smaller. The effect coefficient of wet – screened size of RCC is among 1.01 ~ 1.03, with a average value of 1.02.

3.3.2 Ultimate tension

Theresults of ultimate tension test are shown in table 4. Regardless of Grade I or Grade II flyash, the axial tensile strength and the ultimate tension of fully – graded RCC with layer are lower than those without layer, with the reduction range of 10% ~20%. As the growth of the age, the later layer face effect (180 d) is weakened. The axial tensile strength of fully – graded RCC with layer, which is used of Grade I flyash, the effect coefficient of layer face is among 0.82 ~0.94, with the average value of 0.86. The axial tensile strength of fully – graded RCC with layer, which is used of Grade II flyash, the effect coefficient of layer face is among 0.80 ~ 0.92, with the average value of 0.85. Also, for the layer face effect coefficient of ultimate tension value, Grade I flyash sample is among 0.75 ~ 0.92, with the average of 0.84, and Grade II fly ash sample is among 0.78 ~ 0.94, with the average of 0.87.

The axis tensile strength and ultimate tension of fully – graded RCC without layer is significantly lower than those of their respective wet – screened RCC. The reduction range is 20% ~ 30%. The effect coefficient of wet – screened size of axial tensile strength is among 0.70 ~0.73, with an average of 0.71. The effect coefficient of wet – screened size of ultimate tension value is among 0.70 ~0.79, with an average of 0.74.

Table 4　The results of ultimate tension

Type of flyash	Type of sample	Condition and grading	Axis tensile strength (MPa)			Ultimate tension(10^{-6})		
			28 d	90 d	180 d	28 d	90 d	180 d
Grade I	D_1	fully graded RCC without layer	1.87	2.25	2.79	63	76	85
	D_2	fully graded RCC with layer	1.53	1.84	2.63	47	64	78
	Layer faceeffect coefficient: D_2/D_1		0.82	0.82	0.94	0.75	0.84	0.92
	S	Wet – screened two – graded RCC without layer	2.56	3.19	3.99	80	108	119
	Wet – screen size effect coefficient: D_1/S		0.73	0.71	0.70	0.79	0.70	0.71
Grade II	D_1	fully graded RCC without layer	1.78	2.35	2.82	59	73	83
	D_2	fully graded RCC with layer	1.43	1.95	2.59	46	66	78
	Layer faceeffect coefficient: D_2/D_1		0.80	0.83	0.92	0.78	0.90	0.94
	S	Wet – screened two – graded RCC without layer	2.43	3.32	4.03	77	101	114
	Wet – screen size effect coefficient: D_1/S		0.73	0.71	0.70	0.77	0.72	0.73

3.3.3 Autogenous volume deformation

In the condition of constant temperature and adiabatic humidity, the volume deformation of concrete, which is caused by the hydration of gelled material, is called as autogenous volume deformation. There are two kinds of flyash (Grade I and Grade II). There is no layer face in the molding condition for fully – graded RCC and wet – screened two – graded RCC. And the test results of autogenous volume deformation are shown in table 5 and table 6. The curves of autogenous volume deformation are shown in Fig. 1 and Fig. 2.

Theautogenous volume deformation of fully – graded RCC and wet – screened two – graded RCC with Grade I are shrunk a little, then expanded. The expanding amount of wet – screened RCC is up to maximum at about 60 d, while the expanding amount of fully – graded RCC is up to maximum at about 40 d. The total autogenous volume deformation is microexpanded. The maximum expanding amount is about $(7 \sim 8) \times 10^{-6}$.

The autogenous volume deformation of fully – graded RCC and wet – screened two – graded RCC with Grade II is expanded at the beginning. The maximum expanding amount of two – graded RCC is about 30×10^{-6}, and the maximum expanding amount of fully – graded RCC is about 13×10^{-6}. The expanding amount is up to maximum at 60 d to 80 d, and then shrunk a little. The autogenous volume deformation is microexpanded type.

In general, the autogenous volume deformation of fully – graded RCC is smaller than the corresponding deformation of wet – screened two – graded RCC. The autogenous volume deformation of RCC with flyash of Grade I is smaller than that of RCC with flyash of Grade II.

Table 5　The results of autogenous volume deformation of RCC with Grade I

Time (d)	autogenous volume deformation ($\times 10^{-6}$)		Time (d)	autogenous volume deformation ($\times 10^{-6}$)	
	Fully – graded RCC	Wet – screened two – graded RCC		Fully – graded RCC	Wet – screened two – graded RCC
1	-0.1	-0.1	70	5.2	6.5
2	1.1	-0.5	80	5.5	4.8
3	0.2	-2.5	90	6.3	6.7
5	1.1	-1.8	105	5.2	5.6
7	2.7	2.8	120	5.2	5.9
10	4.2	4.0	135	4.9	4.6
15	4.6	2.7	150	5.0	4.3
20	4.6	2.9	165	4.9	4.8
25	3.6	3.9	180	4.1	4.7
28	3.1	3.2	210	4.8	3.7
35	6.6	5.1	240	4.6	5.2
40	8.1	5.6	270	3.4	5.3
45	7.2	5.1	300	3.3	5.3
50	6.2	5.0	330	3.3	5.3
60	7.7	6.9	360	3.3	5.3

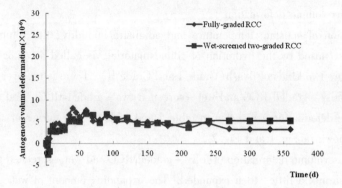

Fig. 1 The autogenousvolume deformation of RCC with Grade I

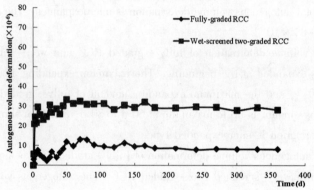

Fig. 2 The autogenous volume deformation of RCC with Grade II

Table 6 The test results of autogenous volume deformation of RCC with Grade II

Time (d)	autogenous volume deformation ($\times 10^{-6}$)		Time (d)	autogenous volume deformation ($\times 10^{-6}$)	
	Fully – graded RCC	Wet – screened two – graded RCC		Fully – graded RCC	Wet – screened two – graded RCC
1	3.1	5.6	70	13.1	32.6
2	0.7	20.9	80	12.8	30.9
3	5.1	24.5	90	10.3	31.2
5	6.2	21.6	105	9.5	31.3
7	7.6	29.6	120	9.1	28.4
10	5.2	25.7	135	11.3	30.5
15	3.9	23.1	150	9.5	29.4
20	2.3	24.7	165	9.9	32.1
25	4.1	30.6	180	8.7	29.4
28	6.9	26.9	210	8.7	29.2
35	4.1	28.7	240	7.9	29.7
40	6.5	26.1	270	8.3	29.5
45	8.3	30.2	300	8.2	28.0
50	11.9	33.4	330	8.2	29.4
60	9.7	31.3	360	8.1	28.2

3.3.4 Creep degree

Two kinds of flyash, Grade I and Grade II are used for creep test. There is also no layer face in the molding condition for fully – graded RCC and wet – screened two – graded RCC. The creep test of RCC includes three loading ages at 28 d, 90 d, 180 d, and the sustained load is observed up to 90 d. The results of creep test are shown in table 7 and table 8.

Table 7   The results of creep degree of RCC with Grade I   (10^{-6}/MPa)

Loading age Sustained loading time (d)	Fully – graded RCC			Wet – screened two – graded RCC		
	28 d	90 d	180 d	28 d	90 d	180 d
0	0.0	0.0	0.0	0.0	0.0	0.0
1	3.5	2.6	0.5	4.2	2.5	2.4
2	5.3	3.0	0.8	5.8	3.1	3.1
3	5.1	3.7	1.1	6.4	3.4	3.1
5	6.6	4.0	1.3	6.9	4.1	3.5
7	7.3	4.4	1.7	8.1	4.5	4.3
10	7.2	4.4	2.0	8.5	4.7	4.5
15	8.4	5.3	2.4	9.2	4.9	4.8
20	8.4	5.5	2.6	9.4	6.2	4.9
25	8.9	5.7	3.1	10.3	5.7	5.0
28	9.3	5.9	3.3	10.4	6.3	5.1
35	9.8	6.0	3.3	11.2	6.5	5.3
40	10.1	6.6	3.5	11.5	6.8	5.5
45	10.8	6.7	3.7	11.6	6.9	5.8
50	10.8	6.7	3.8	11.9	7.2	6.1
60	11.2	6.7	4.2	12.4	7.6	5.9
70	11.4	6.8	4.3	12.6	8.0	6.2
80	11.5	6.9	4.2	13.1	8.1	6.2
90	11.5	7.0	4.2	13.2	8.1	6.3

It can be seen from the results, in addition to the individual loading time, the creep degree of fully – graded RCC with Grade I flyash is smaller than that of wet – screened RCC, while the creep degree of fully – graded RCC with Grade II flyash is smaller than that of wet – screened RCC at the whole loading time. The curves of creep degree of RCC are shown from Fig. 3 to Fig. 6.

At loading ages of 28 d, 90 d and 180 d, the creep degree of fully – graded RCC with Grade I flyash, sustained loaded for 90 d, is from 4×10^{-6}/MPa to 12×10^{-6}/MPa, while the creep degree of wet – screened RCC is from 6×10^{-6}/MPa to 14×10^{-6}/MPa. The average ratio of creep degree of fully – graded RCC to wet – screened RCC at loading ages of 28 d, 90 d and 180 d is 0.88, 0.95, 0.51, respectively. The total ratio of creep degree is 0.78.

Table 8 The results of creep degree of RCC with Grade II(10^{-6}/MPa)

Loading age Sustained loading time (d)	Fully – graded RCC			Wet – screened two – graded RCC		
	28 d	90 d	180 d	28 d	90 d	180 d
0	0.0	0.0	0.0	0.0	0.0	0.0
1	8.7	4.0	2.5	7.5	2.6	2.4
2	9.0	5.0	2.5	9.6	3.3	3.1
3	10.0	5.2	2.3	10.7	3.4	3.3
5	12.7	5.5	1.8	11.6	4.7	4.5
7	14.1	5.6	2.0	14.1	4.2	4.3
10	15.5	5.2	2.5	17.1	4.6	4.0
15	17.1	6.9	4.8	17.4	6.3	4.7
20	18.2	6.6	4.7	17.8	7.1	5.4
25	18.5	7.1	5.0	18.5	7.6	6.1
28	19.0	7.1	5.4	19.6	8.2	6.6
35	19.2	7.5	5.5	20.4	8.8	6.7
40	18.5	7.7	5.4	20.8	10.4	6.9
45	19.2	7.6	5.9	20.9	10.1	7.4
50	19.4	7.9	6.1	23.1	11.6	7.2
60	20.1	7.8	6.3	24.1	12.1	7.6
70	19.7	8.1	6.3	24.7	12.8	8.5
80	20.6	8.6	7.4	25.0	13.1	8.6
90	20.7	8.6	7.5	25.1	13.1	8.6

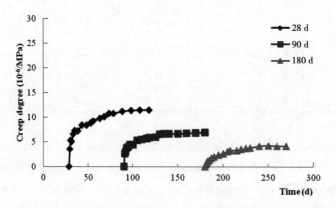

Fig. 3 The creep degree of fully – graded RCC with Grade I

At loading ages of 28 d, 90 d and 180 d, the creep degree of fully – graded RCC with Grade II

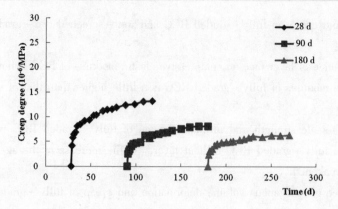

Fig. 4 The creep degree of wet – screened RCC with Grade I

flyash, sustained loaded for 90 d, is from 7×10^{-6}/MPa to 21×10^{-6}/MPa, while the creep degree of wet – screened RCC is from 8×10^{-6}/MPa to 25×10^{-6}/MPa. The average ratio of creep degree of fully – graded RCC to wet – screened RCC at loading ages of 28 d, 90 d and 180 d is 0.94, 0.98, 0.72, respectively. The total ratio of creep degree is 0.88.

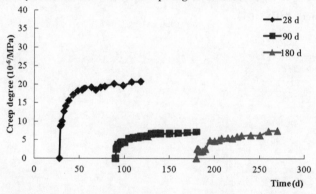

Fig. 5 The creep degree of fully – graded RCC with Grade II

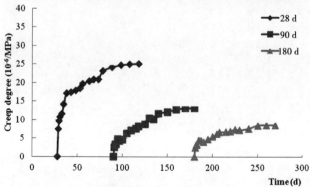

Fig. 6 The creep degree of wet – screened RCC with Grade II

4 Conclusions

This study conducts the tests of compressive elastic modulus, ultimate tension, autogenous volume deformation and creep on fully – graded RCC and wet – screened two – graded RCC. The results indicate that:

(1) The influence of layer face on compressive elastic modulus of RCC is not obvious, and the compressive elastic modulus of fully – graded RCC is a little higher than that of wet – screened two – graded RCC.

(2) The axis tensile strength and ultimate tension of fully – graded RCC with layer face are lower than those of fully – graded RCC without layer. As the increase of the age, the influence of the layer face is weakened.

(3) The degree of autogenous volume deformation and creep of fully – graded RCC is smaller than those of wet – screened two – graded RCC.

References

[1] Fang Kunhe. Research on the deformation characteristics of RCC[J]. Hubei Water Power, 1993,13(2).
[2] Elmo C. Higginson, George B. Wallace, Elwood L. Ore. Effect of Maximum Size Aggregate on Compressive Strength of Mass Concrete[J]. Symposium On Mass Concrete, ACI, 1963, 219-257.
[3] Jiang Rongmei, Feng Wei. Impat of Interlayer and Size Effect on the Mechanicial Peformances of Fully Graded RCC[J]. Water Power, 2007,33(4).

Update of ICOLD Bulletin 126 on RCC Dams

Shaw QHW

(ARQ (PTY) Ltd, South Africa)

Abstract: The original 2003 ICOLD Bulletin 126 on Roller Compacted Concrete dams proved to be the most successful ICOLD bulletin to date, with the highest number of copies published and being translated into more languages than any previous ICOLD bulletin. Over the past decade or so, the number of RCC dams in the world has essentially doubled, with approximately 700 now completed, or under construction. More importantly, some significant developments have occurred in various aspects of RCC technology for dams and to ensure that the current ICOLD bulletin reflects the state of the art, an update of Bulletin 126 was considered necessary. A decision to update the bulletin was consequently taken at the ICOLD annual meeting in Seattle in 2013.

While many of the developments in RCC dam engineering since the publication of Bulletin 126 have been complementary and have come about through experience on RCC structures of ever increasing size and height, several significant developments in our knowledge influence how we should design and build our RCC dams.

The update of ICOLD Bulletin 126 is being prepared by a task team of the ICOLD Committee on Concrete Dams lead by Quentin Shaw and comprising Rafael Ibanez de Aldecoa, John Berthelsen, Marco Conrad, Tim Dolen, Marco Conrad, Malcolm Dunstan, Francisco Ortega, Mike Rogers, Ernie Schrader, Del Shannon and Tsuneo Uesaka.

In this paper, the author describes the developments that will form part of the Bulletin 126 update and discusses the novel manner in which the bulletin update will be published.

Key words: ICOLD, RCC Bulletin 126

1 Background and introduction

Roller compacted concrete dam construction has a history of more than 30 years, with approximately 700 RCC dams completed, or under construction worldwide as of 2015. Despite having evolved to reach a state of some maturity over this period, the technology continues to develop and while certain aspects can be considered to be undergoing a process of ongoing refinement with increasing experience, in other areas necessary changes to current practice remain in the process of being adopted.

The technology for roller compacted concrete dams was first addressed by ICOLD in Bulletin No 75 "Roller – Compacted Concrete Gravity Dams"[1], which was published in 1989, and subsequently in Bulletin No. 126 "Roller – Compacted Concrete Dams: State of the art and case histories"[2] published in 2003. While the latter document saw wide and successful application over the intervening period, a number of important recent developments now need to be addressed to ensure that the current ICOLD publication realistically reflects the contemporary state of the art. With much of the contents of Bulletin No. 126 remaining valid, it was considered that the important developments should be addressed through a bulletin update, rather than a new bulletin.

Particularly important developments in RCC dam technology that necessitated the present update of Bulletin No. 126 include the following:

(1) New developments in the understanding of the early behaviour of different types of RCC that influence design and construction;

(2) The important design differences that relate to the horizontal construction of RCC dams, compared to the vertical construction of conventional vibrated concrete (CVC) dams;

(3) Developments in mix designs and construction techniques, most particularly related to super – retarded, high workability RCC (Fig. 1);

(4) Developments in the design and construction of RCC arch dams; and

(5) Developments arising from the use of RCC in more extreme environments.

In this paper, the author outlines in broad detail the developments in RCC technology since the original publication of Bulletin 126 that will be addressed and included in the update.

Fig. 1 **RCC dam construction**
(illustrating modern, cohesive, non – segregating, high – workability RCC)

2 Bulletin purpose

The purpose of the ICOLD RCC Bulletin is to make available to all engineers a synopsis of current best practice in the use of roller – compacted concrete for dams. The Bulletin update will accordingly present a comprehensive review of the state of the art of the design and construction of RCC dams as at 2015. In principle, the update is a revision of the original document, which will include improvements to aspects through which some increased knowledge and developments in technique have been gained through experience, while adding substantial new sections where significant developments in understanding have changed the basic approaches and methodologies, particularly in respect of dam design.

The Bulletin will address all aspects of roller – compacted concrete for dams, from planning, to design and construction and performance in operation. Materials selection, concrete mixture proportioning and quality control will also be addressed, although the first mentioned will be covered in less detail than previously, as reference can now be made to the recently published ICOLD Bulletin 165 "Selection of Materials for Concrete Dams"[3] for a more exhaustive review of materials selection requirements.

In the process of updating Bulletin 126, the section on Hard – fill will be removed, as the related technology is currently being addressed through a new ICOLD committee on cemented materials dams (CMD), who intend to publish a bulletin on these dam types.

3 Bulletin importance

Various different approaches can be applied for RCC dam design and construction and success is not always easily achieved. Consequently, the ICOLD RCC Bulletin is a very useful and important document in providing information and guidance on best practice and the most successfully applied variations thereof. The particular significance of the RCC Bulletin 126 can be seen in the fact that it is reported to have the highest print numbers and to have been translated into more languages than any other ICOLD bulletin to date.

4 Developments in RCC technology for dams since 2003

4.1 Developments in the understanding of the early behaviour of RCC

Many RCC dams have been comprehensively instrumented and this has allowed significant developments in knowledge and the understanding of the performance and behaviour of the material, in the fresh state, during the hydration process and in its mature form. With this knowledge, it is apparent that RCC can essentially be assumed to behave in a similar manner to CVC only in a mature state. It has been demonstrated that it is not appropriate to assume that the generic rules that were developed to define the stress relaxation creep in CVC during the hydration process can be applied for RCC, with certain low cementitious materials RCCs indicating substantially higher stress relaxation creep and certain higher cementitious materials RCCs (particularly containing fly ash) indicating very substantially lower stress relaxation creep. Furthermore, horizontal construction, with induced, as opposed to formed joints, increases the sensitivity of the structural performance to the level of stress relaxation creep applicable, with structural bridging in narrow valleys far more likelyto develop in low stress relaxation creep RCC than in CVC constructed in independent vertical monoliths.

Researchhas demonstrated such low levels of stress relaxation creep during the hydration process in high workability, fly ash – rich RCC that an upstream movement on a curved dam has been observed during construction (Fig. 2), while high levels of stress relaxation creep have been indicated in lean RCC dams through the measurement of progressively increasing compressive stresses in the surface zones[4]. The evident broad variation in this parameter in different types of RCC implies that the traditional assumptions that are valid for CVC are only appropriate for RCC when no design sensitivity to the level of stress relaxation creep exists.

The above implies that an additional characteristic, or design parameter, should be considered when designing and developing an RCC mix for a specific dam. In this regard, low, or high stress relaxation creep will be a positive characteristic in certain circumstances and a negative in others and careful consideration, investigation and laboratory testing is necessary during the RCC mix development process for dam structures where a design sensitivity to the actual level of hydration – cycle stress relaxation creep exists.

Fig. 2　Changuinola 1 arch/gravity dam – designed on the basis of improved understanding of early RCC behaviour

4.2　Design differences related to horizontal construction

Over its development history to date, in many respects RCC has been treated in a similar manner to CVC and it has been assumed that many "rules of thumb" that apply for CVC dams are equally applicable to RCC dams. The above developments in the understanding of the early behaviour of different RCC mix types, however, have emphasised important structural behaviour differences in RCC compared to CVC. RCC can consequently be designed for a range of behaviours when a single characteristic is often applied in the case of CVC. The reduced early stress relaxation creep in fly ash – rich RCC has also proved very advantageous for the design and construction of more efficient RCC arch dams[4]. The same effects, however, can cause deleterious 3 – dimensional bridging in high RCC gravity dams in narrow valleys[5].

Furthermore, in combination with continuous horizontal construction, the manner in which we currently design induced joints in RCC implies that we are in fact often constructing 3 – dimensional structures. With a practice of 2 – dimensional design for RCC gravity dams that exists due to its heritage in CVC dam technology, the design differences associated with continuous horizontal construction, compared to the separate, vertical monolith construction of CVC dams will be addressed for the first time in the Bulletin update.

Fig. 3 "Older" technology RCC (less cohesive, with more segregation, no set retarder and less paste mobility)

4.3 Developments in RCC mix design

Since the first generation of RCC dams (Fig. 3), a general trend towards high – cementitious RCC types has been apparent, most likely as a consequence of the following:

- While RCC was initially considered as a low strength mass concrete for which design changes might be applicable compared to a traditional gravity dam, developments in the technology have since demonstrated that roller – compaction can be used to produce high quality concrete in large dams;
- High – cementitious RCC is perceived as allowing the construction of gravity dams that are fully equivalent to conventional mass concrete dams;
- Lean RCC is perceived to require modifications in dam design, compared to a conventional mass concrete dam;
- A large dam is a structure with a long design life and designers tend to be conservative, adopting what might be perceived to be a least risk solution and tending not to favour a solution dependent on a geomembrane, or the "separate"[①] approach; and
- Developments in high – cementitious, super – retarded, all – RCC dam construction have produced a methodology that ensures a very efficient, rapid and very cost – effective high quality concrete dam. While this RCC dam type requires cost – effective access to cementitious materials and pozzolans, good aggregates and suitable founding conditions, it undoubtedly currently represents the most efficient approach for the construction of a large,

---

① RCC dams are designed in accordance either with the "Overall" approach, whereby the RCC creates the impermeable barrier, or the "Separate" approach, whereby a separate impermeable barrier is created on the upstream face, usually in the form of a geomembrane (PVC).

high – strength gravity dam.

The last listed of the above implies that High – cementitious RCC will tend often to be applied unless site – specific conditions, such as cementitious materials availability problems, compromise its efficiency. Notwithstanding this fact, all of the listed RCC types represent workable solutions that should be considered within the constraints and opportunities inherent to each specific dam site.

A polarisation of Lean RCC and High – paste RCC occurred as a consequence of distinctly different approaches preferred by various protagonists during the early development of RCC design and construction technology. While a distinct difference is apparent when adopting the "separate" (permeable RCC with an upstream impermeable barrier), or the "overall" approach (impermeable RCC) and while each RCC approach has a number of unique features, the increased use of non – plastic fines in paste has caused more commonality to develop across the various RCC types.

As a consequence of the above, revised terminology will be introduced in the Bulletin 126 update, whereby the previously applied term "High – paste" RCC will be replaced by the term "High – cementitious" RCC. Furthermore, the term "cementitious paste" will be used to define the combination of cementitious materials and water, while "total paste" will refer to all materials in the mix that would pass through a 75 micron sieve. Although the general terminology for RCC indicates a requirement for a cementitious materials content exceeding 150 kg/m^3, modern super – retarded, High – cementitious RCC in fact generally requires a cementitious content exceeding 190 kg/m^3 in order to produce sufficient paste for high – workability modified Vebe times of around 8. While all such mixes contain high percentages of pozzolan, cementitious materials contents of this level usually produce RCC compressive strengths at a one year age exceeding 35 MPa. Concrete strengths of this level are generally unnecessary in gravity dams, unless subject to high seismic loading, and a trend is evident in combining non – plastic fines in slightly lower cementitious materials RCCs (> 160 kg/m^3) to gain the benefits of high – workability, super – retarded RCC while reducing the cost of cementitious materials (Fig. 4).

Depending on the shaping and grading of the aggregates, more than 200 litres of paste will typically be required to produce a cubic metre of high workability RCC. Considering cementitious materials alone, such a paste volume will generally only be possible with contents exceeding 200 kg/m^3. Where non – plastic fines are used to enhance the paste volume, however, the benefits of high – workability, super – retarded RCC construction have been successfully extended to lower strength mixes.

Whether high – cementitious, or Lean, modern RCC is generally more cohesive, less easily segregated and more easily compacted. Modern RCC generally has a softer, or less bony appearance than the original RCC variants, an aspect most noticeable in the high paste RCCs, for which the paste rises to the top surface under compaction (Fig. 4 and Fig. 5).

The "all RCC" dam approach continues to see increasing application, with facings and interfaces being formed in GERCC (Grout – enriched RCC – grout from top), GEVR (Grout Enriched Vibrated RCC – grout on bottom) and IVRCC (Immersion Vibrated RCC). With GERCC and

Fig. 4   Modern RCC construction (high-workability, non-segregating, high-mobility RCC, with moderate cement materials content)

GEVR being variations of grout enrichment to allow the compaction of RCC with an immersion vibrator, IVRCC is RCC that contains sufficient mobility (and paste) to be compacted by immersion vibrator without the need to add grout. Depending on the nature of the aggregates and the RCC, GERCC and GEVR might require the addition of between 50 and 80 litres of paste to enable compaction by immersion vibrator, while, in principle, IVRCC will generally require a paste content of between 230 and 250 litres to allow vibrator compaction, unless particularly well-shaped and well-graded aggregates are used.

Fig. 5   Modern, non-segregating RCC

An additional trend in RCC dam design is a more restrictive aggregate specification, in terms of shaping and void content. With impermeability and density of RCC being possible only when all voids are filled and high workability RCC requiring mobility and excess paste that will rise to the surface under compaction, a tighter specification for the constituent aggregates than is applied for

CVC is specifically beneficial in reducing the total paste requirement and accordingly the cementitious materials content. Consequently, additional attention and expenditure on the aggregate processing plant can often be demonstrated to give rise to a net total RCC unit cost reduction through a lower cement and pozzolan content.

Success has also been achieved in the application of the "overall" approach with the less favoured "medium – cementitious" RCC type. In the cementitious materials content range of 100 to 149 kg/m^3, insufficient paste is generally present to develop impermeability. Consequently, it will often be more economical to increase the cementitious materials, particularly the pozzolan content, or to adopt the "separate" approach, apply an upstream face geomembrane and reduce the cementitious materials content. In certain circumstances, however, particularly when pozzolans are not economically available, adequate impermeability in "medium – cementitious" RCC can be created using non – plastic fines, or milled rock powder.

A range of solutions are now possible with RCC and similar principles and approaches can now be applied for the mix design and construction application of RCC with a significant range of cementitious materials contents.

4.4 Developments in RCC arch dams

In the Bulletin 126 of 2003, only RCC arch dam technology in China was presented. Since 2003, RCC arch dams have been completed in Panama (Fig. 6), Pakistan, Puerto Rico and Turkey. The Bulletin update will cover a more inclusive summary of the full development of RCC arches to date.

It is significant to note that all of the RCC arches outside China to date have been arch/gravity structures, while the majority of examples in China are thin arches. The reason for this situation is probably a combination of the fact that significantly more sites suitable for thin arches exist in China, while RCC loses competitiveness to CVC on sites with difficult access and low volumes of concrete, where the speed advantages of RCC construction cannot effectively be exploited.

Fig. 6 Changuinola 1
RCC arch/gravity, Panama

An interesting benefit that forms a common aspect of the design of several recent arch/gravity RCC dams is the concept of designing for hydrostatic loading as a gravity structure and relying particularly on arch action for stability under seismic loading.

4.5 Secondary benefits of RCC construction speed

Another particular advantage of RCC dam construction has been demonstrated on hydropower schemes, when it is often possible to bring forward the date of scheme commissioning with RCC construction. If the application of an RCC dam might be able to reduce the implementation period for a 500 MW hydropower scheme by six months, the net project benefit is likely to exceed US $ 50 million. Such a benefit of RCC dam construction can substantially influence the selection of dam type and represents a significant factor in the ongoing gr1owth in the application of RCC dams.

5 General discussion on RCC technology

The maximum benefit of RCC construction is realised when full advantage can be taken of the associated high capacity production, delivery and placement equipment and plant. Such an eventuality requires that simplicity be considered as the key element in design and in construction methodology and logistics planning, configuring all works and activities to ensure that all plant can continuously maintain maximum production.

The design and design development for an RCC dam and the optimisation of the RCC mix can be more involved and time consuming than is the case for a traditional, mass concrete dam. Design for an RCC dam is a process of iteration, more similar to that of a fill dam than to that of a traditional mass concrete dam, whereby materials design (cementitious and aggregates) and structural design are developed in parallel to identify an optimal solution. In certain circumstances, RCC mix development can be time – consuming and sometimes it may be advantageous to initiate this process prior to tender, implying a requirement for additional forward planning.

RCC dams can typically be raised at rates exceeding 10 m per month, or more than one 300 mm layer per day. Bond between RCC placement layers can impact the structural and permeability performance of the concrete mass of the dam and different construction methodologies and RCC mix types are applied dependent on the respective performance levels required in terms of a particular dam design.

Although early concern focussed on the apparent low shear strength properties between RCC placement layers, construction techniques have since been developed that can assure high levels of cohesion between placement layers. However, a good RCC mix and stringently controlled construction are required to confidently and consistently realise vertical, inter – layer tensile strengths exceeding 1.5 MPa and it is consequently the vertical tensile strength capacity between placement layers that remains the critical limitation on RCC dam height.

Over its development history to date, RCC has grown from a low strength, mass fill to a material capable of producing low strength, low deformation modulus, high creep concrete to very high strength, high deformation modulus, high density, low creep and impermeable concrete. It is now realistically possible to produce RCCs with strengths ranging from 2 to more than 40 MPa. With appropriate construction techniques and high – workability high strength RCC, it is also possible to achieve good bond, and accordingly vertical tensile strength, between placement layers. Consequently, RCC is now used for the construction of a range of concrete dam types from low stress, prismatic gravity structures to relatively high arch dams.

6 Publication of bulletin update

The original Bulletin 126 was finally published in 2003 after more than 10 years in preparation. With RCC technology continuing to develop, it was considered essential to take every possible measure to avoid such an extended period of preparation for the current Bulletin update. Indeed, this objective was the prime motivation for deciding to proceed with an update of the existing Bulletin, as opposed to compiling a completely new Bulletin.

The ICOLD Committee on Concrete Dams consequently proposed a more modern approach to making the updates available more rapidly. With approval from ICOLD, the update of Bulletin 126 is accordingly being tackled by chapter, with each completed chapter being made available electronically through the ICOLD website on completion. Only once all chapters are completed will the full update be printed in hardcopy. The first drafts of all chapters of the bulletin update are scheduled for completion by the end of 2015.

References

[1] ICOLD. Roller compacted concrete for gravity dams. Pairs; Bulletin No. 75 ICOLD/CIGB. 1989.

[2] ICOLD. Roller compacted concrete dams. State of the art and case histories. Paris; Bulletin No. 126. ICOLD/CIGB, 2003.

[3] ICOLD. Selection of materials for concrete dams. Paris; Bulletin No. 165 ICOLD/CIGB, 2014.

[4] Shaw, QHW. The Beneficial Behavioural Characteristics of Fly Ash – Rich RCC Illustrated through Changuinola 1 Arch/Gravity Dam. Proceedings. New Orleans, USA; USSD Conference. Innovative Dam & Levee Design and Construction for Sustainable Water Management. April 2012.

[5] Shaw, QHW. The Influence of the Low Stress Relaxation Creep on the Design of Large RCC Arch and Gravity Dams. Zaragosa, Spain; 6th Symposium on Roller Compacted Concrete Dams. October 2012.

Features of the Russian Technology of Dam Construction with the Use of Roller Compacted Concrete

Vadim Sudakov, Alfons Pak

(Vedeneev VNIIG, Russian)

Abstract: In the 1970 s during the construction of Toktogulskaya HPP m in layer – by – layer technology, the developers have formed the view that in the inner zone of the dam there would be technologically and economically feasible to lay special concretes and concrete mixes. Their main distinction from vibrated concrete would have been the factor that just after laying and compaction such concrete mixes would be able to bear construction loads and, at the same time, be harmoniously combined with the technology of layer – by – layer concreting

Search and laboratory and field testing of such concretes and mixes took several years and were mainly conducting with superhard concrete mixes.

Key words: dam, roller compacted concrete, technology, construction, hydropower plants

Even during erection of Toktogulskaya HPP of 215 m height with the use of layer – by – layer technology, the developers proposed that it would be useful, from the point of view of techno – economic feasibility, to lay special concretes and concrete mixes into inner zones of the dam. The main difference of such concretes and concrete mixes from the vibrated ones would have been the fact that just after laying such concrete mixes would be able to withstand modern construction loads and, at the same time, harmonically combine with the technology of layer – by – layer concreting.

Research and laboratory and field tests of such concretes and mixes took several years[1, 2]. Concrete mixes of high hardness were mainly researched. The data of the existing experience of their application in native dam construction (Bukharminskaya HPP[3] and others) were taken into account.

In the result in 1985 the inner zones were made from extra – hard low cement content concrete compacted by roller for two small dams: conjugating head wall of 22 m height and 115 m length (Fig. 1) and overflow wall of 18 m height and about 300 m length[4-6].

The conjugating wall was erected in May 1985 when the day temperature was 35 ~ 36 ℃. The temperature of the concrete mix was within 17 ~ 24 ℃ and in the concrete was not higher than 30 ℃[5].

In December the water head on the conjugating wall was 18 m. The tests of cores drilled from the wall in 6 months after laying showed the average strength of 12.7 MPa.

The overflow wall was erected in July to September 1985 in the basement part of the run – of –

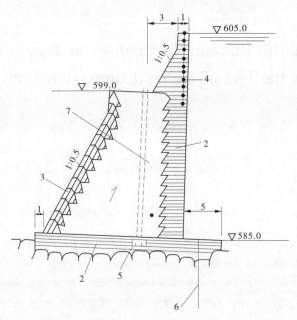

1. roller compacted concrete M 100 (cement consumption 80 ~ 120 kg/m^3);
2. vibrated concrete M 200, B −4 (cement consumption 220 kg/m^3);
3. modular reinforced concrete beam type elements; 4. reinforcing mesh;
5. flume for collecting drainage flows; 6. cement curtain; 7. drainage

Fig. 1 Head wall of Tashkumirskaya HPP

river dam of Kureiskaya HPP[6]. The flood with water discharge of 220 m^3/s was flown over the wall in June 1986.

The erection of 75 m height gravity dam for Tashkumirskaya HPP (Fig. 2) with 400 MW capacity became the next large step in the development of the Russian dam construction with the use of roller compacted concrete.

The concreting technology of Tashkumirskaya HPP as well as other modifications of roller compacted concrete technology is based on concreting by one – layer blocks with the use of extra – hard low – cement content concrete mixes. However it has a number of significant differences from other modifications (Fig. 2).

(1) The technology of concreting of the Tashkumirskaya HPP dam is a combination of the technology of roller compacted concrete laid into the inner zone with the toktogulsk technology of layer – by – layer concreting used to erect outside protection zones.

This combination allowed to unify the concrete laying lines and to concrete all zones of the dam using high – performance machines and mechanisms. With the same thickness of layers of vibrated and roller compacted concrete, this combination allows to get more reliable, dense and strong contact between them.

(2) Concrete mixes were prepared at the continuously operating batch plant.

(3) In the Tashkumirskaya HPP dam the zonal distribution of roller compacted concrete in height is provided according to its stressed state.

Concrete laying means: 1. concrete truck; 2. electrobuldozer M663Б;
3. vibration roller Riomag 200; 4. electric tractor with vibrators

Fig. 2　The dam of Tashkumirskaya HPP during construction

(4) Easy – assembled modular structures are only used to form vertical joints (inter – sectional and joints – cuts).

(5) Horizontal joints in vibrated concrete of the outside zones are prepared by removal of the cement film, surface clearing from dirt and debris, flushing with water and blowing out with compressed air. Underlay grouting is not applied.

The cement film is not removed for preparing the horizontal joints in roller compacted concrete of the inside zones. All other operations are executed as for the vibrated concrete. Underlay grouting is not applied.

(6) Concreting is fulfilled with one – layer blocks to the sections $-35\ m \times 35\ m$.

(7) A vibration roller Riomag (type Bomag BW – 200) was included into the technological line during the construction of the gravity dam for Tashkumirskaya HPP. It gave the possibility to increase the thickness of roller compacted concrete layers to 50 cm.

(8) The surface of the laid roller compacted concrete was constantly moisturized. For the concrete of the outside zones there was organized surface cooling by watering. There are no temperature cracks in the gravity dam.

The experience in the construction of these structures created the necessary prerequisites for successful transition to the design of much more complex hydrosystems with gravity dams made of roller compacted concrete: Bureisk, Katunsk, Turukhansk and others[7-9].

The concrete gravity dam of Bureiskaya HPP of 140 m height, length by crest 780 m and concrete amount of 3.5 mln. m^3 was constructed in the region with the average annular air temperature is -5 ℃, maximum temperature is $+41$ ℃ and minimum is -58 ℃.

Applying the technology of seasonal - separate concreting for this dam, its design is the same for all sections (spillway and station) and consists of two parts - top column and low wedge (Fig. 3).

Fig. 3 Bureiskaya HPP with roller compacted concrete dam

The top column of 14 m width shall provide watertightness of the dam and the low wedge: its stability. The top column was erected by pole blocks of 3~6 m height, the low wedge - by one - layer blocks of 0.4~0.5 m thickness from roller compacted (inside zone) and vibrated (protection zone of the downstream side) concretes.

Field studies showed that the concrete of the head wall, impounded reservoir, is in a compressed state, opening of the horizontal joints on the downstream side is negligible and opening depth does not exceed 5 m.

It is important to point out that the use of such concreting method for the station gravity dam of Bureiskaya HPP with the downstream side slope 0.7 allowed to avoid crack formation and seepage in the horizontal joints.

The main features of the dams from roller compacted concrete and technologies of their erection developed in Russia in relation to the conditions of dam construction are:

(1) All dams has zonal distribution of concrete.

(2) Concrete is laid layer – by – layer with the thickness, as a rule, 0.5 m.

(3) The main means for regulation of the temperature – stressed state of dams is the surface cooling of concrete blocks.

(4) The next higher concrete layer is placed in 3 ~5 days.

(5) the surfaces of the placed concrete layers shall be protected from moisture evaporation and be constantly wetted (before next concrete laying).

The state of the erected dams is under constant in – situ survey (from 1985).

Basing on the existing experience of the design, construction and operation of the dams from roller compacted concrete, the Russian engineers developed normative documents[10, 11] with calculation values of the basic parameters and technical characteristics for vibrated and roller compacted concretes which may be used in the native dam construction.

Reference

[1] ICC on hydroengineering. Low cement content concretes for hydroengineering structures / B. E. Vedeneev VNIIG, L. ,Energy, 1978, – Ed. 121.

[2] ICC on hydroengineering. Ways of gravity dam construction from roller compacted concretes / B. E. Vedeneev VNIIG, L. ,Energy, 1981.

[3] Hydropower & Dams. World Atlas. 2010.

[4] Recommendations P 25 – 85 on application of roller compacted concretes in hydroengineering construction / B. E. Vedeneev VNIIG, – L. ,Energy, 1985.

[5] Shangin V. S. Conjugating head wall from roller compacted concrete / Power construction. – 1986, №1.

[6] Zaltsman O. M. , Anikanov K. A. , Deryugin E. P. , Yagin V. P. Experience of application of low cement content roller compacted concrete for Kureiskaya HPP construction / Power construction – 1986, № 11.

[7] Sudakov V. B. Dam construction from roller compacted concrete, Review / M. , Informenergo. 1988.

[8] Sudakov V. B. , Tolkachev L. A. Modern methods of high dam concreting / M. , Energoatomizdat, 1988.

[9] Sudakov V. B. , Marchuk A. N. , Yepifanov A. P. Features of construction and operation of concrete dams in the regions with severe and particular severe climate / B. E. Vedeneev VNIIG, SPb, 2014.

[10] SNiP 2.06.06 – 85(2012) – Concrete and reinforced concrete dams (SP 40.13330.2012).

[11] SNiP 2.06.08 – 87(2012) – Concrete and reinforced concrete structures of hydroengineering constructions (SP 41.13330.2012).

Design and Quality Control of the IV – RCC Mix USED at Enciso Dam in Spain

Allende M[1], Ortega F[2], Blas C[3]

(1. Ebro River Basin Authority (CHE); 2. FOSCE Consulting Engineers; 3. Técnica y Proyectos S. A. (TYPSA))

Abstract: Preliminary works related with the mix design development for Enciso dam were presented in the last RCC Symposium in Zaragoza 2012 in Spain. By that time the project entered into a stand – by situation due to budget cuts derived from the challenging economical situation. Finally concrete placement could start in the dam at the end of August 2014. The first section with a volume of 170 000 m^3 was placed in the river bed in 2.5 months. After the winter close down placement has resumed and further 180 000 m^3 were placed in 60 working days until middle of June 2015.

This article revisits the immersion vibrated RCC (IV – RCC) mix design development and describes further investigations that have been carried out at the site prior to the start of the placement in the dam. One important pending issue was to define the dosages of the selected retarder admixture under different environmental conditions. In addition further tests were required to confirm the mix properties using the actual materials that were going to be used in the dam.

The information generated by the quality control records during dam constructionis also presented and analysed in this paper. This includes not only fresh and hardened properties measured on specimens but also the results of a preliminary coring and testing programme that has been conducted during the winter break.

Key words: Mix design, IV – RCC, quality control

1 Introduction

Enciso dam is the main component of an irrigation and water supply project located in the region of La Rioja in the North of Spain. The Owner is the Ministry of Agriculture and Environment, who leads the works through a basin organization, the Ebro River Basin Authority. The dam is 103 m high and has a total volume of about 800 000 m^3 of concrete. A dam at this site has been studied since 1950 and evaluation of different typologies concluded that the most suitable option was a straight gravity roller compacted concrete dam. The original design of an RCC dam for Enciso dates from 1994. Initial design considered an RCC mix with 220 kg/m^3 of cementitious materials, split in 50% cement and 50% fly – ash. At that time it was a standard practice the use of a conventional concrete (CVC) mix as skin concrete at the dam faces. The aim was to achieve a good finish and to improve impermeability at the upstream side. This same CVC mix was also used as interface concrete between the RCC and the rock abutments.

Along this period high – paste RCC mixes have evolved towards much more workable mixes were the amount and quality of the aggregate fines have been playing a major role in the optimization of the mix. The extension of the setting time of the RCC mixes incorporating retarder admixtures has also contributed to practically eliminate one of the main points of concern in the in – situ quality of traditional RCC dams, namely bond between successive layers.

Final mix design work was developed at Enciso site between May 2010 and October 2011 including laboratory investigation, and construction and coring of a full – scale trial. Preliminary conclusions were included in the Proceedings of RCC2012 Symposium. A summary of those conclusions is presented in following Sections, including an update of the long term strength test results and the adjustment of the dosage of admixture under different ambient temperatures.

Fig. 1 Construction of the second full – scale trial at Enciso RCC dam: simulation of IV – RCC vibration against rock (left) and typical surface after immersion vibration/compaction (right)

Prior to the start of the RCC placement in the dam a second full – scale trial was constructed with the main objective of providing training to staff and equipment that was going to work on the dam. The same RCC mix has been confirmed after this second trial section.

At a time when almost 50% of the RCC has been placed in the damenough information has been collected by the QC staff and the results can also be presented at this stage. Test results are available from both manufactured specimens and core samples.

2 Design of the IV – RCC mix

Thefollowing parameters have been key – inputs in the design of the IV – RCC mix for Enciso dam:
- Overall and fine aggregate gradation,
- Quality and quantity of fines in the fine aggregate,
- Void content of fine aggregate,
- Initial set,
- Mix consistency, and
- Tensile strength to withstand early surface thermal loads.

Additional material specifications have been adopted to meet the mix design parameters. For example, the amount of non-plastic fines passing ASTM#200 (0.075 mm) sieve could go up to 18% in the fine aggregate and the void content of the compacted fine aggregate should be below 30%. In the final mix that is being placed at Enciso dam the crushed fine aggregate from the limestone quarry has an average percent of fines of 11% and its compacted void ratio is ca. 28%. After some initial adjustments in the crushing plant the gradation of this material is well centred and controlled within the specified limits.

The retarder admixture selected aftertestingvarious suppliers is Conplast RP264 of Fosroc. Initial tests were made at laboratory conditions that were later on confirmed on the field. The dosage has ranged between 1.0% and 1.5% (by weight) of the cementitious materials (cement and fly-ash) modified with changing ambient conditions to ensure initial set above 15~20 h.

The other main aspect that has been evaluated during the mix design development has been the high workability and the ability of the fresh concrete to behave as a cohesive mass in fresh state that does not segregate during handling and transportation. In this regards the optimum mixes were those with the lowest Vebe time, in the order of 8~12 s at the time of compaction.

The final approved mixture proportions for the construction of the damare listed in Table 1 and some important mix parameters are summarized in Table 2.

Table 1 Mixture proportions of the RCC/IV - RCC for Enciso dam

Material	Code	kg/m^3
Coarse aggregate (>5 mm)	[CA]	1 411
Fine aggregate (<5 mm)	[FA]	760 (91<0.075 mm = [FI])
Cement (type I)	[C]	60
Fly-ash	[F]	140
Free water	[W]	93
Admixture	[A]	1.0%~1.5% of [C+F]
Air voids	[AV]	1.0% (by volume)

Table 2 Materials and mix parameters

Material/Mix parameter	Mean value
Nominal/actual maximum size of aggregate	80/50 mm
Proportion of fine aggregate [FA/(FA+CA)]	35%
Water/Cementitious ratio [W/(C+F)]	0.47 (by weight)
Volume of cementitious paste [C+F+W+A] = [P]	169 litres/m^3
Volume of total paste [C+F+W+A+FI]	200 litres/m^3
Cementitious paste/mortar ratio [P/(P+FA)]	0.37 (by volume)

The benefits of the new specification of the fine aggregate and the secondary effect of the admixture as water reducer agent can be appreciated in the relatively low water demand of the mix that has been required to meet the desired range of workability (Vebe times). Thanks to this high workability the concrete could be consolidated by immersion vibration (IV – RCC) using 75mm diameters hand held pokers during construction of the first full – scale trial. At that time a comparison was made with three other facing systems (CVC, grout – enriched and mortar – enriched RCC) and finally IV – RCC was selected due to its simplicity and good finish.

3 Investigation of the dosage of admixture

Once the primary admixture had been selected at the laboratory and its performance confirmed on real scale placement, it was necessary to investigate the adjustments of the dosage for different ambient temperatures. Initial tests were made under relatively warm conditions with mean ambient temperatures of 20 ℃. The optimum dosage for such scenario was fixed at 1.5% by weight of the cement + fly – ash content, i.e. 2.48 litres/m^3. The first and second trial sections were built using this dosage, although in the second trial section different other admixtures were tested that did not give as good results as the selected one in terms of test consistency and layer bonding.

This quantity of 1.5% of the retarder admixture was used in the RCC mix at the start of the placement in the dam under similar ambient temperatures as those of the trial sections. However as the winter was approaching it was necessary to reduce the dosage and adjust it to the lower ambient temperatures. The aim was to keep initial set of the mix above 15 ~ 20 hours also in a colder environment but without extending final set beyond a certain limit that was fixed in about 48 ~ 50 hours. This limitation was necessary to ensure a minimum early concrete strength (above 2.5 ~ 3.0 MPa at 3 days) to ensure stability of the formwork. The results of such investigation in terms of initial set are plotted in Figure 2. The range of dosages that have been used in the dam construction until now has varied basically between 1.2% and 1.5%.

Setting time tests have followed the national Standard which is similar to ASTM C403. Scattered values are usually achieved in this test and simultaneous testing in different moulds with same mix is recommended. Correct preparation of the mortar sieved samples and cured protection under exterior conditions are important to achieve consistency in the results. All penetration tests should be accomplished by well trained personnel that should work in continuous shifts 24 hours/day.

The fresh and hardened properties of the mixes designed with variable dosage of admixture have been also evaluated. In particular VeBe time, fresh VeBe density and compressive strength of specimens with normal curing up to 28 days and under accelerated curing conditions have been tested. The conclusions may differ in other projects with different set of materials. In this case it has not been necessary to make any further adjustments in the composition of the paste (water, cement and fly – ash) to maintain the required workability and level of strength at the desired values.

4 Update of the quality control test results

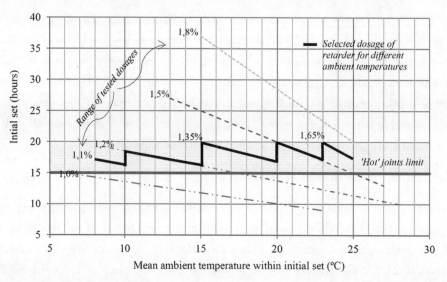

Fig. 2 **Investigation of the dosage of retarder admixture for different ambient temperatures**

4.1 Quality Control Programme

The experience gained during construction of the full – scale trials was very useful in setting up the quality control (QC) programme to be implemented for the dam construction. At that time the-most important key – parameters affecting final in – situ quality of the RCC structure were confirmed and the various checking points and frequency of tests identified.

For QC purposes the layers have been divided in blocks, lanes and sectors. Blocks between layers are identified in the design drawings with typical joint spacing of 20 meters. Size of advancing lanes have been adjusted depending on the rate of placement but typical lanes are placed 8 ~ 10 meters wide. Within this matrix limited by blocks and lanes in each layer, sectors of approximately 100 m^2 each have been identified to keep a systematic register of the RCC properties. Frequency of some tests has been reduced as the available results confirm the anticipated values by a reasonable margin of confidence.

QCand inspection programme includes records of the following items; tests with higher frequency of control that become critical to ensure final quality have been highlighted:

- coarse aggregate: gradation, flat and elongated particles, loss by washing, organic & deleterious substances, clay lumps, specific gravity, absorption, Los Angeles loss,
- fine aggregate: gradation, compacted void ratio, specific gravity, absorption, plasticity index,
- cement and fly – ash: control sheet regularly provided by the supplier including physical and chemical tests, heat of hydration of the combined cementing materials,
- admixture: quality control sheet provided by supplier and setting time tests,
- water: chemical analysis,
- fresh concrete properties: VeBe time, fresh Vebe density, temperature, time between mixing and compaction, number of passes of the vibratory roller, in – situ density (nuclear densiometer), joint treatment and visual inspection,

- hardened concrete properties: density of specimens, cylinder compressive strength at 3 d, 7 d, 28 d, 90 d, 180 d and 365 d with standard curing, cylinder compressive strength after 14 days (7 d normal curing and 7 d at 70 ℃), direct tensile strength of cylinder specimens at 28 d, 90 d, 180 d and 365 d with standard curing, adiabatic temperature rise of RCC, and

- core evaluation: visual inspection and identification of possible joints or planes of failure, in-situ permeability, density, compressive strength, direct tensile strength.

4.2 Fresh concrete properties

Among the tests required to evaluate the fresh properties of a well designed RCC, it is probably the consistency and workability measured with the modified loaded Vebe consistometer the one that has a major impact on the final in-situ quality. The kind of mix that is being placed at Enciso dam is probably among the RCCs withthe lowest VeBe time around the world. Average values recorded at the site at the time of spreading the mix on the lift averages 6 s (Contractor's) and 7 s (Engineer's). When observing the procedures at the site one could argue that the stopwatch is systematically stopped 1~2 seconds ahead of the exact time when paste flows up at the last portion of the perimeter of the disc. Considering this possible deviation from standard practice, it could be assumed that the RCC at Enciso dam is placed with a mean VeBe time of 7~9 seconds. This level of workability is required to minimise segregation while dumping and spreading (Figure 3, left) and to achieve an adequate immersion vibration at the dam faces and other interfaces within the following 30~60 minutes, depending on the ambient conditions. This is also the level of consistency that a high-paste RCC requires to be fully consolidated by roller compaction in less than 4 dynamic passes of the 8~10 t single drum roller. Full consolidation is assumed to be achieved when most of the surface is covered with paste and no major spots of segregation or aggregate are visible.

Fig. 3 Aspect of the cohesive & workable mix after dumping and spreading (left) and tire tracks marked on the surface of the previous 20-hours old layer indicating adequate fresh condition (right)

Statistical records ofover 12 000 observations of the 50% of the volume that has already being placed at Enciso dam shows that the average time between mixing and completion of compaction is about 40~45 minutes and the time between spreading a compaction is just below 30 minutes. Roller compaction is concluded in just 1 static pass followed by 3 dynamic passes in average. After the third dynamic pass, the paste completely covers the surface of the layer and the in-situ density recorded in almost 3 000 tests by the nuclear densitometer is in average 2.481 kg/m^3, i.e. 99,5% of the theoretical – air – free density of the mix. It is remarkable that the immersion vibration of the same mix usually results in a slightly lower density (about 98%) confirming that traditional vibrated concrete has a higher percent of air voids than roller compacted mixes with same materials.

No pre – cooling facilities have been installed at Enciso dam and this has limited the continuity of the RCCplacement throughout the summertime. Different causes have derived into this situation which are not relevant to the subjects discussed in this article. Therefore the placement temperature is directly governed by the ambient temperature. The daily average placement temperature has varied between 14.4 and 21.4 ℃ (average 16.3 ℃).

Another important aspect of the supervision of the dam construction is the control ofthe adequate curing, cleaning and eventually treatment of the joints between layers. The dam construction has been planned with very few long stops creating a 'cold'joint. Up to now, treatments have not been required in 86% of the surfaces ('hot'joint), while 10% have been 'warm'joints and 4% are 'cold'joints. Continuous and uniform curing of the lift joints is a key element to reach a similar setting time on the field as the one measured on the RCC mortar samples. The maturity of the lower layer is visually inspected with the observation of the depth of the tire tracks left by the dump trucks when placing the following layer on top of it. A typical example is shown in the right photo included in previous Figure 3.

Setting time has been controlled once every layer. In average the initial set of the concrete that has been placed in the dam has been 24.3 hours and the final set 38.2 hours.

4.3 Strength of manufactured cylinders

The development with time of the cylinder compressive strength of the concrete mix used in the construction of Enciso dam is plotted in Figure 4. Average results available from two different controls in the dam (Engineer and Contractor) are compared with those that were obtained during construction of both trial sections.

Original dam design specifies a characteristic compressive strength of 15 MPa for the RCC at the age of 90 days. The mix was therefore designed for a mean compressive strength of 20 MPa at that age assuming a allowable percent of failure of 5% (values below design strength) and a coefficient of variation (CV) of 15% (normal control). According to the Engineer's QC records the actual mean compressive strength at the design age is 24.1 MPa and the CV of these records is 15.6%. Based on Contractor's results average strength is 25.7 MPa and CV is 12.3%. The derived characteristic strength in both cases is 17.9 MPa and 20.5 MPa, respectively. At 180 days the strength has raised as shown in Figure 4 and the CV has come down to 14.6% and 10.3% respectively.

Direct tensile strength (DTS) has been measured on manufactured cylinders and the average values vary between 6.2% and 8.3% of the mean compressive strength. It is remarkable the relatively high average value of the cylinder DTS of 2.4 MPa at 180 days.

4.4 Inspection and testing of cores

At the conclusion of the first placement season (2014) cores were extracted fromdifferent locations at the top of the completed section, from the downstream face and from the lower gallery. Un-

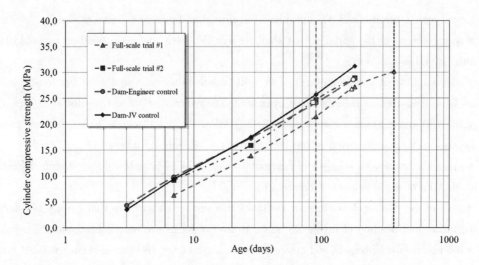

Fig. 4 Development of compressive strength of the RCC mix used in the dam
(60c +140fa +93w)

fortunately only a relatively light and less adequate equipment has been available and just a single corel barrel could be used (Figure 5). These two aspects have had a negative impact in the samples and it must be assumed that all of them have suffered a certain degree of alteration. Based on visual inspection of the cores and the results of the tests it seems like the results of tests of samples cut from the vertical cores are underestimation of the actual in-situ strength.

Fig. 5 Vertical and horizontal cores drilled from the dam using a relatively light equipment

A total of 10 vertical (180 mm diameter, 5 m long) cores and 4 horizontal (150 mm and 1.7 m long) cores have been drilled from the dam. Recovered samples include concrete that was consolidated by both methods: immersion vibration and roller compaction. Despite the drawbacks derived from the type of equipment, it has been possible to draw the following conclusions of this preliminary dam coring programme:

● the in-situ quality is similar regardless of the compaction method,

● air bubbles are appreciated on the surface of the immersion vibrated portions. Roller compacted areas are slightly more dense (between 1% and 1.5%),

● apart from the air bubbles the limit between areas where concrete has been consolidated by different methods is not discernible,

● bond between immersion vibrated RCC and the rock surface is similar to the one that could be achieved with a traditional CVC mix placed against the abutment (Figure 6, left),

● poor cleaning of the surfaces that shall receive IV – RCC decrease bonding as would be the case with any other concrete,

● horizontal direct tensile strength of IV – RCC core samples is similar to the tensile strength that has been measured in manufactured cylinders. The average of 10 test results at an age of 180 days of the 150 mm × 300 mm core samples is 2.4 MPa. This is the same value that was obtained with the specimens manufactured at the laboratory (see Section 4.3 above),

● average compressive strength of horizontal cores is 34 MPa at 180 days, slightly higher than the ca. 30 MPa of the specimens (Figure 4),

● most of the planes of fracture in vertical cores are rough surfaces that cannot be identified with the theoretical position of the joints between layers. Joints are not visible, and

● the test results of tensile and compressive strength of the 180 mm diameter vertical cores is between 50% and 60% of the values obtained with the 150 mm horizontal cores. A maximum of 20% reduction can be attributed to the concrete anisotropy. The rest is only explained through the increasing alteration of the core extracted at lower drilling vertical lengths in a single barrel procedure with less adequate equipment. It is believed that the vertical in – situ strengths are at least 70% of those that could be measured in the horizontal cores. In any case the results of the tests are above the design requirements.

Fig. 6 Detail of an inclined core sample at the contact plane between IV – RCC and the rock (left) and typical vertical core with a breaking plane forced by the drilling operation (right)

In – situ permeability tests have been also made (both Lugeon and Lefranc) and the results are in the order of 1×10^{-9} m/s (coefficient of permeability). It is understood that any potential seepage through this concrete will only be possible in case of uncontrolled cracks are developed at some stage, but not caused by defects in the concrete matrix or across the theoretical position of the lift joints.

5 Thermal considerations

In – situ horizontal static tensile strength is the most critical design parameter of the mix. Such strength is required to support thermal loads and to minimise the risk of subsequent surface cracking. Specially critical are the areas between the gallery and the upstream face. In these areas surface cooling during first winter maydevelop relatively high thermal gradients. Figure 7 shows the development of the temperature records in several locations in one of the dam blocks.

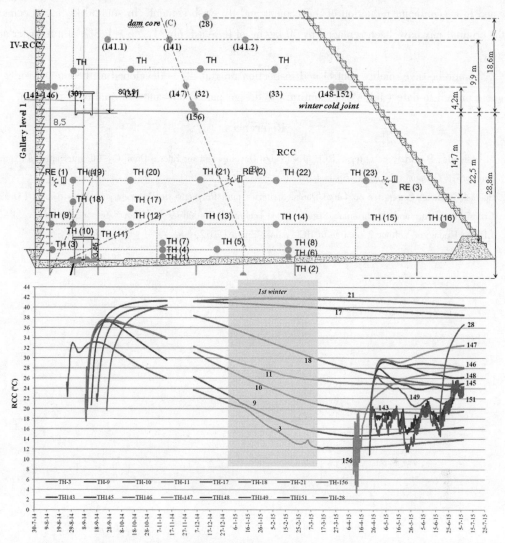

Fig. 7 **Development of temperature in different locations of the dam** (central block)

As can be appreciated the dam and gallery faces are relatively cold at the end of the first winter (8 ~ 10 ℃) whilst the core of the dam (TH#17 - TH#21) remains in near adiabatic conditions at 40 ~ 42 ℃. Consequently high thermal gradients occur along the upstream end of lines (A) & (B). The main point of concern in this scenario is the possibility of cracking in the middle section of the dam blocks that could cause seepage to the gallery. The mean placement temperature has been reduced in 5 ℃ in the second placement season to control this possibility leading to the next winter.

6 Conclusions

Enciso dam mix design represents a further step in the development of high – workability and super retarded RCC. The design andconstruction of the dam benefit from the introduction in recent times of the IV – RCC concept. Improvement in the fine aggregate has led to a much higher quality mix and to a significant reduction of the cement and fly – ash content. In sum 50 kg/m^3 of cement of the original mix have been replaced by 30 kg/m^3 of fly – ash plus 1.2% ~ 1.5% of a retarder admixture.

A comprehensive quality control and inspection programme is developed at the site. Results of concrete placed to date (July 2015) confirm the design parameters and the quality of the dam.

References

[1] M. Allende, D. Cruz, F. Ortega. RCC Mix Design Development for Enciso Dam[C]. 6th International Symposium on RCC Dams, Zaragoza, Spain (2012).

[2] Spanish National Committee on Large Dams. Criterios para Proyectos de Presas y sus Obras Anejas (Design Criteria for Dams and Appurtenant Structures). Technical Guidelines for Dam Safety published by SPANCOLD. Update of Volume 2 relative to RCC dams, Spain (2012).

3D Geomechanical Model Test on Global Stability of Lizhou RCC Arch Dam on Muli River

Chen Yuan[1], Zhang Lin[1], Yang Gengxin[2], Yang Baoquan[1]

(1. State Key Laboratory of Hydraulics and Mountain River Engineering, College of Water Resources and Hydropower, Sichuan University, Chengdu, 610065, China;
2. Shaping Branch of China Guodian Dadu River Hydropower Development Co., Ltd., Leshan, Sichuan 614300, China)

Abstract: Lizhou Hydropower Station is located on the Muli River, a large branch of the Yalong River. The double-curvature RCC arch dam is 132 m high which is the world-class high RCC arch dam. At the dam site, the river valley is narrow and steep with suitable topographic condition for constructing high arch dam. While some weak structural planes including faults, interlayer shear zones and fissure zones and large fissures, are developed in both abutments and affect the global stability of dam and foundation. It is necessary to be studied about this engineering issue. A 3D geomechanical model test is presented in this paper to simulate these adverse geological structures and analyze their negative impacts on stability. Through the overloading destructive experiment, the deformation characters, failure pattern and mechanism of dam, rocks and structure planes in abutment are obtained. The controlling factors and the weak regions impact the stability are revealed. And the overloading safety coefficients of dam and abutment at different destructive stages are determined as follows: $K_1 = 1.4 \sim 2.2$, $K_2 = 3.4 \sim 4.3$, $K_3 = 6.3 \sim 6.6$. The stability of dam and abutment is evaluated. Furthermore, some reinforcement measures for the weakness of dam abutment are advised.

Key words: high RCC arch dam, complicated geological structure, global stability, overloading model test, failure mechanism, safety coefficient

1 Introduction

With the development of hydropower projects and the implementation of strategy of electricity transmission from west to east in China, a number of high dams and large reservoirs are under construction or to be built. These large projects are mainly located on the great rivers in West China, characterized by complicated topographic and geological conditions[1]. These hydropower projects include Xiaowan arch dam (294.5 m high) on the Lancang River, Jinping I arch dam (305 m high) on the Yalong River, Xiluodu arch dam (285.5 m high) and Baihetan arch dam (289 m high) on the Jinsha River, Dagangshan arch dam (210 m high) on the Dadu River, and so on. Analyzing and evaluating the stability and safety of these large projects can meet the current needs of the engineering and academic fields[2,3]. Geomechanical model test is one of the main methods to address this issue[4]. Geomechanical model test is a kind of rupture test. It can scale down the dimensions of prototype project based on the similitude principles to study the engineering and geological issues. Its remarkable characteristics is the actual simulation for complicated geological structures and the visualized test results [5,10].

This paper presents an experimental investigation on the stability of Lizhou RCC arch dam, which is a world-class high-RCC arch dam with a maximum dam height of 132 m. At the dam site, the geological conditions are complicated, and the problem of abutment stability is critical. A 3D geomechanical model test is conducted to study the global stability of Lizhou arch dam. The deformation characters, the failure pattern and the failure mechanism are obtained. And the safety coefficients of dam and abutment at different destructive stages are also determined. These experimental results have been used to comprehensively evaluate the safety of Lizhou RCC arch dam and provide the scientific evidence for project design, construction, and reinforcement.

2 Project description

Lizhou hydropower station is located at Liangshan Yi Autonomous Prefecture in Sichuan Province. It is one of the six cascading hydropower stations along Muli River, with a double-arch RCC arch dam, a reservoir capacity of 189.7 million m^3, and an overall installed capacity of 355 MW. The river valley is narrow and steep with suitable topographic condition for constructing high arch dam.

In the project region, the outcropping strata is limestone. The rock layers incline from right bank to left bank with the inclination angle of 15°~25°. Weak discontinuities mainly in the presence of faults, interlayer shear zones, fissure zones and large fissures are developed in both abutments. In the left abutment, the main geological structures include faults f4 and f5, fissure zones L1 and L2, large fissure Lp285 and interlay shear zones fj1-fj4. In the right abutment, the main geological structures include faults f4 and f5, interlay shear zones fj1-fj4. The project geological structures are shown in Fig. 1 and Fig. 2. As shown in figures, there is a geological asymmetry between both banks. The geological condition of left bank is unfavorable, while the right bank is relatively good. The complicated geological structures and the geological asymmetric characteristics severely affect the stability of dam and abutment. It is necessary to study the global stability of dam and abutment.

3 Geomechanical model test

3.1 Test methods

Geomechanical model test is designed and conducted based on the similitude principles. With the similarity ratio of the model to the prototype, the testing values can be converted to the prototype values. According to the test results, the stability and safety of project can be properly evaluated. Therefore, the geomechanical model test should conform to the similarity relations between the prototype and the model, such as the relations of geometric, physical variable process, mechanics, boundary and initial conditions, etc. The main similarity relations are shown as follows [11]: $C_\gamma =$

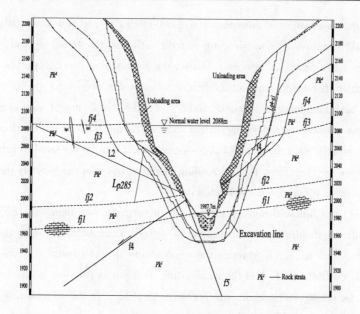

Fig. 1 Geological profile of Lizhou Project

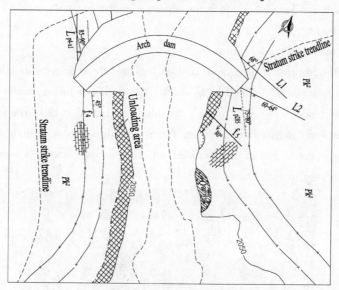

Fig. 2 Geological horizontal section

$1, C_\gamma = 1, C_f = 1, C_\mu = 1, C_\sigma = C_\varepsilon C_E, C_\sigma = C_C = C_E = C_L, C_F = C_\sigma C_L^2 = C_\gamma C_L^3$, where $C_E, C_\gamma, C_L, C_\sigma, C_\varepsilon, C_F, C_\mu, C_\varepsilon, C_f$ and Cc are the similarity factors of Young's modulus, unit weight, geometry, displacement, stress, force, Poisson's ratio, strain, friction factor and cohesion, respectively, and the similarity factor is the ratio of the prototype parameter to the corresponding model parameter.

Geomechanical model test includes three kinds of methods: overloading method, strength reduction method, and comprehensive method [12,13]. Different impact factors on stability are considered by these three methods. The overloading method and the strength reduction method respectively consider the effect of excessive flooding or the parameters reduction caused by leakage, high stress. The comprehensive method combines above two methods, and considers multiple impact factors on engineering stability. The overloading method is the most common method, and it has been widely applied in many practical projects. During the overloading procedure, the upstream water pressure is gradually increasing until model failure to study the overload capacity and evaluate the stability of dam and foundation. Through model test, the displacement development process is measured, and the overloading safety coefficient defined as the ratio of the overload to the normal load is also determined. Moreover, the weaknesses of project can be revealed to provide the conference for reinforcement. Therefore, the overloading model test is adopted to study the global stability of Lizhou RCC arch dam under natural foundation.

According to two Chinese design specifications for concrete arch dam, SL 282—2003 and DL/T 5346—2006, the overloading safety coefficient of arch dam can be assessed by three "water pressure overloading factor" K_1, K_2, K_3 through the geomechanical model test[14,15]. The initial cracking overloading safety coefficient (K_1) is the overloading multiple as the dam heel is initially cracked. The nonlinear deformation overloading safety coefficient (K_2) is the overloading multiple as the downstream dam surface is cracked. The ultimate overloading safety coefficient (K_3) is the overloading multiple as dam and foundation reach the ultimate bearing capacity.

3.2 Research scheme

3.2.1 Geometry scale and model dimension

According to the features of topography and geology and the requirements of test accuracy, the geometry similarity factor is determined as $C_L = 150$, the similarity factor of unit weight is $C_\gamma = 1$, and the similarity factors of stress and Young's modulus are $C_\sigma = C_E = 150$. The model dimension is 2.6 m × 2.8 m × 2 m (longitudinal direction × transverse direction × altitude direction), equivalent to the prototype size of 390 m × 420 m × 300 m in the actual project. The overall view of Lizhou dam model is shown in Fig. 3.

Fig. 3 3D geomechanical model of Lizhou arch dam

3.2.2 Model similar materials and simulation technology

The model similar materials should be developed based on the physical – mechanical parameters converted from the prototype materials by the similitude principles. The main parameters of prototype and model materials are shown in Table 1. In the geomechanical model, there are three series of model material: concrete, rock mass and structural plane (fault and fissure, etc.). The main components of the model material include barite powder, cement, gypsum powder, paraffin, petroleum oil and other additive. The barite powder is the main weighting material to make the mode material has the same density with the prototype material, i.e. $C_\gamma = 1.0$. The cement and gypsum powder are used as the cementing material. Through adjusting the composition and the proportion, different model materials with different physical - mechanical parameters (density, strength, Young's modulus, etc.) can be obtained.

Table 1 Mechanical parameters of prototype and model materials

Material type	Prototype parameters			Model parameters		
	E_p (GPa)	f'_p	c'_p (MPa)	E_m (MPa)	f'_m	c'_m (10^{-3} MPa)
Dam concrete	24	1.2	1.6	160	1.2	10.67
Limestone Pk (moderate weathering)	8	0.80	0.60	53.33	0.80	4.000
Limestone Pk (fresh rock)	12	1.20	1.00	80.00	1.20	6.700
Interlayer shear zones fj1, fj2	—	0.65	0.080	—	0.65	0.530
Interlayer shear zones fj3, fj4	—	0.45	0.030	—	0.45	0.200
Faults f4, f5	3~4	0.45	0.050	20~26.7	0.45	0.330
Fissure zones L1, L2	3~4	0.65	0.060	20~26.7	0.65	0.400
Large fissure L285	—	0.20	0.005	—	0.20	0.033

Note: the subscript m and p indicate the prototype and the model parameter, respectively.

In the geomechanical model, three series of model material are produced by different methods:

(1) Model material of dam: It is produced by pouring method. The liquid raw material is filled into the mold to form the rough blank. As the rough blank is solidified and dried out, it will be finely carved to the design shape and dimension, and finally be accurately located and installed in the foundation.

(2) Model material of rock mass: The powdery raw material is compressed to a small block. Using the masonry technology, the rock mass is constructed by numerous small blocks according to the attitude of rock stratum and the joint connectivity rate.

(3) Model material of structure plane: A kind of soft model material and various thin films with different friction coefficient are used to simulate the weak structural plane. Through adjusting the proportion of the soft model material and the combination of various thin films, the strength parameters of structure plane can be simulate. During the model masonry, the soft model material is firstly plastered on the surface of structural planes and the thin films are paved on the soft model material.

3.2.3 Test procedure

According to the practical operating conditions, the test procedure of Lizhou arch dam model is arranged as follows: First, the water pressure is gradually increased to the normal load, and the operating performance of dam and foundation under normal working condition are investigated. Second, the upstream water pressure is overloaded by the increments of $(0.2 \sim 0.3) P_0$ (P_0 is the design water pressure) until the dam and foundation approach global instability. Then, stop pressing and end test.

4 Test results

4.1 Displacement and strain characteristics of dam body

The displacement of dam body presents the regular charactercs: the upper displacement is larger than the lower displacement, the displacement of arch crown is larger than the displacement of arch end, and the radial displacement is larger than the tangential displacement. Under the normal condition, the symmetry of dam displacement is better. The maximum radial displacement of arch crown is 21.5 mm (prototype values) at El. 2092 m. In the overloading stage, the displacement of left arch is gradually larger than that of right arch with the increase of the overloading multiple K_p, as shown in Fig. 4. In the figure, each curve indicates the radial displacement along the arch ring at certain overloading multiple K_p. Finally, the left arch deformation is significantly larger than the right arch deformation. The dam displacement appears an asymmetric characteristics with deformation downstream and clockwise rotation. This asymmetric deformation is caused by the geological asymmetric characterics and the weakness of left abutment.

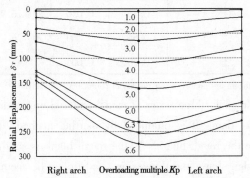

Fig. 4 Distribution curves of radial displacement at EL. 2000 m

Fig. 5 is the curves of horizontal strain on downstream dam surface in arch crown, in which the numbers represent the monitoring points. It shows that the strain on downstream dam surface is mainly compression strain. Under the normal condition, i.e. $K_p = 1.0$, the dam strain is small. In the overloading stage, the dam strain gradually grows with the increase of overloading multiple K_p. As $K_p = 1.4 \sim 2.2$, the strain curves have slight fluctuations, and initial cracking happens at the dam heel. As $K_p = 3.4 \sim 4.3$, the strain curves have obvious fluctuations and form big turning points, and the strain values increase remarkably. At this moment, the dam body is cracked in the left arch. After that, the strain curves frequently fluctuate and the dam cracks further propagate. As $K_p = 6.3 \sim 6.6$, the dam cracks extend from bottom to top, the stress release phenomenon occurs,

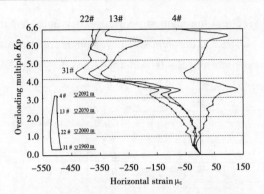

Fig. 5 Horizontal strain μ_ε vs. overloading multiple KP of arch crown

and the dam gradually lose the bearing capacity.

4.2 Surface displacement characteristics of dam abutment

The typical curves of displacement δp vs. overloading multiple K_p are shown in Fig. 6 and Fig. 7. Under the normal condition, the displacements of both abutments are small. The displacement along river trends to downstream, the displacement transverse river trends to valley, and the left abutment displacement is larger than the right abutment displacement. In the overloading stage, the abutment displacement gradually grows as the overloading multiple K_p increases. The displacement near the arch end is the largest, and the displacement downstream gradually reduces. In the left abutment, the displacement of rock at middle – upper part and near interlayer shear zones fj2 – fj3 is relatively large. In the right abutment, the displacement of rock at upper part and near interlayer shear zones fj3 – fj4 is relatively large. Therefore, the interlayer shear zones fj2, fj3, fj4, the faults and the fissures developed in the middle part of left abutment have great effects on

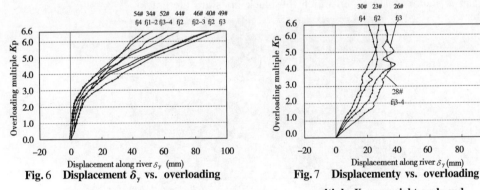

Fig. 6 Displacement δ_y vs. overloading multiple K_p near left arch end

Fig. 7 Displacement vs. overloading multiple K_p near right arch end

Note: 54# fj4 Surface displacement point near fj4; 46# fj2 – 3 – Surface displacement point between fj2 and fj3.

the deformation and stability of dam abutment.

4.3 Impact of structural planes on abutment stability

The typical curves of relative displacement $\Delta\delta$ vs. overloading multiple K_p are shown in Fig. 8. Based on the surface displacement at the outcropping positions and the relative displacement on structural surface, the main structural planes impacted the stability of left abutment include f5,

Lp285, L2, fj2, fj3 and fj4, and the main structural planes impacted the stability of right abutment include fj3, fj4 and f4. Due to the intensive development of structural planes and the broken resisting rock, the structural planes in the left abutment are the critical factors on the deformation and stability of dam abutment.

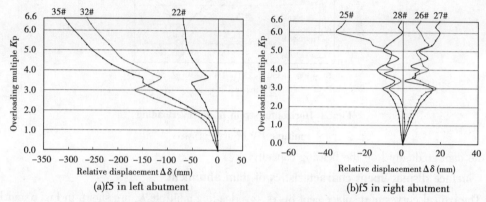

Fig. 8 Relative displacement $\Delta\delta$ vs. overloading multiple K_p

4.4 Failure pattern and failure characteristics

In the overloading stage, two cracks successively appear in the dam body. As $K_p = 3.4 \sim 4.3$, a crack is created on the downstream dam surface of left arch end at El. 2040 m. At last, this crack extends to the dam crest and penetrate the dam body from downstream to upstream. Development of several intersected structural planes in left abutment is the primary cause. In the later overloading stage, i.e. $K_p = 5.0 \sim 6.3$, another crack appears at the dam doe on the right base surface and extend upward to El. 2043 m. The second crack is caused by fault f5. The final pattern of dam is shown in Fig. 9, in which the black lines are the cracks.

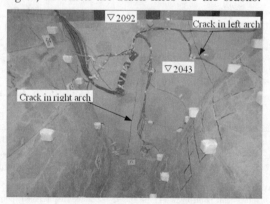

Fig. 9 Final failure pattern of dam model

As the geological asymmetry, the failure patterns of both abutments are also asymmetric. The destruction of left abutment is more severe than right abutment, as shown in Fig. 10. The destroy range of left abutment is approximately 81m (prototype size) long extended from left arch end to downstream. The rock mass and the structural planes (such as f5, fj3, fj4) in the middle - upper area are damaged severely. The destroy range of right abutment is approximately 57m (prototype size) long extended from right arch end to downstream. The rock mass in the upper area is damaged and the interlay shear zones fj3 and fj4 are locally cracked. In the upstream area, the structural planes (such as L1, L2, Lp285) are cracked, and these cracks extend through both banks along dam heel.

(a) Left abutment

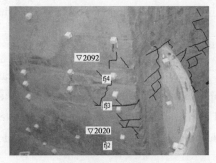
(b) Right abutment

Fig. 10 Final failure patterns of left and right abutments

4.5 Overloading safety coefficients

During the overloading rupture test, as $K_p = 1.4 \sim 2.2$, the initial crack creates at the upstream dam heel; as $K_p = 3.4 \sim 4.3$, the downstream dam surface is cracked in the left arch and the cracks in abutment obviously increase; as $K_p = 6.3 \sim 6.6$, the cracks in dam body extend from bottom to top and the cracks in both abutments propagate and intersect, then the dam and foundation have the tendency of overall instability. Comprehensively analyzing the experiment results, the overloading safety coefficients at different destructive stages are determined as follows: the initial cracking overloading safety coefficient $K_1 = 1.4 \sim 2.2$, the nonlinear deformation overloading safety coefficient $K_2 = 3.4 \sim 4.3$, and the ultimate overloading safety coefficient $K_3 = 6.3 \sim 6.6$.

5 Engineering analogy analysis

Shapai RCC arch dam is located on Chaopo River, Aba Tibetan – Qiang Autonomous Prefecture, Sichuan Province. It is a single – curvature RCC arch dam with a height of 132 m. The project scale of Shapai arch dam is same as Lizhou RCC arch dam. While the geological condition of Shapai Project is much better than Lizhou Project. At the dam site of Shapai Project, there aren't faults and penetrating weak structural planes. The abutment stability is only subject to the groups of joint fissures.

According to the overloading model test of Shapai arch dam, the nonlinear deformation overloading safety coefficient is $K_2 = 4.6 \sim 5.0$[16], which is larger than that of Lizhou arch dam. The test results show that the middle – upper regions of left and right abutments are severely destroyed and become the key controlling factor affect the engineering stability. Thus a large amount of prestressed anchor cables have been adopted to reinforce these weak regions of abutment in the construction of Shapai Project. Consequently, Shapai arch dam and abutment have super capacity of global stability and anti – seismicity. This is the primary reason for Shapai arch dam to withstand the severe test of Wenchuan great earthquake on May 12, 2008.

Considering the complicated geological condition, plenty of adverse geological structures, and the lower safety coefficient, it is necessary to reinforce the main structural planes and the weak regions of abutment for Lizhou Project. The required measures can further improve the global stability of dam and abutment and ensure the project safety.

6 Conclusions

(1) Based on the topographic and geographic features of Lizhou RCC arch dam, a 3D geomechanical model is constructed to conduct the overloading rupture test and study the global stability of dam and abutment. According to the test results, the overloading safety coefficients at different destructive stages are determined as follows: $K_1 = 1.4 \sim 2.2$, $K_2 = 3.4 \sim 4.3$, $K_3 = 6.3 \sim 6.6$.

(2) Test results show that the deformation of rock mass and structural planes in the middle - upper region of left abutment, and the upper region of right abutment is large. The destruction of these regions are very severe. And they are the weakness regions impact the deformation and stability. Some engineering measures should be taken to reinforce the abutment and improve the global stability of dam and foundation, such as concrete replacement, prestressed anchor cable, consolidation grouting, etc.

(3) Contrasting to Shapai RCC arch dam which has the same project scale, the overloading safety coefficients of Lizhou RCC arch dam are relatively lower due to its complicated geological condition. Moreover, the local regions of abutment are destroyed seriously. Therefore, the abutment of Lizhou Project should be paid more attention to reinforce.

Acknowledgements

This work was financially supported by the National Natural Science Foundation of China (Grant No. 51109152, 51379139 & 51409179). All authors would like to express sincere appreciation to Academician Heping Xie, Prof. Chaoguo Li and Prof. Jiankang Chen of Sichuan University, for their continuous supports and valuable suggestions.

References

[1] Jinsheng Jia. Dam Construction in China - A Sixty - Year Review[M]. Beijing: China Water Power Press, 2013(in Chinese).

[2] Yutai Wang, Weiyuan Zhou, Jianquan Mao, et al. Stability analysis of arch dam abutment[M]. Guiyang: Guizhou People's Press, 1983(in Chinese).

[3] Qingwen Ren, Lanyu Xu, Yunhui Wan. Research advance in safety analysis methods for high concrete dam [J]. Science in China (Series E: Technological Sciences), 50 (S1): 62-78, 2007.

[4] Hanpeng Wang, Shucai Li, Xuefen Zheng, et al. Research Progress of Geomechanical Model Test with New Technology and its Engineering Application[J]. Chinese Journal of Rock Mechanics and Engineering, 28 (S1) 2765-2771, 2009(in Chinese).

[5] Jian Liu, Xiating Feng, Xiuli Ding. Stability assessment of the Three - Gorges Dam foundation, China, u-

sing physical and numerical modeling—Part I: physical model tests[J]. International journal of rock mechanics and mining sciences, 40(5): 609-631, 2003.

[6] Fuhai Guan, Yaoru Liu, Qiang Yang, et al. Analysis of stability and reinforcement of faults of Baihetan arch dam[J]. Advanced materials research, 243-249: 4506-4510, 2011.

[7] Weiyuan Zhou, Ruoqiong Yang, Yaoru Liu, et al. Research on geomechanical model of rupture tests of arch dams for their stability[J]. Journal of Hydroelectric Engineering, 24(1): 53-58,64, 2005(in Chinese).

[8] Xiaolan Jiang, Jin Chen, Shaowen Sun. Research on physical model test for high arch dam of Jinping I Hydropower Station[J]. Yangtze River, 40(19): 76-78, 2009(in Chinese).

[9] Jianhua Dong, Heping Xie, Lin Zhang, et al. Experimental study of 3D geomechanical model on global stability of Dagangshan double curvature arch dam[J]. Chinese Journal of Rock Mechanics and Engineering, 26(10): 2027-2033, 2007(in Chinese).

[10] Baoquan Yang, Lin Zhang, Jianye Chen, et al. Experimental study of 3D geomechanical model for global stability of Xiaowan high arch dam[J]. Chinese Journal of Rock Mechanics and Engineering, 29(10): 2086–2093, 2010(in Chinese).

[11] Xinghua Chen. Structure model test for brittle material[M]. Beijing: Water Resources and Electric Power Press, 1984(in Chinese).

[12] Lin Zhang, Jianye Chen. The engineering application of model text about hydraulic dams and foundation [M]. Chengdu:Sichuan University Press, 2009(in Chinese).

[13] Yuan Chen, Lin Zhang, Baoquan Yang, et al.. Geomechanical model test on dam stability and application to Jinping High arch dam[J]. International Journal of Rock Mechanics & Mining Sciences, 76(6): 1-9, 2015.

[14] The Professional Standards Compilation Group of People's Republic of China. SL 282—2003 Design specification for concrete arch dams[S]. Beijing: China Water Power Press, 2003(in Chinese).

[15] The Professional Standards Compilation Group of People's Republic of China. DL/T 5346—2006 Design specification for concrete arch dams[S]. Beijing: China Electric Power Press, 2006(in Chinese).

[16] Liyong Zhang, Lin Zhang, Chaoguo Li, et al. 3D geomechanical model experimental for abutment stability of Shapai RCC arch dam[J]. Design of Hydroelectric Power Station, 19(4): 20-23, 2003(in Chinese).

How Fast should an RCC Dam be Constructed?

Malcolm Dunstan

(Malcolm Dunstan and Associates, U. K.)

Abstract: An analysis of approaching 500 completed RCC dams has shown that there is a very wide range of average rates of placement for RCC dams of similar volumes. For example, for a typical RCC dam with a volume of 1 Mm^3, the average rate of placement varies from circa 25 000 m^3/month (i. e. the 1 M m^3 would be placed in 40 months) to circa 70 000 m^3/month (i. e. the 1 Mm^3 would be placed in just over 14 months, approximately one third of the time). However there are two RCC dams, Upper Stillwater and Olivenhain, both of which had volumes close to 1 Mm^3 at which the average rate of placement was between 120 000 and 125 000 m^3/month, that is five times the lowest average rate of placement. How can it be that two dams, with the same nominal volume of RCC, can be constructed at rates of placement varying by a factor of five?

A detailed study of a number of RCC dams, for which there are considerable data, has shown that there are several factors that impact on the rate of placement and that there is now more than sufficient data to be able to quantify the effect of some of these factors, for example the method used to transport the RCC from the concrete plant to the point of placement.

The paper contains a series of Figures relating the average rate of placement to the maximum month, to the peak day and to the nominal capacity of the concrete plant. In each of the Figures, there is a range of perceived efficiencies from excellent to unacceptable. Given an estimate of the likely efficiency of a Contractor constructing an RCC dam, for which an average rate of placement is required, it is possible to estimate what is the likely maximum month that will be required (from which the stockpiles should be sized, etc.) and what the peak day is likely to have to be (from which the plant on the placement should sized, etc.). In addition it is possible to assess the performance of completed RCC dams by comparing the figures actually achieved with the ranges of efficiencies in the Figures in the paper.

Key words: RCC, dams, rates of placement, efficiency, planning

1 Introduction

The simple answer to the question "How fast should an RCC be constructed?" is "as quickly as practical". Not only will this give an early return on his investment for the Owner/Developer, but rapid construction leads to higher quality and lower costs.

When a Hydropower (and a water supply, and to a lesser extent, irrigation) Project is constructed, the Owner, particularly Private Developers, expect a reasonable rate of return, be that rate 10%, 20% or even more. A completion ahead of schedule means that that return will obviously increase. Conversely any delay to the completion of the Project means that some of that investment will be lost forever.

There are many examples of the benefits of the rapid construction of RCC dams and the benefits that arise from that speed. During a presentation to the Fifth International Symposium on RCC dams in 2007, Mr Shelke, the Chief Engineer of GOMID, the Owner of the Ghatghar Pumped – Storage Scheme (Fig. 1), said:

"Due to selection of RCC dam construction in lieu of conventional concrete or masonry dams, the Project has benefited by three years of early commissioning thereby generation of 1 350 million units (MKWH) the cost of which is 5 200 million INR (Indian Rupees). This is equal to about half the cost of project. This clearly tells the success story of RCC dams and its role in hydro electric Projects." [1]

Thus the return at Ghatghar must have been between 15% and 20% and any delay to the completion would have had a corresponding loss.

Fig. 1 The lower RCC dam at the Ghatghar PSS

There are several examples of early completion of RCC dams leading to significantbenefits. Probably the example with the most power being generated is Longtan[2] (Fig. 2) where the RCC was completed five and a half months early and the Powerhouse (installed capacity 6 300 MW) was accelerated when it became clear that the dam was to be completed early. In the event several million dollars worth of power was generated ahead to schedule. The value of the power was circa $ 2.5M/day and this reflects the benefits of early completion.

Son La Hydropower Project[3] (installed capacity 2 400 MW) was the first large public Project in Vietnam ever to be completed on time and in the first year $ 400M worth of power was

Fig. 2 Completed Stage 1 of Longtan dam

generated. The RCC itself was finished three and a half months ahead of schedule and this exerted pressure on the Powerhouse Contractors to complete on time (in fact three days early).

At YeywaHydropower Project[4] (installed capacity 800 MW) in Myanmar, the RCC was completed five and a half months ahead of the programme. All efforts on the site were then transferred to the Powerhouse that, at the time, was behind schedule. This additional pressure, and an impounded reservoir, was sufficient to accelerate the construction of the Powerhouse and the first unit went on line on programme.

2 Rates of placement achieved at RCC dams

A detailed studyhas been made of the average rates of placement of approaching 500 RCC dams at which the RCC had been placed at an average rate in excess of 1 000 m^3/month. These average rates have been plotted against the volume of RCC in the dam in Fig. 3. Also plotted in the Figure are upper and lower limits that contain 97% of all the points.

At some point the upper (and lower) limits shown in Fig. 3 will become asymptotic to a maximum practical average rate of placement. Based upon the experience of a number of very large RCC dams this figure will probably be circa 250 000 m^3/month. It is possible that the point at which the curve moves away from a straight line will be close to 6 Mm^3.

It is rather disturbing that only 14 out of the 480 RCC (and hard – fill) dams (i. e. circa 3%) in Fig. 3 have average rates of placement within 10% of the upper limit to the range shown in the Fig. 3 Considering only those dams with a volume in excess of 0. 5 Mm^3, only 6 out of 108 (i. e. circa 5. 5%) have average rates within 10% of the upper limit and only 9 (i. e. just over 8%) have rates within 20%. Obviously there will be some dams at which there are very good reasons why it is not possible to achieve high rates of placement, but these should be in a minority. It is possible that over 90% of all RCC dams are not fully utilizing the real advantage of RCC and

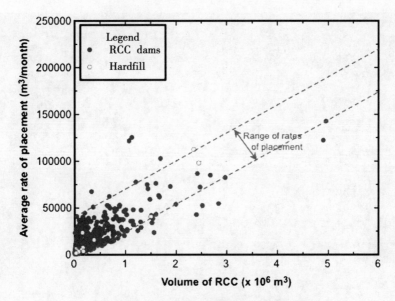

Fig. 3 Average rates of placement of RCC relative to the volume of RCC placed

that is the speed of construction.

What is also of interest in Fig. 3 is that, apart from the two largest hard – fill dams (i. e. Beydag in Turkey and Taum Sauk in the USA), the great majority of hard – fill dams have average rates of placement very close to the lower limit, whereas it might have been expected that they would have been closer to the upper limit given the simplicity of the construction methodology.

3 Factors influencing the speed of construction

The construction of an RCC dam can look easy. For example Fig. 4 shows the RCC being placed at a rate of circa 8 000 m³/day with 13 operatives on the placement.

However it is only easy if very considerable thought has been given to both the design and construction methodology before the start of the placement. For example at Yeywa, investigations of the materials that could be available started four years before the start of the RCC placement. In addition three Full – Scale Trials were conducted to investigate various aspects of the placement procedures as well as the main objective of training all involved on the techniques to be used. All this up front work resulted in the extremely efficient placement shown in Fig. 4.

There are a considerable number of factors that can, and do, impact on the rate of placement that will be achieved at any RCC dam and in most cases it can be predicted what average rate of placement will be achieved prior to the start of the RCC placement. These factors were discussed in a paper presented to ASIA 2013[5] and are shown below:

(1) Design of the layout of the dam;

(2) Design of the interfaces (i. e. the faces of the dam and at the abutments, etc.);

(3) Aggregate supply and stockpiles;

(4) Supply of cementitious materials and silos;

Fig. 4 RCC placement at Yeywa

(5) Design of RCC mixture proportions;

(6) Placement procedures;

(7) Capacity of concrete plant and type of mixers;

(8) Transportation of the RCC from the concrete plant to the point of placement;

(9) Workability of RCC;

(10) Initial (and Final) Setting Times;

(11) Placement procedures (including horizontal joint treatment and bedding mixes);

(12) Contractor's (and Designer's) experience of RCC dams;

(13) Shift patterns;

(14) Condition of equipment and maintenance;

(15) Rainfall and temperature.

It is only by getting most of these factors right that the fullest advantage can be made of the RCC method of construction for dams. Above all, the method of construction should be kept as simple as possible (Fig. 4) because with simplicity will come speed. As probably one of the greatest engineers (and painter, sculptor, architect, musician, mathematician, inventor, anatomist, geologist, cartographer, botanist and writer) who ever lived, Leonardo de Vinci, said, "Simplicity is the ultimate sophistication".

4 Efficiency of the placement of RCC dams

4.1 Introduction

There are now sufficient completed RCC dams so that an assessment can be made of the efficiency of the placement of the RCC at these dams. These "efficiencies" can be used in two ways, first to assess the efficiency of the placement at completed RCC dams, and secondly in order to make a initial estimate of the capacities of the plant that would be needed to complete an RCC dam within a defined period given an estimated level of the efficiency of the placement.

4.2 Assessment of the efficiency of the placement of RCC in dams

Table 1 contains an assessment of the perceived efficiency of the placement of the RCC in dams on the basis of two factors.

Table 1 Possible assessment of the efficiency of the placement of RCC in dams

	Maximum month/Average month	Maximum month/Peak day
Excellent	< 1.80	> 20
Good	1.80 ~ 2.20	18 ~ 20
Average	2.20 ~ 2.60	16 ~ 18
Poor	2.60 ~ 3.00	14 ~ 16
Unacceptable	> 3.00	< 14

The first factor is the Maximum month divided by the Average month; this factor gives an assessment of how the design of the dam impacts on the rate of placement of the RCC. On the assumption that the Contractor will have at least one month in which there is no impediment to the placement of the RCC, the lower this factor is the better will have been the design of the dam in terms of the placement of the RCC. The second factor is the Maximum month divided by the Peak day; this factor gives an assessment of the efficiency of the placement procedures on the dam itself. Again assuming the Contractor has at least one very good day, the greater the number of these "good" days in the Maximum month's placement, the more efficient the actual placement procedures are likely to be.

The average of the first factor – the Maximum month/Average month – for all the RCC dams (for which there are sufficient data) is 2.61 with a range from 1.19 to 6.86, a range of a factor of five. The average of the second factor – the Maximum month/Peak day – is 15.7 with a range from 5.5 to 24.3, a range of a factor of 4.5, i.e. in the worst case the Maximum month was effectively only 5.4 peak days, whereas in the best case the Maximum month consisted of effectively 24.3 peak days, thus an almost continuous placement at the maximum daily rate. The latter was at Upper Stillwater that was completed in the early 1980s and is still probably the most efficiently-placed RCC dam to date[6], the only competitor being Olivenhain[7].

4.3 Utilization of the installed plant

An estimate of the utilization of the concrete plants at RCC dams is shown in Table 2.

Table 2 Possible assessment of the utilization of the plant at RCC dams

Items	Utilization of concrete plant (hours/month)
Excellent	> 150
Good	120 ~ 150
Average	90 ~ 120
Poor	60 ~ 90
Unacceptable	< 60

The factor in Table 2 is simply the average monthly placement (in m^3/month) divided by the nominal capacity of the concrete plant (in m^3/hour). On the assumption that all the other plant will be sized around the concrete plant, that being the most expensive piece of equipment, this will effectively cover all the plant at the site. The overall average figure for all the RCC dams for which there are sufficient data is 94.0 with a range from 13.5 to 195.8, an extraordinary range. Some 150 RCC dams, having average rates of placement in excess of 10 000 m^3/month, were studied. Another 100 RCC dams, for which data existed and which had average rates of placement less than 10 000 m^3/month, had extremely variable results and thus have been neglected from the detailed study of the dams.

Fig. 5 shows the percentages of RCC dams that fall into each of the possible assessments of utilization. As can be seen only circa 10% of all the RCC dams have a utilization that can be considered to be "excellent", whereas nearly 55% have poor or unacceptable utilization of the plant. These figures are consistent with all the other assessments of the level of efficiencies (c.f. Fig. 3 and Section 2). Thus it seems that only approximately 10% of all RCC dams are fully utilizing all the benefits of the RCC dam methodology. That does not mean to say that they are not more economic that other forms of dam construction, but it does mean that they could have been even more efficient and even more economic.

5 Estimation of the plant capacities needed for the construction of an RCC dam

5.1 Average rate of placement

Having determined the average rate of placement required for the construction of an RCC dam and having estimated the level of efficiency that is likely to be achieved, the Peak day's placement can be estimated. This will define the actual (not nominal) capacity of the concrete plant, the transportation to the dam and the transportation and placement equipment on the dam itself. It should be noted that the efficiency of the placement will depend upon all the factors shown in Section 3 - NB practically all Contractors overestimate the level of efficiency that will be achieved, for example it is not generally realised that on average all - truck placement (to which some Contractor's resort when trying to 'save' money) is one third less efficient than a conveyor to the dam loading trucks on the dam - see point (8) in Section 3.

Also shown in Fig. 6 are the average rates of placement (and peak day's placement) at Up-

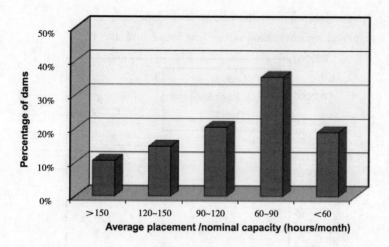

Fig. 5 Percentages of RCC dams that fall within each range of perceived levels of utilization

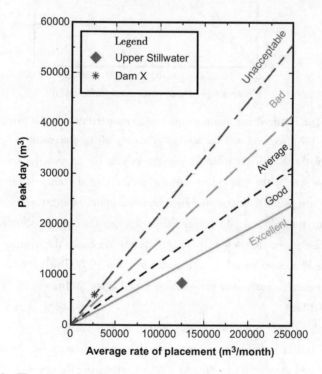

Fig. 6 The required peak day's placement relative to the average rate of placement and the level of efficiency of the placement

per Stillwater, at which the great majority of the factors in Section 3 were got right, and at another dam, Dam X, at which unfortunately most of the factors were got wrong. At Upper Stillwater the average rate of placement was 125 324 m^3/month and the peak day was 8 415 m^3 (a factor of 14.9), whereas at Dam X (albeit a dam with a somewhat smaller volume) the average rate of placement was circa 26 500 m^3/month and the peak days placement was circa 6 000 m^3 (a factor of 4.4, some 3.4 times less efficient than Upper Stillwater).

Fig. 7 is a graphical representation of the first factor in Table 1.

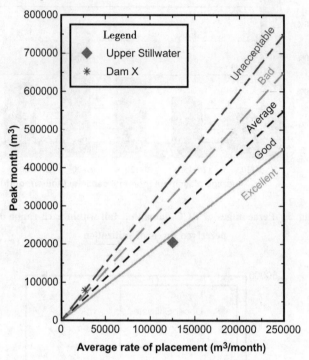

Fig. 7 The required maximum month's placement relative to the average rate of placement and the level of efficiency of the placement

From the Figure the Maximum month's placement can be estimated. The capacity of the aggregate plant and the size of the stockpiles and the delivery of the cementitious materials and the capacity of the silos are usually determined from the Maximum monthly placement (assuming the figures are reasonably typical). Again Upper Stillwater and Dam X are shown, the former with a Maximum month's placement of 204 430 m^3 (i.e. some 1.63 times the average rate of placement) and the latter with a Maximum month's placement of circa 79 000 m^3 (i.e. some 3.00 times the average rate of placement, nearly half the efficiency of Upper Stillwater).

5.2 Capacities of plant

Fig. 8 is a graphical representation of the factors in Table 2.

From the Fig. 8, it is possible to estimate the nominal capacity of the concrete plant (that should also be compared to the data in Fig. 3). When estimating the level of efficiency in the Fig. 8, it should be noted that on average a concrete plant containing a continuous mixer has a realized output some 25% less than a concrete plant containing twin-shaft batch mixers with the same nominal capacity. Also shown again in the Fig. 8 are Upper Stillwater, which had a concrete plant with a nominal capacity of 640 m^3/hour, and Dam X, which had a concrete plant with a nominal capacity of approximately 650 m^3/hour, i.e. essentially an identical nominal capacity. As mentioned above the average rate of placement at Upper Stillwater was 125 324 m^3/month while at Dam X it was circa 26 500 m^3/month, a difference of a factor of 4.7. It is difficult to envisage

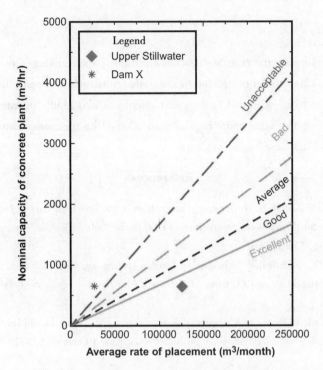

**Fig. 8 The required peak day's placement relative to
the average rate of placement and the level of efficiency of the placement**

that two RCC dams having concrete plants of the same nominal capacity can have average rates of placement that vary by a factor of approaching five. This is an extreme example of the affect of the factors listed in Section 3.

6 Conclusions

Approximately 500 completed RCC dams have been studied in some detail so that an assessment could be made of the efficiency of the placements and the utilization of the installed plant. It was disturbing to find that only circa 10% of all RCC dams are fully utilizing the main benefit of the method of construction, that is the speed of construction. Indeed probably over 50% of all RCC dams are being constructed with an efficiency that might be classified as bad or unacceptable. Nevertheless, even with this level of efficiency, the dams are probably still more economic than other forms of dam construction.

When RCC dams are constructed efficiently, the benefits can be huge with early completion and millions of dollars of power being generated, and/or water supplied, ahead of schedule thus enhancing the rate of return for Owner/Developer.

Although RCC dams can look easy to construct, it is only by getting a considerable number of factors right that the full benefits of the method of construction will be achieved. This generally requires a considerable amount of upfront work even before the award of the Contract for construction.

Acknowledgements

The data used to create the relationships shown in this paper emanate from MD&A's database of the RCC dams that have been completed or are under construction around the World. Acknowledgement must therefore be accorded to the many engineers who kindly update this database each year. Without their inputs, some more regular than others, the tremendous amount of data that is now contained in the database would not exist.

References

[1] V. C. Shelke, S. D. Sapre, M. R. H. Dunstan. Construction of the first RCC dam in India at Ghatghar[C] // Proceedings of the 5th International Symposium on RCC dams 'Celebration for 30 years' application of RCC in dams', Guiyang, China, 2007.

[2] X. Wu. Rapid RCC Construction Technology of Longtan Hydropower Dam Project[C] // Proceedings of the 5th International Symposium on RCC dams 'Celebration for 30 years' application of RCC in dams', Guiyang, China, 2007.

[3] N. H. Ha, M. Conrad, D. Morris, et al.. Speed of construction of EVN's Son La and Lai Chau RCC dams[C] // Proceedings of 7th International Symposium on Roller - Compacted Concrete (RCC) Dams, Chengdu, China, 2015.

[4] W. Kyaw, M. Zaw, A. Dredge, et al.. An Overview of the development of the Yeywa Hydropower Project, Myanmar[C] // Proceedings of the 5th International Symposium on RCC dams 'Celebration for 30 years' application of RCC in dams', Guiyang, China, 2007.

[5] M. R. H. Dunstan. The precedent for the rapid construction of large RCC dams[C] // Proceedings of International Symposium AFRICA 2013 'Water Storage and Hydropower Development for Africa', Addis Ababa, Ethiopia, 2013.

[6] A. T. Richardson. Performance of Upper Stillwater dam[C] // Proceedings of International Symposium on Roller - Compacted Concrete (RCC) Dams, Beijing, China, 1991.

[7] M. R. H. Dunstan, K. A. Steele, G. S. Tarbox, et al.. Value Engineering at Olivenhain RCC dam. USA, Q. 84 - R. 17, XXIIth ICOLD Congress, Vol. 1, Barcelona, 2006.

Earthquake Pipe – Burst Protection of Dam Penstock by Using a VAG Pipe Burst Safety Valve with Integrated Earthquake Sensor

Anton Rienmüller

(VAG Armaturem GmbH, Germany)

1 Introduction

Depending on the level of water a dam has a very high load of potential energy stored. Which is normally used to transport the water to it's destination for treatment or irrigation, or this energy will be transformed to electricity in a hydro power plant.

In any way the pressurized pipes are under the risk of pipe burst for different reasons. In some areas one of the risks is a pipe burst due to an earthquake. A Pipe Burst Safety device (RBS) or so called Safety Valve cannot avoid a pipe burst, but due to his release sensor system, Paddle or Pitot Tube, a Valve will close and protects the dam for uncontrolled emptying. In addition to that common used systems an high sophisticated Earthquake sensor is able to detect the P – wave and close the valve before the destructive S – wave has reached the place. Due to the resulting acceleration forces for a pipe system the risk of a pipe burst for a filled pipe is much higher than of an empty – or partly empty pipe.

So the main advantage of this system is, that the risk of a pipe burst in case of earthquake can be minimized to the lowest level.

2 Pipe break device

The main purposes of dams are to provide drinking water, to generate power, to provide water for industry and agriculture and to protect people from floods. There is an enormous quantity of potential energy captured as water in a dam. Therefore the highest level of safety should be a must in respect of all components which are used in the structure and the mechanical equipment of a dam.

One of the most important components is the water intake and the ongoing pressure pipe though the dam structure down to a hydro power plant or a water supply line.

This pipe has to be designed and calculated against several assumed impacts caused by e. g. hydraulics or by the environment. One of these impacts which have to be considered is the case of

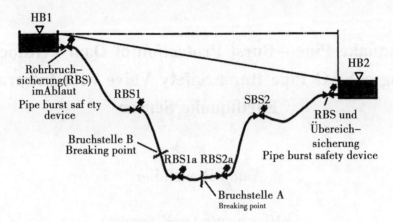

Fig. 1 Locations of pipe break devises (RBS)

an earthquake. Whilst the dam itself can be executed firmly to withstand high magnitudes of earthquake levels, the penstock can't be protected fully against damages or even pipe burst cases. So the common way to protect the pipe is to install safety valves close to the water intake at the dam, the so called "Pipe Break Safety Device" or Safety Valve.

3 Valve type

The type of that safety valve normally is a Butterfly Valve with a lever and weight actuator and a speed sensor as release system. The main duty is to stop absolutely reliable the water flow in case of pipe burst. And it must have a release system which is independent from auxiliary energy as electricity etc.

It is consisting of a Valve with lever and weight, a hydraulic cylinder and an over speed detector (Paddle or Pitot Tube) which is installed in the pipe.

That means, the detecting of the higher flow velocity in case of pipe burst, as well as the release of the closing movement has to be executed without any electricity.

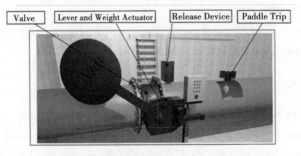

Fig. 2 Pipe break device assembly

4 Closing time

The closing of the Butterfly Valve has not only be independent of the electricity, it has also to follow a closing law, which is the results of a surge analyse.

The closing time of course must be precise adjustable.

As it is depending of the viscosity of the hydraulic oil which is pressed through a control valve the closing time will be affected by the ambient temperature around the Pipe Break Device.

If this would be done by a simple throttle valve, the closing time would not be precise enough all the time and in all environmental conditions.

In that case the closing time would be different under high – or in low temperature ambient conditions respectively in summer and winter seasons.

To guaranty the precise closing time, the hydraulic actuator must work with a viscosity independent capacity control valve to adjust the closing time.

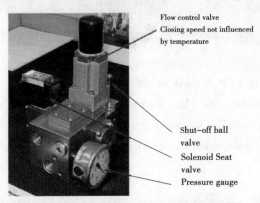

Fig. 3   Flow – , Capacity control valve

The capacity Control Valve guaranties that once a position is adjusted the quantity of hydraulic oil which is passing the valve is always the same, independent of the viscosity of the oil as well independent of the oil pressure. This guaranties always the sam closing time.

5   Earthquake protection

To increase of safety of the pressure pipe (penstock), VAG has developed an integrated and advanced earthquake P – wave detecting and release system in combination with the Pipe Break Safety Device.

The system consists of minimum two separate sensor units, one master – and one slave, which are located on different places in the valve chamber. Only in that case when both sensors will be activated simultaneously and also in the same manner, the alert signal will release the valve actuator and close the valve safely.

The sensitivity of the sensor can be adjusted acc. to the seismologic conditions of the area where it is installed, and the acceleration frequency must contain the typical p – wave character of an earthquake. Therefore it can not be released by local vibrations e. g. traffic etc. in the closer surroundings.

Therefore fail alerts, respectively fail releases are practically impossible.

6   P – wave and S – wave

The different propagation characters of the P – wave and S – wave is the core of the detecting

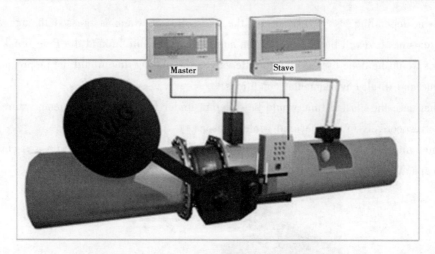

Fig. 4　**Pipe break device with earthquake sensors**

system. Whilst the smoother P – wave has a vertical propagation and the velocity is lower, has the destructive S – wave a horizontal propagation and is much faster.

The P – wave is the signal which can be detected by the VAG Pipe Break device with integrated earthquake sensor.

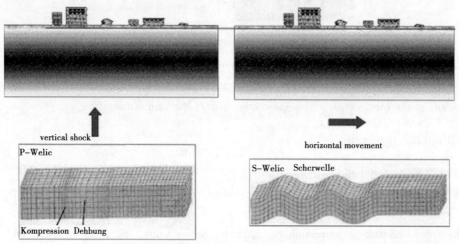

Fig. 5　**P – Wave S – Wave behaviour**

The head – start in time of the P – wave against the destructive S – wave is the important benefit which protects the pipeline system against destroying. When the P – wave has been detected and the Pipe Break Device is closing the flow, the pipe can be drained partly or even totally. Due to the reduced inertia of an empty or partly empty pipe the resulting acceleration – forces towards the pipe are also much lower and can prevent a pipe burst.

The VAG Pipe – burst device with integrated earthquake sensor can avoid pipe burst of main pipes and can maintain the water supply after an earthquake disaster. So It helps to reduce harms of people which are anyway serious affected by an earthquake.

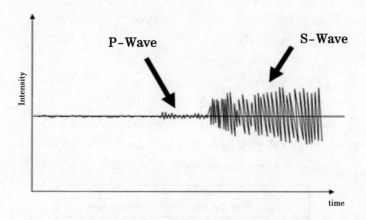

Fig. 6 P – Wave S – Wave intensity

7 Warning time

The warning time is the time between the P – Wave and the S – Wave and of course is depending on the distance of the epicentre (EPZ). All earthquakes around the world are coming up in the same characteristic. First the non destructive and vertical P – Wave appears, and after a while, depending on the distance to the epicentre, the horizontal S – Wave is following with its massive destroying forces.

We dont talk about days, hours or minutes (sometimes it is possible). We have only seconds before the S – wave reaches the place. Till the S wave has reached its total highlight many seconds may pass. But even in few seconds you can do a lot and it can save your and your family's life and as well can avoid damages to structures as penstocks of a dam.

40 km distance to EPZ	8~12 seconds
80 km distance to EPZ	16~24 seconds
120 km distance to EPZ	24~36 seconds
160 km distance to EPZ	32~48 seconds

Fig. 7 Delay time between P – Wave, and S – Wave

The P – Wave detecting release system in the VAG Pipe Break Device is using the delay between P – Wave and S – Wave and close the Valve before the horizontal S – Wave is accelerating the penstock. That means in that time period between P – Wave and S – Wave the pipe has the chance to get completely or partly empty and reduce so the mass of acceleration.

Example: pipe:	DN/PN	2000/10
material:		steel
specific weight:	kg/dm^3	7.85
length:	m	1 000
outside diameter:	dm	20.2
inside diameter:	dm	20.05
wall thickness:	mm	15 medium
water water density	kg/dm^3	1

Weight of a Steel Pipe DN 2000 PN 10, 1000m long		
water filled:	t	3.156
emty:	t	372

Fig. 8 Mass inertia of a filled – and empty pipe

And in respect of pipe load forces under acceleration it's a big difference if filled pipe or an empty pipe will be accelerated.

Attendece of the secty life Patron earthquake warning system[3].

GeoForschungsZentrum Potsdam
Prof. Dr. Zschau - Germany

Friedrich-Schiller-University Jena
Institut for geoscience
Prof. Jentsch - Germany

Federal Institute For Geosciences
And Natural Resources - Germany

Sharif University of Technology
Dr. Bakhshi/ Iran - Teheran

International Institute of Earthquake
Engineering and Seismology
Prof. Mokhtari / Iran - Teheran

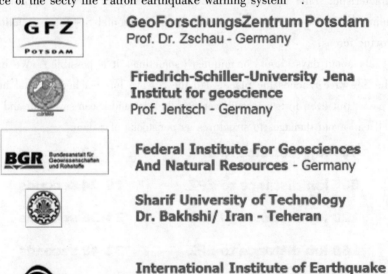

Fig. 9

8 Conclusion

To maintain a basic water supply in earthquake situation, this can be essential for life – saving and to keep up a basic hygienic level.

This assembly is a great opportunity to minimize the risk of pipe break of main pipes in case of earthquake. It is using the delay time period between P – Wave and S – Wave.

Raw Materials Selection and Temperature Control Design for Ludila RCC Dam

Ji Peimin, Huang Tianrun, Wang Tianguang, Yang Xinping, Kang Wenjun

(Power Cchina Northwest Engineering Corporation Limited
Xi'an, Shannxi, 710065, China)

Abstract: The roller compacted concrete (RCC) gravity dam of Ludila Hydropower Project has a maximum dam height of 140 m. The RCC for the main dam was blended with sandstone aggregate, high-quality set retarding superplasticizer, air entraining agent, admixture PL (phosphorus slag powder and limestone powder) and moderate heat Portland cement in an optimal mix proportion. The workability of concrete was improved significantly. Furthermore, through controlling the concrete's ex-mixer temperature and the placing temperature, providing crack-resistant mesh reinforcements at the upstream dam face, taking moisturizing measures to improve microenvironment of concreting faces, employing an intelligent water cooling method, and making 3D stress field simulation and feedback analysis for the dam, the goal of real-time temperature control and cracking prevention for high-performance dam concrete was realized.

Key words: Ludila Hydropower Project, dry-hot valley, RCC dam, real-time temperature control and cracking prevention technology, phosphorus slag, admixture PL

1 Introduction

1.1 Project profile

Ludila Hydropower Project constructed on the midstream of the Jinsha River has a total installed capacity of 2 160 MW. Its main dam is 140 m high to maximum and 622 m long along its crest. Major structures include the left and right water retaining dam sections, the overflow dam section with five surface outlets and bottom outlets in the middle of riverbed, and the headrace and power generation system on the right bank. The underground powerhouse is equipped with 6 sets of turbine-generators. The basic earthquake intensity and seismic fortification intensity of the dam area is Grade Ⅷ and Grade Ⅸ respectively. During the dam construction period, river diversion relied on the diversion tunnel in the low level period and plus flood release through the gaps of the overflow dam section in the flood season. The gross volume of concrete for the dam amounts is 2.008 4 million m^3, of which the RCC accounts for 1.577 1 million m^3.

1.2 Meteorological condition

Ludila project is located in a dry and hot valley of the Jinsha River, where the annual average temperature is 21.9 ℃, the extreme maximum temperature 46.5 ℃, the extreme minimum temperature -2.57 ℃, and the maximum daily temperature difference in May is 20.4 ℃. The annual maximum wind speed is 34.4 m/s. The average annual relative humidity is 63%, and the annual maximum relative humidity 12%. See Table 1 for statistics of meteorological elements recorded by the meteorological station in the Ludila damsite area.

Table 1 Statistics of meteorological elements recorded by meteorological station at Ludila damsite

Month.	Jan.	Feb.	Mar.	Apr.	May.	Jun.	Jul.	Aug.	Sept.	Oct.	Nov.	Dec.	Annual average
Annual average Temperature (℃)	15.6	18	22.7	25.5	25.3	27.5	25.9	24	24	22.2	18.1	15	22
Annual extreme maximum temperature (℃)	27	30.5	33.9	36.4	38.1	40.3	38.2	38.6	38.5	34.7	30.1	27.6	40.3
Annual extreme minimum temperature (℃)	6.3	8.4	10.6	15.9	4	20.4	19.5	19.6	17.2	13.1	8.1	6.1	4
Annual average water temperature (℃)	10.2	12.2	14.8	16.7	18.5	20.8	20.7	20.1	19.8	17.7	13	10	16.2

2 Concrete raw materials

In consideration of the project scale, the total volume of concrete, peak construction capacity and rush hour concreting, as well as the availability and properties of raw materials in the nearby areas of the project site, raw materials for RCC were selected through mix proportion test and alkali reactivity inhibiting test on the principles of utilizing local materials, applying reliable and economical techniques, and satisfying the productivity and quality demands.

2.1 Cement

Moderate heat Portland cement (grade 42.5) produced by Dianxi and Yongbao cement plants in Yunnan province with better performance in control of temperature rise by hydration were selected. With an actual content of MgO of 3% ~4% against the required 3% ~5%, the cement possess a micro-expansion character which is in favor of cracking control for concrete. The hydration heat of the cement in 7 days is 238 kJ/kg and the specific surface area is 343 m^2/kg, which can effectively reduce the heating speed and calorific value of the concrete at its early age.

2.2 Admixture

Based on survey and tests, it was found out that the quality and reserves of coal ash, volcanic ash and micro-slag in the nearby areas can meet the project construction requirement. The probable alkali active reaction of sandstone aggregate can be inhibited when the admixture quantity exceeds 25% of the total weight of the concrete aggregate. However, because of their longer hauling distances, in order to lower the project cost and timely supply admixtures during the construction period, after further investigation and test, the admixture (PL) comprising phosphorous slag (P) from Panzhihua and limestone powder (L) produced by Yonghe cement plant with satisfactory properties and economic indexes was selected for the RCC. Both the phosphorous slag powder and the limestone powder were processed by Yonghe cement plant. The probable alkali reactive of sandstone aggregate can be effectively inhibited when the admixture quantity exceeds 50% of the total weight of the concrete aggregate. The specific surface area of phosphorous slag powder was controlled to be greater than 350 m^2/kg. As for the fineness of rock powder, the 45 μm screen residue was controlled to be less than 20% in weight. The admixture is processed by mixing the two powders in 1:1 ratio. Compared to coal ash with acid property, the admixture PL presents weak alkaline, so its inhibition effect on aggregate alkali reactivity is slightly inferior to that of coal ash, however, it still can gain a satisfactory inhibition effect. The expansion rate of concrete test pieces has been lower than the limit value when the PL admixture quantity exceeds 50% of the total weight of the concrete aggregate. The total alkali content of RCC (at the trial mix proportion) is relatively small, about 0.567~0.685 kg/m^3, which is lower than the value of 2.5 kg/m^3 specified in the Technical Specification for Durability of Hydraulic Concrete (2010 version). See Table 2 for quality control indexes of admixture PL.

Table 2 Quality control indexes of admixture PL

Admixture	Compound ratio	Density kg/m^3	Fineness (residue weight subject to 45 μm screen(%)	Water demand ratio(%)	Compressive strength ratio (28d) %	Water content (%)
PL	1:1	≥2 750	≤25	≤103	≥75	≤1.0
Grad-II coal ash	—	—	≤25	≤105	—	≤1.0

2.3 Aggregate

The aggregates for RCC were artificially processed sandstone (excavated from the underground powerhouse) and syenite from the quarry site. Based on study and test, sandstone would have probable alkali activity reaction, while syenite should be non-active aggregate. In order to inhibit alkali reactivity, besides adding admixture into concrete, syenite was processed and used as fine aggregate, so to reduce the quantity and specific surface area of sandstone aggregate applied in per unit concrete.

2.4 Additive

Basedon test and comparison, JM-II set retarding superplasticizer and GYQ air entraining agent produced by Jiangsu Bote Company were selected as concrete additives, which can control effectively the initial setting time and the peak value of hydration heat of concrete.

2.5 Mix proportion design

After the raw materials were selected, mix proportion of RCC for Ludila project was defined as shown in Table 3 based on the test results of Nanjing Hydraulic Research Institute.

Table 3 Trial mix proportion of RCC (with admixture PL)

No.	Concrete type	No.	Adding quantity of admixture (%)	Water-binder ratio	Water (kg/m³)	Cement (kg/m³)	Phosphorous slag powder (kg/m³)	Limestone powder (kg/m³)	Sand (kg/m³)	Rock (kg/m³)	Water reducing agent (%)	Air entraining agent (‰)	V_C value (s)	Slump (mm)	Air content (%)	Density (kg/m³)
1	$C_{90}25$ (III) RCC	L3RA0	55	0.40	83	93	57	57	684	1 521	1.20	0.45	4.6	—	3.7	2 490
2	$C_{90}20$ (III) RCC	L3RA1	55	0.45	84	84	51	51	689	1 533	1.20	0.45	4.5	—	3.5	2 500
3	$C_{90}15$ (III) RCC	L3RA2	55	0.50	86	77	47	47	691	1 539	1.20	0.45	4.3	—	3.7	2 500
4		L3RB1	60	0.45	84	75	56	56	688	1 532	1.20	0.45	4.3	—	3.6	2 490
5	$C_{90}25$ (II) RCC	L2RA1	50	0.45	100	111	56	56	750	1 392	1.20	0.40	4.0	—	3.8	2 450
6	$C_{90}25$ (II) Conventional concrete	L3NB3	30	0.50	108	151		32	653	1 454	1.00	0.012	—	59	3.8	2 450
7		L3NC2	35	0.45	106	153		41	649	1 445	1.00	0.012	—	63	3.5	2 440
8	$C_{90}20$ (III) conventional concrete	L3NB4	30	0.55	110	140		30	656	1 460	1.00	0.012	—	67	4.0	3 430

3 Finite element theory and simulation result of temperature field and stress field

3.1 Fundamental equation of temperature field

Temperature is taken as a fundamental imposed load in the simulation analysis of a concrete dam. According to the heat conduction theory, for mass concrete with a uniform, isotropy, unsteady temperature filed within a certain region R shall meet the following differential equation and their boundary conditions:

$$\frac{\partial^2 T}{\partial x^2} + \frac{\partial^2 T}{\partial y^2} + \frac{\partial^2 T}{\partial z^2} + \frac{1}{a}\left(\frac{\partial \theta}{\partial \tau}\right) = 0 \tag{1}$$

Boundaryconditions:

$$T = \overline{T}, \quad -\lambda \frac{\partial T}{\partial n} = q, \quad -\lambda \frac{\partial T}{\partial n} = -\beta(T - T_a)$$

Where, τ is time duration (h), λ is heat conductivity coefficient (kJ/(m·h·℃)), β is heat emission coefficient (kJ/(m²·h·℃)), θ is adiabatic temperature rise (℃).

3.2 Fundamental equation of stress filed

After the temperature filed T of concrete is derived, temperature stress of different parts shall be worked out further by calculating first the deformation ε_0 caused by temperature, after that temperature load P_{ε_0} at an equivalent node caused by corresponding initial stress is counted and the node displacement induced by temperature change is counted with common stress calculation method, based on which temperature stress σ can be figured out.

$$P_{\varepsilon_0}^e = \iiint_{\Delta R} B^T D \varepsilon_0 dR \tag{2}$$

$[B]$ is transfer matrix of strain and displacement, $[D]$ is elastic matrix.

The total stress including thethermal stress can be obtained by adding the equivalent node load caused by temperature deformation with other loads.

The initial strain item is included in the stress – strain relation.

$$\sigma = D(\varepsilon - \varepsilon_0) \tag{3}$$

Concrete is characterized by elastic creep, so the influence of creep deformation shall be taken into account in the simulating calculation. Creep compliance of concrete is calculated with the following equation:

$$J(t,\tau) = \frac{1}{E(\tau)} + C(t,\tau) \tag{4}$$

Where, $E(\tau)$ is instantaneous elastic modulus of concrete, $C(t,\tau)$ is creep compliance of concrete.

Calculating the simulation stress withthe increment method by separating duration τ into a series of time periods, $\Delta\tau_1, \Delta\tau_2, \cdots, \Delta\tau_n$, then the strain increment generated in the time period $\Delta\tau_n$ can be calculated out with the following equation:

$$\{\Delta\varepsilon_n\} = \{\varepsilon_n(\tau_n)\} - \{\varepsilon_n(\tau_{n-1})\} = \{\Delta\varepsilon_n^e\} + \{\Delta\varepsilon_n^c\} + \{\Delta\varepsilon_n^T\} + \{d\Delta\varepsilon_n^0\} + \{\Delta\varepsilon_n^s\} \tag{5}$$

Where, $\{\Delta\varepsilon_n^e\}$ is increment of elastic strain, $\{\Delta\varepsilon_n^c\}$ is increment of creep strain, $\{\Delta\varepsilon_n^T\}$ is increment of temperature strain, $\{\Delta\varepsilon_n^0\}$ is increment of self-grown volume deformation, $\{\Delta\varepsilon_n^s\}$ is increment of dry shrinkage strain.

Based on element integration of the above nodeload increments, the integral equilibrium equation can be obtained as below:

$$[K]\{\Delta\delta_n\} = \{\Delta P_n\}^L + \{\Delta P_n\}^C + \{\Delta P_n\}^T + \{\Delta P_n\}^0 + \{\Delta P_n\}^S \qquad (6)$$

Where, $\{\Delta P_n\}^L$ is increment of node load caused by external forces, $\{\Delta P_n\}^C$ is increment of node load caused by creep, $\{\Delta P_n\}^T$ is increment of node load caused by temperature, $\{\Delta P_n\}^0$ is increment of node load caused by self-grown volume deformation, $\{\Delta P_n\}^S$ is increment of node load caused by dry shrinkage.

According to thecorrespondence between $\{\Delta\delta_n\}$ and $\{\Delta\sigma_n\}$, stress at time period τ_n of each element can be obtained by accumulation.

$$\{\sigma_n\} = \sum\{\Delta\sigma_n\} \qquad (7)$$

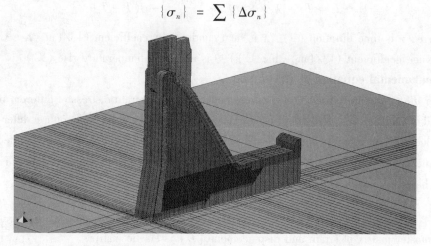

Fig. 1 Finite element calculation model of No. 15 overflow dam section

3.3 Results of simulating calculation

Based on different construction commencement time (February, May and October) for the overflow section of Ludila RCC dam, simulating calculation and analysis were made for controlling cracking due to temperature difference in the dam foundation, surface cracks and vertical cracks on upstream face due to temperature difference between concrete inside and outside, horizontal cracks due to temperature difference between upper and lower concrete layers, and cracks due to flood release in flood season, long-term exposure in air in winter and impact by low-temperature water during the initial impoundment period, and the conclusion was that the maximum temperature of the foundation binding concrete could be limited to not more than 36 ℃ and the maximum temperature of the RCC in the foundation area could be limited below 31 ℃. The maximum stress in the foundation blinding concrete in the flow direction would be 0.8 ~ 1.1MPa or so (the allowable stress is 1.44 MPa) and the maximum stress of RCC in the foundation area would be controlled at 1.0 ~ 1.5 MPa or so (the allowable stress is 1.6 MPa), occurred both when the dam body temperature should have become an equilibrium temperature field; the foundation stress in the flow direction could meet the design requirement for concrete cracking control with all fac-

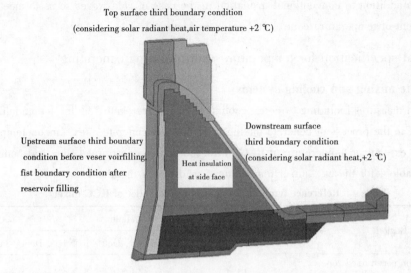

Fig. 2 Schematic diagram of boundary conditions for stable temperature field calculation of No. 15 overflow dam section

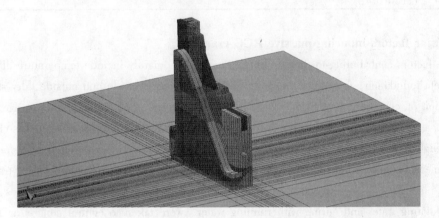

Fig. 3 Finite element calculation model of No. 12 bottom outlet dam section

tors of safety higher than 1.8. The stress in the non-restraint zone in the flow direction could be mostly limited to 0.3 ~ 0.8 MPa under the premise of no long downtime. The maximum axial stress in the upstream face would be 1.2 MPa and the maximum vertical stress would be smaller than 1.0 MPa, the factor of safety would be higher than 1.5 from the perspective of theoretical probable allowable tension stress, which could meet the design requirement for concrete cracking control. In order to lower the concrete cracking risk, the target temperature in the process of flowing through the dam gap and reservoir filling, or before winter coming, should be controlled to 20 ~ 23 ℃. Dam concrete construction should be carried out continuously with short time intervals in lifts of greater than 3.0 m. Effect of cold wave impact on stress would be mostly reflected in deformation of upstream and downstream dam faces as well as surface and laterals of concrete lifts. Excessive stress would occur on the upstream and downstream dam faces in a 3-day cold wave with a temperature fluctuation of 10 ℃. Therefore, external heat insulation measures shall be provided in addition to conventional insulation in the case of cold waves so as to meet the design requirement of temperature control and crack prevention.

4 Technical specification for temperature control during concreting

4.1 Concrete mixing and cooling systems

Technical measures including twice air cooling of coarse aggregate, adding ice andchilled water were taken in the process of concrete mixing to control the concrete's ex – mixer temperature. The minimum ex – mixer temperature of RCC in the foundation constrained zone was controlled at 12 ℃. See Table 4 for the ex – mixer temperatures in different months.

Table 4 Reference temperature values at the outlet of RCC mixer

Region	Month		
	April ~ Sep.	March, Oct.	Nov., Dec., Jan., Feb.
Foundation constrained zone $(0 \sim 0.4)L$	12 ℃	14 ℃	Normal temperature
Non – restrained zone at the upper part (above $0.4L$)	14 ℃	16 ℃	Normal temperature

4.2 Major factors inducing massive RCC cracking

The factors contributing to the cracking in mass RCC mainly include temperature differential in concrete foundation, temperature different between concrete inside and outside, dry shrinkage deformation of concrete, properties of concrete raw materials, mixing proportion, age, elastic modulus, etc. The dry and hot valley of the Jinsha River where the project is located is characterized by high air temperature, big temperature difference and large evaporation. , Therefore, engineering measures such as mixing concrete with chilled materials, heat insulation during transportation and placement, spraying water in concreting places, covering up concreting faces, cooling with circulating water and curing with running water, were taken to control concreting temperature, narrow temperature differential in the dam foundation, temperature difference between concrete outside and inside so as to control concrete deformation due to dry shrinkage. For mass concrete, its core tends to expand due to temperature rise caused by hydration heat, while the concrete surface shrinks because of temperature drop under the influence of ambient temperature. In this strain state of internal swelling and external shrinkage, tension stress will develop on the surface of concrete. In the initial stage of concreting (1 ~ 3 d) when concrete has not become a rigid body, concrete would have a rather small elastic modulus and certain plasticity, this tension stress could be released and no temperature cracking would develop. While in the following 3 ~ 28 days of setting when concrete has turned out to be a completely rigid body, tensile strength and elastic modulus of the concrete would rise with the aggravation of hydration heat inside the concrete mass, but the tensile strength may be insufficient to resist the tensile stress caused due to temperature difference. In this circumstance, cracks are likely to appear on the concrete surface under the action of tensile stress and with cracking development in depth, and deep cracks or penetrating cracks may endanger the safety of the dam. Statistics shows that more than 70% of thermal cracks would come out during this time period. It is concluded that concrete temperature control in the early stage of concreting would play a critical role in control of thermal cracking of concrete.

4.3 Design criteria for concrete temperature control

4.3.1 Concreting temperature

Thetemperature rise of concrete from the mixer outlet to the placed location before being covered up shall not exceed 5 ℃ from April to October and 3 ℃ in other months, respectively. The concreting temperature shall not exceed the values shown in Table 5 to ensure that maximum temperature of concrete should meet design requirements.

Table 5 Concreting temperature of the dam RCC

Region	Month		
	April ~ Sep.	March, Oct.	Nov., Dec., Jan., Feb.
Foundation constrained zone $(0 \sim 0.4)L$	17 ℃	17 ℃	Normal temperature
Non – restrained zone at the upper part (above $0.4L$)	19 ℃	19 ℃	Normal temperature

4.3.2 Temperature difference

(1) Temperature difference in dam foundation

Allowable temperature difference in the dam foundation is shown in Table 6.

Table 6 Allowable temperature difference in dam foundation   (unit: ℃)

Highly constrained zone Allowabletemperature	Gravity dam section		Overflow dam section		Bottom outlet dam section	
	$(0 \sim 0.2)L$	$(0.2 \sim 0.4)L$	$(0 \sim 0.2)L$	$(0.2 \sim 0.4)L$	$(0 \sim 0.2)L$	$(0.2 \sim 0.4)L$
RCC inside the dam / Full working–face concreting	14	16	14	16	14	16

4.3.2.2 Temperature difference between upper and lower concrete lifts

When concreting is conducted on an aged concrete face, temperature of the newly poured concrete within $L/4$ above the aged concrete face shall be controlled based on the temperature difference between upper and lower concrete lifts and limited to 15.0 ℃. The new lift over aged concrete shall rise uniformly with short time intervals so as to avoid resulting in new aged concrete.

4.3.3 Internal and external temperature different and allowable maximum temperature

The allowable maximum temperature defined based on the temperature difference criteria and the dam body stable temperature is shown in Table 7. For individual dam sections with outlets, the allowable maximum temperature of concrete within 15 m below the floor and 15 m above the roof of the outlet should be controlled as per the criterion of 33 ℃. Generally, the height difference of adjacent two dam sections shall not be larger than 10 ~ 12 m, the interval between two lifts shall not be too long and the exposed side faces should be protected in winter.

Table 7 Allowable maximum temperature and temperature difference between inside and outside of RCC (unit: ℃)

Zoning of temperature control	Allowable temperature difference	Equilibrium temperature	Maximum temperature	Allowable temperature difference between inside and outside
Highly constrained zone $(0 \sim 0.2)L$	14	17	31	12
Weakly constrained zone $(0.2 \sim 0.4)L$	16	18	33	14
Non – constrained zone above $0.4L$	18	18	36	16

Note: In case aged concrete is formed at the concreting place, the maximum temperature shall also satisfy the control criteria for temperature difference between upper and lower lifts.

5 Technical requirements and measures for temperature control for concreting dam body

5.1 Technical requirements and measures

The maximum length of RCC blocks was defined as per the full – face placement requirement, i.e. 102 m long. Technical measures applied for temperature control are described as below.

5.1.1 Scheme of RCC placing in different months

(1) From April to September, each lift would be poured and compacted continuously to reach a final thickness of 3.0 m (the lift comprises 10 layers with each compact thickness of 0.3 m), the placing temperature would be controlled not higher than 17.0 ℃, with an interval of 7 days before placing the overlying lift; chilled water (10 ℃) would keep circulating for 20 days for cooling in the initial stage of setting, in addition to water spraying and heat insulation for the work face.

(2) From October to next March, each lift would be poured and compacted continuously to reach a final thickness of 3.0 m (the lift comprises 10 layers with each compact thickness of 0.3 m), the placing temperature would be controlled not higher than 19.0 ℃, with an interval of 7 days before placing the overlying lift; natural river water would keep circulating for 20 days for cooling in the initial stage of setting.

(3) Cooling water tubes are arranged in the pattern of 1.5 m × 1.5 m. The length of one single serpentine circulating tube shall not exceed 250 m. Flowing direction within the tube shall be reversed once every 24 h. Temperature drop of concrete shall not exceed 1 ℃ per day.

5.1.2 Temperature control requirement for the overflowing gap part

(1) In flood seasons, before river water flows through the gaps reserved on the overflow dam

section, the dam concrete within 20 m at least below the overflowing level shall be cooled down to 20~23 ℃.

(2) Before overflowing through the gaps, the concrete within 20 m at least above the water passing level at the internal side of bottom outlets shall be cooled down to 23 ℃.

(3) Two layers of mesh reinforcement are arranged below the overflow surface of the gaps to prevent concrete cracking.

5.1.3 Concrete surface curing

(1) In high temperature seasons, curing with flowing water is applied to concrete surface and the curing shall last 1~2 days when encounter maximum temperature, afterwards concrete can be cured by sprinkling water.

(2) Curing with sprinkling shall start within 6~18 hours after the placing is completed; the curing shall last at least 28 days in dry and hot climatic conditions. Curing of horizontal construction joints shall not stop until placing the overlying lift.

(3) In the process of construction, the RCC placing face must be kept moist. The working face under construction and the working face compacted shall be protected from inflowing of extraneous water.

5.1.4 Surface heat preservation for concrete

(1) Concrete surface shall be covered up promptly with heat insulting material with equivalent heat exchange coefficient of $\beta \leqslant 20$ kJ/(m² · h · ℃) for heat insulation after concreting is done, and the heat insulating cover can be removed gradually until placing or laying cooling water tubes for the overlying lift.

(2) From November to next march, EPE coiled material with $\beta = 20$ kJ/(m² · h · ℃) shall be provided after concrete is placed into the working face or the concrete has been compacted and leveled, and mist spraying shall be applied to the working face. After one lift is done, the concrete surface shall be covered up at once with heat insulating material with equivalent heat exchange coefficient of $\beta \leqslant 8$ kJ/(m² · h · ℃). For concreting in winter with long intervals, all the exposed surfaces of concrete and all outlets and holes shall be covered up with heat insulting material with equivalent heat exchange coefficient of $\beta \leqslant 8$ kJ/(m² · h · ℃). The first stage of cooling with river water shall last for 20 days. In the case of abrupt drop in air temperature, heat insulating material with equivalent heat exchange coefficient of $\beta \leqslant 8$ kJ/(m² · h · ℃) shall be applied as well.

(3) The formwork removal time shall not be earlier than 7 days from the date of concreting completion. Formwork removal is forbidden in the case of sudden drop in air temperature.

(4) Frequency of weather forecasting shall be increased in the season with frequent sudden temperature drops. The surface of newly poured concrete (with age not more than 28 days) and the downstream face of the dam shall be protected with heat insulting material with equivalent heat exchange coefficient of $\beta \leqslant 8$ kJ/(m² · h · ℃) when it is forecasted that a sudden temperature drop is coming.

(5) To prevent vertical cracking on the upstream face induced by temperature fluctuation in-

cluding sudden drop in air temperature, heat insulation material with $\beta \leqslant 8$ kJ/($m^2 \cdot h \cdot ℃$) shall be applied tightly onto the upstream face of the dam all the year around. Polystyrene foam plastic plates of 25 mm thick or so can be stuck onto the heat insulation material if necessary.

(6) From April to September, the ex - mixer temperature of concrete shall be controlled not higher than 12 ℃; EPE coiled material with $\beta = 20$ kJ/($m^2 \cdot h \cdot ℃$) shall be provided after concrete is discharged into the working face or the concrete has been compacted and leveled. Mist spraying shall be applied to the working face in the process of concreting; after the concreting is completed, concrete curing with running water and sprinkling shall be adopted respectively for the surface and side faces of the lifts. The first stage of cooling with chilled water (10 ℃) shall last for 20 days. Shade - shed can be established over the concreting face if necessary.

Fig. 4 View of dam construction (by Jan., 2013)

5.2 Intelligent control of water cooling for concrete temperature and crack - prevention simulation and back analysis

In order to seek a better solution for concrete temperature control and crack prevention in this dry and hot valley of the Jinsha River, an intelligent temperature control system was adopted in an exploratory manner during the construction of the Ludila RCC dam. Since being put into service in December 2012, the system has successfully functioned and provided resultful guidance to the concrete temperature control at site:

(1) Automatic measurement, wireless transmission, automatic input and analysis of water cooling data and concrete temperature real - time monitoring data, in - time early - warning of concrete temperature.

(2) Forecasting of temperature curve; the forecasted curve would be generally consistent with the surveyed curve, which effectively guaranteed the construction quality and safety of the RCC dam.

(3) Intelligent control of water cooling was applied for an RCC dam for the first time.

(4) With the help of the intelligent system, RCC placement in successive rising of 21 m be-

came a reality; concrete temperature could fully meet the temperature control criteria and no harmful cracking occurred.

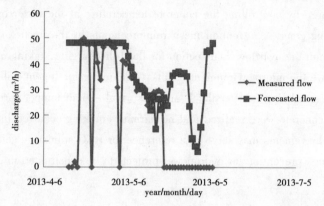

Fig. 5 Sketch of forecasted flow and measured flow of cooling water tube (10#-4-2)

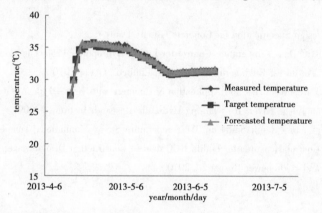

Fig. 6 Sketch of forecasted temperature, target temperature and measured temperature (10#-4-2)

(5) Through gathering relevant data by the intelligent control system, including air temperature, concreting time, concreting temperature, cooling water temperature and cooling time, the real – time relation between flow rate and temperature of cooling water, as well as the concrete block temperature and thermal stress could be well observed. Dynamic simulation and back analysis of temperature stress during construction could be made for real – time calculating the temperature field and stress field of dam blocks, proposing preventive temperature control measures, and timely feedback of adjustments to the temperature control measures, which have yielded a good result in control of harmful cracking, providing a favorable basis for the initial impoundment and final acceptance of the project.

6 Conclusions

Ludila Hydropower Project is located in the dry and hot valley of Jinsha River on the midstream. Sandstone aggregate proposed for the RCC dam construction is characterized by probable alkali active reaction. Applying high – quality set retarding superplasticizer, air entraining agent, admixture PL (phosphorus slag powder + limestone powder) and moderate heat Portland cement, by taking sandstone as coarse aggregate and non – alkali reactive syenite as fine aggregate and through optimization of mix proportion, the working performance of concrete was improved significantly. Further more, by controlling the concrete temperature at the mixer outlet and during its placement, providing crack – prevention mesh reinforcements at the upstream dam face and the gap, which are set on the overflow dam section for flood discharging, taking measures to preserve moisture and improve the microenvironment of lifts, conducting intelligent water cooling and making 3D stress field simulation and feedback analysis, real – time temperature control and crack prevention of dam concrete were realized and no harmful cracking occurred in the dam body. All these works and achievements may serve as a reference for RCC admixture selection and temperature control and crack prevention for hydropower projects to be constructed in dry and hot valleys.

References

[1] Zhu Bofang. Temperature Stress and Temperature Control for Mass Concrete [M]. China Electric Power Press, 1999.
[2] DL 5108—1999 Design Specification for Concrete Gravity Dams.
[3] Design Report of RCC Dam Temperature Control for Ludila Hydropower Project in Construction Design Stage [R]. Powerchina Northwest Engineering Corporation Limited, Oct., 2010.
[4] Lu Cairong, et al.. Report of Test and Application of Concrete with New – Type Admixture PL in Ludila Hydropower Project on Jinsha River[R]. Nanjing Hydraulic Research Institute, 2009.
[5] Zhang Guoxin, et al.. In – depth Study on 3D Temperature Stress Simulation, Temperature Control Criteria and Temperature Control Measures for Ludila RCC dam in Construction Design Stage[R]. China Institute of Water Resources and Hydropower Research, 2010.

Geomembranes on RCC Dams: A Case History After 13 Years of Service

Scuero A[1], Vaschetti G[1], Jimenez M J[2], Cowland J[3]

(1. Carpi Tech, Switzerland; 2. Isagen, Colombia; 3. Carpi Asia Pacific, Hong Kong, China)

Abstract: The paper discusses an RCC dam project where an exposed geomembrane system was installed in 2002 and has been successfully performing for now 13 years. The RCC dam, Miel I, is located in Colombia and when constructed it was the highest RCC dam in the world. 188 m high, the dam is used for power production. The RCC mix has a cement content of 85 to 160 kg/m^3. To meet contractual schedule, the original design of an upstream face made of slip formed reinforced concrete was changed to a Carpi drained exposed PVC geocomposite system, placed on a 0.4 m thick zone of RCC material enriched with cement grout and vibrated. Placement of the geocomposite system in horizontal stages allowed starting dam impoundment while RCC placement was still ongoing at higher elevation. The geocomposite system was completed in 2002, before complementary concrete works were finished. ISAGEN, the owner of the dam, constantly monitors the dam and the geocomposite system. The measured leakage has never surpassed the historical values, or overcome the maximum allowed drainage design values. After 11 years of good performance, in 2013 ISAGEN decided to have a comprehensive monitoring of the general conditions and weathering behaviour of the geocomposite system. Objectives of the investigation were to assess the reason of small cuts that had been detected and repaired (in total 15 in 11 years on a 31 000 square meters surface), to assess the conditions of the PVC geocomposite, ascertaining at which extent its characteristics had changed over service, and to evaluate if the weathering behaviour was in accordance with the expected durability. Inspection at the upstream face and in the drainage galleries was followed by sampling and testing of the exposed geocomposites in the laboratory, to compare the physical and chemical properties of the aged geocomposite sample collected at the dam with those of the same geocomposite as manufactured. Evaluation of test results, based also on most recent research on geomembrane weathering mechanisms, completed the investigation. Results confirmed that the behaviour of the PVC geocomposite is extremely good and fully in line with expectations. The performance of the dam in 2015, after 13 years of service, keeps being extremely satisfactory. The paper will describe the geocomposite system installed at the dam, and then discuss all steps of the investigation, and detail the test results and the conclusions drawn.

Key words: waterproofing, geomembrane, PVC, RCC dam

1 Introduction

The use of geomembranes as upstream water barriers started in 1984, shortly after the first

RCC dam was constructed at the beginning of the 1980s. Information on the available covered and exposed geomembrane systems, and on the advantages that they can provide when adopted at new construction, have been discussed by some of the authors in this same symposium in 2003[3] and 2012[4], and some case histories have been presented. The use of upstream geomembranes has been continued since then, and has been applied also to rehabilitation of existing RCC dams, both in the dry and underwater.

This paper presents the case history of a project completed in 2002, to report on the results of anupstream exposed geomembrane system after 13 years of service and monitoring.

2 Miel I: 2000 to 2002

Miel I, owned by ISAGEN, is a 188 m high and 354 m long Roller Compacted Concrete (RCC) dam used for power production. When constructed in 2002, it was the highest RCC dam in the world.

2.1 The dam

The dam, located at elevation 455 m a.s.l. in Colombia, in a narrow canyon of a hot tropical region with heavy rainfall, is the main component of the Miel I hydroelectric project, which also includes a powerhouse with a capacity of 375 MW and average annual production of 1 460 GW · h. The RCC dam mix is of the low to moderate cement content type, 85 to 160 160 kg/m^3, which was placed in 30 cm high lifts. For more information on the dam design and construction, refer to Marulanda et al.[2].

The design of the dam included a slip-formed reinforced concrete face. To meet contractual schedule, the upstream impervious facing was modified: the reinforced concrete was substituted by a 0.4 m thick zone of RCC enriched with cement grout, on which a Carpi drained exposed Polyvinylchloride composite geomembrane (= geocomposite) system was placed to enhance watertightness. This double water barrier was deemed necessary due to the 188 m height of the exposed face of the dam.

Works on the RCC dam began in April 2000 and were completed after 26 months.

2.2 The waterproofing system

The waterproofing liner is a SIBELON® geocomposite, consisting of PVC geomembrane laminated to a 500 g/m^2 non-woven polypropylene geotextile. In the lowest part of the dam, from elevation 268 m to elevation 330 m, the PVC geomembrane is 3 mm thick, while from elevation 330 m to elevation 450 m it is 2.5 mm thick. The entire upstream face is 31 500 m^2.

The geocomposite is anchored to the dam face by parallel vertical tensioning profiles (Carpi patent), placed at 3.70 m spacing. The assembly of tensioning profiles consists of two main components: the first component is a galvanised steel profile, which was embedded by the main Contractor into the RCC lifts as they were being placed. The second component, placed over the installed geocomposite and connected to the first component, secures and tensions the geocomposite on the upstream face. It consists of a stainless steel profile, placed by Carpi over the installed geocomposite sheets. Coupling of the two profiles, thanks to their geometry and to a three-component connector, secures and tensions the geocomposite on the upstream face. To avoid water infiltration at the connector, the profiles are waterproofed with SIBELON® cover strips (Fig. 1, excerpt of ICOLD Bulletin 135[1]).

From elevation 268 to elevation 358 m a. s. l, the stainless steel profiles have a central reinforcement in order sustain high water pressure.

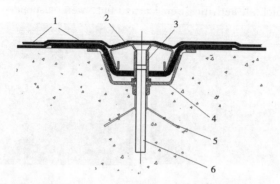

1. PVC geocomposite, 2. Tensioning stainless steel SS profile, 3. Connector,
4. Embedded galvanized steel profile, 5. Galvanised steel anchor wings, 6. Anchor

Fig. 1 Scheme of the tensioning system for face anchorage of the geocomposite (Carpi patent)

Fig. 2 shows installation of the system. This configuration was the state – of – the – art configuration at the beginning of the 2000s, i.e. the U – shaped component of the tensioning profile assembly was attached to the formwork and embedded in the RCC lifts, while in current projects this component is placed after the RCC is completed.

Fig. 2 Left: U – shaped profiles and drainage collector attached to the formworks are embedded in RCC. Middle: U – shaped profiles after embedment appear as vertical drainage grooves. Right: tensioning profiles are waterproofed with PVC cover strips

The geocomposite has a full – face integrated drainage system behind it. The drainage system consists of the gap between the geocompsoiteand the dam face, allowed by the face anchorage system, of the geotextile laminated to the PVC geomembrane, of the vertical conduits formed by the tensioning profiles, of a peripheral collector embedded in the RCC, of the transverse discharge pipes discharging into the gallery, and of ventilation pipes assuring water flow at ambient pressure. The drainage system is divided into 4 horizontal sections (compartments), each discharging in the gallery located at its lower point. Each horizontal compartment is in turn divided into vertical compartments with separate discharge. In total there are 45 separate compartments that allow accurate monitoring of the behaviour of the waterproofing system. At contraction joints, two layers of sacrificial geocomposite provide support to the liner, as shown at left in Fig. 3. The geocomposite system was installed in 6 horizontal sections. A movable railing system was used to install the system concurrent and independent of RCC activities. The railing system was attached to the dam at approximately 90 m above foundation, then moved to some 140 m above foundation. The travelling platforms, from which all activities were carried out, were suspended at the railing system.

Fig. 3 Left: geocomposite installation in lower section, while above the support geocomposite is being installed on the contraction joint from travelling platform at left. Right: installation of the geocomposite with travelling platforms suspended at the movable railing system placed at elevation 407 m, while RCC placement is ongoing above the railing system

Construction of the grouting plinth was made following placement of the RCC. A SIBELON® geocomposite, placed over the completed RCC lifts and over the natural excavation rock, waterproofs the plinth. The liner waterproofing the plinth is watertight connected to the liner waterproofing the upstream face by a mechanical seal. The seal achieves watertightness by compressing the geocomposite with 80 mm × 8 mm stainless steel (Fig. 4) batten strips on the concrete regularised with epoxy resin; rubber gaskets and spice plates assure that compression is evenly distributed. This type of seal, tested at 2.4 MPa, is placed also at crest, to resist water overtopping.

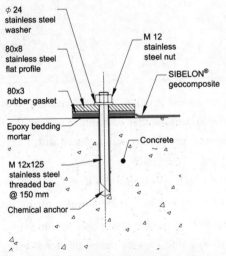

Fig. 4 Scheme of perimeter seal

Installation of 31 453 m² of exposed geocomposite started on October 8, 2001, and was completed on September 7, 2002. Staged installation of the geocomposite waterproofing system allowed early impounding while the dam was still under construction (Fig. 5). The change in design allowed meeting the schedule, and saving several tenths millions US dollars, because of faster completion and earlier power generation.

Fig. 5 Left: the geocompistee installed concurrent with RCC placement reduced times for completion. Right: geocomposite already installed in the lower sections allowed impounding the reservoir and testing the machinery while RCC placement and waterproofing works were ongoing above

3 Miel I: 2002 to 2015

3.1 The monitoring system

As outlined before, the full face drainage system behind the geocomposite was constructed with the main purpose of allowing monitoring the performance of the geocomposite system. The 45 drainage compartments discharge respectively in gallery 9 (compartments of the lowest section), and in galleries 8, 4, and 1 (upper three compartments). In the lowest section, the drainage system is extremely accurate: the central part of the dam has one compartment for each line of vertical tensioning profiles and one compartment for each abutment. In the upper three sections, the central part of the dam has one compartment for each monolith, and one compartment for each abutment.

The purpose of making such unusually numerous separate compartments was to facilitate locating any malfunctioning of the system by monitoring the discharge flow of each of the 45 compartments. Additionally this system, should a failure occur in the geocomposite, guarantees that infiltration water can be detected and managed. Water infiltrated via a damage in the geocomposite is captured by the face drainage system at the upstream face, and from there discharged to the nearest gallery, avoidig that it can infiltrate the dam's body. The possibility of managing infiltration water (defined path where the water can flow) can make the difference between failure and stability in a dam.

The procedure established by ISAGEN for monitoring the geocomposite system was that after establishment of a fairly constant "normal" flow for each discharge, the flow at all discharge pipes would be measured once per month during the dry season at low reservoir levels, twice per month during the rainy season, and twice per week after the reservoir reaches elevation 43 900 m a. s. l.

These routines were defined following the analysis of the inflow and the geotechnical instrumentation behaviour.

3.2 Monitoring results

ISAGEN has been constantly monitoring since 2002 the dam's and geocomposite system's performance. The maintenance group is constantly looking for any deviation from the "normal" flow that is considered a signal of anomaly.

In 2003, after reservoir's impoundment, the recorded rate of leakage from the whole geomembrane system was 3.89 L/s, registered in 2013 with a reservoir level of 446.47 m a. s. l. The average leakage registered since then has been of 2 L/s for the geocomposite's drainage system, and of 25 L/s from the abutments. Such levels are below the design parameters 9.7 L/s from the geomembrane system and 30 L/s from the abutments. The measured leakage has never surpassed the historical values (with stable lecture through 13 years), or overcome the maximum allowed drainage design values defined during the design stage.

Since an abnormal drainage flow does not necessarily mean there is a damage in the liner or in its fastening system (as stated by ICOLD Bulletin 135[1], "··· an abnormal rate of leakage would suggest a defect in the geomembrane, while in reality the drainage system is mainly collecting water which may come from different sources other than a defect through the geomembrane. For instance, water infiltrating through fissures and bypassing the perimeter seal, or infiltrating from foundations, or from crest. Thus it is possible that the amount of water at the discharge point is high, while the dam body is totally dry"), in addition to the inflow rate monitoring, ISAGEN performs an external visual inspection of the geomembrane. The inspection is executed every time that the reservoir level changes of 3 m. This visual inspection allows the maintenance group to detect any problem in the geomembrane. The few small cuts detected occurred mainly during the rainy season, when debris arrive to the dam and bring wood or other materials that can produce the cuts.

3.3 Additional monitoring campaign of 2013

Up to 2013, only fifteen small cuts at the upstream face had been detected and repaired by ISAGEN over a 31.000 m^2 of surface. The largest one was 5 mm wide and 15 cm long. These cuts did not constitute danger or compromise the stability of the geocomposite system. The dam was unaffected due to the full face drainage system. Moreover, measured leakage never surpassed the historical values (stable lecture through 11 years), or overcame the maximum allowed drainage values defined during the design stage.

Nevertheless, in order to have a broader understanding of the behaviour of the system, in summer 2013 ISAGEN decided to perform a more comprehensive monitoring campaign, in addition to the routine one, to assess the general conditions and the weathering behaviour of the waterproofing geocomposite at Miel I. This was made via a site inspection, and with collection and laboratory testing of geocomposite samples in service for 11 years at the dam and of virgin samples, to assess the ageing process of the geocomposite.

In September 2013 visual inspection from a small boat that drew very close to the geocomposite was performed at the upstream face by ISAGEN and Carpi. The inspection was executed when the reservoir was at 428 m a. s. l. (26 meters from the crest). It was agreed that if any problem would be detected an underwater inspection would have been carried out on the rest of the upstream face. During the inspection, the total leakage from the geocomposite system was 0.41 L/s, which is quite low if it is compared with the 2 L/s average that has been recorded over the last 11 years. In addition to vissual and tactile geomembrane inspection, the gallery conditions and the drainage system were verified, and the downstream face inspected. The galleries and drainage system, and the downstream face, were totally dry, confirming that there was practically no water infiltration in the dam. Of the small cuts/holes found and repaired over the years, eight were above water level and fully visible during the inspection. The visual inspection of the repairs showed that they had been performed correctly, and that there were no discontinuities or heat – fusion problems between the patch and the surface of the geomembrane installed in 2002. Watertightness had been perfectly restored. The little leakage measured at the time of the inspection (0.41 L/s) and the dry appearance of the downstream face lead to the conclusion that there were no damages below elevation 428 m a. s. l. With this evidence, it was decided that there was no need to perform an immediate underwater inspection below elevation 428 m.

During the inspection, a 50 cm × 50 cm sample of SIBELON® CNT 3 750 was cut one meter above the water level, in the drawdown zone, at the right end side of the upstream face, near the foundation plinth. After cutting, imperviousness was restored by watertight welding a SIBELON® geomembrane patch over the sampling area. A sample of geocomposite SIBELON® CNT 3750 left over from the time of installation was collected at ISAGEN storehouse. Since this sample was totally protected from UV rays over the eleven years that had passed, its characteristics can be considered identical to those of the virgin material as manufactured. This sample was used for laboratory testing in order to compare the physical and chemical properties of the 11 years old SIBELON® CNT 3750 geocomposite sample collected at the upstream face (Fig. 6) of the dam and SIBELON® CNT 3750 geocomposite as manufactured.

Fig. 6 **Dry downstream face, and geocomposite sampling at upstream face**

3.4 Test results

Testing was performed at the laboratory for geosynthetics of CESI (former Enel. Hydro) in Italy. This company has no affiliation with ISAGEN or CARPI and is considered as an independent source of information. The goals of testing were to ascertain the main physical and chemical properties of the SIBELON® CNT 3750 sample collected at the upstream face of the dam (aged material, 11 years), of the SIBELON® CNT 3750 sample collected at the owner's storehouse (virgin material, and to compare the values obtained in order to understand the weathering performance of the waterproofing geocomposite installed on the upstream face of the dam after 11 years of service. Results are shown in Table1.

Table 1 Test results

Type of test Measured property	Standard	Unit	Sibelon® CNT 3750 aged	Sibelon® CNT 3750 virgin
Tensile Longitudinal direct. Tensile strength Elongation at break Transverse direct. Tensile strength Elongation at break	UNI 8202/8: 1988	kN/m % kN/m %	15.11 244.4 13.05 223.7	13.32 246.4 11.73 234.5
Extraction of plasticisers Plasticisers	UNI ISO 6427: 2001	%	25.73	29.17
Water vapour permeability Permeance Coefficient of permeability	UNI 8202/23: 1988	g/(m² · d) m/s	0.984 1.38×10^{-13}	1.262 1.83×10^{-13}
Thickness Nominal thickness Coefficient of variation	UNI 8202/6: 1988	mm %	2.61 2.48	2.71 0.43
Density Density	UNI 27092: 1972	g/cm³	1.286 1.271	
Shore hardness	UNI 4916: 1984	n°	86.8 81.6	
Cold bending Longitudinal Transverse	UNI 8202/15: 1984	℃ ℃	−40 −35	−40 −40
Dimensional stability Weight variation longitudinal Weight variation transverse	UNI 8202/23: 1988	% %	−2.53 −1.37	−0.92 0.37

The properties most relevant to determine the conditions and efficiency of a PVC geomembrane liner installed on a hydraulic structure are the tensile properties, the plasticisers' content with resulting Shore hardness, the permeability, and thickness and density. The tensile properties of the aged sample are remarkably good, showing an extremely small change: elongation at break of the aged sample is 244.4% (longitudinal) and 223.7% (transverse), while that of the virgin sample is respectively 246.4% and 234.5%. The plasticisers content, which is one of the most important parameters of the ageing process, shows a slight decrease (from 29.17% to 25.73%), with a corresponding slight increase in Shore hardness. The decrease of plasticisers' content documented in a similar application and after the same time of operation (11 years) has been about 80% greater. Other applications have shown that after about 10 years the decrease in plasticisers' content slows down: for example, in the lapse of time between 10 to 23 years and 16 to 29 years, the documented decrease in plasticisers' content was between about 4% and 7%. So in future years performance of the geomembrane in respect to plasticisers' loss should be even better. With regard to the waterproofing properties, the results of the water vapour permeability test show a slight decrease of the permeability coefficient, thus an improvement of the watertightness of the geomembrane. The decrease in thickness is small (from 2.71 mm to 2.6 mm), thickness remaining higher than the nominal minimum thickness of the geomembrane component (2.5 mm). The value of density (from 1 271 g/m^3 to 1 286 g/m^3) also remains substantially stable.

3.5 Final evaluation

The physical appearance of the geocomposite was found to be according to expectations. The changes in properties most relevant to the ageing behaviour of the geomembrane were totally in line with standard behaviour and with the values that are statistically exhibited by exposed SIBELON® CNT 3750 geocomposites installed in exposed positions on other dams. The damages, their number and extent, if compared to the number of years of service and to the total surface of the upstream face, must be considered extremely low, and do not represent any risk for the service life of the geomembrane. Concerning the cause of damage, the location and appearance of the small cuts seem to indicate the possibility that they could have been caused by mechanical events such as falling of loose part of concrete from the spillway, or impact by large sharp floating objects. There is no reason or evidence that they can be relevant to a significant alteration of the properties of the geocomposite. Inspection of the repairs confirmed that the methods followed by the owner's maintenance personnel is totally adequate to guarantee locating any flaws and performing correct repair

4 Miel I today, and conclusions

Since 2013, no additional repairs have been made. Inspection of the galleries, piezometers close to the upstream face, and other dam instrumentation, do not show wet areas, signs of leakage or of any problem.

This case history is a further confirmation of the effectiveness of exposed geocomposite systems as watertight facings in RCC dams.

References

[1] ICOLD - International Commission on Large Dams, Bulletin 135, Geomembrane sealing systems for dams - Design principles and review of experience, Paris, France, 2010.

[2] A. Marulanda, A. Castro, N. R. Rubiano. Miel I: a 188 - m high RCC dam in Colombia, Hydropower & Dams, Issue three, 2002.

[3] A. M. Scuero, G. L. Vaschetti, Synthetic geomembranes in RCC dams: since 1984, a reliable cost effective way to stop leakage, Proceedings of the 4th International Symposium on Roller Compacted Concrete (RCC) Dams, 2003.

[4] A. Scuero, G. Vaschetti, Three recent geomembrane projects on new RCC dams, Proceedings of the 6th International Symposium on Roller Compacted Concrete (RCC) Dams, 2012.

Inga Project in Drc – design Considerations to Facilitate the Construction in Two Stages of the Bundi Dam

Arnaud ROUSSELIN

(ELECTRICITE DE FRANCE (EDF) Hydro Engineering Center, FRANCE)

Abstract: The hydropower potential of the Congo River at Inga site has been identified since the beginning of the 20th century. The Inga site is a 32 km stretch of the Congo River, located approximately 280 km downstream of Kinshasa in the Democratic Republic of the Congo (DRC). This stretch of river naturally drops by 97 m in elevation, creating an impressive succession of natural cascades. With a mean annual flow of 40 800 m^3/s and a gross head that may be raised to 150 m, the hydroelectric potential of the site is considerable and the total installed capacity can exceed 44 000 MW. Today, two power plants, Inga 1 and Inga 2, built in the 70's and 80's, provide a respective capacity of 350 MW and 1 400 MW.

The present article is based on the results of the Feasibility Study completed in 2013 by the AECOM – EDF Consortium. This study proposes a phased development of the site in order to adequately respond to the projected short and long term energy demands of the DRC and countries of the African sub – region.

One major component of the Grand Inga Project is the Bundi RCC dam, used to close the Bundi Valley on the right bank of the Congo River. Considering the size of the dam, it will be constructed in two stages.

In the first step of development of the site (Low Head Stage), about 5 millions of m3 of concrete are required to erect the Bundi dam to a crest elevation El. 172.5 m. The reservoir will be fed by a 12 km long channel diverting up to 6 000 m^3/s into the Bundi Valley. The dam will incorporate the water intake for Inga 3 powerhouse, which is located at the toe of the Dam and equipped with a capacity of 4 800 MW.

In the High Head Stage of development of the site, the dam will be heightened by 37.5 m in order to create the final reservoir for Grand Inga Project, featuring a total of seven powerhouses totalizing a generating capacity exceeding 40 GW.

At its final crest elevation El. 210 m, the Bundi dam total concrete volume will reach 9 millions m^3, becoming the largest RCC dam in the world.

The paper will focus on the technical design features of the dam to facilitate the future heightening of the dam and to minimize the impact of the works on the generation of Inga 3.

Key words: DRC, INGA, Bundi Dam, heightening

1 Introduction

Since the first studies where performed in the 1950s, it was apparent that successive phases of development, executed as a function of the energy demand, were required for the development of a site with a hydopower potential exceeding 40 GW. The crucial phase toward harnessing the full potential of the site, the Grand Inga Phase, is the construction of the dam crossing the Congo River. This phase is the Gordian Knot of the project, creating a roadblock for its realization due to the massive investments and large energy market required to make it feasible in terms of financing.

2 Context of the study

The first two phases of the development of the Inga site, Inga1 (350 MW) and Inga 2 (1 400 MW), were completed in 1971 and 1982, respectively. Since then, no major new power plant has been constructed in the country. As a result, the DRC faces an energy deficit today, which is accentuated by the current dilapidated state of the equipment and sedimentation problems in the intake channels of Inga1 and Inga2. In order to resolve this problem and to jumpstart its economy, the Congolese State intends to launch the next phase of development of the Inga site, appropriately named Inga 3. It is in this context that the AECOM – EDF Consortium was mandated by the Société Nationale d'E'lectricité (SNEL), with financing from the African Development Bank (AfDB), to reevaluate the existing studies and propose an optimized scheme for the long-term development of the site.

3 Characteristics of Inga site

3.1 Site location

The location of the Inga site on the Congo River is shown in Fig. 1.

The catchment area of the Congo River covers approximately 3.7 million square kilometers, of which a large centre portion is equatorial forest.

3.2 Topography

The topography of the Inga site, shown in Fig. 2, is characterized by three major depressions, upon which the layout schemes are centered: a) the low-water channel of the Congo River, b) the three valleys (Grande Vallée, Sikila Valley and Nkokolo Valley) that form the flood plains of the Congo River, and c) the Bundi River Valley, which is a tributary on the west bank of Congo River. The digital elevation model of the area shows the hills and valleys that will be flooded by the projected Grand Inga reservoir.

3.3 Hydrology

Given the immense size of the Congo River watershed, and the fact that it is well distributed across the equator, the discharge at the Inga site is naturally regulated along the year. Based on one hundred years of hydrometric data, the mean annual flow is 40 800 m^3/s and the extreme minimum and maximum historic flows are respectively 22 300 m^3/s and 85 100 m^3/s.

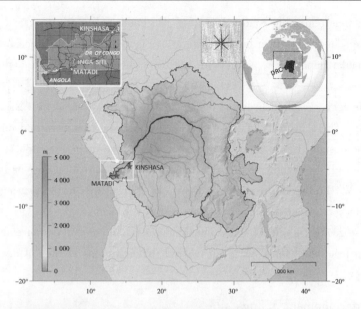

Fig. 1 Catchment basin of the Congo River (from Wikipedia) and Inga site location

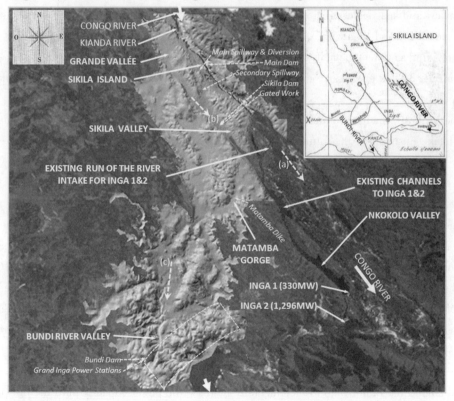

Fig. 2 Topography of the Inga site and location of the main hydraulic structures

3.4 Geology

The bedrock at the Inga site area is composed primarily of five types of rock: granite, granitic gneiss, greenstone, rhyolite schist, and micaceous & sericite quartzite. These rock formations are found in bands, aligned in a N.E. − S.W. direction. The Congo River carved its bed in the least solid rock, thus remaining confined between two bands of granitic rock.

The rhyolite schist (shown in the photograph of Figure 4) and the greenstone in the S. W. area are of particular importance, as the hydroelectric structures will be founded on these formations.

The rhyolite schist is found in layers 0.5 to 1.0 m thick, oriented in a NE – SW direction, with a S. W. dip. The schistosity of the rock has always been a concern and has been carefully considered in the design. In the Bundi Valley, the greenstone transitions from a schist formation near its interface with the rhyolite in the N. E. to a more solid formation near its interface with the granite in the S. W. Furthermore, concerns were raised in regards to the bearing capacity of the greenstone and the construction of concrete dams on this formation.

Due to these geological concerns and considerations, the construction of aboveground powerhouses and open headrace canals has been favored in the various studies of development of the site.

4 Proposed development of Inga site

Review and analysis of previous studies and geological dataconfirmed the necessity to start the development of Grand Inga with a Low Head Stage in order to minimize the cost of the first phase, while providing sufficient capacity for an export project. The scheme proposed by AECOM – EDF, shown in Fig. 3, is a progressive development of the site in successive phases that are economically competitive.

The general principle behind the scheme proposed by AECOM – EDF is to first construct a lateral intake upstream of the future main dam site, which will divert water via a transfer canal to the low head RCC Bundi dam and Inga 3 powerhouse constructed in the Bundi Valley. The installed power capacity of Inga 3 will depend on the size of the transfer canal; the reference scenario currently retained (called Inga 3 Low Head Stage) will guarantee 4 800 MW in order to meet the forecasted energy demand. In the Low Head phase, in absence of any dam across the Congo River, the Bundi reservoir level will correspond to the natural river level at the lateral intake, and Inga 3 will be operated as a run – off river plant with a reservoir at a mean level of El. 160 m.

The subsequent phase of development (called Inga 3 High Head Stage), involves the heightening of the Bundi Dam, the closure of the Congo River by a large earthfill dam and the raising of the reservoir level to its final elevation of 205 m. Inga 3 capacity will automatically increase by 2 500 MW of guaranteed power without any new turbine generator unit.

In the following phases of development, the Inga 4, 5, 6, 7, etc. powerhouses will be constructed around the Bundi Dam to incrementally increase the power capacity until the full potential of the site is reached.

Although Inga 3 units must be designed to operate from the Low Head to the High Head stages, subsequent plants of Grand Inga would be designed only for the High Head Stage.

At the final stage of development of the site, the cumulative installed capacity will exceed 44 GW and the annual energy produced will reach 335 TWh.

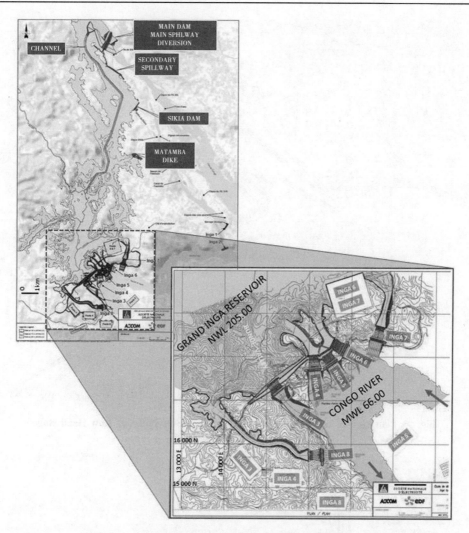

Fig. 3 Scheme overall layout proposed by AECOM – EDF

The present paper focuses on the Bundi Dam, a very large RCC dam constructed in two stages.

The plan view of the Bundi Dam at Low Head Stage is shown in the following figure.

The typical sections of the Bundi Dam across the water intake and across the RCC dam are provided in the following figures, for Low Head and High Head stages.

5 Design considerations to facilitate the heightening of the Bundi Dam

The Bundi RCC dam will be constructed in two stages. In the Low Head Stage, the dam will be erected up to the crest level El. 172.5 m and in the High Head Stage the dam will be heightened up to El. 210 m. The dam will be heightened by 37.5 m, which represents 34 % of the dam initial height. A substantial period of time may last between the two phases.

The dam and its water intake have been specially designed to minimize the impact of the heightening of the dam on Inga 3 generation. The turbines, alternators, and all the auxiliary sys-

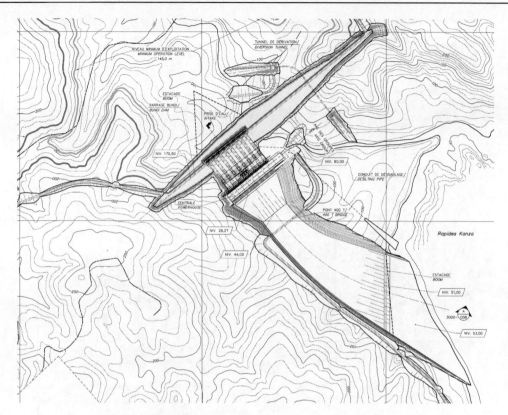

Fig. 4 Plan view of the Bundi Dam and Inga 3 Power house at Low Head stage

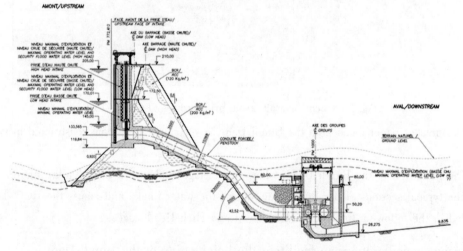

Fig. 5 Cross section of Inga 3 water intake and power house at Low Head and High Head stages

tems of the powerhouse are designed to operate on the complete range of water head which will apply during the dam heightening, up to the final level.

Following is a summary of the main provisions envisaged in order to ensure a continuous generation during the dam heightening works.

5.1 Dam profile

The dam profile has been selected in order to ensure the dam stability for both stages, while

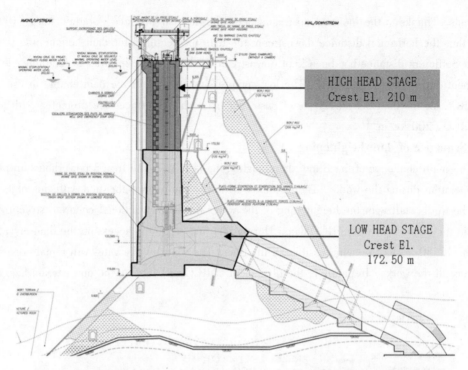

Fig. 6 Cross section of Inga 3 water intake at Low Head and High Head stages

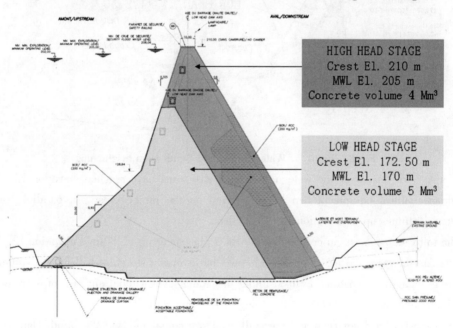

Fig. 7 Cross section of Bundi dam at Low Head and High Head stages

minimizing the overall concrete volume. The interface between rock and RCC was assumed to be cohesionless with a friction angle equal to 39°. Theupstream face is inclined $(0.92(H)/1(V)$ in the low part and $0.525\ (H):1(V)$ above El. 128 m) in order to keep the compressive stresses on the greenstone foundation below 2.5 MPa, and to take benefit of the stabilizing effects of the sediments. Thanks to the inclined upstream face, the downstream face is rather steep (0.6H / 1V), thus the horizontal distance downstream the dam toe required for dam heightening is minimized. Sufficient distance has been kept downstream of the dam toe in order to facilitate the RCC placement during dam heightening. The low compressive stresses in the dam enable to use a low paste RCC for the dam body (120 kg of paste per m^3) while the faces are protected with a high paste RCC (200 kg/m^3).

5.2 Sequence of dam heightening

The main challenge of the Bundi dam heightening is to keep the Inga 3 powerhouse in permanent operation during the works. The sequence of works has been elaborated with this objective.

The works start with the heightening of the water intake conventional concrete structure, located in the center part of the RCC dam. During this first step, the access on the dam crest at elevation 172.50 m shall remain possible all the time. The water intake gates will remain operational during all the works. In parallel, the placement of RCC will start on the downstream face of the dam.

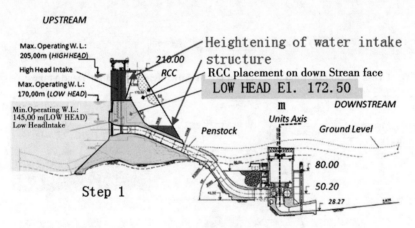

Fig. 8 Step 1 – Water Intake and Bundi Dam heightening

In a second step, once the water intake civil structures have reached elevation El. 210 m, the hydromechanical equipment can be transferred on the intake crest. In parallel, the RCC placement continues up to elevation El. 172.50 m.

The third step consists in completing the RCC placement up to its final elevation El. 210 m. Once the dam has reached elevation El. 177 m, the diversion structure will have a sufficient capacity to evacuate the construction flood under this water level without any risk of overtopping the Bundi Dam, and the final closure of the Congo River can start.

Once the Congo River Dam and the spillway 1 are completed and the Bundi Dam is heightened up to El. 210 m, the diversion gates can be closed and impounding of the reservoir up to El. 205 m can start.

At this moment, the units of Inga 3 powerhouse will start generating under the high head and Inga 3 capacity will automatically increase from 4 800 MW to 7 300 MW.

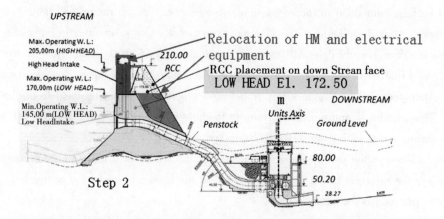

Fig. 9 Step 2 – Water intake and Bundi Dam heightening

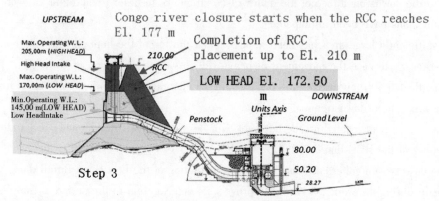

Fig. 10 Step 3 – Bundi Dam heightening

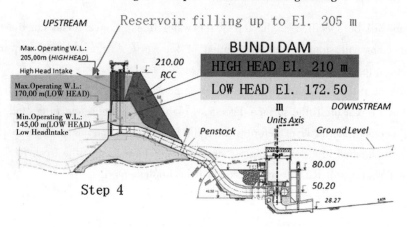

Fig. 11 Step 4 – Bundi Dam filling

5.3 Specific provisions for water intake hydromechanical equipment

During the heightening of the water intake, the gantry crane will be dissassembled. Temporary lifting equipment will be required for the cleaning of the trashracks, or for installation of the stoplog elements, if required. The intake gate hoisting system, located in a chamber above El. 170 m, will remain fully operational during the works. The electrical cabling of the gate hoisting system, will be fully embedded in the concrete above elevation El. 170 m and will not be impacted by the works. After completion of the concrete structures of the water intake and of the RCC placement on the dam, two bridges will connect the intake crest with the dam crest at elevation El. 210 m. As a first step, the gantry crane will be reassembled on the crest. The electrical and mechanical equipment will be progressively transferred, so that they can be used from elevation El. 210 m. This transfer of equipment will require to shutdown the units one by one and to install the stoplogs during the modification of the hoisting system of the water intake gates.

After completion of the modifications and testing of the mechanical and electrical equipment of all the units, the reservoir filling can start as long as the upstream works are completed.

5.4 Specific provisions for civil works

The most important aspect with respect to the heightening of the RCC dam is to ensure that the access to the downstream face of the dam will be efficient. In order to facilitate the works, the following provisions have been anticipated:

– A sufficiently large area has been kept between the toe of the dam and the usptream side of the powerhouse, to ease the movement of heavy vehicles and installation of conveyor.

– A dedicated 400 ton bridge crossing the tailrace channel will permit the access of trucks to the working area, from downstream.

– The access road located on the left bank will be progressively heightened, in parallel with the heightening of the RCC dam.

– The downstream cofferdam, used for the construction of the Low Head Bundi dam, will be required again during the works. In order to prevent its deterioration in the period of time between the Low Head and High Head phases, the cofferdam will be protected against erosion by a backfill.

– Connectors will be specified at the top of the water intake conventional concrete structures. Such connectors will simplify the installation of the vertical reinforcement required for heightening the concrete structures.

– The treatment of the RCC surface at the interface between the low head dam and the fresh RCC will consist in a repair of all the surface defects, a high water pressure cleaning, followed by the placement of a cement mortar layer before placing the fresh RCC. The concrete parapet on the dam crest will be demolished in order to create a horizontal platform to start the fresh RCC placement. The recommendations of ICOLD Bulletin 64 will be implemented.

Acknowledgements

The author would like to pay homage to the Congolese State and its Representatives for their guidance and assistance during the mandate. They express their gratitude to the African Development Bank (AfDB), who financed this study. The author also addresses special thanks to his colleagues of the AECOM – EDF Consortium; this project would not have been possible without the hard work of their dedicated personnel (engineers, technicians and support staff).

The First Two RCC Arch/Gravity Dams in Turkey

Shaw QHW

(ARQ (PTY) Ltd, South Africa)

Abstract: While Turkey was a late starter in RCC dam construction, with the first dams of the type only completed during the middle of the last decade, the country has since been catching up very quickly. Current data suggest that more than 50 RCC dams have been completed, are under construction, or are in an advanced stage of planning or design in Turkey, with the highest completed dam 177 m in height.

Two of the Turkish RCC dams are arch/gravity structures. The Kotanli and Köroğlu dams are currently being constructed by Ünal Construction for EBD Enerji on the Kura River in the north east of the country, close to the borders with Georgia and Armenia. RCC placement for the 72 m Kotanli dam will be completed in June 2015, with power generation at the 50 MW station to be commenced in July. RCC placement at the 95 m Köroğlu dam was commenced in May 2015, with the pace of placement to be increased once the RCC plant at Kotanli becomes available to augment the new plant at Köroğlu. Completion of RCC placement at Köroğlu Dam is scheduled for the end of 2016, allowing generation at the 80 MW power station to be commenced during early 2017.

In this paper, the author describes the design and construction of these two dams, with Köroğlu Dam indicating a higher degree of arch function than Kotanli Dam. Particular challenges in the dam design and construction relate to the relatively extreme climate of the area, which required interruption of RCC placement between November and March each year and innovation in the arch design to accommodate a significant temperature drop load.

Key words: RCC, Arch/Gravity, Turkey

1 Background and introduction

1.1 RCC arch dams and RCC dams in Turkey

After the first RCC arch/gravity dams were constructed in the late 1980s, it was only in China that this technology received significant attention and development, with more than 40 RCC arch and arch/gravity dams constructed in that country in the subsequent two decades; the highest being 147 m[1]. The early part of the second decade of the 21st century, however, saw the completion of RCC arch dams outside China, with the Changuinola 1 Dam in Panama, Gomal Zam Dam in Pakistan and Portugues Dam in Puerto Rico.

Although Turkey was a late starter in RCC dam construction, thesheer number of dams constructed in the country in recent years, with a great many being for hydropower production, has implied that that RCC dam construction has now become relatively commonplace. With approximately 25 RCC dams in operation, as of 2015 and much of the hydropower development in Turkey being on the deeply incised rivers in the north east of the country, a number of dam sites are topographically well-suited to arch and arch/gravity dams.

1.2 Köroğlu & Kotanli HEPs

Two such sites were identified on the Kura River, close toArdahan and a short distance upstream of the point at which the river flows into Georgia. At this location, the river flows through the rolling topography in a gorge of approximately 200 m depth, offering significant opportunities for dams for power generation. The Köroğlu and Kontanli hydropower schemes were originally a single scheme, with one dam and a long tunnel, but a significant fault was discovered on the alignment of the tunnel and it was decided to separate the scheme into two separate projects, each comprising a dam with generation at a power station located at the toe of the structure.

With both schemes designed for a discharge of approximately 100 m^3/s, the higher head of the Köroğlu HEP allows an installed capacity of 80 MW, compared to 50 MW at Kotanli.

2 Climatic conditions

The maximum operating elevation of Köroğlu Dam is at 1 723 m above sea level, while that of Kontanli is at 1 635 m above sea level and the climate of the associated region of Turkey experiences quite extreme cold during winter, with daytime temperatures commonly not rising above freezing between November/December and March.

Indicative average monthly temperatures are listed in Table 1.

Table 1 Köroğlu Dam site typical monthly temperatures

Month	Jan.	Feb.	Mar.	Apr.	May	Jun.	Jul.	Aug.	Sep.	Oct.	Nov.	Dec.	Ave.
Mean Temp (°C)	-9.9	-8.3	-2.4	5.8	10.2	14.0	17.8	17.8	13.8	7.3	0.3	-6.4	5.0

On the basis of the above, it is clear that RCC placement can only realistically occur between April and November each year, with the cold also generally preventing night-time placement during the later half of November. These cold temperatures also obviously substantially impact the design of an arch dam, with a thermal analysis typically indicating a long-term winter temperature at the core of the upper structure between 5.5 and 7 °C and in the lower core of the structure of approximately 8 °C. Applying a realistically achievable maximum RCC placement temperature of 15 °C, a net structural temperature drop of between 7 and 10 °C is accordingly apparent, before the effects of stress relaxation creep are considered.

3 Factors of influence on dam design

3.1 Kotanli

3.1.1 Topography

Kotanli Dam is located in an asymmetrical section of the Kura River valley, with a gradually inclined right abutment and a very steeply – inclined left abutment. The hill wash – covered slope of the lower left abutment provided a good location for the power station. The final dam arrangement indicates a total crest length of 201 m and a height of approximately 72 m and while the low ratio of crest length/height might suggest an opportunity for an efficient arch structure, the width of the valley at the bottom and the geotechnical rock mass conditions were not similarly favourable (Fig. 1).

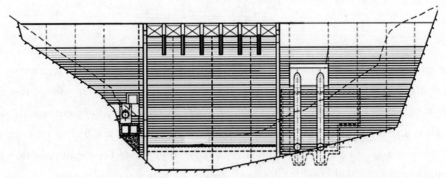

Fig. 1 Developed downstream elevation of Kotanli Dam

3.1.2 Geotechnical conditions

The geology of the Kotanli HEP site comprises volcanic rock types, primarily in the form of Basalt; with alternating layers of porous and fractured Basalt and a particularly low strength Red Basalt towards the top of the right abutment and approximately half way up the left abutment. The project geotechnical investigations indicated a deformation modulus of approximately 6 GPa for both the porous and fractured Basalt, while the equivalent estimated value for the Red Basalt was just 1.28 GPa.

3.2 Köroğlu

3.2.1 Topography

Although the topography at the Köroğlu site indicates better symmetry, a crest length/height ratio of approximately 4 (372 m/93.5 m) suggests a reduced suitability for an efficient arch dam. Furthermore, the gentler slopes of the flanks imply less space to accommodate the power station.

3.2.2 Geotechnical conditions

At Köroğlu, the Kura volcanites were differentiated into two specific geotechnical units. Unit 1 can be described as a closely jointed Basalt rock mass, while Unit 2 is a widely spaced jointed to massive Basalt rock mass. In general, the river section of the dam site comprises Unit 2 rock mass, while both abutments comprise Unit 1. Geotechnical investigations indicated a deformation modulus exceeding 17 GPa for the Unit 2 rock mass, but a modulus of only 3.3 GPa for Unit 1 (Fig. 2).

3.3 Induced joint grouting in RCC dams

Joints induced in RCC have apparently been successfully grouted in China and on Gomal Zam Dam in Pakistan. The experience in South Africa, however, produced less absolute results. The induced joints at Wolwedans Dam were grouted in winter 3 years after completion of the dam,

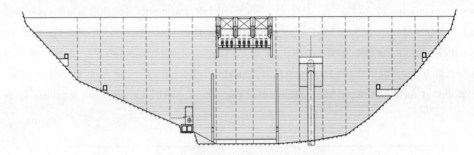

Fig. 2　Developed downstream elevation of Köroğlu dam

during which time the water level was reduced with the intention of eliminating arching. At the time of grouting, due to particularly low stress relaxation creep of the RCC, none of the important central induced joints had opened right up to crest level. Furthermore, the technique used for placing the RCC and the conventional vibrated concrete (CVC) facing resulted in an interface zone of permeable concrete. Consequently, although an induced joint had been split open through pressure grouting on the full scale trial for the dam, during the dam structure grouting, it was not realistically possible to build up adequate pressure to open closed induced joints, as grout simply entered the permeable zone between the RCC and the facing concrete.

At Wolwedans Dam, only 5 of the 27 induced joints opened significantly, with only a joint towards the top of each abutment opening all the way to the crest and several joints opening towards the base of the structure, where constraint had caused stress relaxation creep to be greatest. Grouting of the induced joints gave rise to two observed effects; the leakage originating in the permeable zone between the CVC skin and the RCC was comprehensively sealed, while the open parts of the induced joints were filled. The primary structural purpose of the joint grouting, however, was not achieved. The latter conclusion is supported by the fact that the downstream displacement of the dam structure during subsequent winters was unchanged following grouting, while the upstream displacement during summers was significantly increased[2].

Since the construction of Wolwedans Dam, the application and installation of the induced joint grouting system has been substantially improved and, of course, the use of GEVR has completely eliminated the problem related to a permeable zone immediately beneath the dam faces. Although the system installed in Changuinola 1 Dam was only locally grouted in zones that would not be accessible after impoundment, no problems were experienced in sustaining high pressures with this grouting arrangement.

3.4　The impacts of low stress relaxation creep

Although low stress relaxation creep gives rise to significant benefits in reduced long – term mass gradient temperature drop loadings, a critical medium term temperature load condition can be developed. With some expansion during hydration, as cooling of the typically thinner crest zone occurs more rapidly than the more massive base, vertical tensions can be developed over the upstream face of the loaded structure if the upper arch joints are not grouted before 2 to 3 years after dam completion. Conversely, if the joints are grouted at this time, compressions on the important parts of the structure for arching will increase with subsequent cooling of the lower mass.

4 Cementitious materials

In the north east of Turkey, no flyash is produced and its use consequently requires very significant distances of transportation; in certain cases with road – ship – road transfers. The volcanic geology of the region, however, implies that good sources of natural pozzolans exist and while one such source was located within close proximity of the dam sites, its development was not economically viable on the scale of the two dams in question. High quality Trass pozzolan is used as part of the cement manufacturing process at the Aşkale cement plant at Erzurum and this material is available for bulk delivery. Aşkale Trass indicates a relatively low specific gravity of 2.38, a high active SiO_2 content and a very high pozzolanic strength activity index of 9.6 MPa. Although milled to a Blaine value of 5 650 cm^2/g, the material, however, does not indicate the same workability benefits inherent to a good quality fly ash.

For both theKotanli and Köroğlu dams, the Trass was blended with an Aşkale CEM I (42.5) cement, which indicates a Blaine value typically of 3 800 cm^2/g. Using 85 kg/m^3 cement with 130 kg/m^3 Trass and 114 to 118 litres water, testing has reflected approximately equivalent accelerated cure and 365 day compressive strengths exceeding 20 MPa.

5 Dam structural designs

5.1 General

With both the Kotanli and Köruğlu sites indicating less than ideal geotechnical conditions and topography for an efficient arch dam and with the inherently unfavourable climatic conditions, the dam configuration optimisation exercise sought to identify worthwhile savings, assuming that some additional pre – and post cooling of the RCC would probably be required for an arch configuration. While the curved alignments created additional space to accommodate the power station and the spillway, the key benefit to be realised through arching was a reduction in total concrete quantities, with an associated reduction in input costs and benefit in an earlier commencement in power sales revenue.

For truly successful RCC dam construction, simplicity is the key element and a significant objective of the dam design for Kotanli and Köruğlu was accordingly to ensure simplicity of construction. This in turn requires that the benefit in a reduced concrete volume in conversion from a gravity dam to an arch must not be compromised through any requirement for complicated construction, or intricate post – cooling systems. Furthermore, the benefit gained through earlier completion of RCC placement should not be lost to delays in impoundment consequential to a requirement that the dam structure cools, or is artificially cooled to allow joint grouting. These objectives have to date been successfully achieved on arch dams in relatively mild climatic conditions through the beneficial use of low stress relaxation creep RCC and placement temperature control. While RCC construction on a gentle curvature and including the installation of groutable joints has been demonstrated to incur no programme, or cost impacts, in a severe climate, the long – term mass gradient temperature differentials to be accommodated are significant and consequently, the issues and requirements for post – cooling and joint grouting become of greater importance and influence.

5.2 Seismic loads

The applicable MDE pga values for the Kotanli and Köruğlu sites were established as 0.341g and 0.310g, respectively. While these earthquake accelerations are quite manageable in terms of dam design, considering the fact that the gated spillways ensure minimal floodwater rise, it was obvious that MDE seismic loadings would represent the critical structural design loading case for both dams.

5.3 Temperature loads

In the severe cold experienced at theKöroğlu and Kotanli sites, both surface gradient and mass gradient temperature effects are critical and while the problems associated with surface gradient effects are no different at any RCC dam, the consequences of mass gradient temperature effects are of particular significance in the case of an arch dam.

Assuming a 15 ℃ maximum RCC placement temperatureand an adiabatic hydration temperature rise of 20 ℃ (measurement has suggested < 18 ℃), a thermal analysis was undertaken over a n equivalent period of 20 years, with a monthly time step and applying external air and water temperatures estimated from local records. The analysis demonstrated that the Köruğlu dam structure will require a period of approximately 3 years and 6 months (Fig. 3) to completely dissipate all hydration heat and to reach a final stable equilibrium temperature condition, with subsequent temperature variations controlled only by external climatic conditions.

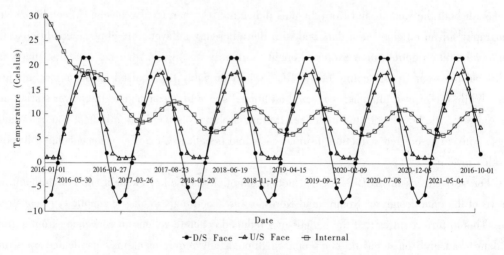

Fig. 3 Typical internal cooling for Köroğlu Dam

Such analyses were completed to establish the maximum structural temperature range which various parts of the dam structures can be expected to experience, enabling an evaluation and optimisation of pre – cooling, post – cooling and joint grouting. The thermal analyses illustrated that the minimum dam core temperatures are experienced in the month of March.

5.4 RCC early stress relaxation creep behaviour

It was intended from the outset that the smaller Kotanli HEP would be constructed first and that construction would immediately continue to Köroğlu. This would allow information to be collected through instrumentation on the stress relaxation creep behaviour of the applicable RCC mix with Trasspozzolan, using the smaller dam, which would be less sensitive to a more conservatively design approach. With the information gained, more confidence in anticipated behaviour was developed, allowing a more ambitious design for the larger, and consequently more quantity – sensitive, Köroğlu dam.

Experience has nowallowed a reasonable understanding of the typical stress relaxation creep behaviour during the hydration cycle that can be anticipated for a fly ash – rich RCC. No such information, however, is yet available for RCCs with high contents of natural, or other pozzolans. With a database of strain performance under hydration for various different RCC types, performance measured for the RCC mix at Kotanli Dam was reviewed, in conjunction with measured behaviour from other RCC dams in the region constructed with RCC mixes using the same cementitious materials. These evaluations demonstrated the RCC to take up compression strain under hydration temperature rise and to indicate no subsequent strain relaxation during extended periods at constant temperature. This observation provides a strong indication that the RCC of both Kotanli and Köruğlu dams demonstrates low stress relaxation creep, to the extent that no differentiation could be made with the behaviour of a fly ash – rich RCC, which is now known to demonstrate particularly low stress relaxation creep. Consideration, however, must be given to the fact that only strain can be measured and not stress and that no related laboratory testing of RCC paste, or RCC manufactured with the particular pozzolan/cement blend has yet been undertaken. Accordingly, it was considered expedient to retain a degree of conservatism in all design assumptions in respect of stress relaxation creep behaviour.

5.5 Kotanli

Considering the relatively poor geotechnical conditions, the unexceptional topography, the cold climatic conditions, the uncertain stress relaxation creep behaviour of the RCC incorporating Trass pozzolan and the relatively small dimensions of the dam site in conjunction with the fact that the dam was to be the first RCC arch dam in Turkey, a relatively conservative design approach was adopted for Kotanli Dam Fig. 4. To this end, a design configuration was applied that only really relies on arch action for an additional factor of safety and for stability under seismic loading.

The stability was analysed and confirmed in 2 – dimensions for all hydrostatic loadings and the operational basis earthquake loading. The stability and structural function of the dam was subsequently analysed using finite elements (FE) in 3 – dimensions under all applicable load cases, but with a specific requirement to meet design criteria for the maximum design earthquake (MDE). Furthermore, the dam structure was analysed under a significant structural temperature drop, considering the long – term winter temperature distribution related to the placement temperatures and allowing an assumed stress relaxation creep of 50 microstrain, but assuming no joint grouting.

Fig. 4　Kotanli arch/gravity dam － May 2015

In addition, the spillway crest for Kotanli Dam was constructed in CVC to a height of 3.5 m and over a length of approximately 88 m and the associated concrete was post – cooled with steel cooling pipes and the circulation of cool river water. The joints in between each CVC placement were constructed with shear keys and re – groutable joint systems and consequently, the joints were grouted after the concrete was cooled, creating a structural strut in a significant part of the dam crest that will not shrink with subsequent dam cooling. This facility further ensures an ability to grout the upper sections of the arch structure once cooling of the RCC has progressed sufficiently. With this additional facility, the dam structure was demonstrated to easily meet all design criteria under all possible conditions and in the case of all possible stress relaxation creep development.

A simple gravity dam on theKotanli site would have indicated a total concrete volume of approximately 290 000 m^3. Converting to an arch, it was possible to save a little more than 50 000 m^3, or approximately 17% of the concrete volume. With an annual construction season limited to 8 months and an average monthly RCC placement of approximately 25 000 m^3, the RCC volume reduction further allowed completion and commissioning two months earlier than otherwise would have been the case, with a consequential net revenue benefit.

5.6　Köroğlu

Neither the topography, nor the geotechnical conditions favoured a thin arch at the Köroğlu site and the optimal arrangement proved to be a simple structure with a vertical upstream face, an upstream face circular radius of 200 m (= 2.14 H) and a downstream face slope of 0.6(H):1 (V). The arch configuration applied for Köroğlu Dam resulted in a substantially reduced concrete volume; with the 580 000 m^3 required for implementation representing almost 30% less concrete than necessary for an equivalent gravity dam. This benefit translates into construction completion almost one year early, or the requirement for an RCC plant, conveyance and placement system with a 30% lower capacity.

The structural configuration for Köroğlu Dam relies on 3 – Dimensional arch function under

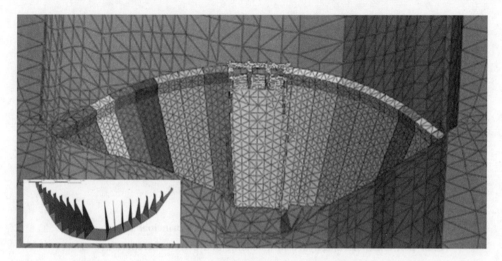

Fig. 5　Köroğlu Dam – basic layout illustrated through FE model
(including joint elements)

all operational and safety – related loading cases and accordingly, all structural analyses were completed using FE analysis and a full 3 – Dimensional model, with interface elements to represent the induced joints between blocks, as illustrated in Fig. 5.

As can be seen in Fig. 6, the Köroğlu structure demonstrates efficient arch function, although it is evident that little arching occurs across the lower sections of the taller central cantilevers.

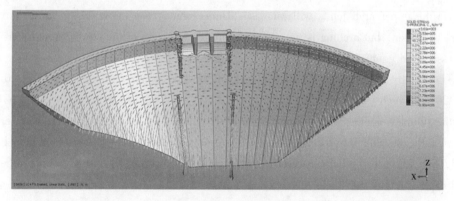

Fig. 6　Köroğlu Dam – downstream face principal stress vector plot under
normal hydrostatic loading

The structural and thermal analyses demonstrated that the arching action becomes compromised by a temperature drop equivalent to that applicable for a 15 ℃ RCC placement temperature, in conjunction with 50 microstrain stress relaxation creep and at the long – term temperatures in the month of March. As illustrated in Fig. 7, the area of compression stress is raised significantly higher up the structure and the the stress is carried down much more vertically in the lateral cantilevers.

While it is significant to note that the structural analysis reached convergence in a non – linear analysis under this load case, the associated structural behaviour cannot be considered ideal,

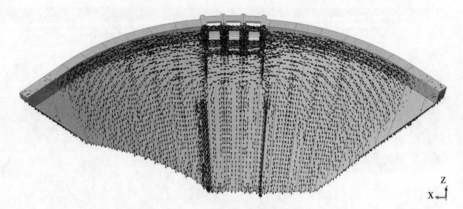

Fig. 7　Köroğlu Dam – downstream face principal stress
vector plot under normal hydrostatic loading and extreme temperature drop

with maximum crest displacement increased by almost 120%. Consequently, it was considered preferrable and necessary to include some post – cooling and grouting in the important upper section of the arch structure to ensure a strong arch function immediately on dam impoundment and under the eventual maximum cold conditions without the requirement for excessive displacements.

Analyses demonstrated that the required condition could be achievedthrough strategic post – cooling and joint grouting at a temperature of 16 ℃. Cooling only the upper part of the dam structure before impoundment will open joints in this section, where arching is required, while the lower section remains heated and expanded. As the lower section subsequently cools, compression stresses in the upper arch joints will increase. Once cooled, the joints in the lower part of the structure will open, but arching is not required through this area of the structure and consequently, no associated grouting is foreseen.

It is currently planned that the RCC placement for the Köroğlu dam will be completed by November 2016, which will imply that pre – cooling to 15 ℃ will be required in the RCC placed during the summer months of 2016 and that post – cooling must be undertaken in a strategic pattern, generally allowing approximately 4 weeks hydration before cooling. A thermal model will be applied, simulating the actual construction programme in real time, to ensure maximum benefit of thermal expansion and maximum effectiveness post – cooling.

6　Summary & conclusions

A significant benefit can be seen to have been demonstrated in implementing both the dams presented as arch/gravity structures. It must be recognised, however, that a particularly important part of the design of Kotanli and Köroğlu dams was the retention of simplicity, in order to ensure that the inherent speed benefits of RCC dam construction were not compromised, or lost. Avoiding formed joints, the comprehensive implementation of post – cooling systems and the need to delay impoundment to complete post – cooling and joint grouting eliminates any consequential impact on the RCC construction programme and cost and correspondingly ensures a realisation of the full benefit of the RCC arch dam.

With a concrete volume reduction in the case of Kotanli Dam of just 17%, it was considered essential to ensure that none of the inherent construction simplicity and speed was lost as a consequence of the conversion to an arch. This objective was successfully achieved, with no secondary cost, or time impacts compromising the benefits of a reduced concrete volume. In the case of Köroğlu Dam on the other hand, the benefits gained were substantially greater, while the consequential adaptations to ensure an effective technical performance were similarly more significant. Applying a strategic approach, however, and only including cooling and grouting where structurally necessary, the impacts of the necessary additional aspects were significantly mitigated.

References

[1] Dunstan, M. R. H. (2014). 2014 World Atlas & Industry Guide. International Journal of Hydropower & Dams. Aqua – Media International. Wallington, Surrey, UK. September 2014.

[2] Shaw, QHW. A New Understanding of the Early Behaviour of Roller Compacted Concrete in Large Dams [D]. University of Pretoria, South Africa, 2010.

Practical Application of Super High – Volume Flyash Roller – compacted Concrete

Liu Xianjiang

(Sichuan Ertan International Engineering Consulting Co., Ltd. Chengdu, 611130, China)

Abstract: Mamaya I Hydropower Station dam is roller – compacted concrete gravity dam. During the construction, the dam concrete material and mix proportion are adjusted: adjusting Grade II C20 roller – compacted concrete (flyash content 50%) in the permeation resistance area above EL. 489.5 to Grade III C20 roller – compacted concrete (flyash content 60%); adjusting Grade III C15 roller – compacted concrete (flyash content 60%) in the dam interior to Grade III C15 roller – compacted concrete (flyash content 70%). The higher flyash content and lower cement content reduce the concrete material cost and decrease hydration heat temperature rise of concrete greatly. This paper compares and analyzes the difference of mechanical properties and permeation and frost resistance before and after the adjustment of mix proportion, based on the test indexes of the properties above – mentioned and adiabatic temperature rise of hydration heat monitored by the thermometers embedded in the dam during the super high – volume flyash concrete construction period.

Key words: roller – compacted, superhigh – volume flyash, property

1 Project overview

Located in the lower and middleBeipanjiang River in Guizhou Province, Mamaya I Hydropower Station is the second hydropower station of the trunk stream cascade. Main structures of the hydro project include the roller compacted concrete gravity dam, the crest overflow outlet, the emptying bottom outlet in dam, the headrace on the left bank and the underground power house on the left bank. The top of the roller compacted concrete gravity dam is 592.0 m, with maximum dam height being 109.0 m, dam crest width being 10.0 m, and maximum dam base width being 100.5 m. The dam is divided into eight sections from the left to the right, between which are transverse joints. The concrete volume of the main dam structure is about 712 600 m^3, with 265 600 m^3 of normal concrete and 447 000 m^3 of roller – compacted concrete. During the construction, through specific test demonstration, the dam concrete material and mix proportion are adjusted: adjusting Grade II C20 roller – compacted concrete (flyash content 50%) in the permeation resistance area above EL. 489.5 to Grade III C20 roller – compacted concrete (flyash content 60%); adjusting Grade III C15 roller-compacted concrete (flyash content 60%) in the dam interior to Grade III C15 roller-compacted concrete (flyash content 70%).

In the early 1990s, Puding Hydropower Station gained certain achievements about superhigh-volume flyash of its "the 8th Five-year National Science and Technology Attack Plan" arch dam project. The Grade II flyash content which is in cementitious material of internal dam is about 60% and that of external dam is about 50%, which improved the hydraulic concrete material technique in our country to a higher level. The hydropower technicians in our country have been studying the application of super-mix-percent flyash to roller-compacted concrete. The maximum flyash content in roller-compacted concrete in Longtan Hydropower Station is 66%. The maximum flyash content in roller-compacted concrete at elevation of EL. 745.50 ~ EL. 748.90 m on dam monolith 13# ~ 16# in Guizhou Guangzhao Hydropower Station is 65%. Both of the two stations have witnessed satisfactory effects.

Super high-volume flyash concrete is used to all the dam at the elevation of EL. 489.5 ~ EL. 590.5 m in Mamaya I Hydropower Station and the flyash content in the dam interior reaches 70%, which exceeds the stipulated value and is the first case in China.

2 Key technical indexes of roller-compacted concrete

As specified in Construction Technical Requirements on Dam Concrete of Mamaya I Hydropower Station, the key design indexes of roller-compacted concrete are shown in the Table 1.

Table 1 Design indexes of roller-compacted concrete used for Mamaya I Hydropower Station Dam

Concrete type	Construction position	Strength grade	Grade	Permeation resistance grade	Frost resistant grade	Unit weight of concrete (kN/m^3)
Roller-compacted concrete	Upstream face	$C_{90}20$	2	W8	F100	≥2 350
Roller-compacted concrete	Upstream face	$C_{90}20$	3	W8	F100	≥2 350
Roller-compacted concrete	Dam interior	$C_{90}15$	3	W6	F50	≥2 350

3 Property test of raw materials

3.1 Cement

The cement used for dam concrete of Mamaya I Hydropower Station is "Tai ni" P · O 42.5 ordinary portland cement produced by Guizhou Anshun Cement Co. Ltd. Before the construction, the physical and mechanical properties of the cement are tested, with the result shown in Table 2.

Table 2　Test result of physical and mechanical properties of "Tai ni" P · O 42.5 cement

Item	Normal consistence (%)	Specific surface (m²/kg)	Set time (min)		Flexural strength (MPa)		Compressive strength (MPa)		Stability	Specific gravity
			Initial set	Final set	3 d	28 d	3 d	28 d		
Measured value	25.5	351	154	195	6.1	8.5	29.1	50.2	Qualified	3.04
Standard P · O 42.5	—	⩾300	⩾45	⩽600	⩾3.5	⩾6.5	⩾17.0	⩾42.5	Qualified	—

The spot test result of "Tai ni" P · O 42.5 cement produced by Guizhou Anshun Cement Co. Ltd shows that the tested properties can satisfy the requirements in *Common Portland Cement GB 175—2007*.

3.2　Flyash

The flyash (Class F flyash) used for dam concrete of Mamaya I Hydropower Station is Grade II flyash produced by Guizhou Zhuosheng Fengye Environment - friendly Material Development Co. Ltd. Before the construction, the physical mechanical and thermal properties of the flyash are tested, with the result shown in Table 3.

Table 3　Test result of mechanical properties of "Zhuosheng" flyash

Test standard	Fineness (%) (45 μm)	Water demand ratio (%)	Loss on ignition (%)	Moisture content (%)	SO₃ (%)	Specific gravity
"Zhuosheng" flyash	10.0	95	7.2	0.4	1.43	2.52
DL/T 5055—2007 Grade I flyash requirements	⩽12	⩽95	⩽5	⩽1	⩽3	—
DL/T 5055—2007 Grade II flyash requirements	⩽25	⩽105	⩽8	⩽1	⩽3	—

During the initial mix proportion test, the thermal properties of cement with different flyash content (0%, 30% and 60%) are tested, with the result shown in Table 4.

Table 4　Test of thermal properties of cement with different flyash content

Test item	Hydration heat (kJ/kg) (hydration heat ratio)		
	3 d	7 d	28 d
100% cement	279(100%)	305(100%)	339(100%)
70% cement + 30% flyash	233(84%)	263(86%)	314(93%)
40% cement + 60% flyash	154(55%)	176(58%)	233(69%)

The test result of Grade II flyash produced by Guizhou Zhuosheng Fengye Environment-friendly Material Development Co. Ltd shows that the tested properties can satisfy the requirements on Grade II flyash in *Technical Specification of Flyash for Use in Hydraulic Concrete* DL/T 5055—2007.

3.3 Aggregate

The aggregate used for dam concrete of Mamaya I Hydropower Station is artificial limestone aggregate produced by the aggregate stockyard on the right bank of the project, which is non-alkali-active. There are four grades of aggregate produced on site: large aggregate (80~40 mm), middle aggregate (40~20 mm), small aggregate (20~5 mm), and artificial sand. The property test result is shown in Table 5, Table 6 and Table 7.

Table 5 Test result of artificial sand particle grade

Item	Screen size (mm)								Stone powder content (%)	Fineness modulus
	>5	2.5	1.25	0.63	0.315	0.160	0.08	<0.08		
Sectional retained percentage (%)	0	18.50	21.90	22.40	13.80	7.00	8.10	8.30	16.40	2.82
Accumulated retained percentage (%)	0	18.50	40.40	62.80	76.60	83.60	91.70	100		

Table 6 Test result of artificial sand physical properties

Test item	Saturated surface dry density (kg/m³)	Saturated surface dry percent sorption (%)	Clod content (%)	Sturdiness (%)	Bulk density (kg/m³)	Compacted density (kg/m³)	Organic content (%)	Mica content (%)
Test result	2 650	1.9	0	1.8	1 530	1 764	Lighter than the normal color	0
DL/T 5144—2001 requirements	—	—	Not permitted	≤8	—	—	Lighter than the normal color	≤2

The artificial aggregate is identified as middle aggregate, based on the test result of the aggregate produced by the aggregate system on the right bank.

Table 7 Test result of artificial coarse aggregate physical properties

Test item	Saturated surface dry density (kg/m³)	Saturated surface dry percent sorption (%)	Mud content (%)	Clod content (%)	Oversize (%)	Undersize (%)
Small aggregate (5~20 mm)	2 710	0.65	0.25	0	7.8	1.1
Middle aggregate (20~40 mm)	2 710	0.50	0.06	0	7.3	0
Large aggregate (40~80 mm)	2 710	0.38	0.02	0	4	0
DL/T 5144—2001 requirements	≥2 550	≤2.5	Small and middle aggregate: ≤1, large aggregate: ≤0.5	Not permitted	<5	<10

Test item	Crush index (%)	Elongated and flaky (%)	Sturdiness (%)	Bulk density (kg/m³)	Compacted density (kg/m³)	Organic content (%)
Small aggregate (5~20 mm)	7.6	4.0	0.6	1 425	1 640	Lighter than the normal color
Middle aggregate (20~40 mm)	7.6	2.8	0.2	1 396	1 610	Lighter than the normal color
Large aggregate (40~80 mm)	7.6	6.0	0.1	1 344	1 560	Lighter than the normal color
DL/T 5144—2001 requirements	≤12	≤15	≤8	—	—	Lighter than the normal color

The test result of artificial coarse aggregate properties shows that the properties can satisfy the requirements in *Specifications for Hydraulic Concrete Construction* DL/T 5144—2001, except that the oversize percent of small and middle sample aggregate is over the normal value.

3.4 Additive

The air-entraining agent used for dam concrete of Mamaya I Hydropower Station is GK-9A air-entraining agent produced by Shijiazhuang Chang'an Yucai Construction Material Company. The mix proportion is in line with the construction mix proportion. The water reducing agent is HLC-NAF naphthalene-based super water-reducing agent produced by Nanjing R&D High Technology Co. Ltd. The mix proportion is in line with the construction mix proportion. Corresponding test results are shown in Table 8 amd Table 9.

Table 8 Test result of Gk – 9a Air – Entraining agent properties

Name of air – entraining agent	Content	Water reducing ratio (%)	Air entrainment (%)	Bleeding ratio (%)	Set time difference (min)		Compressive strength ratio (%)		Durability index
					Initial set	Final set	7 d	28 d	
Chang' an Yucai GK – 9A	0.8/ 10 000	6.1	3.3	53	+33	+9	100	110	Relative dynamic elastic modulus is 88.2% after quick – freeze for 200 times
GB 8076—2008 requirements	—	≥6	≥3.0	≤70	−90 ~ +120		≥95	≥90	Relative dynamic elastic modulus ≥85% after quick – freeze for 200 times

Table 9 Test result of HLC – NAF Water – reducing agent properties

Name of water – reducing agent	Content (%)	Water reducing ratio (%)	Air entrainment (%)	Bleeding ratio (%)	Set time difference (min)		Compressive strength ratio (%)	
					Initial set	Final set	7 d	28 d
NanjingR&D LC – NAF naphthalene series	0.8	21.7	1.7	38	+857	+844	162	155
GB 8076—2008 requirements	—	≥14	≤4.5	≤100	> +90	—	≥125	≥120

The test results of air – entraining agent and water – reducing agent show that GK-9A air – entraining agent and HLC-NAF water – reducing agent (naphthalene – based) can satisfy the requirements in *Concrete Addmixtures* GB 8076—2008.

4 Mix proportion for roller – compacted concrete construction

The construction mix proportion listed in Table 10 is put forward by the construction unit based on the key design indexes of the dam roller – compacted concrete and the test results, and has been approved by the Engineer.

Table 10 Construction mix Proportion of dam Roller – compacted concrete

No.	Concrete type	Grade	Mix proportion parameter					Material consumption per unit volume (kg/m³)							V_C (S)
			Water – binder ratio	Flyash content (%)	Sand ratio (%)	Water – reducing agent (%)	Air – entraining agent (%)	Cement	Flyash	Sand	5~20 (mm) G_S	20~40 (mm) G_m	40~80 (mm) G_L	Water	
1	$C_{90}15W6F50$	III	0.55	60	35	0.7	0.06	56	84	780	446	595	446	77	4~6
2	$C_{90}20W8F100$	II	0.50	50	39	0.7	0.08	86	86	839	539	808	—	86	4~6

During the roller – compacted construction of the dam, through laboratory test demonstration, the design unit adjusts Grade II C20 roller – compacted concrete (flyash content 50%) in the permeation resistance area above EL. 489.5 to Grade III C20 roller – compacted concrete (flyash content 60%), adjusts Grade III C15 roller – compacted concrete (flyash content 60%) in the dam interior to Grade III C15 roller – compacted concrete (flyash content 70%), and proposes the construction mix proportion after testing. Refer to the following Table 11 for details.

Table 11 Construction mix proportion of super high – volume flyash roller – compacted concrete

No.	Concrete type	Grade	Mix proportion parameter					Material consumption per unit volume(kg/m^3)							V_C (s)
			Water – binder ratio	Flyash content (%)	Sand ratio (%)	Water – reducing agent (%)	Air – entraining agent (%)	Cement	Flyash	Sand	5~20 (mm) G_S	20~40 (mm) G_m	40~80 (mm) G_L	Water	
1	$C_{90}15W6F50$	III	0.50	70	33	0.8	0.06	42	98	735	458	610	458	70	4~6
2	$C_{90}20W8F100$	III	0.46	60	33	0.8	0.08	60.9	91.3	727	453	604	453	70	4~6

5 Property test of Roller – compacted Concrete

5.1 Tests of compressive strength and split tensile strength

The Engineer has carried out sample tests on the compressive strength split tensile strength of the concrete before and after the adjustment of mix proportion. Refer to the following table 12 for details.

Table 12 Test result of the compressive strength and split tensile strength

No.	Concrete type	Grade	Flyash content(%)	Strength	Age (d)	N	min	max	$\bar{x}$
1	$C_{90}15W6F50$	III	60	Compressive strength	28	25	8.1	14.3	11.3
2	$C_{90}15W6F50$	III	60	Compressive strength	90	25	16.5	23.9	20.6
3	$C_{90}15W6F50$	III	60	Split tensile strength	90	5	1.70	2.20	1.97
4	$C_{90}15W6F50$	III	70	Compressive strength	28	52	9.9	21.9	12.7
5	$C_{90}15W6F50$	III	70	Compressive strength	90	45	16.1	27.2	20.0
6	$C_{90}15W6F50$	III	70	Split tensile strength	90	14	1.42	2.98	1.80
7	$C_{90}20W8F100$	II	50	Compressive strength	28	8	18.2	22.8	20.9
8	$C_{90}20W8F100$	II	50	Compressive strength	90	6	23.1	28.6	24.9
9	$C_{90}20W8F100$	II	50	Split tensile strength	90	2	2.15	2.99	2.57
10	$C_{90}20W8F100$	III	60	Compressive strength	28	20	13.4	23.0	17.8
11	$C_{90}20W8F100$	III	60	Compressive strength	90	17	20.7	30.2	25.1
12	$C_{90}20W8F100$	III	60	Split tensile strength	90	15	1.66	2.82	2.13

The statistical result analysis of the table 12 shows that the mechanical properties (compressive strength and tensile strength) of super high - volume flyash roller - compacted concrete are only slightly different from those of common high - volume flyash roller - compacted concrete. The two types of concrete are of the same level, have no obvious differences and can meet design requirements.

5.2 Property test of permeation and frost resistance

The Engineer has carried out sample tests on the permeation and frost resistance of the concrete before and after the adjustment of mix proportion. Refer to the table 13 for details.

Table 13 Test result of the permeation and frost resistance

No.	Concrete type	Grade	Flyash content (%)	N	Permeation resistance		Frost resistant	
					Designed	Measured	Designed	Measured
1	$C_{90}15W6F50$	III	60	1	$\geq W6$	$> W6$	$\geq F50$	$> F50$
2	$C_{90}20W8F100$	II	50	1	$\geq W8$	$> W8$	$\geq F100$	$> F100$
3	$C_{90}15W6F50$	III	70	3	$\geq W6$	$> W6$	$\geq F50$	$> F50$
4	$C_{90}20W8F100$	III	60	3	$\geq W8$	$> W8$	$\geq F100$	$> F100$

Through analyzing the above statistics, the comparison between the permeation and frost resistance before and after mixing the super high - volume flyash can not be conducted, since the Engineer does not have sufficient samples. Nonetheless, the test result shows that the permeation resistance and frost resistance of super high - volume flyash roller - compacted concrete can meet the design requirement. In addition, during the inspection on the drainage hole in the permeation resistance area after impoundment, there is little converge in the hole, which shows that the permeation resistance can satisfy the design requirement.

5.3 Property test of adiabatic temperature rise

Adiabatic temperature rise refers to the temperature change and maximum temperature rise of concretecementing materials under adiabatic conditions. A test has been designed and carried out about the adiabatic temperature rise of common high - volume flyash roller - compacted concrete and super high - volume flyash roller - compacted concrete. The result is shown in Table 14.

Test result of laboratory adiabatic temperature rise shows that with the same quantity ofcementing materials, the adiabatic temperature rise of super high - volume flyash roller - compacted concrete is about 1.5 ℃ lower than that of common high - volume flyash roller - compacted concrete.

Table 14 Adiabatic temperature rise of common high – volume flyash roller – compacted concrete and super high – volume flyash roller – compacted concrete

No.	Concrete type	Grade	Adiabatic temperature rise (℃)								Fitting formula
			1 d	3 d	5 d	7 d	10 d	14 d	21 d	28 d	
1	$C_{90}20W8F100$ (flyash content 50%)	II	6.3	9.7	12.1	14.3	16.4	18.0	19.1	19.8	$T = 22.3d/(d+3.6)$
2	$C_{90}15W6F50$ (flyash content 60%)	III	4.2	7.0	9.4	11.3	13.4	15.0	16.0	16.8	$T = 20.3d/(d+4.7)$
3	$C_{90}20W8F100$ (flyash content 60%)	III	4.3	6.8	9.4	11.5	13.9	15.3	16.4	17.0	$T = 20.1d/(d+4.9)$
4	$C_{90}15W6F50$ (flyash content 70%)	III	2.9	5.7	8.2	9.9	12.1	13.4	14.8	15.4	$T = 19.0d/(d+6.2)$

Through statistical analysis of the temperature changes of the thermometers imbedded inside the dam, the average temperature rise at different elevations inside the dam is shown in table 15.

Table 15 Statistics of temperature rise of thermometers at different elevations inside the roller – compacted concrete dam

No.	Elevation	Quantity	Average temperature rise (℃)	Construction period	No.	Elevation	Quantity	Average temperature rise (℃)	Construction period
1	483	5	12.8	January	5	500	10	9.9	June
2	485	3	11.9	April	6	506	10	11.3	July
3	490	3	11.3	May	7	512	7	10.6	July
4	495	2	10.8	May	8	518	7	11.4	August

Although, the temperature rise of the embedded thermometers depends on various factors including placing temperature, placing time and the adiabatic temperature rise of concrete, the analysis on external construction conditions at the same elevation of EL. 485 ~ EL. 500 m shows that the internal temperature rise of super high – volume flyash roller – compacted concrete is generally lower than that of the common one.

6 Economic applicability of the technique

(1) Due to the technique of superhigh – volume flyash for full – face Grade III construction, the construction mix proportion adjustment reduces 14 kg cement, increases 14 kg flyash and

saves RMB 3.22 for each cubic meter of C15 roller-compacted concrete inside the dam. The construction mix proportion adjustment reduces 25.1 kg cement, increases 5.3 kg flyash and saves RMB 9.73 for each cubic meter of C20 roller-compacted concrete on the upstream face of the dam, and the direct investment for the project reduces RMB 1.65 million.

(2) Owing to thetechnique of super high-volume flyash construction, the temperature control measures are simplified. Previously, the cooling pipes inside the dam are arranged with interval of 1.5 m × 1.5 m. After the simplification, the cooling pipes at EL.530 ~ EL.560 m and EL.560 ~ EL.578 m are arranged respectively with interval of 2.0 m × 2.1 m and 3.0 m × 3.0 m. The simplification of cooling pipe arrangement reduces the disturbance during roller-compacted concrete construction and reduces the workload of water injection and pipe maintenance.

(3) Thetechnique of super high-volume flyash construction can reduce cement consumption and make better use of solid waste in large quantity, which will contribute to significant ecological and environmental benefits such as energy conservation and emission reduction.

7 Conclusion

Mamaya I Hydropower Station successfully realized diversion closure and impoundment in November 10, 2014 and the water level at the upstream reservoir area has kept above EL.580 m ever since. Inspection on the drainage hole inside upstream dam shows that the permeation resistance at upstream dam is satisfactory. Upon the construction completion, the super high-volume flyash area is drilled for sample cores. 12.50 m-long and 15.38 m-long sample cores are continuously taken from J2-2 hole. Through testing, the properties of the sample cores including unit weight, compressive strength, split tensile strength, modulus of elasticity, ultimate tension and permeation resistance can meet the design requirements. At the beginning of the year 2015, the super high-volume flyash roller-compacted concrete construction technique passed the evaluation by the experts from Power Construction Corporation of China.

Based on the property test results, compared with commonhigh-volume flyash roller-compacted concrete, super high-volume flyash roller-compacted concrete has economic advantages: similar mechanical properties and endurance quality, slightly lower hydration heat temperature rise, and quicker roller-compacted concrete construction. Therefore, super high-volume flyash roller-compacted concrete is more advantageous than common high-volume flyash roller-compacted concrete, and is worth of popularization and application in industrial practice.

Design and Construction of Cerro Del Águila Gravity Dam in Peru

Sayah S. M.[1], Bianco V.[2], Ravelli M.[1], and Bonanni S.[2]

(1. Lombardi Engineering Limited, Switzerland; 2. Astaldi SpA, Italy)

Abstract: Cerro del Águila project in Peru represents the final step of the Mantaro River major hydropower scheme cascade development. This 520 MW hydropower scheme, presently under construction, will include an 88 m high and 270 m long RCC gravity – arch dam equipped with 6 mobile gates providing a total capacity of the surface spillway of around 7 000 m^3/s. The bottom outlets increase the total discharge capacity of the dam up to 12 000 m^3/s. The volume of the dam is approx. 0.5 Mm^3 with a relatively high percentage of vibrated concrete due to the fact that the 6 bottom outlets are large and require high strength concrete. The bottom outlets are in fact designed to allow an easy annual flushing of the reservoir knowing that the sediment yield is around 3.5 Mm^3/year. A special attention was also given to the design of the upstream face of the dam in order to confirm its behavior during earthquakes as the site is considered seismically very active. This dam is equipped with galleries for the foundation consolidation grouting. These galleries were designed in order to accelerate the construction process allowing consequently the treatment of the base of the dam without interfering with the construction activities. Several other concepts were also applied that allowed rapid placement of the RCC paste. Concrete placement is mainly carried out using a high capacity blondin combined with transport trucks and steel chute. The mix design of the RCC was optimized using full scale test sections. With a low percentage of pozzolanic content up to around 20%, temperature rising during the hydration of the concrete was a concern. Extensive investigations were thus considered in order to reduce the cement content while respecting the design strength. In this paper an outlook of the main design features of the dam is provided together with some major highlights regarding the optimized mix design, the construction process, and concrete placement.

Key words: Gravity Dam, RCC mix design, bottom outlet

1 Project background

Cerro del Águila project in Peru represents the third step of the Mantaro river major hydropower scheme cascade development. It is situated downstream of SAM/Restitución Hydroelectric Plants. Originally, another project was planned to develop the second curve practically down to its confluence with the Apurimac river. This project would have included a long low pressure tunnel and a 250 m high dam situated downstream of the Colcabamba river confluence in order to include the discharge provided by the additional catchment area and the additional head between the SAM

HPP tailwater level and the Mantaro river.

Taking into consideration the very large landslide in the Mayunmarca area that occurred in 1974 and the apparent vulnerability of the valley side stability, the idea of implementing a very high dam was abandoned and partially substituted by the following:

(1) The Restitución Project, developing the 250 m remaining head between the SAM HPP tailrace and the Mantaro river, and

(2) The Guitarra Project, developing at short distance downstream of the denominated Guitarra curve. After almost 25 years of consideration, this Guitarra Project (1983) has been redefined and renamed Cerro del Águila Project.

Fig. 1　The 77 m Tablachaca gravity dam, situated at around 100 km upstream of Cerro del Águila Dam on the Mantaro River, constructed in the 70's

The project is being developed by Kallpa Generación S. A., a subsidiary of IC Power. It falls under an Engineering Procurement and Construction (EPC) scheme. In November 2011, Astaldi S. p. A and their joint venture partner Graña y Montero SAA won the contract. Lombardi SA was chosen later as the project designer. In the beginning of 2012 the excavation of the access roads started. In the mid of 2014 the dam foundation preparation ended allowing the start of the dam construction.

2　Dam geology

Cerro del Águila Dam is located in a higher mountainous environment with steep valley slopes (average 30°, locally exceeding 60°), some 50 km upstream of the Amazonian area, and located in the main bedrock lithologies formed of granites/granodiorites (Villa Azul batholiths). Mantaro

River has eroded its current river bed directly into the bedrock. The left – hand and the right – hand side slope shows different quaternary deposits: The left – hand side is steeper and directly shows bedrock under a thin cover of colluvium deposits and locally present rockfall/debris flow deposits (Fig. 2). On the right – hand slope the dam site is on the terrace of a fan of mixed origin and terraced recent alluvial/moraine deposits. Due the erosive environment of Mantaro River, only very little current alluvial deposits are present around the river bed.

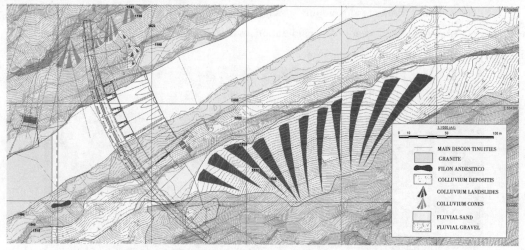

Fig. 2 Geology of Cerro del Águila dam site

3 Design of Cerro del Águila Dam

3.1 Dam characteristics

The new dam main characteristics are given below and illustrated in Fig. 3.

(1) Hydrology and geomorphology:
- Catchment area: 28 096 km^2; Drainage at intake: 9.04 L/(s·km^2)
- Project flood (at the dam site): Q_{1000} = 6 125 m^3/s
- Assumed sediment yield in the new reservoir: 1 to 3 ~4 Mm^3/y

(2) Artificial reservoir and dam:
- Dam type: RCC gravity dam (arch form)
- Dam height: 88 m from foundation (El. 1 560.00 El. 1 472.00 m asl)
- Dam crest length: 270 m; Dam crest elevation: El. 1 560.10 m asl
- Exceptional operating water level: EL. 1 560.00 m asl
- Normal operating water level: EL. 1 556.00 m asl
- Total impoundment volume: ~37 Mm^3
- Bottom outlet: 6 × 2 slide gates $b \times h$ = 4.60 m × 6.00 m; sill level: El. 1 495.00 m asl
- Spillway: 4 × radial gates $b \times h$ = 12.40 m × 16.00 m; 2 flap gates $b \times h$ = 12.00 m × 5.20 m
- Spillway crest sill level: El. 1544.50 m asl (radial gates); El. 1 551.50 m asl (flap gates)

• River diversion: 340 m long pressure tunnel; capacity of 715 m³/s.

3.2 Dam layout and typical section

A typical section of the new dam is illustrated in Fig. 4. The dam is an RCC concrete gravity dam with 18 independent blocks, each around 16 m long. The 270 m long dam has a slightly curved planimetric axis ($R = 400$ m), with a crest elevation of 1 560.10 m asl. At the lowest point, the base elevation of the upstream toe of the dam is 1 472.00 m asl. The maximum dam height measured at foundation is 88 m. The normal operating water level is at El. 1 556.00 m asl, allowing a total storage volume of around 37 Mm³. The typical cross section of the dam is designed with inclined faces in order to ensure the required stability during a seismic event (MCE = 0.4.g; MDE = 0.25.g). The upstream dam face has an inclination of 1:0.1 ($V:H$), and the downstream face correspond to an inclination of 1:0.75 ($V:H$). The maximum width at foundation is around 70 m. The typical width of the crown blocks is 6.2 m. The dam crest is 6.5 m wide and equipped with a concrete parapet at the upstream face. The total volume of concrete of the dam is around 450 000 m³.

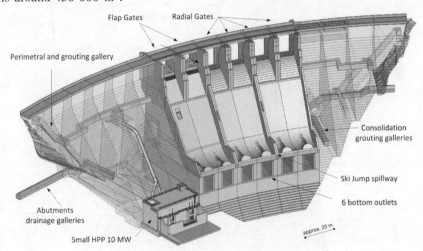

Fig. 3 **General layout of Cerro del Águila Dam**

The release of extreme flood events is guaranteed by a gated surface spillway equipped with 4 radial gates and 2 flap gates, as well as 6 bottom outlets equipped with slide gates. With a total capacity of 7 000 m³/s for the gated surface spillway and 5 000 m³/s for the bottom outlet, the combined total flood discharge capacity of the scheme amounts to 12 000 m³/s.

The floods are discharged downstream at the dam – integrated 45 m long chute spillway equipped with a ski jump. A series of 3 m wide and 3 m high deflectors are foreseen at the end of the ski jump in order to open the hydraulic vein and facilitate the air entrainment, favoring the energy dissipation of the water jet before impacting the plunge pool.

3.3 Big bottom outlets for sediment flushing

The operational conditions of the Cerro del Águila Dam in the long term will significantly depend on the proper management of the sediments yield carried by the flows of the Mantaro River and deposited at least partially in the reservoir during the flood season. And average annual sediment yield of around 2 to 4 Mm^3 entering the reservoir was estimated. Therefore an annual flushing will be required. In order to optimize the flushing procedure and make it shorter in time thus avoiding significant loss in energy generation, 6 big bottom outlets were proposed (Fig. 5).

Each is equipped with two slide gates that can be operated partially for partial sediment flushing. When they are completely opened it is possible to empty the reservoir and carry out a complete flushing of all the deposited sediments. Based on physical and numerical modeling, a flushing period of several days it was estimated as sufficient. All the bottom outletsare steel lined and completely aerated allowing a free surface flow during the emptying of the reservoir (Fig. 5).

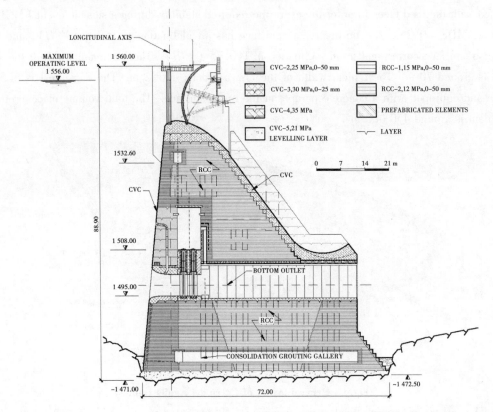

Fig. 4 **Typical section of the central blocks**

3.4 Dam RCC – CVC zoning

The 3D stress analysis of the typical central block of the dam is illustrated in Fig. 6. This analysis carried out for typical seismic events (in the present case an MDE event), showed that the upstream face of the dam exhibits positive tension up to around 2.5 ~ 3 MPa while at the toe of this face tensile stresses might rise to values as high as 6 ~ 7 MPa. In order to cope with these high values, it was decided to apply a high resistance conventional vibrated concrete with a compressive strength equal to 25 MPa. An RCC concrete was applied to the remaining sections of the dam. Two different RCC strength were applied depending on the stress distribution inside the dam body. In the core of the dam body a strength of 12 MPa was adopted, and, at the downstream toe and below the slide gates, a 15 MPa compression strength was applied. Applying the zoning con-

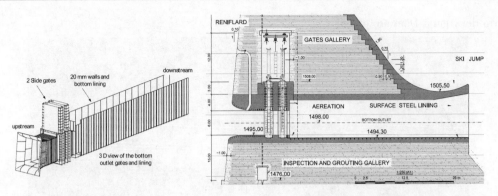

Fig. 5 Steel liner concept (ATB – Riva Calzioni S. p. A.) (left) Bottom outlet design (right)

cept it was possible to avoid thermal treatment of the main massive concrete (RCC – 12 MPa). Furthermore, this optimization allowed a significant reduction of the cement content of the concrete.

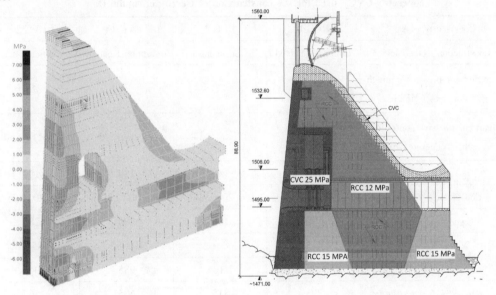

Fig. 6 Stress Analysis of the typical central block for the MDE earthquake (left) and zoning of the Dam body as a function of the concrete type and strength (right)[1]

4 Construction of the dam

4.1 Mix design and full scale test sections

In the beginning of the 2014, a comprehensive study of the mix design for both the CVC and RCC concrete started. Fig. 7 illustrates the full scale test section of the RCC concrete (left) which was around 4 meters wide and 10 meter long. This figure also illustrates the core extraction procedure carried out for the systematic and followed – up laboratory analyses. Typical laboratory testswere carried out, such as compressive and tensile strength (on the main body concrete and lift joints), permeability, density, vebe number, thermal conductivity, heat of hydration, etc. In Table 1 the main characteristics of the different mix design as it was adopted for the construction of

Cerro del Águila Dam are given.

Fig. 7 Full scale test section for the study of the dam concrete mix design

Table 1 Mix design of the Roller compacted concrete (RCC) and the conventional vibrated concrete (CVC) used for the construction of Cerro del Águila Dam[2]

Concrete typology	RCC 1	RCC 2	CVC 1	CVC 2
Construction methodology	Roller compacted	Roller compacted	Conventional massive	Conventional massive
Specified compressive strength of concrete (MPa)	15	12	15	25
Mean compressive controlled strength of concrete (MPa)	$15 < f_{cj} < 19$	$12 < f_{cj} < 14$	$15 < f_{cj} < 19$	$25 < f_{cj} < 30$
Age (days)	180	180	180	180
Consistency / slump	Vebe(20 ± 5) sec.	Vebe(20 ± 5) sec.	Slump(100 ± 25) mm	Slump(100 ± 25) mm
Cement (kg/m^3)	130	100	200	280
Water (kg/m^3)	130 ± 2	132 ± 2	190 ± 5	185 ± 5
Sand (kg/m^3)	1 095	1 104	985	948
Aggregate (25~5 mm) (kg/m^3)	795	815	605	565
Aggregate (50~25 mm) (kg/m^3)	275	275	400	380
Admixtures	Set retarder/ water reducer	Set retarder/ water reducer	Set retarder/ water reducer + Superplasticizer	Set retarder/water reducer + Superplasticizer
Theoretical density (kg/m^3)	$2\ 430 \pm 20$	$2\ 430 \pm 20$	$2\ 380 \pm 20$	$2\ 380 \pm 20$

4.2 Concrete transportation, placement and compacting

Concrete placement was carried out using mainly a blondin device designed by the Italian companyAgudioS. p. A with a transportation capacity of around 9 m^3 (Fig. 8). The average placementvolume registered during the construction of the central blocks of the dam varied between 100 and 120 m^3/h knowing that the geometry of the dam galleries is quite complex. After its placement and extension, a 12 t vibrating roller drum carried out the compaction of the RCC with an average of 7 passes. In order to increase the transportation and placement rate, a steel chute was also mounted on the right abutment and used as a complementary device to the blondin. This allowed reaching on several occasions a placement rate of around 210 m^3/h. While a layer of 30 cm thickness was applied for the RCC, it was considered sufficient to adopt a 60 cm thickness layer for the CVC. As illustrated in Fig. 4 and Fig. 8, a 1 m wide perimetral vibrated concrete layer (CVC 15 MPa) was applied around all the galleries of the dam and the RCC was placed later against this layer. This prevents having the RCC in direct contact with air and provides a better finish of the walls and base of the galleries. In order to avoid placingformworks for the construction of the roof of the galleries, several types of prefabricated reinforced beams (Fig. 8) were used depending on the width of each gallery (the largest roof beam was 9 m long placed at the bottom outlets). By this manner it was possible to avoid long construction delays.

Fig. 8 Concrete placement using the blondin for the placement of the perimetral CVC of the diagonal access gallery to the gates chamber

4.3 Lift Joint treatment procedure

A special attention was made for the treatment of lift joints between two consecutive layers. Due to the complex dam design and the concentration of galleries having sometimes a complex geometry, it was unavoidable at several occasions to bring to a halt the concrete placement for more than 24 hours. Since the lift joints are considered very critical for RCC dams regarding the tensile and cohesion strength, it was considered wise to apply a specially designed bedding mortar after a sound cleaning with high pressure water/air of the old layer surface before placement of the consecutive layer (Fig. 9). This procedure was judged successful, specially after carrying out concluding laboratory tests made on drilled cores extracted from several locations of the dam.

Fig. 9 Special treatment of the RCC lift joints

4.4 Consolidation grouting and grout curtain

All the works regarding the consolidation and contact grouting of the dam foundation were carried out using special galleries foreseen in each block of the dam in the upstream – downstream direction. These galleries have a section of 3 m × 3 m (see Fig. 10). With this manner it was possible to carry out the grouting works independently of the construction and concrete placement of the dam body thus avoiding any delay.

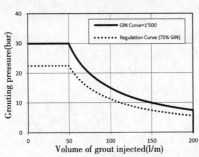 

Fig. 10 GIN curve applied for the grout curtain (left); consolidation and contact grouting galleries foreseen in each dam block (right)

In Fig. 11 it is illustrated the grout curtain of the dam for primary and secondary grouting. All the grouting works follow the GIN curve (Fig. 10) with a GIN number of around 1 500[3]. This was considered sufficient taking into account the average rock quality of the dam foundation. The grouting pressure applied in each hole varies between a maximum of 30 bar (around 40 m below the dam base) and a minimum of 5 bar (close to the dam base).

5 Conclusions

Presently the dam construction is at around 30% from completion (Fig. 12). Consolidation and contact grouting of the foundation are almost terminated. The grouting curtain works are ongoing. The first filling of the dam is foreseen by the end of 2015.

Acknowledgements

The authors acknowledge the work of G. Rotundo head of the technical on site office of

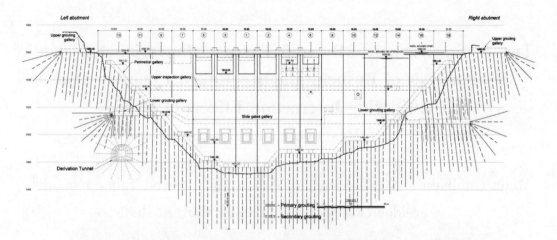

Fig. 11　Grout curtain of Cerro del Águila Dam

Fig. 12　Cerro del Águila Dam under construction (photo taken on July 2015)

Astaldi S. p. A, the contribution of M. D'arrigo with regards to the construction procedures, the participation of F. Andriolo in the definition of the mix design, the strong involvement of A. Ricciardi, F. Tognola and J. Arbolí in the design work of the Dam, the support of M. Braghini, head of the hydraulic section of Lombardi Eng. Ltd, and last but not least the support of J. Monaco, owner's project manager.

References

[1] Lombardi SA. Dam safety evaluation, General Stability and Stress Analysis, 2013.
[2] CRM SA. Informe técnico características de los concretos colocados en la presa, 2015.
[3] G. Lombardi, D. Deere. Grouting design and control using the GIN principle, 1993.

Two – dimensional (2D) and Three – dimensional Dynamic (3D) Analysis of Bushehr Baghan Rolled Concrete Dam by Considering Interaction Effects of Dam, Reservoir and Lake

Majid Gholhaki[1], Ali Mohammad Mahan Far[2], Seyed Taher Esmaili[3], Fereidoon Karampoor[4], Seyed Morteza Rad[4]

(1. Civil engineering college in Semnan University, Iran;
2. Islamic Azad University, Iran; 3. Absaran Consulting Engineers, Iran;
4. Bushehr Regional Water Authority employee, Iran)

Abstract: With regard to importance of stability and safety of dams against incoming loads, it is very important to choose suitable constructions as dam foundation. Bushehr Baghan rolled concrete dam has two different Bakhtiari constructions under its foundation.

Existence of many differences in two constructions towards each other, increasingly pay attention to dam dynamic analysis importance and to results comparison between 2D and 3D analysis. Results analysis indicate that there is meaningful difference of maximum horizontal displacement on the top of dam under final loading in two constructions.

2D and 3D behavior comparison shows that in similar loading, because of tension distribution and adjustment, replacement and foundation sinking and tensions in 3D state is lower and smaller than 2D state.

Key words: Rolled concrete dam, dam interaction, two – dimensional and three – dimensional dynamic analysis

1 Introduction

Always, one of the most important human needs and requests has been water and mankind has endorsed constructing or building some obstacles in the course of flowing as a dam in order to store water in droughty conditions and periods and to use produce able energies by water. Dams have unique or single position for many reasons such as the importance and significance of construction goals and also severity and intensity of dangers and hazards and damages arising from their possible demolition.

In the past, analysis and designing of concrete dams was usually carried out statically and without considering lake trance in calculations and accounts. As first step, Westergard analytically solved 2D problems by considering water compressibility and dam rigidity theorem and assumption and he carried into account hydrodynamic load factor in his calculations by solving Helmholtz equation in order to design a dam in U. S. In the year 1959, Kotsubo indicated that Westergard solution is only true for vibrations having frequencies lower than normal frequency of reservoir. Also Zangar(in the year 1952) invented a quick and fast method for relating to weight dams by taking into account superior position to calculate hydrodynamic load. In this method, it was considered that liquid is incompressible and it was extended from superficial wave influence and hydrodynamic load by Laplace equation. Hydrodynamic load distribution two – dimensional problem, by assuming dam rigidity, was continued by other researchers such as Chopra. Chopra generalized Westergard solution and showed that hydrodynamic load response to earth horizontal vibrations generally depends on different values.

First non – linear analysis of concrete dams´ limited elements was carried out by Pal in the year 1974. In the year 1982 Haul and Chopra designed and modeled lake in limited element method. Aforementioned methods included dam – lake interaction in 2D models. In the year 1983 Hall and Chopra continued their studies (which were carried out in the year 1982) as extending 2D method of limited elements for 3D analysis. Bhattacharjeeand Le'ger(in the year 1993 and 1994) carried out extensive studies to analyze non – linear weight – related concrete dams and used "distributed crack model" to examine dam failure response and they also used added mass for lake. In the year 1995, Tan and Chopra extended the presented method in EACA – 3D program in order to consider dynamic interaction effect of foundation. In the year 1990 Ghaemian and Chopra presented 2D non – linear vibratory analysis of weight – related concrete dam by considering dam and lake interaction. In this research, distributed crack model has been used to indicate the response of dam and its frame cracking. In the year 2002 and 2003 Le'gerand has coworkers accomplished stability analysis of weight – related concrete dams using computer software such as CADAM and RSDAM. In the year 2003, Chuhanand Guanglunand Shaominpresented results obtained from laboratory tests on rolled concrete samples and non – linear failure analysis of rolled concrete dams. Ghaemian and Moeini (in the year 1382) presented non – linear vibratory analysis of concrete dams consists of lake and dam interaction. Ghaemian and Fazeli (in the year 1383) presented parametric studies of distributed crack non – linear model and failure mechanic in weight – related concrete dams analysis. Ghaemian and Mazloomi (in the year 1385) presented non – linear vibratory analysis of kooyan and pineflat weight – related concrete dams and Jegin rolled concrete dam by considering lake and dam interaction.

In accomplished studies, the direction of place – finding for selecting dam establishment and construction place, is selected locally and stone quality and type in dam foundation not only is desirable and high – quality but also has monotone construction along the dam. But in all cases the existence and presence of incompatible foundation together with different constructions under building is an in avoidable concern. So in this paper we examine the effect and influence of different constructions on rolled concrete dam frame in Baghan under 2D and 3D dynamic analysis.

2 Modeling (making – model)

Baghan rolled concrete dam is placed on two different Bakhtiari constructions in the east of Booshehrprovince with an altitude of 57 m. BK1 construction has coarse – grained to mean and average grained conglomerate type, with calcareous and argilous cement that inside or interior sand layers are stony that is seen in right leaning, foundation and lower figures of left leaning. Also BK2 construction has young and very weak coarse – grained conglomerate with argillous cement and silty and sandy matrix and grain – size ranges from gravel to bolder. This unit is extended in left leaning of dam and in syncline axes. The slope of dam superior position is 0.05 perpendicular to horizontal 1 and the slope of lower position is 0.85 perpendicular to horizontal 1.

For Modeling (making – model), Baghan dam, four models have been used that consist of three 2D models and one 3D models. 2D models contain over flowing section placed on BK1 construction and two non – over flowing sections placed on BK1 and BK2 construction. In order to select non – over flowing sections, the provision and condition of largest latitude or height of dam in this section has been considered. In two – dimensional models, in order to consider dam – foundation – reservoir (storage) interaction, in addition to dam frame or body, foundation and water reservoir have also been modeled. In order to modeling of dam frame and foundation, we use CPE4R and for modeling of liquid we use AC2D4 element that is an acoustic element (in ABAQUS software). Wave distribution only exists in compressible liquid and whereas this wave spreads, it will require a suitable boundary condition to absorb wave at the reservoir extremity and its bottom. In these models, compressibility effects and influence of lake water, has been considered by describing Balk model and liquid density that will result in speed up sound in water. Also, foundation having mass represented in this model in order to consider the property and quality of wave distribution in foundation.

In 2D model, foundation depth has been prolongatedabout 2.5 times of dam height and latitude (150 m from the lowest dam frame and body level) and its length has also prolongated similarly towards upwards and downwards of dam (300 m). Baghan dam 3D model with foundation depths and its length at the X axis direction, is 2.5 times of dam latitude and foundation size towards dam axis is two – fold of dam tip length and also lake length is two – fold as much dam length. Also reservoir bottom has been considered without any slope. In three – dimensional model, both constructions have been considered as making – model foundation. Fig. 1 ~ Fig. 3 shows 2D and 3D of dam frame and foundation.

3 Relations dominant in dam and reservoir system

Dynamic balance equations are dominant on constructions and lake under vibration loads, according to relations (1) and (2) we have:

$$[M]\{\ddot{u}\} + [C]\{\dot{u}\} + K\{u\} = \{f_1\} - \{M\}\{\ddot{u}_g\} + \{Q\}\{P\} = \{F_1\} + \{Q\}\{P\} \quad (1)$$

$$[G]\{\ddot{P}\} + [C']\{\dot{P}\} + [K']\{P\} = \{F\} - \rho[Q]^T(\{\ddot{u}\} + \{\ddot{u}_g\}) = \{F_2\} - \rho[Q]^T\{\ddot{u}\} \quad (2)$$

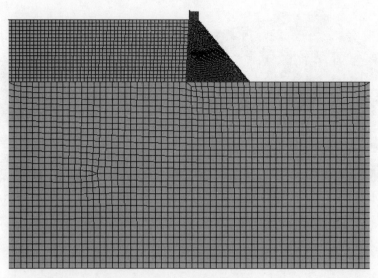

Fig. 1 Two-dimensional finite element's model of Baghan dam non-overflowing section

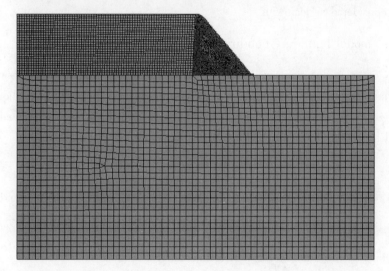

Fig. 2 Two-dimensional finite element's model of Baghan dam overflowing section

In which $[M], [C], [K]$ are mass, damping and hardness or in flexibility of constructions and foundation matrices respectively and $[G], [C'], [K']$ are respectively mass equivalent, damping and hardness or inflexibility of lake. $[Q]$ is couple matrix and $[F_1]$ is body and frame power vector and hydrostatic power and $\{P\}, \{U\}$ are hydrodynamic load of lake node and construction nodes of displacement vectors and $\{\ddot{U}_g\}$ are earth speed and acceleration vector.

Dynamic equations dominant on dam and lake interaction system can be stated in two types and manners lie Lagrange - Lagrange and Oiler - Lagrange. In the Lagrange - Lagrange method we use equation 1 for reservoir and construction and we consider contemporary and synchronizes solution of equations 1 and 2 in Oiler and Lagrange method. In other words, in Lagrange - Lagrange method, dam and reservoir degrees of freedom, both of them are of replacement kind but

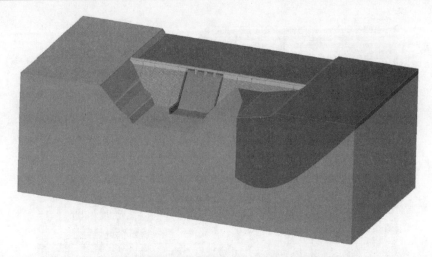

Fig. 3　Baghan dam three – dimensional model

in Oiler – Lagrange method, dam degrees of freedom are of replacement kind and reservoir degrees of freedom is considered as load king.

4　Making – model boundary condition

Generally equation dominant on liquid boundary is known as Navier – Stocks relation. The relation dominant on reservoir or storage known as wave equation, supposing that liquid compressibility is presented as relation (3) using connection relation and considering speeds with small amplitude and supposing that the liquid is non – rotational.

$$\Delta^2 P - \frac{1}{c^2}\ddot{P} = 0 \tag{3}$$

In above relation known as Helmholtz relation, C is load wave speed in liquid, P is liquid load and $\ddot{P}$ is the second derivative relative to time ($\frac{\partial^2 P}{\partial t^2}$). In order to solve above equation, it is necessary to consider condition of geometrical boundary.

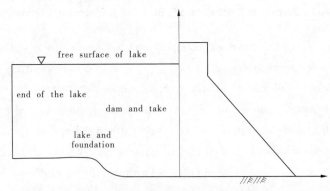

Fig. 4　Dam and Lake system borders

4.1 Lake and dam boundary provision and condition

Considering that dam frame and body is floodgate and there isn't any flowing in the border between dam and reservoir, similar speed in liquid and construction perpendicular to common border is established and formed ($v_n^s = vn = v.n$). This subject expresses impermeable condition in this surface in concrete dams. By derivating above relation and by considering a_{nd}^s as a speed or acceleration perperdicularto construction surface in common section of dam and lake, we can apply relation(4) as a boundary provision and condition as superior surface of dam in which n indicates normal unit vector on dam superior surface towards lake:

$$\frac{\partial p}{\partial n} = -\rho a_{nd}^s \qquad (4)$$

4.2 Boundary provision and condition of lake and foundation

Provided that reservoir bottom or bed is suspected to be rigid and hard, we can again apply relation 4 as boundary condition and provision between storage and reservoir and surroundings' wall and bottom. Rigidity and hardness of reservoir bottom means that there is any wave absorption or permeability of water inside and about the reservoir but we should pay attention that always there is similar deposits or sediments in reservoir that cause percentage absorption of hydrodynamic waves. Utilized boundary condition for this section of reservoir is discussed as relation(5) supposing that after clashing or colliding waves to reservoir bottom, only vertical longitudinal waves are spread in foundation:

$$\frac{\partial p}{\partial n} = -\rho a_{nr}^g - q \frac{\partial p}{\partial t} \qquad (5)$$

Excess and additional sentence and term in relation(5) a scompared with relation 4 expresses hydrodynamic wave absorption. q is wave absorption coefficient that its value (in terms of α) is wave reflection coefficient that is retroversion hydrodynamic wave amplitude and scope to initial wave scope ratio. From relation(6) we can obtain:

$$q = \frac{11 - \alpha}{c1 + \alpha} \qquad (6)$$

Theoretically, wave reflection coefficient value ranges between $+1$ and -1. It is evident that in hardness and rigidity bed presumption case, wave absorption coefficient value will be equal to zero. Anyhow, selection of suitable α depends on reservoir bed deposits'condition as generally we use $0.9 \sim 1$ values for new dams and use $0.75 \sim 0.9$ values for dams that passing of time from building time, causes deposits and sediments occur.

4.3 Boundary condition of lake free surface

Boundary condition of reservoir free surface is applied as relation(7) irrespective of shallow and superficial waves:

$$P = 0 \qquad (7)$$

4.4 Boundary condition of extremity of around lake

Dam and reservoir interaction is considered as semi - infinite element to restrain and prevent waves' reflection because of loading waves'creation and transmission of these waves (in compressible condition) so in this condition, we simply use analysis method in frequency limit and if we want to examine and inspect it in time limit we should consider a suitable boundary condition for dam discontinued and cut extremity. In such a manner that those waves which get away from construction, they are completely absorbed in far extremity. The most common and current waves is Summerfield boundary condition that ispresented in relation(8):

$$\frac{\partial p}{\partial n} = -\frac{1}{c}\frac{\partial p}{\partial t} \tag{8}$$

In physical interpretation, Summerfield boundary condition is such that a number of amortizing factors of loading waves have been placed on reservoir superior border.

5 The characteristics of construction materials of dam frame and body and foundation

The properties and characteristic of Baghan dam frame and foundation construction materials have been used in making mode as in table 1.

Table 1 Properties of building materials of Baghan dam foundation and frame

Properties of building materials dam	unit	value
Volumetric mass	kg/m^3	2 400
elasteticity module		26.3
Poason coefficient	GPa	0.2
Damping coefficient		3
Loading resistance rolled consrete cylindrical 90 days	MPa	14
90 days loading resistance of surface concrete cylindrical	MPa	25
Property of building materials of foundation and dam	unit	value
Poison coefficient		0.3
BK1 construction transition module	GPa	3
BK2 construction transformation module	GPa	1

6 Loading conditions

In order to analyze models' tension, we have applied or used three loading combinations according to table 2.

Table 2 Loading combinations used in the dynamic analysis of dam Baghan

Input load type	1	2	3
Dam mass	*	*	*
Water hydrostatic load in normal level		*	*
Sediment load		*	*
Load on action		*	*
Vibration lateral load in DBE surface in superior position	*		
Vibration lateral load in DBE surface in lower position		*	
Vibration lateral load in MCE surface in lower position			
Loading type — Normal			
Loading type — Abnormal		*	
Loading type — Final	*		*

For non-linear dynamic analysis of Baghan dam, we use Elcentro and Tabas earthquake hodograph for DBE and MCE earthquake levels respectively and they have been compared with maximum horizontal speed (0.24 g) of Baghan dam construction for earthquake of DBE level and maximum horizontal speed 0.46 g of dam construction for earthquake MCE level.

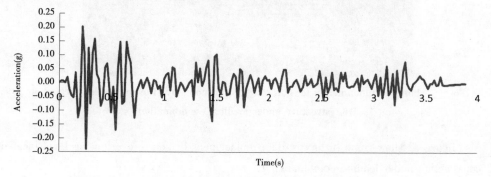

Fig. 5 Horizontal composer of speedometer in Elcentro earthquake

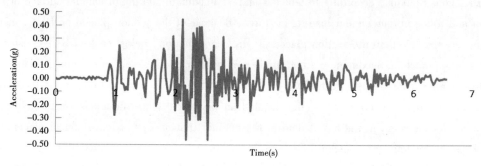

Fig. 6 Horizontal composer of speedometer in Tabas earthquake

7 Analysis results

Tractional tension contours σ1 have been shown in figures 7 – 11 non – overflowing and over – flowing levels located on BK1 and BK2 constructions under loading combination 3 (Tabas earthquake MCE level).

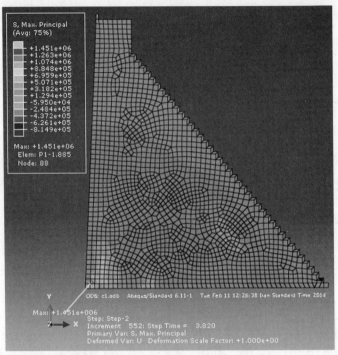

Fig. 7 Main and fundamental tension condition in non – overflowing section located on BK1 structure under loading 3 combination

Also figure 12 shows time history of dam replacement in non – overflowing level located on BK2 construction under loading combination 1.

With comparison whit dynamic analysis results of non – overflowing sections located on BK1 and BK2 constructions, according to tables 3 and 4, maximum sinking of contact surface of foundation and frame in dam similar loading mixtures, because of the section placed on BK1 construction is very massive than the section placed on BK2 construction, being weak of BK2 construction causes in approaching maximum sinking numbers in similar loading. As an example, this value in loading number 3 for non – overflowing section located on BK1 construction equals to 0. 010 3 meters and for non – overflowing section located on BK2 weak construction is very noticeable in maximum displacement in the top of dam in the manner that maximum horizontal displacement of top of dam in loading state number 3 for two aforementioned sections equals to 0. 031 4 and 0. 048 3 meters for BK1 and BK2 constructions respectively that indicate negative influence of BK2 construction weakness on obtained results. With comparison with results of 2D and 3D sections it was determined that maximum loading and tractional tensions, dam contact surface sinking in founda-

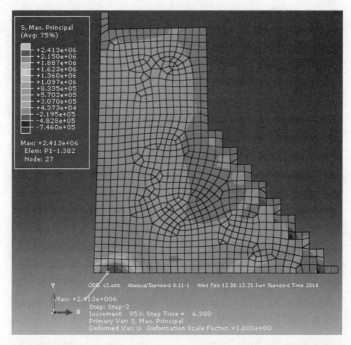

Fig. 8 Main tension condition in non-overflowing section located on BK2 structure under loading 3 combination

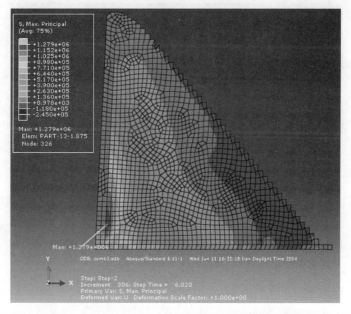

Fig. 9 Main and fundamental tension condition in overflowing section located on BK1 structure under loading 3 combination

tion and displacement and speed of top of dam, in all three loading cases in 3D sections has been decreased in 2D sections. As an example according to table 5, maximum loading and tractional tension in overflowing section respectively equals to 1.346 and 3.02 Mega-pascal in 2D condi-

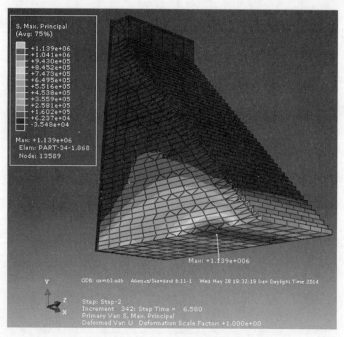

Fig. 10　Main tension condition in non – overflowing section located on BK1 structure under loading 3 combination

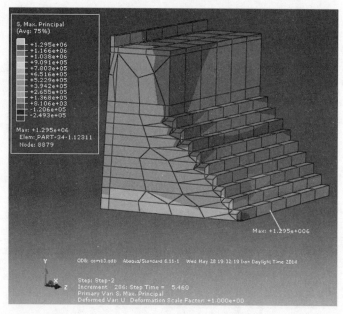

Fig. 11　Main tension condition in non – overflowing section located on BK2 structure under loading 3 combination

tion and 1.14 and 1.49 MPa in 3D section for loading1. The reason of value decrease in 3D section are considered interaction effect and input loading distribution in neighbor section in 3D sections. As explained, with regard to greater dimensions of non – overflowing section located to

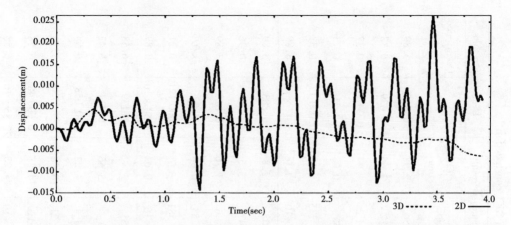

Fig. 12 Temporal history of dam crest displacement in non-overflowing section BK2 under loading 1 combination

BK1 construction relative to BK2, it is expected that initial and first section analysis results is very more than second section that from comparing results, contrary of this case is seen. The effect and influence of BK2 construction weakness on three-dimensional behavior of dam alike to 2D section, with comparison with maximum torsional tension in two non-overflowing sections BK1 and BK2 is seen that respective equals to 1.354 and 1.331 MPa for two BK1 and BK2 constructions.

8 Conclusion

In this paper, BoosherBaghan rolled concrete dam has been modeled by using ABAQUS finite element software and they have analytically been analyzed under loading mixtures. Dynamic analysis was carried out as timing history and while considering interaction effects of dam-reservoir and lake.

Nonlinear dynamic analysis results situated on two 2D dimensions on BK1 and BK2 constructions showing weak effect of BK2 construction on increasing tensions and displacement values of dams.

With comparison with 2D and 3D analysis results, in similar and equal loading, decrease of tensions and displacement values were seen in 3D section analysis results because of input tension distribution to lateral section and considering 3D behavior on dynamic analysis. So 3D analysis of dams especially in dams having asymmetrical foundation is considered necessary because of different response potential as a result of different constructions in dam's foundation.

Table 3 Dynamic analysis results of non-overflowing section in Baghan dam located on BK1 construction

Section name	Construction type	Loading	Description	Unit	2D analysis results		3D analysis results	
					Value	Occurrence time	Value	Occurrence time
Non-overflowing	BK1	1	Maximum tension - torsion between frame and dam	MPa	1.354	2.98	0.885	3.89
			Maximum loading tension of frame and dam	MPa	2.812	3.66	0.639	0.38
			Maximum horizontal displacement of tip of dam	m.1	0.011 2	1.94	0.007 2	0.37
			Maximum horizontal speed of tip of dam	m/s^2	4.52	0.08	2.51	0.73
			Maximum contact surface sinking of dam and foundation	m.1	0.003 735	1.18	0.000 17	0.37
			Maximum cutting power of contact surface of dam and foundation	N	7.328×10^6	1.6	13.47×10^6	3.89
		2	Maximum torsional tension of frame and foundation	MPa	1.402	2.8	0.883	3.89
			Maximum loading tension of frame and foundation	MPa	5.147	3.6	0.51	0.36
			Maximum displacement tip of dam	m.1	0.018 6	3.76	0.025	3.89
			Maximum horizontal speed of tip of dam	m/s^2	-16.53	0.08	3.02	0.025
			Maximum contact surface sinking of dam and foundation	m	0.002 581	3.7	0.000 121	0.21
			Maximum cutting power of contact surface of dam and foundation	N	6.54×10^6	0.02	13.58×10^6	3.89
		3	Maximum torsional tension of frame and dam	MPa	1.451	3.82	1.139	6.58
			Maximum loading tension of dam frame	MPa	10	3.66	0.411	6.42
			Maximum horizontal displacement of tip of dam	m.1	0.031 4	4.22	-0.058	6.58
			Maximum horizontal speed of top of dam	m/s^2	-11.21	0.08	3.77	1.14
			Maximum contact surface sinking of dam and foundation	m.1	0.010 3	4.88	0.000 48	6.58
			Maximum cutting power of contact surface of dam and foundation	N	18.88×10^6	4.28	22.85×10^6	6.58

Table 4 Dynamic analysis results of non – overflowing section in Baghan dam located on BK2 construction

Section name	Construction type	Loading	Description	Unit	2D analysis results Value	2D analysis results Occurrence time	3D analysis results Value	3D analysis results Occurrence time
Non – overflowing	BK1	1	Maximum tension – torsion between frame and dam	MPa	1.331	3.46	1.156	3.89
			Maximum loading tension of frame and dam	MPa	0.907	3.36	1.475	0.7
			Maximum horizontal displacement of tip of dam	m.1	0.026 3	3.44	0.007	3.89
			Maximum horizontal speed of tip of dam	m/s²	−6.35	1.44	2.11	0.75
			Maximum contact surface sinking of dam and foundation	m.1	0.002 3	2.96	0.000 12	0.71
			Maximum cutting power of contact surface of dam and foundation	N	1.1×10^6	2.84	6.69×10^6	3.86
		2	Maximum torsional tension of frame and foundation	MPa	1.344	2.94	1.15	3.89
			Maximum loading tension of frame and foundation	MPa	1.483	3.56	0.06	3.89
			Maximum displacement tip of dam	m.1	0.023 6	3.62	0.025	3.89
			Maximum horizontal speed of tip of dam	m/s²	6.43	1.44	1.74	0.23
			Maximum contact surface sinking of dam and foundation	m.1	0.002 961	2.44	0.009 9	0.23
			Maximum cutting power of contact surface of dam and foundation	N	1.82×10^6	3.44	7.91×10^6	0.37
		3	Maximum torsional tension of frame and dam	MPa	2.413	6.38	1.295	5.46
			Maximum loading tension of dam frame	MPa	3.769	6.38	1.37	6.58
			Maximum horizontal displacement of tip of dam	m.1	0.048 3	4.44	−0.058	6.58
			Maximum horizontal speed of top of dam	m/s²	−16.54	4.46	−2.7	1.03
			Maximum contact surface sinking of dam and foundation	m.1	0.007 593	4.9	0.003	6.58
			Maximum cutting power of contact surface of dam and foundation	N	3.42×10^6	5.56	4.52×10^6	6.58

Table 5 Dynamic analysis results of overflowing section in Baghan dam located on BK1 construction

Section name	Construction type	Loading	Description	Unit	2D analysis results		3D analysis results	
					Value	Occurrence time	Value	Occurrence time
Overflowing	BK1	1	Maximum tension – torsion between frame and dam	MPa	1.346	3	1.14	0.7
			Maximum loading tension of frame and dam	MPa	3.02	3.66	1.49	0.76
		2	Maximum horizontal displacement of tip of dam	MPa	1.198	1.74	1.02	0.22
			Maximum horizontal speed of tip of dam	MPa	5.51	3.62	1.22	0.36
		3	Maximum contact surface sinking of dam and foundation	MPa	1.28	6.02	1.273	6.58
			Maximum cutting power of contact surface of dam and foundation	MPa	1.27	6.38	3.54	3.84

References

[1] Westergard, H. M. (1933). Water pressure on dams during earthquakes. ASCE.

[2] Kotsubo S. (1960). Dynamic Water Pressure on Dams During Earthquakes. PROC. 2ND. World Conf. Earthquake ENG.

[3] Zangar C. M. 1953. Hydrodynamic Pressure on Dams Due to Horizontal Earthquakes. PROC. Soc. Exper. Stress Analysis Vol. 10.

[4] Chopra A. K. 1967. Hydrodynamic Pressure on Dame During Earthqukes. PROC. ASCE, Vol. 93 NO Ems.

[5] Pal, N. Seismic Cracking of Concrete Gravity Dams. Journal of the Structural Division, ASCE, 102(9), 1976,1827-1844.

[6] Fok, K. L., Hall, J. F. Chopra, A. K. EACD-3D: A Computer Program for Three-Dimensional Analysis of Concrete Dams. Report No. UCB/EERC 86/09, University of California, Berkeley, 1986.

[7] Bhattacharjee S. S., Leger P. (1994). Application of NLFM models to predict cracking in concrete gravity dams. Journal of Structural Engineering, ASCE, 120(4), 1255-1271.

[8] Ghaemian, M., Ghobarah, A. Nonlinear Seismic Response of Concrete Gravity Dams with Dam-Reservoir Interaction. Journal of Engineering Structures, Vol. 21, 1999,306-315.

[9] Leclerc M, Le'ger P, Tinawi R. RS-DAM Seismic rocking and sliding of concrete dams, USERS Manual. Department of Civil Engineering, E'cole Polytechnique, Montre'al, Quebec, Canada, 2002.

[10] Leclerc M, Le'ger P, Tinawi R. Computer aided stability analysis of gravity dams - CADAM, USERS Manual. Department of Civil Engineering, E'cole Polytechnique, Montre'al, Quebec, Canada, 2002.

[11] Leclerc M, Le'ger P, Tinawi R. Computer aided stability analysis of Gravity Dams - CADAM. Int J AdvEng Software 2003,34:403-20.

[12] Chuhan, Z., Guanglun, W., Shaomin, W., Yuexing, D. Experimental Tests of Rolled Compacted Concrete and Nonlinear Fracture Analysis of Rolled Compacted Concrete Dams. Journal of Materials in Civil Engineering, Vol. 14, No. 2, 2002, 108-115.

[13] Tan, H. and Chopra, A. K. EACD-3D-96: A Computer Program for Three-Dimensional Earthquake Analysis of Concrete Dams, Report No. UCB/SEMM-96/06, University of California, Berkeley, 1996.

[14] Mirza-Bozorg, Hassan. 3D nonlinear seismic analysis of concrete gravity dams by considering interaction effect of dam lake. Doctoral dissertation in construction, sharif university, Establishing and derievilent engineering college.

[15] Rastegarfar, Asghar and Moeini, Syavash and Kiamanesh, Hassan and Ayazi, Mohammad Hassan, 2011. The comparison of 3D and 2D analysis method of flat tension and 2D flat bowing down in light-weight concrete dams; national congress os construction, roadways, architecture of Ghaloos Azad university.

[16] Moeini, Mohsen and Ghaemian, Mohsen and Mohammadi-Shoja, Hossein, 2003. Analysis of non-linear seismic response of concrete gravity dams consisting dam and lake interaction by using damage mechanic. Fourth seismology international conference and seismic engineering.

[17] Fazeli, Meisam, and Ghaemian, Mohsen. 1383. Parametric study of distributed cracking non-linear model in non-linear analysis of concrete gravity dams due to earthquake. Second national congress about civil engineering.

[18] Mazloomi-Balsini, Arash. 2006. Nonlinear seismic analysis of rolled concrete dam consisting of dam and lake interaction by using distributed cracking method. Senior expert thesis about hydrolic construction, sharif university, civil engineering college.

[19] Mahmoodianshooshtari, Mohammad, and Sadeghi Chikani, Pooriya. 2013. Non-linear seismic analysis of

concrete gravity dams consisting of dam and reservoir – interaction emphasized on stability criteria. Seventh civil engineering national congress, Sistan and Baloochestan university, Shahid Nikbakht engineering department.

[20] Omidi, Rolled concrete dam analysis.

[21] Cheraghi, Nader. 1998. Concrete gravity dams dynamic analysis by considering dam and reservoir interaction, senior expert thesis about construction. Sharif university, civil engineering department.

[22] Moradi – Moghadam, Mohammad – Reza. 2004. Foundation effect on arciform concrete dams' seismic behavior by considering dam and lake interaction. Senior expert thesis about hydrolic construction, Sharif university, civil engineering department.

[23] Heirani, Zahra, and Razm – khah , Arash and Vosooghi – far, Hamid – Reza. 2005. Necessary arrangement and preparations for, designing RCC and RMD seismic dams. The first seismic international congress and construction style – making, Ghom university engineering – technical department.

[24] Esmail, Nia Omran, Mohammad and Mahdilu – Torkamani, Hamed. 2011. The evaluation of chegin rolled concrete dam stability compared to statical and dynamical loads. The first international congress and the third national congress of dam and water – electric power – producer.

[25] Farhad – Abadi, Chia, and Hassanloo, Mahmood. 2011. The application of ANSYS and ABAQUS finite element software and inspection of their optimum influence in dynamical gravity dams analysis. The first international congress and the third national congress of dams and water – electric power – producer.

[26] Bagheri, Seyyed – yaser, Ghaemian, Mohsen. 2012. Dynamical non – linear rolled concrete dams analysis having low cement. The ninth international congress of civil engineering, Esfahan industrial university.

[27] Moghaddam, Haman. Dynamic analysis method. Sharif university booklet.

[28] Mahan – Far. 2014. The inspection and investigation of asymmetrical foundation on RCC dams 3D behavior and comparison and analysis of 2D and 3D behavior of Boosher Baghan dam.

Thermal Stress Analysis for the Lai Chau RCC Gravity Dam

Marco Conrad[1], Marko Brusin[2], David Morris[1], Pham Van Trong[3]

(1. AF – Consult Switzerland Ltd, Switzerland;
2. AF – Consult Energy d. o. o. Beograd, Serbia;
3. Power Engineering Consulting JSC (PECC1), Vietnam)

Abstract: The Lai Chau HPP is the uppermost hydro power project in the cascade of power plants on the Da River in North West Viet Nam. It includes a 131 m high RCC gravity dam and a 1 200 MW power station. The technical design started in 2010, just after completion of the downstream Son La RCC dam, construction of the Lai Chau RCC dam with 1.884 Mm^3 of RCC began in March 2013 and the RCC placement was complete in May 2015. The paper describes the 2D thermal stress analyses carried out during the Technical Design Stage, when only limited data of the RCC properties representing the RCC mix to be used in the dam were available. The methodology of RCC properties estimation and thermal stress models setup is presented to determine the allowable maximum RCC placement temperature to be specified. The paper further describes the 3D thermal stress analyses performed during the Construction Drawings Design Stage, when the RCC mix used for the dam was refined and finally defined and according RCC properties were available, as well as the RCC construction programme finalised. The 3D thermal stress analyses considered a relaxation of the specified RCC placement temperature from 18 to 20 ℃ and confirmed that a reduction of RCC monoliths joint spacing to 20 m for the largest part of the dam had to be implemented. It was seen in the analyses and through embedded thermistors that one of the most important and effective temperature control measures during the hot seasons, apart from passive and forced cooling, was a rapid RCC placement to reduce ambient heat gains of the fresh RCC.

Key words: RCC gravity dams, thermal analysis, temperature control, early age behaviour

1 Introduction

The thermal analysis of an RCC dam is carried out to determine the temperature variations within the dam, during and after construction to determine the thermal stresses, confirm transverse joint spacing between dam monoliths, evaluate the requirement for a longitudinal joint and specify possible temperature control measures (maximum permissible RCC placement temperature and measures to control it during construction of the RCC dam) with the aim to sensibly reduce thermal cracking potential. As is well known, RCC temperatures increase after placement as a result of the hydration of the cementitious materials. Cooling within the dam body commences due to the difference between the internal temperature and the external ambient temperature and continues until the centre of the dam reaches a stable mean temperature or temperature cycle, whereas the faces will follow the ambient temperature cycles over the year. Cooling causes a volume reduction and under restraint this will cause tensile stresses in the RCC. Cracking, either as surface or mass cracking, will occur if tensile strains higher than the tensile strain capacity of the RCC occur or stresses exceed the tensile strength.

Due to the transient behaviour of tempertaure fields, evolution of RCC properties and the layered RCC construction process, the thermal stress analysis of a RCC dam is rather complex and requires the consideration and implementaton of a number of different domains such as the rock foundation, the RCC dam monolith geometry including number and locations of galleries, the RCC construction programme, ambient conditions and other boundary conditions. In addition, the RCC mix along with its heat generation characteristics and development of material parameters with time should be known to assist in arriving at reasonable and representative conclusions regarding temperature control measures.

In early design stages when the RCC trial mix programme has just started, especially the RCC properties and their development with RCC age are usually not fully known and confirmed. In some cases this also refers to the RCC construction programme, which has a major impact on the temperature and stress fields in the RCC dam body. It is therefore recommended that the degree of complexity of the thermal stress analysis methodology should be selected sensibly based on the available knowledge about the input parameters and on the number of parametric variations to be studied (table 1).

Table 1 Thermal stress analysis concepts

Project	Analysis methdology	Complexity of analysis	Estimate / prediction of missing inputs	Remarks	
Design progresses / Knowledge about input parameters becomes increasingly available	All spreadsheet analysis	Less complex	More conservative	e.g. according to USACE [1]	Estimate of a possible RCC maximum placement temperature (envelope) for a certain monolith joint spacing
	2D Finite Element Model of dam cross section			May be accompanied by mass gradient cracking comparison analysis as spreadsheet analysis	Allows to study a larger number of parameter variations to understand impacts on thermal stresses
	3D Finite Element Model of full monolith	Very complex	Less conservative		Confirmation of RCC maximum placement temperature for designed monolith joint spacing. Limited set of parameteric variations due to considerable computational efforts

In the case of the Lai Chau RCC gravity dam the above procedure has been followed through the course of the Technical Design Stage (Tender Design Stage), in which the range of possible maximum allowable RCC placement temperatures was first established in a spreadsheet analysis and the temperature control measures further refined with a 2D Finite Element Analysis (FEA), and continued into the Detail Design Stage, in which the temperature control measures were confirmed in a 3D FEA of full monoliths. The spreadsheet analysis narrowed the possible maximum allowable RCC placement temperature to between 16 and 20 ℃, and 18 ℃ was finally determined in the 2D thermal stress FEA. Since various important parameters could not be confirmed during the Technical Design Stage, such as the actual RCC placement programme and RCC strength and elastic properties, the 3D thermal stress analysis was envisaged for the early phase of the Detail Design Stage, when the relevant parameters were confirmed. The Higher Authorities decreed a maximum allowable RCC placement temperature of 20 ℃, higher than what was recommended earlier, and this led to an adjustment of the maximum monolith joint spacing to 20 m for the majority of monoliths. The full block 3D thermal stress analysis was therefore carried out to confirm the specified maximum allowable RCC placement temperature of 20 ℃ for the monolith joint spacing of 20 m.

Temperature control measures specified from both, the 2D and the 3D thermal stress analyses included the installation of wet belts to cool the coarse aggregates to about 10 to 11 ℃ at the outlet of the cooling tunnel, the use of chilled batching water and the provision of flake ice to allow continued RCC placement even through very hot periods of the year (average daily ambient air temperatures varied between 13 and 33 ℃ over the year). To control thermal gradients at exposed surfaces, temporary surface insulation by styrofoam boards was introduced in parts of the dam.

2 Technical Design Stage 2D FEA

2.1 FE model and selection of analysed dam cross – sections

The finite element analysis was carried out using the software package FEnas (Finite Element Non – linear Analysis Software by Walder&Trüeb, Switzerland), which particularly addresses the features required for a detailed thermal stress analysis of an RCC dam. A two – dimensional geometry of the dam sections considered was modelled as a typical and sufficient approximation for the thermalstress analysis of a gravity dam with a significantly longer dam axis than the upstream – downstream dimension of the cross – section. Each individual RCC layer was modelled with one iso – parametric, eight nodes element in height. The consideration of each single RCC layer is necessary for the realistic simulation of heat gains and losses of the RCC during its exposure to ambient conditions until being covered by the subsequent RCC layer. The modelling methodology was phased. The transient temperature fields resulting from the RCC hydration and according start and boundary conditions during dam construction and operation were computed first. In a second transient computation the thermal stress fields resulting from the temperature fields at each time step were calculated with the plane strain condition and a linear – elastic material model, taking into particular consideration the evolutionary character of the RCC modulus of elasticity of each RCC layer. The time steps for both analysis phases were 6 hr until one year after completion of the RCC dam section and 2.5 d during reservoir operation. The simulation covered 25 yr.

For the selection of the dam cross – sections to be considered for the thermal stress analysis, the placement schedules of different dam sections in conjunction with the ambient temperatures were studied with particular focus on planned placement breaks, the corresponding seasons of their exposure and on the season of starting RCC placement in the dam sections considered. From this analysis two dam cross-sections were identified for study (Fig. 1):

(1) Non-overflow section Right of Intakes: Placement break at low dam elevation, spanning cold season and including cold spell (increased surface cracking risks);

(2) Spillway section: Started RCC placing during hottest time of the year (increased mass gradient cracking risks) and placement break at mid-height including a cold spell (increased surface cracking risk).

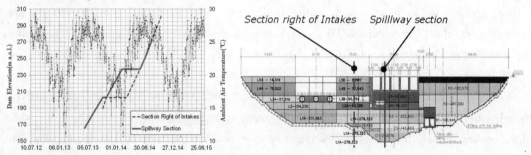

Fig. 1 Initial RCC placement schedule of the two modelled cross sections in relation to ambient temperatures available in Technical Design Stage

(note that RCC block layout and programme were adapted during approval process of the Technical Design)

2.2 Prediction of some important input parameters

At the time of carrying out the Technical Design Stage thermal stress analysis only some of the required model input parameters were available, such as past records of diurnal ambient air temperatures, cylinder compressive and direct tensile strength results up to 182 days of age from the Stage 1 RCC trial mix programme and up to 91 days of age from the Stage 2 RCC trial mix programme, both programmes using aggregate which did not represent the actual material from the prospected quarry. Since it was anticipated that the same cementitious materials would be used for Lai Chau as were used for Son La (But Son Portland cement PC40, treated fly ash from Pha Lai), the RCC testing results from Son La were available for predicting missing RCC properties for the Lai Chau thermal stress analysis. The prediction of the Lai Chau RCC properties anticipated a mix proportion containing 60 kg/m^3 But Son PC40 and 160 kg/m^3 treated fly ash. The predictions could later be compared against the actual tested properties from the final trial mix programme stages and from in-situ testing results obtained from the Lai Chau full scale trial embankments, prior to the Detail Design Stage 3D thermal stress analysis.

Long-term core compressive strengths were derived on basis of past experience and in particular on the basis of the RCC strength development observed at Son La. An exponential type ageing function was derived, also using the experience from other RCC projects, which correlated the early-age and long-term strengths to the compressive strength at 91 days (Fig. 2). The regression showed that 365-day strengths would rise to 130% and 1,000 day strengths to 150% of the 91-day strength. Fig. 2, right, shows that the predicted core compressive strengths were slightly higher than the initial core testing results.

The predicted in-situ vertical direct tensile strengths were based on the same ageing funtion as derived for the core compressive strengths and a ratio of 6.3% between the in-situ parent vertical direct tensile strength and the in-situ compressive strength was assumed, based on the experience made at Son La. Fig. 3 shows a good agreement between the prediction and the initially available core testing results. Since the thermal stresses have to be compared against the horizontal direct tensile strength for surface cracking and mass gradient cracking perpendicular to the dam axis (transverse cracking), the vertical direct tensile strengths were increased by a factor of 1.14 to consider the anisotropic behaviour of RCC as experienced at Son La.

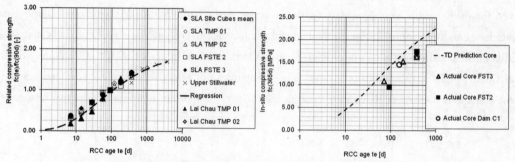

Fig. 2 Derived ageing function for strength development and comparison of predicted with actually tested core compressive strength

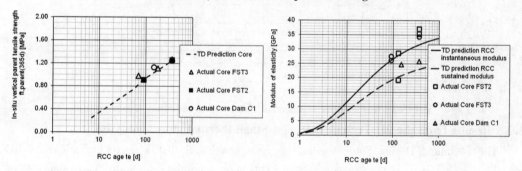

Fig. 3 Comparison of predicted with actually tested coreparent vertical direct tensile strength and modulus of elasticity

The prediction of the modulus of elasticity is comlicated by the absence of any test results. Limited 91-day results from cylinder testing were available, along with the experience from Son La and other RCC projects. An ultimate modulus of elasticity was estimated by using multi-composite material models (see e.g. Mindess[2]) and conservatively estimating the instantaneous modulus to 34 GPa at 365 days of RCC age. An empirical ageing function was used to correlate the 1-year modulus to other RCC ages. Fig. 3 shows that the prediction more or less met the actual properties tested from cores from the full scale trials. In order to account for creep-relaxation effects in the RCC dam body, a sustained modulus of elasticity was implemented in the thermal stress analysis as an effective modulus. The usual conventional assumption to date was that the sustained modulus would be in the order of 2/3 of the instantaneous modulus, but recent research[3] showed that creep-relaxation in a high cementitious high volume fly ash RCC is less pronounced and that this ratio should be considered to be larger. In agreement with comparisons made at Son La the ratio between sustained and instantaneous modulus was increased to 3/4.

Solar radiation was implicitly considered in the thermal stress study as a supplemental temperature to the diurnal ambient temperatures. The temperature supplement was derived from a geometric model[4] which takes the spatial orientation and the geographic location of the dam into account. Due to the spatial orientation of the dam and the different inclinations of the facings, different temperature supplements normally apply for the upstream and downstream faces as well as the horizontal surfaces of exposed RCC layers. The solar radiation temperature supplement is derived for each month and is added to the average daily ambient air temperatures as the final boundary condition input in the thermal model. It has been seen on other RCC projects that solar radiation heat gains can be as high as 5 ℃ and a maximum supplement of 3.5 ℃ was derived for Lai Chau (Fig. 4). Maintaining a rapid placement rate and effective curing to minimize ambient heat gains and the loss of RCC pre-cooling are therefore important temperature control measures.

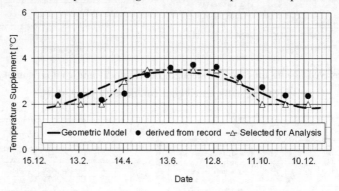

Fig. 4 Lai Chau temperature supplement due to solar radiation(horizontal surface)

2.3 Results from the 2D Technical Design Stage thermal stress analysis

The Technical Design Stage thermal analysis concluded that a maximum allowable RCC placement temperature of 18 ℃ was specified. This was concluded from both, the surface gradient and mass gradient point of view. Maximum interior RCC temperatures of 42 ℃ were assessed in the upper part of the Spillway Section. Maximum tensile stresses of 1.4 MPa in upstream-downstream direction occurred during the construction period from steep thermal gradients at the exposed surfaces and particularly at the extended placement break in the bottom part of the Non-overflow section (Section at right of Intakes). This resulted in the requirement of an insulation (0.5 m sand layer) during the exposure of the horizontal layer surface. In the long-term the temporary high tensile stresses at the placement break turned into compressive stresses with long-term tensile stresses developing above the placement break. Long-term maximum tensile stresses occurred at the bottom part of the Spillway Section and reached 1.3 MPa in upstream-downstream direction. Maximum transverse tensile stresses were predicted at 0.9 MPa in both sections.

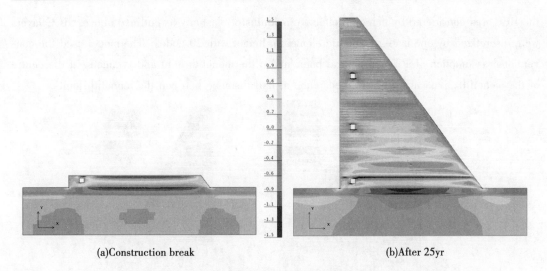

Fig. 5 Non-overflow block - longitudinal thermal stresses during construction break (left) and after 25 yr (right)

3 Detail Design Stage 3D FEA

3.1 FE model and procedure of analysis

The full 3D transient thermal stress analysis was carried out with an improved information on input parameters (in particular the RCC placement programme, ambient and boundary conditions, RCC properties) and used the opportunity that no parametric study had to be performed. The 3D thermal stress analysis solely considered the specified maximum RCC placement temperature of 20 °C and the as-designed conditions for the two RCC monoliths simulated. Again, each individual RCC layer and its exposure to ambient conditions was modelled in accordance with the construction programme.

The simulation was carried out in a number of steps to make sure that all relevant loads were correctly considered and results as realistic as possible. The first stage was the transient incremental calculation of temperature fields in accordance with the actual construction boundary conditions. The second stage was the transient incremental calculation of thermal stresses as a result of the spatial and temporal temperature differences calculated from the temperature fields. The third stage represented the transient incremental calculation of the stress-strain state due to the self-weight in accordance with the construction programme. The final stage represented the linear first order calculation of the stress-strain state due to the hydrostatic and uplift loads under normal operation conditions. The final combined stresses resulted from the superposition of the individual stress computations for each time step. Time steps were slightly increased compared to the 2D analysis, with 8 hr from start of RCC placement until one year after the completion of the RCC monoliths modelled and 3.0 d from then on until 25 yr after start of RCC placement (end of finite element simulation). All analyses used linear elastic isotropic materials, whereas the thermal stress and self-weight analyses implemented a time variable modulus of elasticity to simulate the increasing RCC stiffness over time (hardening process). Creep, respectively stress relaxation of the RCC was considered by using an effective modulus of elasticity as outlined above. RCC layers were discretized by one iso – parametric elementin height with 20 nodes. The model used the conventional assumption of a free structural boundary at the monolith joint and symmetry at the centre of the monolith, i. e. a fully restrained plane at half distance between the monolith joints.

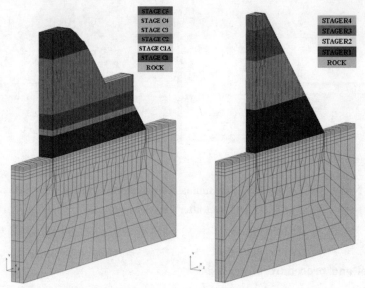

Fig. 6 3D FE models of Spillway Section (left) and Non – overflow Section (right)

3.2 Results of 3D thermal stress FEA

Fig. 7 shows the combined (static + thermal) long – term cross valley stresses and longitudinal stresses in the Spillway Section in the symmetry plane, i. e. in the monolith centre. Maximum RCC temperatures at the bottom and top part of the section reached 43 ℃ with a placement temperature of 20 ℃. The long – term cooling results in maximum cross – valley stresses of around 0.6 to 0.9 MPa in zones of around 5 m from the facings, where tensile stress zones seem to be trapped and appear cyclically during dam operation. Long – term cross – valley stresses in the section interior are limited to about 0.5 MPa after 25 yr. Long – term longitudinal stresses are in the same order and reach their maximum close to the dam – foundation interface and below a construction break with a season spanning exposure into autumn 2013. Minimum safety factors against transverse and longitudinal cracking below or close to the desired safety factor of 1.5 mostly occur during the construction time and at exposed surfaces with steep thermal gradients.

3.3 Comparison of 3D thermal stress FEA results with dam instrumentation

A number of concrete effective stressmeters and long – base strain gages in an arrangement proposed in[5] were installed in some cross sections of the Lai Chau RCC dam. While the agreement between computed and measured RCC temperature histories is very good, it is the stress comparison which is of interest. There are only very few RCC dams with thermal stress monitoring to some extent and such monitoring is extremely helpful to understand the early – age behaviour of RCC especially regarding the creep – relaxation behaviour, respectively regarding the effective modulus of RCC. Figure 8 compares the computed and measured longitudinal and cross – valley stresses in the centre of the modelled Spillway Section. The comparison includes results from a much shorter 3D check simulation of the same cross section with constraints parallel to the dam axis at both lateral ends (i.e. deformation restraint at the monolith joint assuming full bond at the monolith joint).

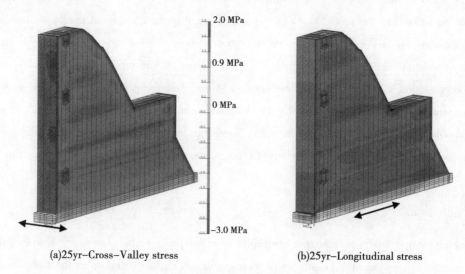

(a) 25yr-Cross-Valley stress (b) 25yr-Longitudinal stress

Fig. 7 3D FE models of Spillway Section (left) and Non-overflow Section (right)

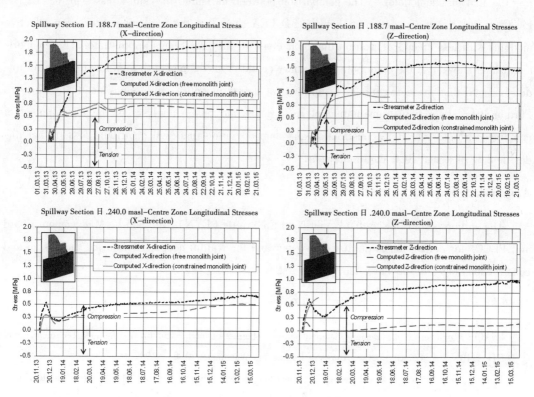

Fig. 8 Spillway Section, comparison between measured and 3D computed stresses

There is significant disagreement between the measured and computed long – term longitudinal stresses close to the dam – foundation interface, which most probably is related to boundary conditions not fully considered in the model (e.g. rock foundation topography and large working platforms filled at the upstream and downstream faces whose influences on deformation restraint were not considered in the model). Early – age and longitudinal stresses at the upper part of the section (where boudary conditions are more certain) show a good agreement. The deformaton boundary condition at the monolith joint has a great impact on the cross – valley stresses and the results suggest that a better agreement between measured and computed cross – valley stresses at both elevations is obtained by restraining the deformation at the monolith joint parallel to the dam axis. A more long – term monitoring from the stress meters is required to allow further conclusions, especially covering the period when stresses move toward tension and when monolith joints open.

4 Conclusions

Thermal stress analyses require a multitude of input parameters. The level of knowledge about those parameters and the possibility of their prediction should be taken as the basis for selecting the analysis methodology. The comparison between measured in – situ stresses in the RCC at Lai Chau and the stresses computed in the 3D FEA show that the implementation of an effective modulus with less creep – relaxation than conceived in the past seems appropriate. The comparison further shows that the deformation boundary condition at the monolith joint has to be better understood and that the conventional assumption of a free monolith joint seems not to be correct. Further stress – strain instrumentation of RCC dams and post – analyses of the development of combined stresses over RCC construction and dam operation time will contribute to a further understanding of actual RCC dam behaviour.

Acknowledgements

The authors would like to thank Viet Nam Electricity (EVN), the Son La Management Board and PECC1 for permission to publish this paper and the fruitful cooperation during the design and construction stages of the Lai Chau HPP.

References

[1] USACE. Engineering and Design: Thermal studies of mass concrete structures. Engineer Technical Letter ETL 1110 – 2 – 542, U.S. Army Corps of Engineers, Washington D.C., 1997.

[2] Mindess, S., Young, J.F., Darwin, D. Concrete. Second edition, Prentice Hall, Upper Saddle River, 2003.

[3] SHAW, Q.H.W. A New Understanding of the Early Behaviour of Roller Compacted Concrete in Large Dams. PhD Thesis. University of Pretoria, South Africa. 2010.

[4] Van Breugel, K. , Koenders, E:A:B. Solar radiation. Report BE96 - 3843/2001:31 -1, Improved Production of Advanced Concrete Structures (IPACS), Luleå University of Technology, 2001.

[5] Conrad, M. , Shaw, Q. H. W. , Dunstan, M. R. H. Proposing a standardized approach to stress - strain instrumentation for RCC dams. 6th Int. Symp. on Roller Compacted Concrete (RCC) Dams, Zaragoza, 2012.

Simulation Analysis on the Thermal Field and the Stress Field for Gongguoqiao RCC Dam

Chen Hongjie[1], Wang Zhaoying[2]

(1. Huaneng Lancang River Hydropower Ltd, Technology R.&D center, Kunming, 650214, China;
2. Miaowei · Gongyuoqiao Hydropower Project Administrationof HuanengLancang
River Hydropower Inc. Dali, 672708, China)

Abstract: Gongguoqiao dam is a concrete gravity dam compacted by rollers with sand and slate as concrete aggregate. Due to the poor grain shape with needles and the poor properties of the concrete, the crack resistance of this dam is relatively low. It makes the project having more complicated temperature control, which leads to the notch flow in flood seas on. To better optimize design and construction, there is a great need of simulation analysis of the thermal stress field for the Gongguoqiao RCC dam. A review of features and research on temperature control of the RCC is introduced briefly firstly. Followed by the basic finite element equations of the thermal stress field, computational method of simulation analysis is carried forward in the paper. The research is focused on the temperature control standard and stable stress control standard of the concrete according to the "Design specification for concrete gravity dams". The simulation calculation of the thermal and stress field during the whole construction period and long-term operation are presented in the paper. And based on the simulation calculation results, the thermal field and stress field are judged to meet the temperature control standard or not. Compared with the thermal and stress field of the non-water cooling and the water cooling conditions, temperature control measuresare proposed at last.

Key words: RCC dam, Gongguoqiao hydropower station, the thermal field, the thermal stress, the temperature control standard, simulation analysis

1 Introduction

Roller compacted concrete is one kind of dry concrete in manner of compaction by rolling[1]. Temperature rise of massive RCC is relatively slow at the beginning and fast in the later stage. And the period to reach the maximum temperature and duration is a little long. Owing to construction complicated and lasting temperature effect, how to dynamically simulate temperature field of the whole process is an important prerequisite to achieve the temperature control of RCC dam. RCC dam takes the continuous pattern for pouring. Heat dissipation of internal is slow and temperature rising is high. All of this will result in great temperature stress. Due to the poor tensile properties of concrete and dam cracks at the over-wintering surfaces, temperature control plays a prominent role in the process. It always last several years to complete the construction for a RCC dam construction, which leads to the difference between temperature of internal dam and that of upper - lower over - wintering intermittent surface[2]. Temperature control calculation is involved in the whole process dynamic simulation method by considering changes in various factors at different stage. And finite element method is one of the most commonly used measure for RCC temperature field calculation at present.

Gongguoqiao dam is a roller compacted concrete gravity dam with design parameter as 356.0 m in total length in, 1310.0 m in crest elevation, 105.0m in maximum height. The concrete strength grade of this dam is relativity low, $C_{180}20W10F100$ for upstream anti - seepage RCC and $C_{180}15W4F50$ for middle and downstream RCC. The concrete aggregate composed of sand and slate and due to the poor grain shape that has many needles and the poor properties of the concrete, the crack resistance of this dam is relatively low. It makes the temperature control of this project much more complicated, which is at the risk of the notch flow in flood season. There is a great need of simulation analysis of the thermal stress field for the Gongguoqiao RCC dam.

2 Calculation model

The finite element method is adopted to this simulation calculation. The entire dam along with dam axis is selected to be the calculation model. Coordinate origin of the global coordinate system is on the left side of the dam heel. The right bank of the dam direction is positive x - axis, the downstream direction is positive y - axis, the vertically upward direction is positive y - axis. Calculation model in the dam depth direction is 100 m, 100 m in the upstream direction and 100 m in the downstream direction.

Boundary conditions of calculation: Bottom foundation, four flanks and contraction joints of the dam to be adiabatic boundary; upstream and downstream surfaces above the water level to be solid - air boundary; upstream and downstream surfaces below the water table to be solid - water boundary. And solid - air boundary is the third boundary condition while solid - water boundary is the first boundary condition. Stress Field calculation takes the entire dam. Bottom foundation takes as fixed support, foundation along upstream and downstream takes as fixed support along Y direction and the rest is free boundary. Calculation model of the spillway section is shown in Fig. 1.

The bedding concrete takes as part of foundation and its elastic modulus is 33 GPa. The paper aims at studying temperature control measures upon constructing bedding concrete. RCC dam takes the continuous pattern for pouring and construction joint research is not considered.

3 Calculation condition

This paper only selects two scenarios, and compares water cooling with non - water - cooling in the first stage through simulated analysis. The working conditions are presented in table 1.

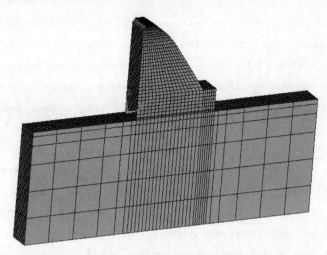

Fig. 1　Calculation Model of the spillway section

Table 1　Calculation conditions for unstable temperature field and temperature stress

Combination	Water cooling in the first stage
Condition 1	No
Condition 2	10 ℃ water cooling for the whole section of 1 205 ~ 1 288 m

4　Analysis of the simulation result

4.1　Emulation analysis of unstable temperature field

Typical temperature field distributions of working condition 1 (without water) and 2 (with water) are shown in Fig. 2 ~ Fig. 4.

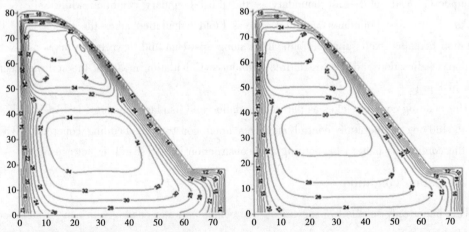

Fig. 2　temperature isogram at the final stage of construction period under working condition 1

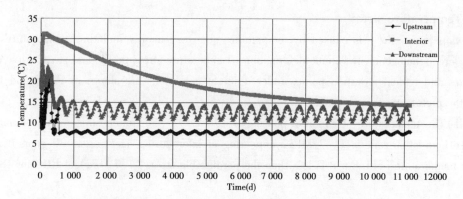

Fig. 3   Temperature duration curve of the classic points of the horizontal plane with the height of 1 214 m under working condition 1

We can see from the calculation of the construction and operation periods of the overflow dam that:

(1) During the construction period, the temperature was high in high – grade RCC on upstream and downstream faces and in the area of normal concrete around wear – resisting layers of spillway surfaces, while the temperature was relatively lower in the main RCC. The main reason is that the adiabatic temperature rise in high – grade RCC on upstream and downstream faces and in the area of normal concrete around wear – resisting layers of spillway surfaces was high while that in the main RCC was low.

(2) Compared with plan 2 with no water supply, under the condition that supplying cooling water of 10℃, maximum temperature on the inside of the dam was lowered by 4.5 ~5℃ in plane 1.

(3) The surface temperature of the dam clearly has a positive correlation with the outside temperature. During the construction and operation period, the interior temperature of the dam merely is linked with temperature within or outside of concrete during pouring and agingin particular, measures for supplying water and cooling the temperature. After 30 years of operation, the interior temperature of the dam basically reaches the quasi – stationary temperature field.

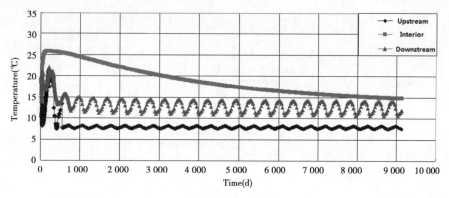

Fig. 4   Temperature duration curve of the typical points of the horizontal plane with the height of 1 214 m under working condition 2

4.2 Temperature stress

Temperature Stress is caused by temperature changes, mainly including temperature differences, creep, autogenously volume deformation. Temperature stress profile and stress curve of typical sites and typical elevation points under different conditions are shown in Fig. 5 and Fig. 6. From the pictures we can see that:

(1) The temperature of concrete pouring is 19 ~ 21 ℃. Maximum stress of the strong – restrained foundation zone is 1.43 MPa, appearing at later curing age (up to 30 years) close to steady temperature field. But it is less than the permissible stress (1.7 MPa) of RCC for 180 d ages.

(2) Maximum stress of weak – restrained foundation RCC zone (height of 1 223.0 ~ 1 238.0 m) is 1.12 MPa, 0.95 MPa for none – restrained RCC zone (height of 1 238.0 ~ 1 244.0 m), 0.85 MPa for affected intermittent surface RCC zone (height of 1 244.0 ~ 1 268.0 m), 0.64 MPa for none – restrained RCC zone (height of 1 268.0 ~ 1 280.0 m) and 0.80 MPa for none – restrained normal concrete zone (height of 1 280.0 ~ 1 288.0 m). All of them are less than the maximum permissible tensile stress.

(3) Maximum stress along the surface stream of 1 256 m intermittent surface is about 0.6Mpa after overflow, which basically meet the central section of the RCC 28 days' permissible tensile stress. But if the curing time cannot reach 28 d, the tensile stress will exceed the permissible tensile stress. There is a need to perform the first or second cooling and surface cooling near the intermittent surface.

If the first age cooling is carried out around the affected intermittent surface and the second age cooling before overflow, the tensile stress near the height of 1 256 m will be reduced obviously. Maximum stress after overflow is about 0.2 MPa.

Maximum stress on vertical direction of spillway sections will greatly be reduced after self – weight is considered. The changes in surface heat release coefficient is taken into consideration in the simulation of this paper while the influence of water load is not. Essentially, this will be beneficial to the temperature control.

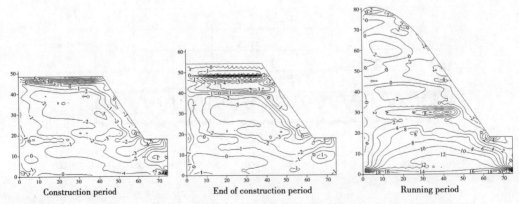

Fig. 5 **Isogram of temperature stress σ_y of typical period under working condition 1**

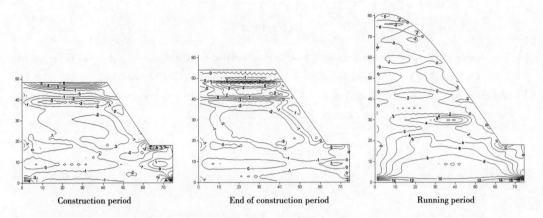

Fig. 6 Isogram of temperature stress σ_y of typical period under working condition 2

5 Conclusions

(1) Maximum temperature of internal dam can be lowered at 2 ~ 5℃ by laying cooling water pipe at local site during construction period. Thus it is an effective way to take the cooling water pipe to lower the internal temperature. Maximum temperature of strong – restrained zone under condition 1 temperature control are 33.2℃ and 34.6℃. The corresponding foundation temperature differences are 19.7℃ and 20.1℃. Both of them are bigger than the suggested temperature difference standard.

(2) In addition to that stress of dam heel and dam toe edge is a little high, normal stress in each condition of the spillway section all meet final permissible tensile stress standard. The key control issues are early stress, overflow impact stress and dam heel and dam toe stress.

(3) The stress of 1 256 m intermittent surface after overflow suddenly increases from 0.2 to 0.6 MPa under non – water cooling condition, which basically satisfy 28 days' stress control standards. But it may beyond the standard if it overflows at early age (earlier than 28 d).

(4) The curing age of long intermittent surface concrete reaches to 150 d before recovering pouring, which belongs to old concrete. Affected by this concrete, the restraint stress of upper concrete along the river increases. The maximum value is 0.65 MPa, but it still remains within the scope permitted. By laying cooling water pipe at 1/4 of the upper and lower of the intermittent surface to carry out the first stage cooling, second stage cooling and surface cooling, the maximum stress after overcurrent is 0.2 MPa, which will meet the early stress control standard.

(5) The construction of the spillway restrained foundation zone is mainly arranged at low temperature season. According to temperature stress calculation, it is not necessary to lay the cooling water pipe.

(6) The principle stress of dam heel and dam toe area is high. And there is a need to take the filling stability measures in front of the dam. The simulation calculation of the thermal and stress field during the whole construction period and long – term operation are presented. Compared with the thermal and stress field of the non – water cooling and the water cooling conditions, temperature control measures are proposed and the result shows it is an effective method.

References

[1] Tang Xinwei, Li Penghui, Zhang Chuhan. Entire process simulation of temperature and stress fields for RCC dams[J]. Journal of Yangtze River Scientific Research Institute, 2007, 24(3): 50-53.

[2] Zhu Zhengyang, Qiang Shen, Wang Haibo, . et al. The research of winter construction of RCC dam temperature control[J]. Journal of hydropower energy science, 2011, 29(1):45-47.

Thermal Study and Implemenation of Short Joints at Upper Paunglaung RCC Dam

U Aye Sann[1], U Zaw Min San[1], Christof Rohrer[2], Marko Brusin[3]

(1. Ministry of Electric Power, DHPI, Myanmar;
2. AF – Consult Switzerland Ltd., Switzerland;
3. AF – Consult Energy d. o. o., Serbia)

Abstract: The paper gives a brief overview of the layout of the Upper Paunglaung Project in Myanmar and the RCC Dam and concentrates then on the study of the short joints.

Due to changes of construction materials, RCC mix composition and construction equipment, the risk of thermal cracking of the present design had been increased and mitigating measures had to be taken. As the detail/construction design had been progressed and construction already started, many restrictions had to be taken into account when looking for a solution.

A simple and economic way to overcome the issue of thermal cracking risk was to introduce short joints in the middle of the RCC blocks at the upstream face of the dam body. A FE thermal analysis of two RCC dam Blocks has been performed, first as normal 20 m wide dam blocks, and then with introduced short joint. The results of the finite element analysis showed a reduction of the thermal stresses at the u/s face of the dam, so that the risk of thermal cracking at the upstream face could be reduced. The results of the thermal analysis will be presented here.

In the last part of the paper it is presented how the short joints had been established at construction site, and how the final completed RCC Gravity dam looks like at end of construction.

Key words: RCC Gravity Dam, Short Joints, Thermal FE Analysis

1 Introduction

The Upper Paunglaung Hydropower Project is located on the Paunglaung River, about 50 km east of the Capital Nay Pyi Taw, in Mandalay Division, Myanmar. The Project is owned by the Ministry of Electric Power with the main purpose of power production for the national grid. The 140MW Upper Paunglaung Hydropower Project comprises a RCC dam of gravity type with a height of 103 m and with a crest length of 530 m. The Gravity dam is divided into 27 concrete blocks of maximum 20 m width, and contains a total concrete volume of 950 000 m^3. The Upper Paunglaung dam is the second RCC dam project in Myanmar and the first built by local contractors. The RCC dam construction has been completed end of 2013, and the plant installation has been completed by end of 2014.

2 The project at design stage

The RCC Dam has been designed with 27 dam blocks of maximum 20 m width. In the RCC Dam incorporated are the Bottom Outlet structure, the Power Intakes and Penstocks, as well as the Spillway. Furthermore, the layout of the RCC Dam with a length of 530 m is characterised by 4 kinks. The RCC Mix has been designed in several stages in the laboratory before construction. The design RCC mix included 4 proportions of aggregates with a maximum size of 40 mm, 70 kg/m^3 cement, 160 kg/m^3 natural Pozzolan, and 120 kg/m^3 of water to meet the requirements of the design strength. It has been considered to reuse the batching plant, transporting systems and equipment for placing and forming of the RCC from the previous project Yeywa.

2.1 RCC dam block width

The maximum RCC Dam block width has been assessed based on thermal analysis, and on experience made with other project designs developed in similar climatic conditions and with similar RCC mixes. The individual blocks of the RCC dam at Upper Paunglaung were defined based on the boundary conditions of the overall layout and presence of kinks as well as the incorporated structures. The layout of the RCC dam with the block arrangement is shown in the following Fig. 1.

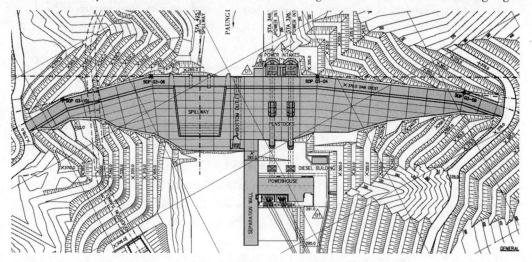

Fig. 1 Layout of the Upper Paunglaung RCC Dam

The four kinks which ensure the RCC dam is founded on good rock, the defined width and location of the Power Intakes and Penstocks, the location of the Spillway, Bottom Outlet and Separation wall, left only small range for variation and optimisation of spacing of the Block Joints. Some of these restrictions are described in the following subchapters.

2.2 Power intakes, penstocks and powerhouse

The Powerhouse with Unit spacing of 20 m matches ideally with the maximum RCC dam block spacing and therefore the spacing of the Power Intakes/Penstocks – Dam Blocks waskept to 20 m. With the determination of the block widths to 20 m, the Power Intakes, Penstocks and entry into the Powerhouse are in straight axis, and no additional bends of the Penstocks are required.

2.3 Spillway

The spillway shall be placed in the centre of the river for an ideal reservoir approach flow on the upstream side and for good energy dissipation downstream. As the valley is narrow and the required spillway width is occupying most of the impact area, the location of the spillway is fixed. The Spillway design, discharge performance and energy dissipation had been confirmed in the hydraulic model tests, and therefore should not be changed.

2.4 Separation wall and bottom outlet

The separation wall shall be adjacent to the Powerhouse with determined width based on the stability safety. The space between powerhouse and spillway is very narrow and the separation wall as well as the bottom outlet had to be fit between both structures. Therefore the bottom outlet was designed to be integrated into a 15 m wide RCC block. The bottom outlet is equipped with two maintenance (bonnet) gates and two service (radial) gates.

2.5 Foundation of the RCC dam

The RCC dam had to be moved and kinked during the design and early stage of construction due to unexpected geological conditions. The RCC dam is founded on a granite intrusion at the left bank lower part, and at the upper left abutment as well as in the riverbed and on the right abutment on metasandstone. At the upper level near surface the metasandstone had been found very blocky, disturbed, and moderate to highly weathered. Only at deeper excavation level the foundation improved too slightly to moderately weathered metasandstone. To limit the excavation depth the RCC dam had been moved exactly to the ridge of good foundation.

3 Changes in the mix properties at late stage of design

At late stage of RCC dam design and with already started construction of the project structures, it had been concluded that the 365day design strength of the RCC using the actual quarry material produced by the actual crushing plants could not be reached. The performance of the mix in the lab from early stage could not be reproduced with the produced materials on site and the specified mix proportions. Since the quarry was developed, the crushing plant installed, commissioned and production started, the materials could not be changed, only slightly improved in quality, gradation and shape.

It had been concluded that there was no option such as changing the dam shape, or changing the quarry or crushing plants, to overcome the strength problem in acceptable time. To reach the design strength of the RCC, the cement content had to be increased by 20 kg/m^3.

The consequence of the increase of cement content is a higher heat of hydration of the RCC. To minimise the risk of thermal effects several alternative measures had been discussed.

One of the possible solutions for compensation of a higher heat of hydration is to lower the placing temperature of the RCC. But at that late stage of design and already installed batching and ice plant on site with limited capacity in ice production, the placing temperature of the RCC could only be lowered in a small range.

The width of the RCC blocks depends on the thermal behaviour of the RCC. If the cement content in the RCC mix and therefore the maximum temperature in the mass concrete is increased, the risk of thermal cracking is increased as well. Therefore it had to be either proven that the block width is acceptable for the new RCC mix, or the Block width had to be reduced to minimise the tensile stresses due to the thermal load.

The reduction of the block width was not a practical solution at that time of the project. Too many restrictions such as construction started, process of manufacturing the equipment started, and design issues as explained above, did not allow reducing the block width of the RCC dam.

The only practical and economical solution was the possibility to introduce short joints at the upstream face in the centre line of the RCC blocks to reduce the thermal tensile stresses.

4 Thermal analysis of RCC blocks

The objective of the thermal analysis was to find out if the stresses of the 20 m wide RCC Dam blocks would be within the limits applying a mix with 20 kg/m^3 more cement. As it was expected that this will not be the case and the stresses will exceed the maximum design strength, thermal analysis with short joints included in the model were considered.

4.1 The thermal analysis methodology

The thermal and stress simulation of actual 3D monolith model had been used to allow realistic analysis of cross – valley stresses which potentially cause exceedance of tensile strength. The Analysis was carried out in four stages:

(1) Incremental calculation of temperature fields that are developing in the dam body over time and in accordance with actual boundary conditions applied.

(2) Calculation of thermal stresses in the dam body based on the spatial and temporal temperature differences resulted in the first stage.

(3) Calculation of stress – strain state due to applied self – weight loads.

(4) Calculation of stress – strain state due to hydrostatic pressure and uplift at normal operation conditions.

The first three individual phases were performed by an incremental transient procedure, using sufficiently small time steps. All the three analysis steps used linear elastic isotropic materials, whereas the thermal stress and self – weight analysis implemented a time variable modulus of elasticity to simulate the increasing RCC stiffness over time (hardening process). Creep, respectively stress relaxation of the RCC was considered by using an effective modulus of elasticity or sustained modulus. The analyses make use of the death and birth option of activating elements in the dam body to simulate the dam construction schedule and to activate heat conductivity or rigidity as well as the heat capacity at the right time of calculation.

The fourth phase was carried out as linear static analysis for the fully constructed dam.

Thermal stresses were superimposed with stresses from self – weight at appropriate time steps in order to get a complete stress – strain state during the construction of the dam. Superposition of stresses due to hydrostatic loads is carried out only with results of the previous three analyses at time steps that are corresponding to the time after which the water level in the reservoir has reached elevation for normal operating conditions.

4.2 The thermal 3D model

The finite element mesh for the dam was generated with an element thickness of 30 cm corresponding to the RCC layer thickness placed in the dam. This concept of discretization requires the application of quadratic, respectively iso – parametric elements with 8 nodes in the 2D case and 20 nodes in the 3D case. A finer mesh is applied in the area of upstream and downstream faces to allow simulation of steep thermal gradients and resulting stresses due to temperature changes at the facings, a coarser mesh is applied at the core of the dam body.

The foundation rock mass is modelled as a sufficiently large domain to cope with the heat generated by the RCC and conducted into the foundation (minimization of erratic heat accumulation), as well as to adequately consider deformation restraint in the dam – foundation interface. The UPL model considers a rock domain of 40 m from the upstream as well as 40 m from downstream face and 82 m below the dam – foundation interface, using a coarser mesh than applied for the dam body to reduce the number of elements.

Fig. 2 represents the 3D numerical models used in the thermal and stress analysis of modelled blocks of the UPL RCC dam.

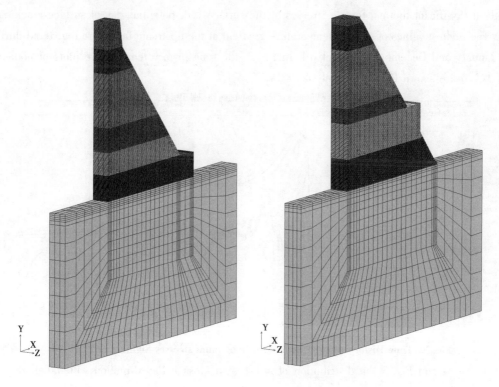

**Fig. 2 3D models of Block 14 (left) and 18 (right),
including dam body construction stages and foundation**

4.3 The modelling of the short joints

The numerical 3D models used the symmetry properties of the monolith in central plane perpendicular to the dam axis to optimize computational efforts by modelling only a half of the block. This results in free deformation of the RCC monolith at the contraction joint, and a fully restrained deformation at the plane of symmetry in lateral direction parallel to the dam axis. To model the short joints which are located at the centre line of the block, simply the type and condition of supports at specific location had to be changed.

For typical non-overflow Block 14, the second step of above mentioned calculation of thermal stresses was therefore conducted two times with different boundary conditions.

4.4 The results

This paper focuses on the risk of development of surface cracks, which is highest during the construction time of the dam. Temperature drops at exposed faces under internal restraint may cause surface cracks.

The temperature time histories (see Fig. 3) show that the temperatures close to surface quickly cool down when exposed to the ambient conditions and approach the ambient temperature cycle. Due to higher and relatively constant (slowly decreasing) temperature further away from the surface, steep thermal gradients towards the cooled surface occur. These gradients generally result in significant thermal tensile stresses at the surface with potential risk of surface cracking.

The highest values of surface temperature gradient at the upstream face were registered during the January and December for Block 14. In Fig. 3, the time history temperature plots of nodes at El. 341.0 are shown for the normal Block 14.

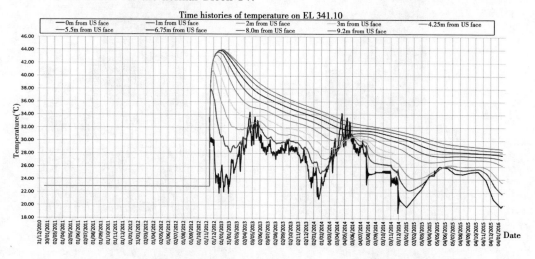

Fig. 3 Time History plots of nodes with maximum stresses Sigma z at Block 14

In Fig. 4 and Fig. 5 the distribution of stresses in cross valley direction with labelled peak values that have been observed at the upstream face are presented for normal Block 14 and for Block 14 with short joints.

It can be seen that the introduction of short joints reduced the tensile stress area and the maximum tensile stress value.

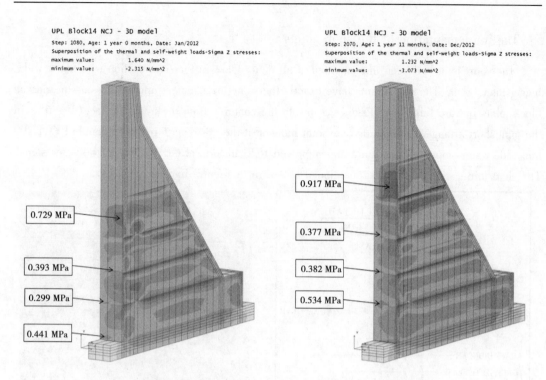

Fig. 4 Maximum observed tensile stresses on upstream surface – Block 14 (without short joints) during critical construction stages

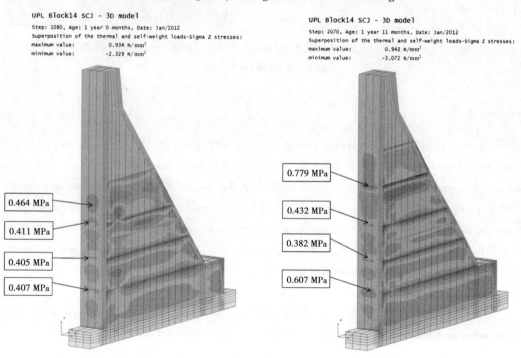

Fig. 5 Maximum observed tensile stresses on upstream surface – Block 14 (with short joints) during critical construction stages

5 The implementation of the short joints

The short joint had been implemented at all dam blocks in the river section with block width bigger than 18 m. The short joints have exactly the same arrangement and features as the normal block joints but are limited in distance 3 m into the concrete from the upstream face(Fig. 6). In the joint short arrangement included are joint inducer at the upstream formwork, double PVC, 300 mm wide water stops, formed drain and improved RCC in form of GEVR around the water stops. The short joints were installed as mentioned before in the centre line of the blocks.

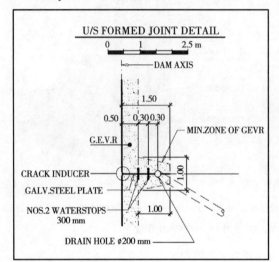 

Fig. 6 Detail of U/S formed joint valid for short and normal joint, and Photograph with indicated short joints

6 Conclusions

The thermal analysis had shown during late planning phase of the Upper Paunglaung RCC dam(Fig. 7) project, that the introduction of short joint reduced the thermal caused tensile stresses at the upstream face of the dam, and minimised the risk of surface cracks. The implementation of the short joint for construction was fast and easy because no major impact was made to the design of the structures and RCC. The costs for implementation purchase of materials and execution of the work were low as all materials, equipment and know how was the same as for the normal block joints. Last but not least the solution did not delay or slow down the placing of RCC in any aspects.

Now, two year after completion of the Upper Paunglaung RCC Gravity Dam no leakage could be detected and no short joint propagated to the inspection galleries.

Fig. 7 **The Photograph of completed Upper Paunglaung RCC dam** (taken 2015)

Acknowledgements

Gratitude must be accorded to MOEP (Ministry of Electric Power) and AF – Consult Switzerland Ltd. for permission to publish this paper, but the views expressed in this paper are purely those of the authors. Many thanks are also due to the DHPI (Department of Hydropower Implementation) engineers on the construction site who made this project to a success and to our RCC Expert M. Dunstan.

References

[1] Ch. Rohrer, et al. "Design and Construction Aspects of the Upper Paunglaung RCC Dam", Paper of ASIA 2012 Conference – Water Resources and Renewable Energy Development in Asia, of Hydropower and Dams, 2012.

[2] Ch. Rohrer, et al. "Experience made with retarder admixturesat two RCC dams in Myanmar", Paper of 6th International Symposium on Roller Compacted Concrete (RCC) dams, Zaragoza, 2012.

[3] Ch. Rohrer, et al. "Completion of the Upper Paunglaung RCC Dam", Paper of ASIA 2014 Conference – Water Resources and Hydropower Development in Asia, of Hydropower and Dams, 2014.

RCC Construction Aspects and Quality Control of Spring Grove Dam (South Africa)

Nyakale J[1], Badenhorst D[2], Mohale N[2], Trümpelmann M[2], Ortega F[3]

(1. TCTA (Trans – Caledon Tunnel Authority);
2. AECOM Africa; 3. FOSCE Consulting Engineers)

Abstract: The Spring Grove Dam is located in the Mooi River near Rosetta in the KwaZulu – Natal province of South Africa. It is a composite dam with a concrete gravity section in the river and an earthfill embankment on the right bank. The concrete gravity structure has a maximum height of 37m and the embankment is 11 m high.

The concrete dam has been constructed with an innovative optimised Roller Compacted Concrete (RCC) mix. Preliminary works related with the development of the mix design were presented in RCC2012 Symposium in Zaragoza (Spain). The simplification and advantages of this very workable RCC mix will be commented in this paper. The same mix could be either compacted in the dam with very few passes of the 10t vibratory rollers or vibrated against formwork and in confined areas with 50 mm poker needle vibrators without any additional grout (called IV – RCC = Immersion Vibrated RCC).

RCC placement in the dam started in July 2012 and partial impoundment started in March 2013. This article focuses on the dam construction issues and will analyse the quality control and instrumentation records, from early construction times to final impoundment. First spilling of the stepped spillway located on the central section of the RCC dam took place in March 2015.

Key words: Efficient construction, quality control, very workable RCC, IV – RCC

1 Project information

The 37 m high roller compacted concrete Spring Grove Dam is located in the Mooi River near Rosetta in the KwaZulu – Natal province of South Africa. Spring Grove dam is the main component of Phase 2 of the Mooi – Mgeni Transfer Scheme Phase 2. Water from the dam will be transferred from the Mooi River catchment to the Umgeni River catchment to augment water supply by 60 million m³ per year for about five million domestic and industrial users in the Durban and Pietermaritzburg regions in KwaZulu Natal province.

The composite dam has an RCC gravity section and an earthfill embankment on the far right flank. The concrete section comprises an ogee stepped spillway across the river channel with non – overspill sections on both sides and an outlet works upstream of the concrete structure, on the right side of the spillway. The concrete gravity structure was constructed using an optimised RCC mix and the outlet works using conventional concrete.

The dam was designed on laboratory tested results obtained from sampled rock. To ensure stability of the dam structure the upstream side of the dam was designed with a slope, thus formwork on the upstream side had to be modified to accommodate the slope.

Thehigh paste RCC mix was optimized to the minimum cementitious content (cement and fly – ash) with similar hardened properties to conventional mass concrete to guarantee impermeability and acceptable concrete compressive strengths (⩾6.5 MPa at 7 days and ⩾15 MPa at 365 days) and tensile strengths (⩾1 MPa in the mass and 0.4 MPa across the horizontal joints). In countries such as South Africa, where any kind of pozzolanic material (fly – ash, natural pozzolan, slag, etc.) is available using such materials is the most economical way of building RCC dams.

The design mix was first tested in a laboratory using the materials proposed by the Engineer during the tender phase, and then tested in a laboratory again using the materials proposed by the appointed Contractor. Three (3) different trial sections were constructed on site to assess mixes and placing methods.

The third mainfull – scale trial test section was constructed using the design mix developed in the laboratory for further optimization before construction of the RCC gravity structure commenced. The workability of the mix, nominal segregation and construction techniques were tested in the on-site conditions. Coring, sampling and testing of the RCC trial section were also carried out to compare the characteristics of the RCC trial section with the design requirements.

The new developed RCC mix consisting of 160 kg/m^3 (50 cement +110 fly – ash) of cementitious content met the requirements for a workable, impermeable mix with a smooth finish on the external faces and does not segregate during handling. The mix also guaranteed that enough paste rises to the surface of the layers required for good bonding between layers using vibratory rollers. A high proportion of a set retarder admixture in the RCC mix improved the workability and ensured a mass which is not set before placement of the following layer aided by continuous curing of the compacted surface. RCC placement in the dam started in July 2012 with partial impoundment on 26 March 2013. The dam first spilled on 13 March 2015, two years from partial impoundment (Fig. 1).

2　IV – RCC concept

The traditional use in RCC to ensure a smooth face is 'skin concrete', which is approximately a 600 mm wide layer of conventional concrete (CVC). A separate CVC mix is required for the 'skin concrete, takes longer to place and is more labour intensive. There is also a potential of discontinuity between the 'skin concrete' and the RCC, apart from a different thermal – elastic behaviour that has caused cracking is some cases.

With the optimization and development of a high paste RCC mix at Spring Grove dam, the RCC mix developed for the body could also be placed on the external faces and could be successfully compacted by poker vibrators. The Immersion Vibrated Roller Compacted Concrete (IV –

Fig. 1 General view of the completed Spring Grove dam

RCC) provides a good finish on the external face and also save construction time and other resources. There was no need to add grout (to improve workability and impermeability) to this RCC mix prior to immersion vibration. The use of the IV – RCC in different areas of the dam is shown in Figure 2. The typical finish of IV – RCC at the dam faces and inside the gallery can be appreciated in Fig. 3.

Fig. 2 IV – RCC along formwork faces, around cast – in items and along abutments

3 Experience during construction

3.1 Placement and compaction

RCC was transported from the batching plant to the works by trucks. To prevent contamination of the fresh RCC, two fixed frame cantilever conveyors were positioned at pre – determined strategic positions to feed fresh RCC onto trucks inside the RCC placing area.

Fresh RCC concrete was spread and levelled with a dozer to the required layer thickness (ie. 350 ~ 400 mm) and then compacted with a smooth 10 t single drum vibratory roller to the design thickness of 300 mm. The highly workable RCC was compacted against formwork and in confined areas with 50 mm handheld needle vibrators.

The vibratory roller was making an average of 2 to 3 overlapping passes until the expected RCC density and sufficient mix paste uniformly covered the top of the compacted layer. Small

Fig. 3 Typical finish of Immersion Vibrated RCC(IV – RCC) at dam faces and in the gallery

smooth vibratory rollers were used to compact RCC concrete adjacent to confined areas, formwork faces (600 mm from the formwork), abutments, other concrete structures and also immersion vibrated RCC (IV – RCC) interfaces.

Various RCC mix designs have been adopted from the new developed mix to accommodate for various weather or climatic conditions. The differences between them are a slight variation in the water content and the adjusted dosage of retarder based on ambient temperatures.

3.2 Formwork

Facing systems (ie. formwork) for RCC at Spring Grove Dam included 900 mm high vertical stepped formwork for the formation of the downstream steps and also smooth inclined vertical formwork for the upstream face. Special shutters for the upstream and downstream side were designed to accommodate unset wet paste roller compacted concrete which exerts higher forces onto the shutters and counteract displaced and/or uplifted forces by hydrostatic pressure from the high paste RCC. The shutters extended for three steps of concrete.

The downstream 900 mm steps were formed with steel formwork panels with 900 mm high soldiers and jacks for alignment. The horizontal steps were floated with a mechanised steel trowel.

The inclined upstream face was formed with conventional steel formwork panels with steel soldiers and recoverable dywidag bars fitted with anchor plates. The panels were fitted with manually adjustable jacks to align the formwork. It was fitted with walkways which permitted hand raising of formwork and finishing of the RCC while placement was is in progress.

The formwork performed satisfactorily without any problems.

3.3 Induced contraction joints and waterstops

Vertical contraction joints are spaced at 12 m intervals along the Spring Grove dam wall to accommodate contraction mainly during the cooling period of the RCC. The contraction joints have been induced successfully with a combination of crack inducers and crack directors respectively across the dam wall.

A 275 mm single centre – bulb natural rubber water stop has been used on the upstream face of the vertical contraction joints in the RCC with continuous HDPE crack directors on both the upstream and downstream of the water stop respectively. The purpose – made brackets were designed to firmly and simultaneously hold the water stop and crack directors in place during RCC placement as shown on Figure 4.

Fig. 4 Waterstop support bracket holding the waterstop, crack director in position during construction (left) and 250mm HDPE sheets inserted to act as crack inducers (right)

Continuous crack inducers have been inserted from the upstream face through to the downstream face in every second (2nd) layer by forcing the folded 250 mm wide HDPE sheets in freshly compacted RCC.

The poker vibration of the RCC around the waterstop was carried out with care, making sure that the paste occurs around the waterstop for sealing purposes. This was done successfully except at two places where minor leakage was observed during the filling of the dam.

3.4 Horizontal joints

Hot, warm and cold horizontal joints have been defined as follows. An RCC concrete surface or joint was considered to be hot before initial set of the compacted RCC layer beneath has occurred as measured according to ASTM C403. A compacted RCC concrete surface or joint was considered to be warm for the duration of the period between initial set of the RCC and the time at which the surface/joint is considered cold. An RCC concrete surface or joint was considered to be cold when it is judged that one of the following conditions has been met:

- Little or no penetration of the aggregate from the new layer into the previously compacted layer will occur, or
- The final set of the compacted RCC layer beneath has occurred as measured according to ASTM C403, or
- A modified maturity factor of 1 200o C×hour has been exceeded. The modified maturity factor is the product of exposure time (or time between placements of successive layers of RCC) in hours and the average ambient temperature in °C at the placement modified in +12 °C.

Due to the relatively cool conditions during most of the RCC placing programme and the effectiveness of the set retarder, cold joints did not hamper the integrity of the concrete or the rate of construction. By the end of production and placement of RCC, a total of eight (8) cold joints were encountered. This is twice the originally planned total number of four (4) cold joints.

No seepage was observed through these joints which showed good quality assurance.

3.5 Gallery formation

The longitudinal gallery along the body of the dam wall with two (2) perpendicular downstream entrances, respectively on the right and left hand side of the spillway chute walls, was formed using the ordinary formwork panels. RCC concrete was respectively placed on either side of the formed channel at different times. The placement of RCC was interrupted at floor level of the central horizontal gallery so that the conventional formwork for the gallery walls could be put in place. Once complete, RCC placement continued to the gallery soffit level before being stopped so that the precast roof soffit could be installed.

The walls of the inclined sections of the gallery were formed with vertical formwork, and 300 mm thick precast roof soffits were placed horizontally and tied in with the RCC layers, which produced a neatly formed stepped roof soffit in the inclined sections of the gallery, permitted continuous RCC placement during forming of the inclined gallery sections, and accommodated changes in the slope and landings in the galleries. Once the RCC had cured (50 days) and placement progress had been well above the gallery, the gallery formwork was removed through the gallery entrance.

Construction of the left bank gallery access obstructed the movement of RCC equipment between the RCC delivery point on the left bank and the central RCC work area. Sandbags and sand filling were used to gradually form the gallery access walls while the RCC progressed lift by lift. This sand void former was removed afterwards by pneumatic breaker and hand work and finished with conventional concrete.

3.6 RCC Heat of Hydration

Themain full-scale RCC placement commenced on 7 July 2012, during the South African winter. The winter night temperatures occasionally dropped to -5 ℃ and due to the cold conditions the temperature of fresh RCC sometimes was as low as 8 ℃ at night. Strings of thermistors were installed at a number of positions and the RCC temperatures were continuously monitored.

During the first few days of placement the cold conditions delayed initial set by 48 hours, which restricted the raising of the stepped formwork. The lowest RCC temperature at the time of placing was approximately 8 ℃.

It took approximately one week for the RCC temperatures to reach 15 ℃, after which RCC placement achieved resulted in more than one layer per day. Due to the cold conditions in the winter months and effectiveness of the set retarder unplanned cold joints were seldom an issue. The total increase in RCC temperatures due to heat of hydration seldom exceeded 10 ℃. The bulk of the RCC in the lower third of the wall was placed in winter and the temperature of this RCC did not exceed 22 ℃.

During the warmer months the RCC section was narrow and dissipation of heat was effective. The specified maximum placing temperature was 25 ℃ and the temperature increase did not exceed 34 ℃.

4 Review and evaluation of quality control parameters

4.1 **Aggregate and Materials**

All RCC concrete aggregates were produced from a Dolerite Quarry located approximately 65 km from Spring Grove Dam. RCC crusher – run dolerite aggregates, 37.5 mm, 19 mm and 9.5 mm, were routinely and continuously monitored for grading, elongation, flakiness, water absorption, relative density, dust content at the quarry and on arrival at the site.

The sum of Flakiness (FI) and Elongation (EI) whose specified maximum value was to be less than or equal to 25% was a challenge. Adjustments had to be made at the crusher plant to produce aggregates according to BS 812 (i.e. cubical shaped aggregates). The shape of coarse aggregates is critical to the workability and tendency to segregation of RCC mix and it directly also affects the water demand.

Crusher sand was monitored for grading, void ratio, fineness modulus, etc. Both void ratio and fineness modulus are critical parameters forfine aggregate. The final product could not meet the maximum specified void ratio of 32% and the maximum value was relaxed to 33%, thus allowing for a certain increase in water content. This was a balanced compromise as the long – term strength results were well above those required in the design. The site laboratory monitored concrete quality during production and placement. Monitoring points were flexibly located at the batch plant, placement area and the site laboratory.

Despite the random aggregates quality deviations during the production, which were appropriately dealt with as they appeared, the required integrity of the concrete and a workable impermeable structure with smooth external faces were achieved.

4.2 **Roller compacted concrete**

The qualitycontrol of following parameters was monitored for the RCC:

- ambient and concrete temperature;
- VeBe times (workability);
- in – situ RCC concrete densities;
- initial & final setting times;
- heat of hydration (using installed instrumentation);
- compressive strengths;
- direct tensile strengths, etc.

A team at the batch plant tested the workability (VeBe Time), concrete and ambient temperatures and checked the batching variances per load and was the first indication of the quality of the RCC.

At the placement area the laboratory representative also checked the concrete and ambient temperatures, respectively, to establish any temperature gain or loss during RCC concrete haulage. Another laboratory team was recording the concrete loads as they reached the conveyor belts, to correctly give the volume of placed RCC concrete per layer and/or per shift.

The density of the roller compacted mix was 2 650 kg/m^3, which is more than the accepted value of 2 400 kg/m^3 for conventional vibrated concrete. The achieved cube compressive strength exceeded the specified 15 MPa at the age of 90 days and also exceeded the 20 MPa target design strength at the age of 180 days as opposed to the design age of 365 days. The direct tensile strength of 1 MPa was achieved at the design age of 365 days. The cube compressive and tensile strengths achieved with this developed mix is shown Fig. 5.

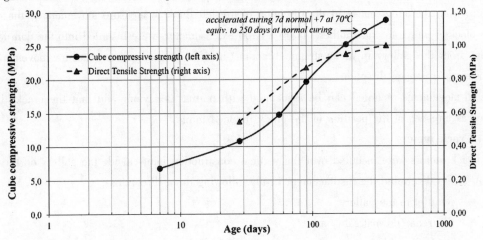

Fig. 5 **Development with time of the RCC cube compressive and direct tensile strengths**

The maximum reached heat of hydration along the thicker section of the dam wall was 34 ℃ with a minimum of 8 ℃ during the colder winter weather.

5 Dam behaviour after impoundment

The dam behaviour from partial impoundment (26 March 2013) to the first spilling (13 March 2015) was monitored. The behaviour of the dam was assessed using the instrumentation installed at the dam and visual observations.

5.1 Induced contraction joints and vertical face drains

The induced contraction joints and vertical face drains in the gallery were visually inspected during impoundment period and the following were observed. Most of the vertical drainage holes through the concrete wall above the gallery have no leakage except for very low leakage at three positions where the following respective conclusions at each position can be made:

- (Pos. 1) Leakage is most likely from an ineffective water stop at the joint.
- (Pos. 2) Leakage observed during impoundment has not increased as the dam water level increased.
- (Pos. 3) Leakage has increased as the dam water level increased and is frequently monitored.

No significant leakage was observed at the position where slight movement of the dam was encountered.

5.2 Seepage

A V-notch measuring structure was installed in the dam's outlet valve house to measure all seepage water collected from the dam. The result is that the seepage water increases as the dam water level increases but is not significant seepage indicating that the IV – RCC was correctly and effectively applied. At full supply level the total seepage measured with the V – notch structure is approximately 0,7 ℓ/s.

5.3 Rod Extensometers

Four (4) rod extensometers were installed across the horizontal crack directors in the ogee spillway from the outlet house in the direction of the left flank to measure movement of the concrete blocks relative to one another. Two (2) rod extensometers were installed into the dam foundation, from the gallery in an upstream direction (at 45° inclination) to measure movement in downstream/upstream direction.

All movements recorded can be associated with normal behaviour and that the cracks have formed as foreseen at the position of the crack induced joints.

5.4 Crack meters

Crack meters were installed over the vertical construction joints inside the gallery on the upstream faceto determine the displacement in the following three directions:

- parallel to the gallery,
- upstream/downstream, and
- vertical up/down.

The larger portion of crack meters shows a displacement of less than 0.2 mm. The plot for the data range for each individual crack meter showed that the displacement slightly increases when the dam water level increases.

6 Recommendations and Conclusions

"A two – point quality control" of fresh concrete should be maintained to avoid costly (ie. with respect to time and eventually money) removal of failing concrete at the placement area. Concrete sampled and tested first at production point (ie. batch plant) proved to be the best approach and it should be implemented on all construction projects.

Despite the random aggregates quality deviations during the production, which were appropriately dealt with as they appeared, the required integrity of the concrete and a workable impermeable structure with smooth external faces were achieved.

The RCC concrete design strength could conservatively be targeted at the age of 180 days instead of 365 days as per the obtained project compressive strength results.

In countries such as South Africa, where any kind of pozzolanic material (fly – ash, natural pozzolan, slag, etc.) is availableusing such materials is the most economical way of building RCC dams. The optimised concrete mix design with lower cementitious content and optimized fly – ash utilization produced significant cost savings.

The dam behaviour of the concrete gravity RCC structure focusing on the seepage through the foundation and concrete structure and relative movement are normal.

Acknowledgements

Thank you to the Trans Caledon Tunnel Authority (TCTA) for the consent to publish this paper.

References

[1] Nyakale J., BadenhorstD. B., Ortega F.. The Optimisation of the RCC Mix Design of the Spring Grove Dam in South Africa[C]. The 6th International Symposium on Roller Compacted Concrete (RCC) Dams, Zaragoza, Spain, 2012.

[2] AECOM (Pty) Ltd. Spring Grove Dam: Construction Completion Report. 2014.

Kahir RCC Dam Thermal Analysis

Araghian H. R. , Hajialikhani M. R. , Jafarbegloo M.

(Jahan Kowsar Co. , Iran)

Abstract: Kahir Dam is an RCC dam in Iran which is located in South east of Iran in a semi dry area over a wild / heavy flooded river. It is FSHD type with 54.5 m height and crest length of 380 m and with RCC volume of about 500 000 m^3.

In this paper, 2-D Kahir thermal analysis within construction by using ANSYS finite element software is presented. Based on the dam construction methodology, right and left parts has different construction methodology. So two diffetent construction priority has been analyzed and temperature histories have been calculated in different nodes upto 6 000 days after start of construction. Effect of internal heat of concrete on the rate of hydration has been considered in the model.

Finally, according to the calculated temperatures in different points of the dam, the potential of thermal cracks in concrete dam body are investigated.

Key words: Kahir RCC Dam, Thermal Analysis, FSHD, Heat of hydration

1 Introduction

Due to gradual construction of massive concrete structures and RCC dams, calculation of heat distribution in mass concrete is a complicated matter. Also, additional to thermal properties of concrete and its initial temperature, some other parameters like as time interval between lifts and mean ambient temperature act on heat distribution in massive concrete structures.

Totally, solving the heat differential equation is necessary to obtain adequate and accurate results. Nowadays, because of computing developments, these equations are solved by finite element or finite difference methods. The equation of heat distribution is expressed like as Equation (1):

$$k_{xx}\frac{\partial^2 T}{\partial x^2} + k_{yy}\frac{\partial^2 T}{\partial y^2} + k_{zz}\frac{\partial^2 T}{\partial z^2} + w = \rho c \frac{\partial T}{\partial t} \qquad (1)$$

In Equation (1), the heat distribution in a mass is a function of time; but if a steady heat distribution is envisaged, Equation (1) changes to Equation (2) – where T is the mass temperature, k is coefficient of heat transmission, ρ is the mass density, c is specific heat and w is heat generation in the mass.

$$k_{xx}\frac{\partial^2 T}{\partial x^2} + k_{yy}\frac{\partial^2 T}{\partial y^2} + k_{zz}\frac{\partial^2 T}{\partial z^2} + w = 0 \qquad (2)$$

2 Effective parameters on hydration

Hydration process depends on various parameters such as chemical composition of cement, water to cement ratio, fineness of cement particles and cement particle size distribution.

According to scientific observations, hydration does not exceed more than 70 to 80 percent. Also theoretically, the highest rate of probable hydration is about 80 percent.

Hydration process is accelerated by heat like the most of chemical and physical processes (Fig. 1). This effect can be seen in cement hydration in temperatures above 20 ℃. So, for correct simulation of heat generation in concrete mass, it's necessary to model the temperature effect on hydration. Majority of math models for determining the temperature effect are based on Fourier differential equation.

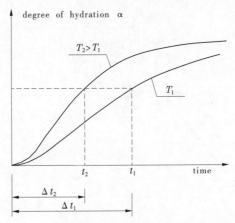

Fig. 1 **Temperature effect on concrete hydration**

Equation (3) indicates the temperature effect on hydration rate:

$$H(T) = e^{\frac{E_a}{R}(\frac{1}{T}-\frac{1}{T_0})} \qquad (3)$$

Where T is concrete temperature at the calculation time and T_0 is the initial temperature of reaction equal to 293 Kelvin. E_a is the activation energy of Portland cement which is equal to 33.5 kJ/mol and gas constant, R is equal to 8.31 J/mol · K. In the Kahir thermal analysis the effect of heat on hydration of cement is considered.

3 Calculation of cracking risk

In Kahir project, cracking risk calculation is based on strain method. ACI 207.2R expresses this relation as a function of L/H which presents in Equation (4):

$$\varepsilon = \alpha \times K_R \times K_F \times \Delta T \qquad (4)$$

Equation (4) shows the created strain in concrete influenced by temperature difference where, α is the coefficient of thermal expansion of concrete and ΔT is temperature difference. Also, K_R is the internal restraint degree as a result of structure and foundation geometry which differs from 1 to 100 percent and obtained from ACI 207.2R. It should be mentioned that the reduction of restraint with heigh has been neglected in the calculation of thermal restraint and K_R was considered equal 1.0 conservatively.

4 General project information

Kahir RCC dam was designed in south-east of Iranand is under construction. The foundation level is 13.5 MASL and the foundation width is 83 m. Kahir dam typical section has been shown in Fig. 2.

Some major project data are as below:

Type of dam: RCC gravity dam (FSHD type)

Crest length: 380 m

Crest width: 5 m

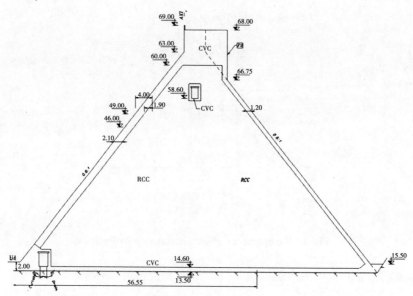

Fig. 2 Typical cross section of Kahir RCC dam

Spillway width:	160 m
Dam height from foundation:	54.5 m
Reservoir Volume:	314 million m^3
RCC volume:	500 000 m^3
CVC volume:	180 000 m^3
RCC required compressive strength:	70 kg/cm^2 @ 180 days
Diversion system:	One Tunnel with 6m diameter & 280 m length

5 Ambient conditions

Kahir dam is located in a semi dry region. Sinusoidal curve of mean monthly ambient temperature is shown in Fig. 3. The site average annual temperature is 27.7 degrees celsius.

At present, RCC mix proportion is designed at the dam local laboratory. Cement content in

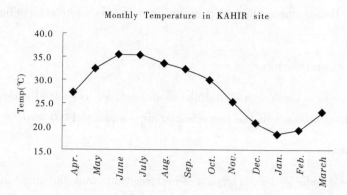

Fig. 3　Mean monthly temperature in Kahir dam region RCC mixture

the current mix design is equal to 110 kg/m³. The required specified strength is 70 kg/cm².

6　Heat generating of RCC

Khash pozzolanic cement with 20% Natural pozzolan is utilized in preliminary mix program. The heat of hydration of Portland pozzolanic cement is considered in the analysis. and the heat hydration from 3 to 90 days is shown in Fig. 4.

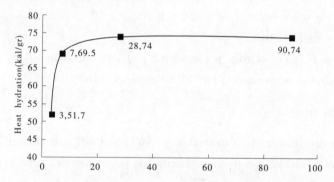

Fig. 4　Heat of hydration of Khash pozzolanic cement

7　Coefficient of thermal expansion

The coefficient of thermal expansion of RCC is a function of expansion coefficient of the aggregates and consequently, the petrology of concrete aggregates (Sandstone, Limestone and Volcanic). Due to nature of RCC aggregates – in this project – and considering tables 2.9.1 and 2.9.2 of ACI – 209, the coefficient of thermal expansion is determined equal to 10.2×10^{-6} (1/℃).

8　Coefficient of heat diffusivity

This coefficient is dependent on the type of concrete aggregates. The greater the coefficient of heat diffusivity, the more the transmitted heat per unit time within concrete. Considering the Kahir borrow area investigations, the aggregates are a combination of Sandstone, Limestone and Volcanic aggregates. Hence, the coefficient of heat diffusivity of concrete is identified equal to 0.117 m^2/day.

9 Coefficient of convection

When concrete is in touch with a fluid like as air, concrete is cooled through convection. Coefficient of heat transmission between concrete and air is equal to 11.6 $kcal/m^2 \cdot hr \cdot °C$

10 Specific heat

The concrete specific heat is an amount of heat needed to raise the temperature of one unit of mass of concrete by one degree of centigrade. The concrete specific heat increases with temperature rise – which is not counted for confidence – and determined equal to 950 $J/kg \cdot °C$ according to ACI 207.2 R.

11 Structural properties

The tensile strain capacity is an amount of strain which concrete can suffer without cracking and indeed, is the quotient of division of tensile strength by modulus of elasticity. In slow – rate loading, the creep effect is also considered and so, the capacity is increased. Supposing such a condition, the tensile strain capacity of concrete is equal to 60 μ strain, which is considered due to the low cement content of RCC.

The modulus of elasticity of RCC is considered 12 GPa. Foundation Rock deformation modulus is about 1GPa, but it is assumed to be 1.5 GPa conservatively for calculation of restraints.

12 Construction time schedule

Based on the dam construction methodology, right and left parts has different construction methodology. So two diffetent construction priority has been analyzed. Due to the flood seasons, right part of dam construct up to level +33 first, then left part construct up to this level, and finally the whole part of the dam construct to the final level.

Furthermore, the rate of concrete pouring has been calculated for two different conditions. First, the roller compacted concrete of dam will be constructed during 17 months and second, the roller compacted concrete of dam will be constructed during 13 months. According to two time schedules, the concrete volume at the end of each monthin different levels is mentioned in table 1.

Table 1 Two different construction time schedules

Month	Concrete Level MASL		Concrete Level MASL	
	Right	Left	Right	Left
1	18.0	14.5	18.0	14.5
2	21.5	14.5	23.0	14.5
3	25.5	14.5	30.5	14.5
4	31.5	14.5	33.0	18.8
5	33.0	18.0	33.0	24.0
6	33.0	22.3	33.0	29.0
7	33.0	26.2	34.0	
8	33.0	30.0	37.5	
9	33.6		41.2	
10	36.0		45.2	
11	39.0		50.0	
12	42.0		57.0	
13	45.0		60.0	
14	48.3		—	
15	52.5		—	
16	56.0		—	
17	60.0		—	

13 Calculation of placing temps

Placing Temperature of Fresh RCC has been calculated based on Usarmy method. So the Sinusuidal equation of placing Temperatures is as equation(5):

$$T(t) = 28.8 + 7.3\sin(\pi(t - 27)/365) \tag{5}$$

Which T is placing temperature and t is the day number from beginning of the year (persian year).

14 Finite element model

The finite element program Ansys 5.4 and the finite element PLANE 77 is used to build a finite element model for thermal analysis of Kahir dam body (Fig. 5). Overall algorithm of the cited model is like below:

(1) Perform the 1st lift with the specified initial temperature and the upper surface in touch with the air.

(2) Increase the degree of heat of hydration considering the cement content.

(3) Read the temperatures calculated in clause 2 and calculate the new heat generation.

(4) Calculate the air temperature in respect of the elapsed time from beginning of the previous lift construction.

(5) Re – compute the clauses 2 to 4, considering that the upper lift is in touch with the air during the lift construction cycle.

(6) Remove the surface transmission of concrete with the air and make the next lift elements alive.

(7) Re – compute the clauses 2 to 6 for the other lifts.

The heat generation rate of concrete is dependent on the time and temperature. So, these effects are taken into account in thermal analysis and a nonlinear analysis is performed. It must be mentioned that the analysis time step is based on the results′ convergence and heat changes inside the structure and in all cases, the ANSYS program automatically calculates and checks the convergence of results. Based on construction methodology and time schedules, there are four models in Ansys containing left and right part and two priority.

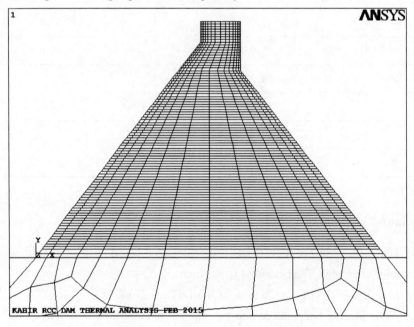

Fig. 5 **Ansys finite element model**

15 Results of thermal analysis

Temperature distribution have been presented in different times up to 6 000 days after start of dam body placement (Fig. 6). Also for better understanding of details of graphs, temperature history of the nodes has been presented up to 6 000 days (Fig. 7).

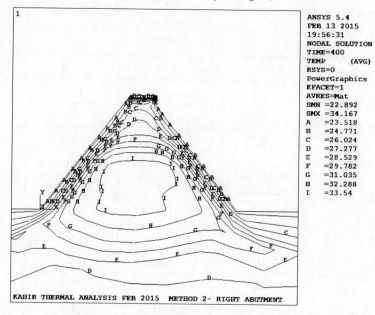

Fig. 6 Sample of isothermal contours – 400 days after start – Right part – Second priority

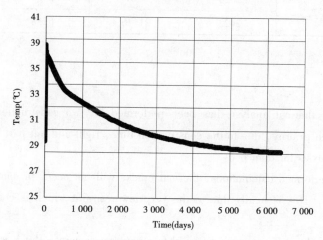

Fig. 7 Sample of temperature history – Node 530 – Left part – First priority

In Table 2, risk of cracking in different Nodes is shown for first priority of construction methodology. Results show that there will be no crack in dam body caused by thermal stresses.

Generally parts of the dam which has been poured in the warm months (summer) are more vulnerable to crack. In this months, both of RCC placing temperature and ambient temperature are higher than the other months of the year.

Table 2 Calculation of risk of cracking type 2 for Kahir dam — For First priority

No.	Priority – Part	Node Number	Max. Temperature change(C)	Produced Strain (μ)	Status
1	First Priority – Right part	155	5.09	12.4	No Crack
2		310	5	12.1	No Crack
3		530	5.3	12.9	No Crack
4		903	9.42	22.9	No Crack
5		1 630	2	4.9	No Crack
6		2 026	10.15	24.6	No Crack
7		2 422	21.77	52.8	No Crack
8	First Priority – Left part	155	11.84	28.7	No Crack
9		310	10.83	26.3	No Crack
10		530	9.7	23.5	No Crack
11		903	4.97	12.1	No Crack
12		1630	2.01	4.9	No Crack
13		2026	9.81	23.8	No Crack
14		2422	21.76	52.8	No Crack

16 Conclusion

In this paper, thermal analysis has been performed for two construction priority of Kahir FSHD dam. In each priority, due to the construction time, right and left part seperately analysed so there is four analysis for dam body.

• Based on performed calculations, thermal cracks will not occur in dam body in both first and second priority. This is manly due to the very low restraint which is come from foundation properties.

• Calculations show that there is no need to a contraction joint. But it was advised to place some movement joints in 36 m spacing to prevent cracking due to the probable differential settlement.

• Because the performed analyses are based on continuous placement of RCC. In condition of stop of placing, two meters of RCC should be placed in a low rate.

References

[1] ACI 207.2 R –95 – Effect of restraint, volume change, and reinforcement on cracking of mass concrete.
[2] ACI 207.1 R – Mass concrete.

[3] ACI 207.4 R – Cooling and insulating systems for mass concrete.
[4] ACI 209.R – 92 – Prediction of creep, shrinkage, and temperature effects in concrete structures.
[5] US Army – ETL 1110 – 2 – 542 – Appendix A: Techniques for performing concrete thermal studies.
[6] ASTMC 1074 – Standard practice for estimating concrete strength by the maturity method.
[7] ANSYS 5.4 ADPL, User's manual.
[8] F. R. Andriolo. The use of roller compacted concrete.
[9] Bentz, de Larrard. Prediction of Adiabatic temperature rise in Conventional and high Performance Concretes using a 3 – D Microstructural model, cement and concrete Research.
[10] F. Rueda, N. Camprubí, G. García. Thermal cracking evaluation for La Breña II dam during the construction Process.

Design Feature of the Nam Ngiep 1 Hydropower Project

Makoto ASAKAWA, Mareki HANSAMOTO

(Kansai Electric Power Co., Inc. Japan)

Abstract: A main dam and a re-regulation dam are constructed for power generation of the Nam Ngiep 1 hydropower project ("Project"), located in the Lao People's Democratic Republic. The Project consists of a main power station and a re-regulation power station. The main dam with 148 m in height and the main power station creates the reservoir by which the main power station generates the power of 272 MW. The re-regulation dam with 20.6 m in height and the re-regulation power station is planned to re-regulate and stabilize the maximum plant discharge of 230 m^3/s released from the main power station for the safety to the downstream area of the re-regulation dam. The design features of the Project including hydrology, seismic analysis, dam stability analysis, and spillway design are introduced in this paper. The Probable Flood discharge is calculated by using a frequency distribution curve of "Log Peason III". 1,000 years return period of flood discharge is adopted as the design flood which is calculated to be 5 210 m^3/s. Additionally, Probable Maximum Flood is calculated to 9 050 m^3/s. The dam stability on earthquake is considered under Operating Basis Earthquake (OBE) and Maximum Credible Earthquake (MCE). The design seismic intensity of OBE is decided based on Global Seismic hazard Assessment Map (GSHAP). At the point of NNP1 the intensity estimated to be ranged between (0.041 ~ 0.082)g from the map. Accordingly the design seismic intensity as OBE of NNP1 is set at 0.10g conservatively. For the calculation of MCE, hypocenter modeling based on the historical earthquake data are carried out. Peak ground acceleration (PGA) is calculated based on attenuation relation modeling and hypocenter modeling and MCE is calculated to be 0.2 g. RCC is adopted for the main dam. Both rigid static analysis and dynamic elastic analysis are carried out to confirm the stability of the dam. In case of MCE, open crack and sliding failure are quite limited and the stability of the dam is confirmed. RCC trial mixing tests are underway and the design are to be revised based on the results of the RCC trial mixing tests. The spillway types of the main dam and the re-regulation dam are ski jump type and labyrinth type. The details of each structure are determined based on the hydraulic model tests. Applicability of both types has been confirmed.

Key words: Flood Analysis, Seismic Analysis, Hydraulic model test

1 Introduction

The site of the Nam Ngiep 1 hydropower project ("Project") is located along the Nam Ngiep River, 145 km northeast of Vientiane, the capital of Lao PDR and 50 km north of Paksan city as shown in Fig. 1.

The Project consists of a main dam and a re-regulation dam. The crest length and dam height for the main dam, RCC gravity dam, are 530 m and 148 m respectively. The reservoir created by the main dam will store water of around 10 billion m³ to generate the electricity of maximum output of 272 MW for exporting to Thailand. The re-regulation dam, concrete gravity dam with labyrinth spillway, are located 6.5 km downstream of the main dam, and its crest length and dam height are 252.6 m and 20.6 m respectively. Main feature of the project is shown in Table 1.

Fig. 1 Project location

Table 1 Main features of the project

Facility	Items	Unit	Specifications
Main Reservoir	Effective storage capacity	10^6 m³	1 192
	Catchment area	km²	3 700
	Average annual inflow	m³/s	148.4
Main dam	Type	—	Concrete gravity dam Roller-Compacted Concrete
	Dam height	m	148.0
	Crest length	m	530.0
	Dam volume	10^3 m³	2 245
Spillway (Ski jump type)	Gate type	—	Radial gate
	Number of gates	—	4
	Design flood	m³/s	5 210(1 000 – year)
Turbine and generator	Maximum plant discharge	m³/s	230.0
	Effective head	m	130.9
	Rated output	MW	272 at Substation

2 Hydrology

2.1 Design flood

Probable flood discharges for various return periods are calculated based on the observed discharge data at the Muong Mai G/S (R3) for the period from 1978 to 2000. Annual daily maximum discharges at the Muong Mai G/S are plotted in Fig. 2. Peak flood discharge is estimated by using a frequency distribution curve of "Log Peason III" which is the most adaptable curve among all the frequency distribution curves. 1 000 years flood is selected as the desing flood and calculated to be 5 210 m³/s.

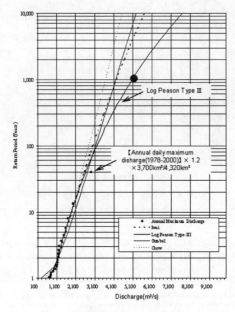

Fig. 2 Frequency distribation curve

2.2 Probable maximum flood

Probable Maximum Flood (PMF) are calculated by using unit hydrograph method. Base flow of the rainy sesason is estimated to be 267 m³/s. For the caluculation of PMF, flood analysis of PMF is required to add to the Base flow. For the calculation of Probable Maximum Precipitation (PMP), Hershfield method is applied with the rain fall data obtained at Paksan rainfall station located around 50 km south of the Nam Ngiep 1. The estimated result of the point PMP with Hershfield method is 1 072 mm (72 h). This point rain fall is converted into entire Nam Ngiep basin rain fall by multiplying the regional factor of 39 %. The reginal factor is calculated based on the camparison between Paksan rainfall data and Thiessen – method in which the rainfall data is collected broadly around Nam Ngiep catchment area. Flood analysis is carried out by unit hydrograph method with lag time of 18 hours calculated by using Modified Snyder equation. This value is verified with time lag between the peak rain – fall at upstream area which is almost centre of the Nam Ngiep basin and peak water flow at the gauging station near the dam site. PMF is calculated to 9 050 m³/s.

2.3 Dam height and freeboard

The crest elevation of the main dam is determined to add a freeboard based on USBR to the high water level of EL. 320.000 m during 1 000 - year flood of 5 210 m³/s. The wave height due to wind is obtained by generalized correlations of significant wave height (Average of the highest of 1% of the waves) with related factors, where the effective fetch is 1 312 m and the 1 minute average wind velocity is 20 m/s at Pakxan; $h_w = 0.93$ m is obtained. The formula used for the wave height due to earthquakes is, $h_e = (1/2)(KT)/\pi(gH_o)1/2$, and when design seismic intensity is 0.10 g, earthquake period is 1 sec, and H_o: depth of main dam is 148.0 m; $h_e = 0.61$ m is obtained. Now the dam crest elevation is set at EL. 322.000 m by adding a freeboard consisting of wave due to wind and earthquakes. The dam crest elevation for the case of a design flood and an earthquake is high sufficiently, even if wave due to wind and earthquakes is considered. In addition, the main dam is confirmed not to overflow even in case of the PMF.

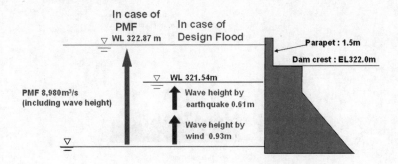

Fig. 3 Calculation of wave height

3 Seismicity

The dam stability on earthquake is considered under Operating Basis Earthquake (OBE) and Maximum Credible Earthquake (MCE). OBE is defined as the serviceability of the dam remains without larger damage after earthquake. MCE is defined as the dam may be damaged but a catastrophic damage equivalent to an uncontrolled outflow of water must be avoided after earthquake.

3.1 Operational basis earthquake

The design seismic intensity of OBE is decided based on Global Seismic hazard Assessment Map (GSHAP) as shown in Fig. 4. The map shows the distribution of earthquake with 475 years of return period which means 10% chance of being exceeded in 50 Years corresponding to OBE. Inside Lao PDR, seismic intensity becomes high in the south area and low in the north area. At the point of NNP1 the intensity estimated to be ranged between(0.041 ~0.082)g from the map. Accordingly the design seismic intensity as OBE of NNP1 is set at 0.10g conservatively. Additionally, seismic intensity at NNP1 is estimated by some attenuation models. Okamoto, Esteva, Iwasaki and Fukushima - Tanaka models are used for practical models. Design seismic intensity of OBE set at 0.10g (98gal) is much larger than any other estimated seismic intensity as shown in Fig. 5.

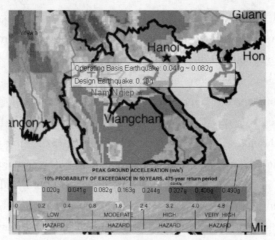

Fig. 4

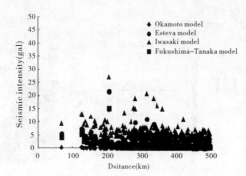

Fig. 5

3.2 Maximum credible earthquake

Seismic hazard analysis is carried out to predict the behavior of a dam by using relationship between magnitude and occurrence probability of earthquake. MCE is adopted for ground motion at certain occurrence probability based on the seismic hazard analysis. Required seismic capacity in MCE should be confirmed to avoid a catastrophic dam failure. The dynamic analysis is conducted to confirm whether the required seismic capacity is to be secured or not. The procedure of seismic hazard analysis are shown as followings.

3.2.1 Data collection

Seismic data with the magunitude of 4.5 to 8.0 around the project site within 500 km radius are collected. The data for the seismic hazard analysis are extracted latest 31 years, relatively reliable data to prevent from adverse impacts on the probabilistic analysis.

3.2.2 Frequency and magunitude

The seismic activity such as magnitude and frequency occurrence within a specific period is quantified by using the Gutenberg Richter relationship (G—R relationship).

$$\text{Log}(N) = a - bM \tag{1}$$

where, N is the frequency occurrence of earthquake with magnitude M, and a and b are constant coefficients.

The values of a and b are set at 6.59 and 0.94 respectively based on the total 225 of observed historical earthquake data as shown in Fig.6.

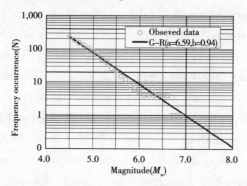

Fig.6 G—R relationship

3.2.3 Hypocenter modeling (random trial)

The seismic source location is basically distributed by means of random numbers in the area of 500 km radius and 20 km depth at the center of NNP1 site. 1 000 000 trials are carried out to estimate seisumic source at the low frequencey.

3.2.4 Peak ground acceleration and seismic hazard curve

The calculation of peak ground acceleration (PGA) is required to estimate the maximum acceleration at NNP1 site using the attenuation relationship of Atkinson and Boore (2006). As MCE at NNP1 site, the 10 000 years of return period, which is reciprocal number of annual exceedance probability, is applied. To estimate the stationary and random occurrence of earthquake, the Stationary Poisson Process is applied as probabilistic model. The annual exceedance probability of each PGA is calculated as follows:

$$P = 1 - \exp(-\nu) \tag{2}$$

where, $\nu = (1 - n_1/1\,000\,000) \times (N/\text{Sampling period})$, n_1 is Frequency occurrence in case of acceleration, N is Total observation data (225), sampling period is 31 years.

The result of calculation of seismic hazard curve is shown in Fig.7. The PGA of return period of 10 000 years is corresponds to 192.2 cm/s². Accordingly, the design intensity as MCE of NNP1 is set at 0.20 g.

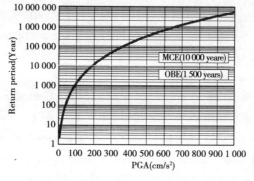

Fig.7 Seismic hazard curve

4 Dam stability

The main dam is designed through stability analysis for a typical section of dam having an apex on the top of upstream side. The design of dam body (downstream slope, slope of fillet, inflection and apex elevation of fillet) is studied by trial – and – error method to simulate the most economical concrete volume. The downstream slope angle is determined to be 0.73. The dam body is basically designed considering four load combinations such as usual and unusual (flood, OBE, MCE). The rigid static analysis is applied to usual, flood and OBE with the proper safety factor for each. The dynamic analysis is applied to MCE. A series of dynamic analysis flow under MCE is shown in Fig. 8 to confirm the inelastic behavior without a catastrophic failure as explained below. In addition, the main dam is confirmed to secure a sufficient safety from structural viewpoint even in case of PMF with the water level of EL. 322.944 m and one gate out of operation with the water level of EL. 320.290 m. The analysis is conducted based on test results of geological survey and RCC trial mix.

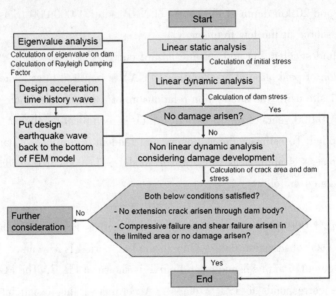

Fig. 8 Dynamic analysis flow under MCE

Both rigid static analysis and dynamic elastic analysis are carried out. In this paper the dynamic non – liner analysis is focused. The following items are examined on the required seismic capacity in MCE by means of non – linear dynamic analysis considering damage development:

(1) Cracks are not extended thorough dam.

(2) Arisen compressive failures and/or shear failures are limited in the local area.

Firstly, the linear static analysis is conducted to calculate the initial stress without earthquake wave. Secondly, the linear dynamic analysis is conducted whether some damage arises within dam body or not. If some damage arises, non – linear dynamic analysis considering damage development is conducted to confirm the crack area within dam body as next step. Both dynamic analyses are conducted by means of time history response analysis using the simultaneous input of the horizontal and the vertical design acceleration time history wave. The typical non overflow cross section of the dam at maximum dam height is considered as 2D finite element method. The analysis model is extended as far as sufficient large area of dam foundation so as to consider the dam behavior on dam foundation appropriately. The dam body is divided into two zones consist of zone – 1 with cement 80 kg/m^3 and flyash; 90 kg/m^3 and zone – 2 with cement 70 kg/m^3 and flyash; 80 kg/m^3. Analysis model and material propertiesare shown in Fig. 9 and Table 2 respectively. Three different types of historical wave are adopted. In all cases tensile stress exceeds more than tensile strength of 2. 18 MPa of Zone – 1 at the upstream dam foundation. Thus, the non – linear dynamic analysis is conducted on the conditions same as linear dynamic analysis taking into account the non – linear constitutive law. The maximum tensile stress after the re – distribution of concrete stress in the non – linear dynamic analysis is smaller than tensile stress of the linear dynamic analysis because of the non – linear constitutive law. On the other hand, the opening crack area may arise as a result of the accumulation of tensile strain even though the re – distributed tensile stress is less than the tensile strength.

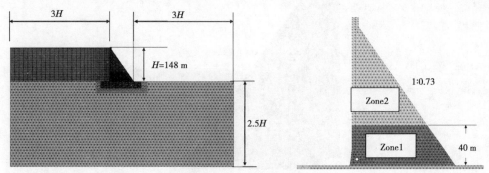

Fig. 9 G—R relationship

Table 2 Material property

Items	Concrete		Foundation Rock (CM class)	Note
	Zone – 1 Cement 80 kg/m^3 Fly ash 90 kg/m^3	Zone – 2 Cement 70 kg/m^3 Fly ash 80 kg/m^3		
Unit weight	2 300 kg/m^3		2 590 kg/m^3	
Dynamic modulus of elasticity	27 459 MPa	21 575 MPa	12 750 MPa	
Dynamic poisson's ration	0. 20	0. 20	0. 30	Conventional value
Damping factor	10 %	10 %	5 %	Conventional value
Static compressive strength	28. 4 MPa	17. 1 MPa	—	Estimated value at 365 days Not applicable of the ratio dynamic/static tensile strength
Static tensile strength	2. 18 MPa	1. 32 MPa	—	
Cohesion	2. 84 MPa	1. 71 MPa	2. 01 MPa	
Internal friction angle	45 degree	45 degree	42. 5 degree	
Fracture energy	248 N/m	174 N/m	—	

The maximum tensile stress after the re-distribution of concrete stress in the non-linear dynamic analysis is smaller than tensile stress of the linear dynamic analysis because of the non-linear constitutive law. On the other hand, the opening crack area may arise as a result of the accumulation of tensile strain even though the re-distributed tensile stress is less than the tensile strength. However, the simulated area of opening crack is quite limited at the upstream dam foundation as shown in Fig. 10. In all cases, the maximum compressive stress is less than the compressive strength of 28.4 MPa. As for the shear failure, the area that the safety factor is less than 1.0 is limited in the upstream dam foundation as shown in Fig. 11. Accordingly, the dam stability is hardly influenced judging from the limited area of shear failure. Even in case of MCE, the crack extension through dam body, is not arisen and the shear failure is limited in the local area. RCC trial mixing is underway, and the design of the dam will be updated through the laboratory

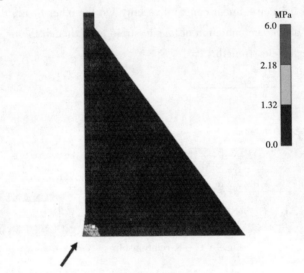

Fig. 10

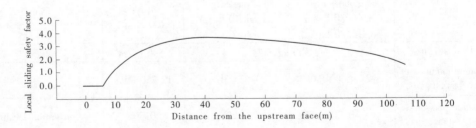

Fig. 11

5 Spillway

A ski jump type of spillway is adopted for the project, in light of the characteristics of the feature of topography at the site. The design of the spillway is carried out based on hydraulic model tests as shown in Fig. 12.

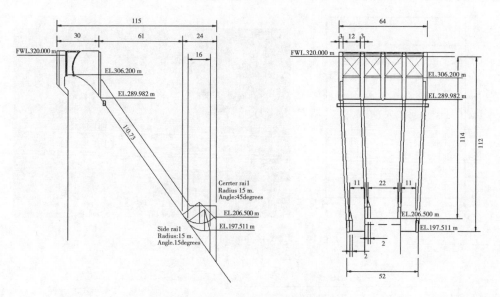

Fig. 12 Current design of spillway

The hydraulic model includes the reservoir, main dam, spillway, and downstream riverbed as described in Fig. 13. The scale of the model is 1∶65 which is calculated based on the Fluid rule.

Fig. 13 Model structure

5.1 Capacity

Four radial gate of 12.25 m wide and 16.9 m high are mounted on the spillway and the capacity of the over crest portion is checked to spill the design flood of 5 210 m^3/s smoothly. When setting the high water level of 320 m, the sufficient distance of 3 m between the trunnion and the water surface is secured as shown in Fig. 14.

5.2 Negative pressure and aerator

The negative pressure and cavitation, which may affect concrete structure, are checked at the over crest and the flip bucket. The measurement points are shown in Fig. 15. Though slight negative pressure is found in the case of partial open of the gate, it is quite acceptable level. To prevent the negative pressure, aerator is to be set.

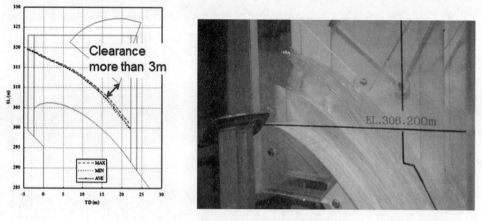

Fig. 14 Flow regime at over crest

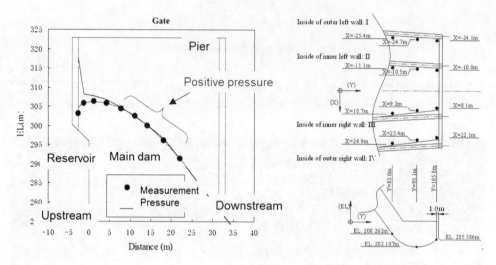

Fig. 15 Negative pressure measurement

5.3 Multi bucke type

Movable bed tests for both normal ski jump and Multi Bucket Type (Fig. 16) are carried out to evaluate the scoring the riverbed, deposition of the riverbed, diving points, impacts to the tail water level, and impacts to the current topography. Large volume riverbed deposits are scored at the diving points in the case of normal ski jump type as shown in Fig. 17. In the case of the Multi Flip Buckets Type, outside lanes are set to disperse the diving point and results in reducing the total volume of scored riverbed and impact area of the diving water. It follows that such adverse effects due to the diving water point as the surging of tail water level and the risk of the failures of both banks of downstream can be reduced as shown in Fig. 18. In the case of the Multi Flip Buckets Type, the tail water level, the average river water level and the elevation of river bed are lower by 6.7 m, 3.0 m and 0.5 m respectively compared to those of the normal ski jump type. The effects of the energy dissipater are confirmed by measuring the water velocity of the downstream. The water flow at the downstream of the main dam are sufficiently dissipated and the velocities are equivalent to the results of the non-uniform flow calculation.

 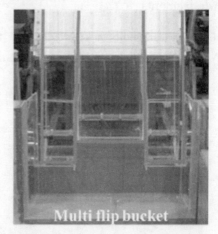

Fig. 16 Ski jump type

Fig. 17 Hydraulic model test

Fig. 18 Comparison between normal ski jump and Multi Flip Bucket Type

6 Closing

Nam Ngiep 1 hydropower plant is now under construction. RCC placing will start in 2016. Operation of the power plant will start in 2019. During the construction, the detail design may be modified.

Acknowledgements

The authors gratefully acknowledge that the kind cooperation of the Owner of the Project (Nam Ngiep 1 Power Company) to prepare and present this paper.

Construction Challenges of the Portugues Arch – Gravity RCC Dam in Puerto Rico

Rafael Ibáñez – de – Aldecoa, David Hernández, Eskil Carlsson

(DRAGADOS USA, USA)

Abstract: At the end of 2013 the construction of the Portugues Dam near Ponce, in Puerto Rico, was completed by DRAGADOS USA Construction Company. The project was designed and procured by the U. S. Army Corps of Engineers, and is owned by the USACE and the Puerto Rican Department of Natural and Environmental Resources. Upon completion, Ponce residents have now much needed flood damage protection during the intense rainy season, sometimes including tropical storms and hurricanes. The work included construction of a 67 – m high roller – compacted concrete single centered arch – gravity structure with a crest length of 375 m and some 270 000 m³ of RCC. Appurtenant structures include integral spillway, intake structure, and a control and valve house located on top of a 18 000 m³ RCC buttress. The project also included foundation rock excavation, foundation treatment work, developing an onsite quarry for aggregate production, and an access road running to the dam crest and to the valve house. The Portugues Dam is the first arch – gravity (or thick – arch, term preferred by USACE's designers) RCC dam built in the United States, and the first RCC dam in Puerto Rico. As such, it presented unique design and construction challenges. Focusing primarily on the construction process, this paper addresses particular constraints of very different nature that had to be overcome, among which include: not availability of local personnel with experience in RCC; limited availability of certain materials, equipment, subcontracting and managerial required resources; very seasonally hot and wet climate with frequent river floods during the wet season (reason for which only winter windows were allowed for RCC placement, preventing a steady workforce and requiring retraining of personnel throughout the entire dam structure construction period); high temperatures for placing concrete even during the winter windows; complex geology at the dam foundation leading to extensive foundation treatments; etc.

Key words: arch – gravity, flood protection, RCC placement windows, knowledge transfer

1 Introduction

The Portugues Dam is the first Roller – Compacted Concrete (RCC) dam ever built in Puerto Rico and the first single – centered thick – arched RCC dam in the United States. The project was awarded to Dragados USA in 2008, was completed in December 2013 and the dedication ceremony was conducted in February 2014.

This dam completes the last phase of the Portugues and Bucana Rivers Project to protect the city of Ponce from floods. The city of Ponce is subject to frequent and serious flooding from recurrent torrential rains, tropical storms, and hurricanes, especially when these rains occur in the mountainous terrain to the north of the city. This complex climatology also greatly affected the construction of the dam.

Being the first RCC project in Puerto Rico, local experience was not available and an extensive training program on RCC was implemented to be able to build the dam, meet the multiple requirements and achieve the project objectives.

The fact of the location of the project in an island, greatly conditioned supplies of all types (materials, spare parts, etc.) that not being available on the island had to come from the mainland.

A summary of the main site specific challenges are presented in the document below.

Fig. 1 Night view of the completed dam, from downstream

2 Background

2.1 Project location

The Portugues Dam is located in the central region of Puerto Rico near the southern Caribbean cost, less than eight miles from the mouth of the Portugues River and approximately three miles northwest of the city of Ponce. Ponce is the second most populous city in Puerto Rico.

2.2 Project history

The mountainous terrain north of Ponce is subject to sudden and rapid runoff of the existing rivers that caused frequent floods in the downstream urban areas. These floods often resulted in extensive property damages and even loss of life in the city of Ponce.

The Portugues Dam was originally designed by the United States Army Corps of Engineers (USACE) to be its first three – centered double curvature thin – arch dam, as a traditional arch concrete dam. In the early 2000s, the cost of the single construction bidder, significantly exceeded the government's cost estimate and USACE decided to revise its original design for a more cost effective alternatives, finally selecting the current RCC design.

The Portugues Dam, completed in December 2013, is the last of the two dam structures built as part of the Portugues and Bucana Rivers flood control project that was authorized in 1970. The first dam, the Cerrillos Dam, was completed in 1992. The entire project consists of two multipurpose reservoirs with downs – stream channel improvements from the dams to the river outflows as shown in Fig. 2. The Puerto Rican Department of Natural and Environmental Resources (DNER) is the ultimate owner of the project.

Fig. 2 **Project location map**

2.3 Project description

The Portugues Dam was built as a flood protection. The Portugues Dam is a 67 – m high single centered thick – arched RCC dam with a crest length of 375 m. The project was built with over 270 000 m^3 of RCC for the dam, over 7 600 m^3 of conventional concrete on the spillway, intake and top of dam, nearly 15 000 m^3 of mass concrete for foundation and dental concrete and over 18 000 m^3 of RCC for the foundation buttress of the valve house. The dam has a vertical upstream face and a stepped downstream face. The downstream steps are 1.2 m high (four 0.3 m thick layers) resulting in equivalent downstream slope of 0.35h:1v. Even though the RCC layers were built concurrently, the body of the dam had 18 sections created by installing contractions joints spaced 21.3 m apart. An internal gallery (2.1 m wide and 3.7 m in height) was built in the RCC approximately 6 m above the design foundation level. An upstream slopped grout curtain and vertical foundation drain holes were drilled from inside the gallery. The dam has a 42.7 m wide un – gated ogee – crested and flip bucket spillway which was constructed with conventional concrete in the central section of the dam, above the original river channel.

The permanent outlet works consists of two 1500 - mm diameter regulating outlet pipes traversing the dam through a conventional concrete block built between the dam foundation and the RCC structure. The outlet pipes extend from the sluice gates at the intake structure just upstream of the dam to the fixed - cone valves located at the valve house just downstream of the dam.

The slow strength developmentpozzolan - cement used in the RCC was designed to achieve a compressive strength of 31 MPa after one year, but generally exceeded 41 MPa. The slow strength development nature of the RCC allowed continuous RCC placement on "hot" surface joints (i. e. relatively "fresh" concrete surfaces) for a period up to a maximum of 24 h between layers. Partially set concrete surface or "warm" joint surfaces (up to 72 h) and "cold" joints surfaces (over 72 h), required special joint surface treatments before the following layer was placed.

Fig. 3 Aerial view of the dam under construction and RCC batch plant facilities

2.4 Construction facilities

The project required the installation of several facilities to support the construction of the dam. The principal support structures included:

(1) An aggregate production plant to produce three types of coarse aggregates (NMAS: 50 mm, 20 mm, and 10 mm) and fine aggregate (sand). The different size aggregates allowed the production of a wide variety of diverse concrete.

(2) A sand washing plant to remove fines of a portion of the processed sand.

(3) An on - site double concrete plant able to produce up to 4 m^3 per batch each of RCC, conventional concrete, or mortar, supported with:

①Two 1 000 t cement silos

②One 1 000 t flyash silo

③Three coarse aggregate silos of 2 000 t each

④Two fine aggregate silos of 1 000 t each (one for washed sand and another for unwashed sand)

⑤A plant for fabricating ice flakes capable of producing 200 t/day

⑥A water cooling plant for the "wet belts" tunnel and chilled mixing water

⑦Two 120 m long "wet belts" located inside an aggregate cooling tunnel to lower the temperature of the coarse aggregate to below 10 ℃ and help meet the maximum average RCC placement temperature of 15.5 ℃.

(4) An off-site portable batch concrete plant to produce conventional concrete and mortar.

(5) A conveyor system, with a length of up to 298 m, to transport the concrete from the plant to the central portion of the dam.

(6) Miscellaneous support facilities such as the mechanical, electrical, welding and carpentry shops.

The construction of the Portugues Dam did not have access to the electrical grid system and all the on-site facilities were powered with generators.

Fig. 4　Aerial view of the RCC batch plant and pre-cooling facilities

3　Construction challenges

3.1　Site and subsurface conditions

The project is located in heavily fractured and in areas highly weathered sandstone, siltstones and metamorphic volcanic conglomerates intersected by shears and diorite dikes. The highly fractured foundation geology made it difficult to obtain a regular foundation surface. The aggregate material was excavated from a quarry located just over 1 mile north from the dam site.

Heavy excavation equipment was permitted to be used by contract up to the theoretical foundation level, leaving a significant portion of the final cleanup to be performed by extensive use of hand tools and occasional very light excavation equipment (see Fig. 5). This resulted in a very labor intensive and time consuming dental treatment process that was often impacted by heavy rain and sudden floods.

The mountainous terrain of the site created difficulties for the layout of the support facilities and forced the division of the aggregate stockpiles in multiple locations. Likewise, the required temporary construction accesses to different levels of the dam were difficult to deploy.

Fig. 5 Labor intensive foundation preparation

3.2 RCC placement requirements and supplies constraints

The Portugues Dam was constructed with a series of very restrictive requirements. These requirements, although not unusual for many RCC projects and many are common USACE and USA standards, presented special challenges during construction of the project. As indicated earlier, this project is the first RCC structure in Puerto Rico, hence many of these requirements were not previously used in Puerto Rico. This impacted particularly services, materials and equipment. Because of these requirements, many of the local companies in the island could not provide services, materials and equipment to the project. Many of the materials and services had to be provided either by selected qualified local suppliers or by off-island companies.

Producing RCC is basically an industrial process, where continuous RCC plant operation is vital for a consistent product. Minor problems and small breakdowns can cause costly disruptions and even stoppages of the production operation where hundreds of people would be sitting idle due to spare part delivery dependencies.

Fig. 6 General view of the RCC placement

Puerto Rico is an island that does not have similar facilities to the ones used for the construction of this dam, so the vast majority of the specialized spare parts and other equipment had to be brought from off – island suppliers, resulting in a critical logistical challenge to minimize the impact of delivery, maintenance and repair delays. To mitigate these high risks, Dragados USA established a special program to stockpile on site a significant amount of spare parts and long lead items, not commonly stockpiled in other projects, to reduce the potential impact on the schedule. Despite this, on several occasions, spare parts and materials had to be shipped by air from the mainland US. In addition, Dragados USA had on its payroll a larger than normal team of mechanical and electrical personnel working 24 h per day, 7 days per week. These measures increase the cost of a project, but are essential to maintaining the project on track.

Some RCC placement contract requirements included:

(1) Place RCC at an average temperature below 15.5 ℃.

(2) RCC placement windows allowed only during the cooler months of the year.

(3) Cover placed RCC with new layer within 24 h (maximum exposure time between layers); otherwise special joint surface treatment and placement of mortar required between layers (warm joint treatment).

(4) Produce, transport, place and compact RCC within a 45 minute window.

(5) Extensive and labor intensive RCC surface cleaning requirements between layers.

3.3 Limited access to the crest of both abutments

The only permanent access road to the dam crest is from the downstream side towards the right abutment. But to connect this road with the dam crest there is a bridge which logically could not be built until the end of the upper part of the right abutment.

Therefore, to have access for the auxiliary equipment to the proximity of the upper portion of the right abutment while constructing the RCC, and to remove the RCC placement equipment from the top of the dam, and move it to build the upper part of the left abutment, a difficult temporary road was arranged at the upstream side of the right abutment.

Likewise and due to the left abutment does not have a permanent access road, to place the RCC placement equipment into the left abutment in order to finish the upper part of it, and to remove all the equipment from the dam after completion of the dam portion to the left of the central spillway, another very difficult temporary road, because of the steep natural slope conditions of the left bank of the valley, had to be built at the downstream side.

Both accesses, as were no part of the project design, were removed upon completion of their role.

3.4 Weather inclement

Puerto Rico is one of the islands that separate the Caribbean Sea from the Atlantic Ocean. It is located in an area that is prone to heavy tropical storms and hurricanes. As indicated earlier, the project was built to protect the city of Ponce from frequent and dangerous floods. The project itself is located in a narrow valley with steep slopes in the mountainous terrain just north of the city. The watershed area of the Portugues River receives on average a precipitation exceeding 2000 mm per year. The frequent tropical rains often come down heavily and suddenly, giving little time to protect the work areas, removing equipment and safely relocate personnel from the incoming river floods. The steep terrain required the temporary diversion system to be designed for overtopping conditions. The work also had to stop until the water in the inundated access road and low laying work areas receded. Floods were a constant threat during the early construction stages, requiring a vigilant eye on the weather. When rain occurred during RCC placement, often there was not sufficient time to protect the work in progress. This resulted in extensive repairs and stoppages that exceeded the allowed time between layers for regular placement and required extensive joint surface treatment before resuming construction.

Overtopping of the diversion cofferdam during heavy rain events, divided the construction site and limited access to the quarry, RCC plant, and right abutment area.

The frequent rains also created continuously varying and non – homogeneous moisture content of the fine aggregates. RCC is very sensitive to moisture variations and slight changes in moisture had an impact on the consistency of the RCC. This required that a portion of the sand be stockpiled and handled two or three times into covered sheds to maintain a more consistent moisture content for optimum RCC production.

The contract required that RCC be placed below a daily average temperature of 15.5 ℃. The average temperature in Ponce, even during the winter months, is above 24 ℃. To meet the project requirements, all coarse aggregates used for RCC production were chilled to a temperature below 10 ℃ and ice chips were used in lieu of most of the batch water to lower the RCC temperature. The relatively high moisture content of the fine aggregates caused by the frequent rains, in combination with the fact that the coarse aggregate leaving the cooling tunnel was wet, limited the amount of ice that could be added to the batches to cool down the RCC, so maintaining the fine aggregates dry was very important to achieve the low placement temperatures.

Fig. 7　One of the several floods experienced during the dam construction

3.5　Work force availability and experience

The construction of RCC dams is a continuous operation and is essential to maintain a 24 hours – 7 days per week work schedule. This required the availability of a large pool of skilled construction labor force. During peak construction periods, the Portugues Dam employed up to 800 personnel. The city of Ponce and its vicinity could not provide all the needed experienced field personnel for the project, so a significant portion of the labor force had limited or no construction background.

Additionally and as indicated earlier, the Portugues Dam was the first RCC project constructed in Puerto Rico. Therefore, finding local technical personnel with experience in RCC technology was not possible and bringing experienced personnel from outside the island had to be limited due to cost constraints. Therefore, most personnel had to be locally trained in RCC production and placement technology.

Most of the activities in the construction of RCC are very time sensitive. This required that all activities be carefully coordinated with specialized crews for the success of the project; from the time between fabrication and placement of the RCC, to the maximum time allowed between placement of consecutive layers, beyond which special joint surface treatment is required. Some of the specialized activities that the different crews had to accomplish include:

(1) Dental concrete treatment.
(2) Drilling and grouting.
(3) RCC fabrication.

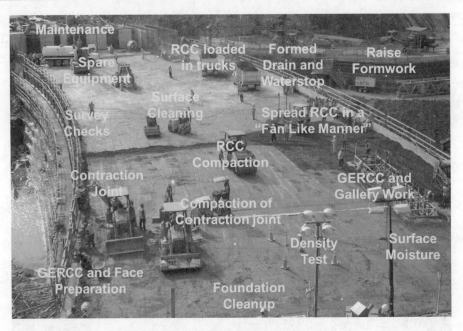

Fig. 8 Typical RCC placement activities

(4) RCC conveyance.

(5) RCC placement and consolidation.

(6) Cleaning and preparation or RCC surface.

(7) Moisture control of RCC surface.

(8) Raising and preparation of formwork.

(9) Grout – enriched RCC consolidation near exposed surfaces.

(10) Contraction joint cutting.

(11) Instrumentation installation and monitoring.

(12) Inspection and testing.

(13) Equipment maintenance.

Poor execution or delays in any of the above activities posed a significant risk of stopping RCC placement that required costly and time consuming remedial work to resume construction after "warm" or "cold" joint treatment. So a reliable and well trained team is essential.

A sizeable portion of the available resources did not have the required construction experience and could not easily adjust to the continuous shift work. This shift work is generally very demanding and resulted in a recurrent turnover of the work force that had to be replaced and re-trained.

The fact that RCC placement was limited to a six month window in the winter months, complicated the experience retention even further, since the majority of the RCC workforce had to be dismissed during the summer months. For the next placement window, an attempt was made to re-hire the same personnel; however, a significant portion of the workforce did not returned and the previously acquired local knowledge was lost and new personnel had to be re – trained and a second learning curve of the field personnel had to be absorbed by the project.

3.6 Safety

Safety is of paramount importance to USACE and Dragados USA. An extensive safety program was implemented in the project with dedicated safety personnel available in the project at all times during the construction activities.

The time sensitive nature of RCC construction requires that equipment and personnel work fast and in close proximity posing a serious safety risk. Furthermore, RCC dam construction uses concrete conveying systems that have to be raised regularly as the dam height increases, demanding personnel to work from ever increasing elevations.

The continuous changing nature of the workforce, the limited construction experience of some of the available personnel hired for the project and the construction requirements described above, also posed considerable risks of serious injury to the field personnel that had to be mitigated throughout the construction period.

The safety requirements and expectations from USACE and Dragados USA for the project were significantly more stringent than the typical safety requirements for construction projects in the region. This made the implementation of the safety plan more challenging and prompted Dragados USA to implement an extensive safety awareness and safety culture program through continuous safety training. It is important to note that even though the project experienced a series of minor incidents, it did not suffer a fatality or other serious accident.

3.7 Quality program

The overall safety of a dam is greatly dependent on the quality of its construction. The Portugues Dam was built to the highest quality standards. Quality Control (QC) was the responsibility of Dragados USA and Quality Assurance (QA) the responsibility of USACE. During peak construction periods, the project quality personnel had a staff of over 30 QC and 20 QA personnel.

A wealth of information was collected and analyzed during the construction process. Some highlights of the quality checks are presented below:

(1) Submittal process: All activities were thoroughly documented through the submittal process. Over 4,500 submittals were prepared for this project

(2) Construction tolerances: Were very demanding. For example the spillway required that the finish grade of the 42.7 m wide ogee, chute and flip bucket be built to within 3.2 mm

(3) Testing frequencies: Vebeconsistency, RCC placement temperatures, RCC density, compressive strength, aggregate gradation, were just a few of the numerous tests and other quality data points collected during the construction of the dam. Many of these tests were performed at high frequencies, for example, most of the over 2 000 conventional concrete truck loads delivered to the project were tested for slump, air content, temperature, and unit weight

The Dam was constructed using the USACE's three step Quality Control and Quality Assurance Process strengthened by Dragados USA's extensive RCC dam experience. First, all definable features of work of the project required to a formal planning session prior to its execution, where its execution plan, as well as all safety, quality, and environmental requirements were thoroughly reviewed and planned by the Dragados USA's construction and quality teams and approved by the

USACE's QA team. Then, before any activity was started, the supervisor reviewed all safety, quality and construction requirements with the construction crew to ensure alignment. Finally, the execution was continuously inspected and verified to ensure compliance with the requirements.

Fig. 9   USACE's and Dragados USA's personnel celebrating the completion of RCC placement

Before RCC placement begun, Dragados USA used the initial RCC test section and other RCC test strips to train the different field crews, select best RCC placement equipment and optimize placement methodologies. For the second RCC placement window, Dragados USA constructed a second test section to refresh the knowledge of existing employees, capture lessons learned from the previous placement session and help train new employees. This second test section was critical to achieve the anticipated quality of the RCC placement during the second window.

4   RCC technology knowledge transfer

As indicated earlier, the Portugues Dam resulted in the first single – centered, thick – arch RCC dam designed and constructed in the United States and the first RCC dam in Puerto Rico.

The dam has served as training ground for the next generation of USACE dam engineers through the "Dam Safety University" program. Using this program USACE trained many of their engineers in workshops and temporary field assignments in dam engineering. The project was also used to transfer RCC technology knowledge to the local engineering community through workshops and temporary employment.

5   Conclusion

The Portugues Dam was successfully completed in December 2013. However, the combination of the project being located in an area where delivery of materials and availability of spare parts for specialized equipment is limited, with lack of experienced personnel in RCC technology, can present a significant challenge to the project without sufficient planning and a strong team experienced in the RCC construction technology.

To mitigate these difficulties, an extensive training program must be implemented to ensure all personnel can work safely and effectively with the numerous time sensitive activities that must be coordinated to achieve a quality RCC dam structure.

Fig. 10 **Aerial view of the completed Portugues Dam**

Finally, it is very important that the owner, designer, and contractor teams work cooperatively to resolve the difficulties that are sure to occur during construction of complicated projects like the Portugues Dam.

Acknowledgements

Grateful thanks must be accorded to the USACE, Jacksonville District, the designer and temporary owner of the Portugues Dam project, for permission to publish this paper. Nevertheless USACEcannot be held responsible for the views expressed in this paper, they being solely those of the authors.

The authors also wish to publicly thank USSD for allowing the use of the information and figures that were previously published in the Proceedings of the 35th Annual USSD Conference held in Louisville, Kentucky, in April 2015.

References

[1] USSD, 2013, Achievements and Advancements in U. S. Dam Engineering.
[2] P. Vázquez, A. González, Moving Successfully from a Conventional Concrete into an RCC Design for Portugues Dam. Proceedings of the 6th International Symposium on RCC Dams, Zaragoza, Spain, October 2012.
[3] R. Ibáñez – de – Aldecoa, D. Hernández, E. Carlsson, Highlights and Challenges of Constructing the Portugues Thick – Arch RCC Dam in Puerto Rico. Proceedings of the 35th Annual USSD Conference, Louisville, Kentucky, USA, April 2015.

The Key Technology of Cement Grouting for Concrete Interlayer Crack Seepage or Leakage of RCC Dam

Li Yan, Chen Weilie, Li Geng, Yue Mingtao

(Gezhouba Group Testing Company, Yichang, 443002, China)

Abstract: Concrete interlayer crack seepage (leakage) is one of the common diseases and defects of hydraulic RCC dam. To resolve this problem, there are a variety of solutions, which include "cutting – off curtain" method and interlayer grouting at upstream face with grooving sealing method and the combination of the two formers, among which the third method is widely used. Hereby, the author has come to a summarized and narrative introduction and description for the key technology in these methods in accordance with own experience of several years' work, and raised tips for attention to provide technical reference for the design and construction of similar project.

Key words: RCC, layer, seepage (leakage), treatment, key technology

1 Introduction

Concrete interlayer crack seepage (leakage) is one of the common defects, with its reason including mix design, construction process and operation management. The defect usually occurs during operation, and according to forms of leakage, it can be divided into point leakage (concentration leakage), line leakage (curtain leakage) and jet leakage (including point jet leakage and line jet leakage), and etc. Leakage quantity is related to the factors such as head pressure, leakage path length, and cross – sectional area.

Dam interlay leakage can cause greater seepage pressure in the interior of concrete, and sometimes even affect the stability of the building safety. In the case of erosive water, it may cause damaging erosion and reduce concrete strength. Even in the case of water without erosiveness, long – term leakage may cause dissolving of $Ca(OH)_2$ in cement stone and CaO in silicate hydrate may also dissolves gradually, and more leakage may cause more and bigger voids, which may causes loose structure, lower strength and long – term leakage, and eventually lead to the destruction of the entire structure. In cold areas, frozen ice from leakage water may piles on outcrops (See Fig. 1), which may cause freezing and thawing damage of concrete.

In short, interlay seepage (leakage) of a compacted concrete dam is a type of non negligible hazard to the safe operation of the dam, which shall be promptly treated.

Fig. 1　Ice piled on outcrops of leakage water

2　Main technical solutions

There are a variety of solutions for interlay leakage of a RCC dam, which include "cutting – off curtain" method (Fig. 2), interlayer grouting at the upstream face with grooving sealing method (Fig. 3), and the combination of the two formers.

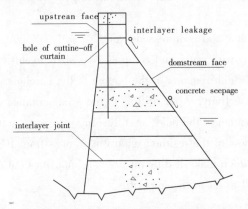

Fig. 2　Diagram of Cutting – off Curtain

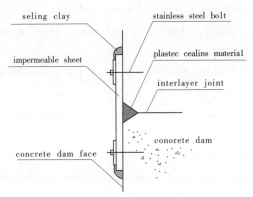

Fig. 3　Diagram of Interlayer Sealing at the Upstream Face

(1) "Cutting – off curtain" refers to the continuous, complete impervious structure with lower

permeability than concrete, which is constructed in the RCC dam to cut off seepage and is like a band in the plane and a stage curtain in the three-dimensional space. Grouting method can be adopted to form this structure, referred to as "cutting-off curtain" method.

(2) When the reservoir is emptied, drill borehole along the interlayer joints, and then conduct controlled chemical grouting at the upstream face, and finally chisel grooves along the joints and conduct joint sealing with sealing material.

(3) In the case of reservoir emptying and permitted construction period, the combination of the two former methods can be considered.

3 Key process

3.1 "Cutting-off curtain" method

3.1.1 Drilling, flushing with water and compressed-air, simple Lugeon test and drainage

3.1.1.1 Drilling

Rotary drilling rig and diamond or carbide drill bits shall be used for drilling and non-core drilling construction method is not allowed. When drilling, leaking observation must be carried out.

3.1.1.2 Flushing with water and compressed-air

The purpose of flushing with water and compressed-air is to flush powder, debris and other impurities out of the hole, which remains in the bottom of the borehole and adheres to the wall of the hole, and to check connectivity. After drilling to the predetermined depth and coring, put the drilling bit down into the bottom of the hole, and flushing with water until the backwater becomes clear, and when the thickness of the residual impurities ranges from 10 to 20 cm, end flushing. In the case of deep hole, which make it difficult to flush impurities out of the hole, cleaning pipe can be mounted on the drill to clean the impurities.

3.1.1.3 Simple Lugeon test

Simple Lugeon test can be conducted before grouting as required. In the case of obvious leakage, grouting can be conducted directly. Lugeon test aims to understand the interlayer seam leakage including leakage quantity, connectivity and so on. Dyed water shall be adopted for the test. Test requirements are as follows:

(1) Test pressure is 80% of grouting pressure, and not more than 1 MPa. Groundwater level is assumed to be a dry hole.

(2) Test times is 20 min, and take the flow rate reading once every 5 min, and take the final flow rate as calculation flow rate.

(3) Test results are expressed with permeability (q) with the unit of Lugeon (Lu).

It shall be noted that due to long time flushing or Lugeon test injecting large amounts of water it is harmful and shall be avoided.

3.1.1.4 Drainage

After completing Lugeon test, connect the inlet pipe to the compressor, and intermittently start air compressor to drain the water from the hole through the return pipe. Time interval shall be determined according to the field observations, usually shorter early and extended later. When draining water, neighboring air leakage phenomenon shall be observed.

3.1.2 Grouting, lifting observation and sealing

"Cutting – off curtain" grouting shall be conducted in the order of holes stipulated by design requirements. In the case of multi – hole grouting, firstly grout in the holes of the downstream row, and then that of upstream row, and finally that of middle row; and for the same row of holes, order II ~ III is usually adopted.

3.1.2.1 Grouting

(1) According to the project requirements opening packer, top – down segmenting or hole packer can be adopted.

(2) Grouting segment length shall be determined according to the thickness of the poured concrete and leakage elevation. In the case of no interlayer leakage, if the top – down segmenting and opening packer method is used, end hole shall be located between compacted layers, and if the top – down segmenting and hole packer method is used, packer shall be located 1/2 of compacted layers. In the case of interlayer leakage, the leakage joint shall be in the grouting segment.

(3) Grouting pressure shall be determined according to the design requirement or boundary conditions such as hydrostatic pressure of grouting segment and pouring quality of concrete.

(4) Water – cement ratio shall be determined according to the design requirements or through field test. Grouting pressure shall be controlled through the gauge installed on the return pipe with its fluctuation less than 20% of grouting pressure. Loggers shall record the average and maximum of grouting pressure with the average being 90% peak pressure. The pressure shall rise step by step to prevent lifting the concrete structure.

(5) The criterion to end grouting shall be in accordance with the design requirements.

(6) Treatment for special circumstances. In the case of s grout leakage, according to the specific circumstances the measures such as caulking, surface sealing, low pressure, thick grout, limited flow, limited quantity, intermittent holes, leaving to set and so on, may be used. In the case of grouting leaking between adjacent holes, if conditions permit the leaking holes for grouting, conduct grouting at the same time, otherwise seal the leaking holes and after completing grouting of grouting holes, conduct cleaning, flushing and grouting for the leaking holes. If grouting is interrupted for more than 30min, flush the holes before re – grouting. In the case of re – grouting, if injection rate is similar to that before interruption, initial grout ratio shall be used, otherwise the measures such as thicker grout and cleaning holes may be used. In the case of thickening grout, adjust the grout ratio to the initial one, and if there are no obvious effects, re – mix the grout.

3.1.2.2 Observation of absolute lifting

During Lugeon test and grouting, lifting values shall be less than the design values. The device for lifting observation is shown in Fig. 4.

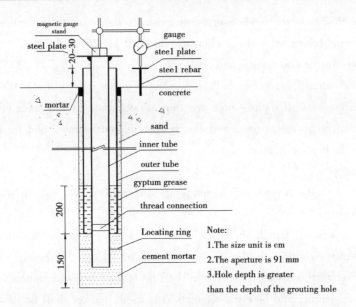

Fig. 4 Diagram of device for lifting observation

3.1.2.3 End to grouting

Criterion to end grouting: at the design pressure, when grouting – absorption capacity is not more than 0.4 L/min for a single hole or 0.8 L/min for group holes, continue grouting for 40 min to end grouting.

3.1.2.4 Sealing

Thick grout and total pressure method shall be used for sealing.

3.1.3 Grouting material

3.1.3.1 Cement

Cement for grout shall be fresh and free from lumps with strength grade of 42.5 MPa, which may be ordinary portland cement or portland dam cement or ultrafine portland cement. Fineness (retaining percentage through 80 μm square – hole sieve) for ordinary portland cement and portland dam cement shall be less than 5%, and specific surface area for ultrafine portland cement with strength grade of 42.5 MPa shall be greater than 6 500 cm²/g, with $D_{50} = 8 \sim 12$ μm, $D_{max} \leqslant 40$ μm.

3.1.3.2 Grout mixtures

1) ordinary portland cement grout mixtures

Grout mixtures may be made of ordinary portland cement with strength grade of 42.5 MPa and water – cement ratio being 1:1, 0.8:1 and 0.5:1.

2) Wet ground cement grout mixtures

Grout mixtures may be made of ordinary portland cement with strength grade of 42.5MPa ground by three wet grinders. Mix procedures: dispensing – high speed stirring – wet grinding – ordinary stirring. Fineness requirements in the case of colloid grinders are $D_{97} \leqslant 40$ μm and $D_{50} \leqslant 12$ μm. Grout should be screened before using, and wet ground cement grout shall be used within 2h from mixing. Water – cement ratio shall be determined according to site requirements.

3) Ultrafine cement grout mixtures

Grout mixtures may be made of ground ordinary portland cement with strength grade of 42.5 MPa by high speed mixer. In the case water ratio not greater than 1:1, superplasticizer shall be introduced in aqueous solution with its dosage determined through indoor test. Marsh funnel viscosity should be about 30 s. Water – cement ratio may be 3:1, 2:1, 1:1 and 0.5:1 (mass ratio), or adjusted according to site requirements.

3.2 Interlayer grouting at the upstream face with grooving sealing method

Treatment procedure: grouting – grooving – infilling with sealing material – covering with impermeable sheet.

3.2.1 Grouting

Drilling and header embedment shall be conducted along the interlayer joints of upstream face corresponding to the leaking joints of downstream face. Controlled grout shall be adopted, and grout material may be chemical material such as polyurethane, and in cases where there are overhead concrete, cement grout may be used. The spacing of grouting holes shall be determined through grout take with the principle of more grout take, greater spacing. Empirical grouting pressure generally ranges between 0.2 ~ 0.5 MPa.

3.1.2 Grooving

After grouting is complete, conduct grooving as per Fig. 2. Grooves may be "triangular" or "U" type, in both cases where the depth of grooves should be greater than the width.

3.2.3 Infilling with sealing material and covering with impermeable sheet

After completion of grooves, infill with sealing material and cover with impermeable sheet in accordance with the design requirements. The process may refer to "*Technical code for Repair and reinforcement of Hydraulic Concrete structures* (DL / T 5315—2014)".

3.3 Combination of the formers

The treatment procedure for the solution: after grouting for the crest or corridor is complete, empty the reservoir, and proceed to conduct interlayer grouting at the upstream face and grooving sealing method. When leakage is detected near during water (air) pressure test for the crest, it shall immediately be treated as leaking between adjacent holes.

4 Project example analyses

A RCC hyperbolic arch for a hydroelectric power station, crest elevation: 620.0 m, maximum height: 84.5 m, top width: 6.0 m, bottom width: 21.5 m, dam arc length: 143.49 m, ratio of height to thickness: 0.24, installed capacity: 2 × 15 MW, engineering grade: III.

After normal reservoir impounding in November 2006, leakage occurred on the local downstream face of the non – overflow monolith, and free calcium precipitated near the leaking area.

Grouting holes for the cutting – off curtain were arranged in a single row, order III, spacing of 1 m, with deflection rate not greater than 1%. For the holes of order I, II ordinary portland cement with strength of 42.5 was used, with water cement ratio being 2, 1, 0.8, 0.6 (or 0.5). For the holes of order III, if water injection rate was greater than 15 L/min, the former grout mixtures was used, otherwise ultrafine cement was used, in which case water cement ratio is 2, 1, 0.8, 0.6.

Opening packer method was used for grouting. Segment length was generally 3 m and segment length for final holes was less than to 5 m. During drilling when backwater was obviously getting less, stop drilling and then conduct grouting for a single segment.

In the case of no water in holes, grouting pressure was less than 0.5 MPa for holes of order I and II and less than 0.6 MPa for holes of order III. In the case of water in holes (reservoir impounding), raise the grouting pressure accordingly.

Before grouting, each grouting segment was flushed, and simple water pressure test was conducted concurrently, and test pressure was 80% of grouting pressure and less than 1 MPa.

Criterion to end grouting: at the design maximum grouting pressure, when injection rate is not more than 1 L/ min, continue grouting for 40 min to end grouting.

Hole Sealing: total pressure grouting sealing method was used.

After grouting average velocity of acoustic test was greater than 3 750 m/s, and the rate of measuring points with acoustic velocity less than or equal to 3 000 m/s was within 3%. Except for a few damp phenomenons were in sight, visual dam leakage points disappeared. Concrete permeability met the requirement that at least 85% (or 80%) shall be not greater than 3 Lu and the left should be less than 3 Lu. The project has safely operated for nearly 10 years after completion.

5 Conclusions

(1) There is another solution for interlayer seepage (leakage) treatment of RCC dam, which is pouring permeable panels or painting or spring water – proof material (such as polyurea, and etc.) on the entire upstream face. Compared with the above – mentioned methods, this solution is feasible in technology, however due to needing to empty the reservoir, it is not practical economically. Therefore, cutting – off curtain method is the priority solution. In addition, except cement grout, chemical grout mixtures are also used. Given limited space, they are not involved.

(2) For the interlayer grouting at the upstream face with grooving sealing method, its disadvantage is the same as that of pouring permeable panels or painting or spring water – proof material solution, therefore it is also not practical. When conditions permit, the author think controlled is the most critical process since it involves holes spacing, grouting pressure, grout mixtures (cement or chemical grout), criterion to end grouting and so on.

(3) Cutting – off curtain method is the priority solution for interlayer seepage (leakage) treatment of RCC dam. The author thinks flushing with compressed water and air is the most critical process. Only through flushing with compressed water and air, it is possible to find the relationship between leaking paths including upstream and downstream faces. And then customized program of grouting can be developed.

(4) According to "*Technical code for Repair and reinforcement of Hydraulic Concrete structures* (DL/T 5315—2014)", concrete seepage (leakage) treatment shall follow the principle "blocking at the upstream face and drainage at the downstream face, and combination of blocking and drainage". Due to limited space, how to "drain" is not involved in this paper.

Concrete Production in the Required Quality, Temperature and Output Capacity, Considering the Different Climatic and Geographic Conditions

Reinhold Kletsch

(Dam Projects Liebherr – Mischtechnik GmbH, Germany)

Introduction

Concrete is playing an increasingly significant role in today's state – of – the – art methods of dam construction and in particular for the RCC dam. The demand for outstanding mix quality, correct concrete temperature and necessary production output is rising continuously. If these high demands are to be met, very specific and meticulous planning of the entire concrete mixing plants as well as the handling and transport of the concrete is of utmost importance. Geographic factors and climatic conditions must also be considered.

Furthermore, reliability of all installed equipment is tremendously important as every discontinuation in the supply of concrete can prove extremely problematic, both financially and with regards to agreed deadlines.

1 Planning of the mixing plant with aggregate and binder storage

1.1 Site plan

It must be ensured when planning mixing plants, storage for aggregates and binders, as well as cooling systems, that the transport routes for aggregates and concrete are kept as short as possible, taking geographical factors into consideration. As it is evident in our example, the Son La dam project in Vietnam, irregularities in height over rough terrain can actually be used as an advantage to optimise the material flow, as long as planning is carried out correctly. Material discharge, gravel storage, cooling system and mixing plant for this project have all been arranged at various heights, allowing maximum utilisation of the terrain and ensuring that the necessary transport routes are kept as short as possible. Shorter transport routes also minimised the heat up of the concrete.

1.2 Storage of aggregate and binders

Due to the huge demand for additives (different sized aggregates) and binders (cement, fly ash or pozzolana), sufficient storage capacity must be provided directly at the mixing plant. The channels of supply or the required provision times for the individual components are absolutely de-

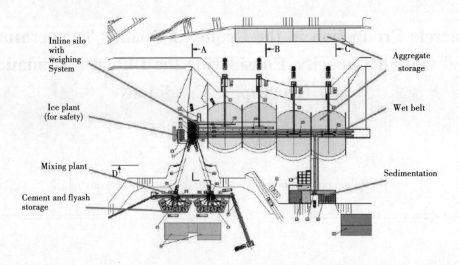

Fig. 1 Layout plan for Son La Dam Project in Vietnam

cisive for the extent of the storage capacity. To prevent supply shortcomings, the storage capacity must include a stockpile directly at the mixing plant sufficient to ensure concrete production for at least 2 ~ 3 days.

The advantages of extensive aggregate storage are the high storage capacity, a cooler aggregate temperature of up to 5 ℃ within the stockpile, and in particular the additional drainage for sand. It also provides the possibility of lowering the aggregate temperature by shading the stockpile (sand storage) and thus reducing the required cooling capacity to a minimum.

The binders are stored in silos. The storage volumes for these silos are determined by the channels of supply or the provision times for the binders.

2 Cooling methods

Depending on the required concrete temperatures or the aggregate temperatures, as well as the ambient temperature, a cooling system may have to be provided for concrete production.

Essentially, one of the following cooling methods will be applied: Cooling of the mixing water, evaporation cooling, ice cooling, aggregate cooling with water or air or a combination of these systems. Cooling with liquid nitrogen may only be applied under certain circumstances.

2.1 Cooling of the mixing water

The simplest method of cooling concrete is to use cooled mixing water, whereby approx. 33 litres of cold water (5 ℃) is required to cool 1 m^3 concrete by 1 ℃. As the aggregates being used already indicate their own moisture, often only minimum amounts of water can be added, this method of cooling on its own will not suffice in most cases.

2.2 Evaporation cooling

Another possibility of cooling the concrete is evaporation cooling. For this cooling method the aggregate stockpile is sprayed with uncooled water. As the water then evaporates, heat is drawn from the stones thus cooling the stones down. It must be ensured with this method, however, that

Fig. 2 Storage of aggregate and binders

the water is clean, and that only the exact amount of required water is used. If too much water is applied the moisture content of the aggregates increases, thus restricting the amount of cooled water or ice which can be added. If applied correctly, this method will cool the aggregates down to approx. 2 ℃ above the Wet Bulb temperature, and is used essentially in very hot and dry regions. In most cases, however, cooling of the aggregates achieved with this method will prove insufficient and an additional cooling method will be needed. Moreover, incorporation of this cooling method into the cooling system is only possible under certain circumstances due to potential deviations in the Wet Bulb temperature.

2.3 Ice cooling

The most frequently applied cooling method is ice cooling. Pre-cooled water is frozen in the ice machine and then a rotating scraper works the ice into shards with a thickness of not more than 1.5~2 mm. The ice shards (ice temperature −5 ℃) have a tremendous cooling potential due to the conversion energy and melt very quickly due to their size. It is extremely light and is thus easy to store and to transport.

As an alternative, plate ice is sometimes also used. Financially, this proves much cheaper but also takes longer to melt, effecting both a decrease in plant output and an increase in wear.

Approx. 7.5 kg of ice is required to cool 1 m^3 of concrete by 1 ℃. One disadvantage of this cooling system is the high power consumption required to produce the ice.

2.4 Cooling with liquid nitrogen

For this cooling method liquid nitrogen with a temperature of approx. −196 ℃ is supplied directly to the concrete by way of a lance or conveyed to the binder (cement) via a feeding device.

Nitrogen cooling is extremely expensive and is therefore applied for minimum quantities of concrete only. Acquisition of liquid nitrogen can also prove difficult.

2.5 Pre-Cooling of coarse aggregates with water (water-cooling)

Water cooling of the aggregates (pre-cooling) is a method applied whenever significant amounts of cooled concrete are required. For this cooling method, aggregates >5 mm are sprayed with cold water (3 ℃) and subsequently cooled to the required aggregate temperature in accordance with cooling time and aggregate size. The additives can thus be cooled to approx. 6 ℃. Cooling can be carried out either in a cooling silo or on a wet belt.

2.5.1 Cooling silo

The cooling time in a cooling silo can be influenced relatively easily via the storage volume. Aggregate is sprayed with water in the silo. Cold water cools the aggregate to the required temperature as the water flows through the silo, before being separated from the additives at the base of the silo via drainage sieves and supplied to the sedimentation. The advantages of the cooling silo for aggregate sizes greater than 50 mm are the minimum special requirements and the low investment costs.

Following drainage of the aggregates, the aggregates are then transported to the working silo via conveyor belts.

2.5.2 Wet belts

Cooling times are influenced by the wet belt in accordance with belt speed and conveyance distances. As the aggregates are being transported from the aggregate storage to the mixing plant, they are sprayed with cold water on the conveyor belt and thus cooled to the required aggregate temperature. At the end of the wet belt, the water is separated from the aggregates via drainage sieves and then supplied to the sedimentation.

Fig. 3 Complete Liebherr plant with cooling silo, Se San III Dam Project, Vietnam

Following drainage of the aggregates, the aggregates are transported to the working silo via conveyor belts. The necessary storage volume of the working silo is determined by the number of transport lines or cooling lines. If each aggregate is provided its own transport line or cooling line, the material can be supplied directly to the working silo, thus allowing the necessary storage volume to be reduced to a minimum ('just on time').

Fig. 4 4 Liebherr Wet - Belt

2.5.3 Sedimentation

Direct cooling of the aggregates with water causes fine fraction to be flushed out of the stones. To ensure that the necessary cooling output is retained at an absolute minimum, the water which was warmed ever so slightly during the cooling procedure is purified and returned once more to the cooling process.

Purification of the cooling water is resulted in three stages. In the first stage, all aggregates greater than 1.5 mm are separated from the water via drainage sieves. In the second step the dirt water will be pumped together with flocculant into the settling tank, where the fine fraction can settle down. The size of the sedimentation tank is dependent upon the water flow, as well as the rate at which the fine fraction settles. And finally the sludge from the settling tank with fine fraction from the aggregates, approx. 1%, can be drained once again in the Chamber Filterpress.

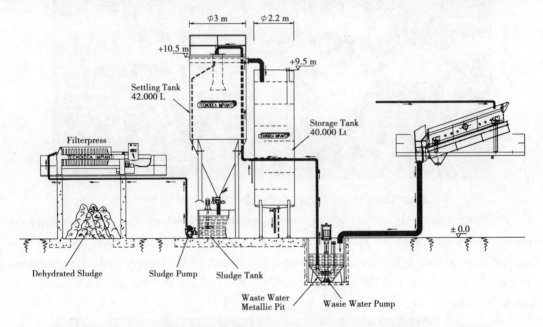

Fig. 5 Sedimentation system

2.6 Pre – cooling of coarse aggregates with air (air – cooling)

For classic air cooling, air which has been cooled to a temperature of $-20\ ^\circ C$ is blown into the working silo. This cooling method allows the aggregates which have already undergone cooling with water to be cooled still further. It must be noted here, however, that fine fraction is no longer present within the aggregates and that storage volume of the working silo is determined in accordance with the cooling time.

3 Batching plants

Modular and semi – mobile systems have been implemented within our mixing plant technology. The advantages of these plants are:

  High flexibility in aggregate and binder storage.

  High flexibility in mixer size (from $2.25\ m^3$ up to $6\ m^3$).

  Short delivery time.

  Standard, well proven components.

  High availability of components and spare parts.

  Short erection time and low erection costs.

  Fast dismantling.

Low foundation costs.

High reusability in subsequent projects.

To meet the strict requirements for operational safety and redundancy for RCC – production, we recommend that these plants be configured as double plants. The utilisation of 1, 2 or 3 double plants will guarantee plant production outputs of 1 000 m³ of ready – mixed concrete or more every hour.

With the realisation of this high production output with optimum quality standards, we can assume that the plant components, and in particular the calibratable microprocessor control, are being subjected to tremendous operational stresses. The microprocessor control allows monitoring, regulation and protocolling of all procedures. Moisture of the aggregates for example, can be determined as the aggregates are being batched, and this can then be taken into consideration when water and sand correction is initiated for the same batch. Precise weigher measured values are evaluated, protocolled and, if necessary, adjusted by the control. Quick and meticulous batching of the components is quintessential if consistent mixes with high production outputs are to be achieved. This system guarantees unchanging consistency.

Once all components (aggregates, binders, water, additives etc.) have been batched, they are then conveyed to the 'heart' of the mixing plant, the mixer. The mixer should be of rugged design and be capable of high production outputs. Mixing performance of the mixing system is decisive for a high production output. A sound mixing action will reduce mixing time to an absolute minimum (e.g. 30 sec.) and ensure outstanding plant production output, as well as reducing wear to an absolute minimum. Additionally, the mixer has been fitted with highly wear – resistant tiles to guarantee an extensive life – expectancy.

Concrete temperature must be measured constantly throughout the mixing procedure. Any deviations in mix temperature are detected immediately within the mixer and an adjustment initiated for the following batch.

Fig. 6　Liebherr mixing plant, Son La Dam Project, Vietnam

4 Concrete transportation

Upon conclusion of the mixing procedure, the cooled concrete must be transported as quickly as possible to the incorporation site. Any delays during this transport period will allow the concrete to warm up again.

Various means of transport can be used to convey the concrete to the incorporation site (dump trucks, belts, vacuum chutes and cable crane).

4.1 Dump trucks

Depending on the respective location of the mixing plant, as well as the quantity of concrete required and various construction stages, dump trucks are often introduced to transport the concrete to the dam. When using dump trucks, however, it must be ensured as early as the planning stages that clear travel lanes to the individual construction levels are feasible.

Moreover, it must be considered that cooled concrete is going to warm up by 2~4 ℃ during transport, depending on the respective distance which must be covered. This must be taken into account during the production stages.

The concrete is then discharged from the dump truck into a chute or onto a belt before being conveyed to the incorporation site. Any subsequent transport, or distribution around the dam, is then undertaken by the dump trucks provided.

4.2 Vacuum chute

On steeper slopes, it can be extremely problematic to convey the concrete using dump trucks or belts, and a vacuum chute provides the ideal solution. Up to 80 m in length and featuring an inclination of 45°, the vacuum chute is installed directly onto the slope. To ensure a continuous flow of material, a so-called surge hopper must be placed at the top of the chute. Dump trucks are filled at the bottom of the chute to continue conveyance of the concrete. The chute is then shortened as construction progresses (Fig. 7).

Taking consistency, mix formula and conveyance distance into account, the vacuum chute should allow an efficient and cost-effective transportation of concrete on the steepest of slopes without separation of the mix occurring.

4.4 High speed conveyor belt with swivel belt

High speed conveyors transport the concrete from the mixing plant to the dam by way of a series of conveyor belts. Standard conveyance capacities are 250 m^3/h, 350 m^3/h and 500 m^3/h in hardened concrete. The belts are driven at speeds of up to 3.8m/s to ensure that the stipulated conveyance outputs are maintained with the shortest possible transport time. Shorter transport times also ensure that the temperature of the concrete is maintained at a max. 2~4 ℃ and does not have an opportunity to warm up again.

An optimal, continuous filling of the conveyor belts is guaranteed by way of a surge hopper installed directly after the mixing plant.

This surge hopper also facilitates the filling of dump trucks in the event of the conveyance line breaking down (Fig. 8).

Fig. 7 Vacuum chute, Yewa Dam Project, Myanmar

At the end of the conveyance line a swivel belt distributes the concrete into dump trucks, which then undertake subsequent transport to the dam.

As the dam continuously increases in height, the conveyance line can be repositioned using the climbing equipment installed in the belt supports.

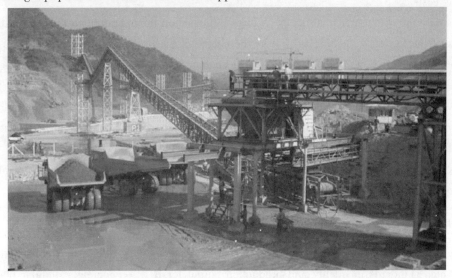

Fig. 8 Conveyor system with surge hopper, Yewa Dam Project, Myanmar

5 Concrete production

Successful incorporation of concrete or high production output of optimum quality concrete depends essentially on the system components and the operating personnel.

As early as the production process, the quality of every batch must be constantly monitored at the mixing plant by measuring moisture and temperature, as well as protocolling all batched aggregate components (gravel, binders and water). These batch inspections ensure that any deviations in moisture and temperature, as well as batching irregularities during concrete production, are recognised immediately and corrected accordingly.

Professional training of the operating and maintenance personnel is also an essential factor for quality assurance. A training schedule must therefore be drawn up for the operating and maintenance personnel, covering commencement of plant assembly and start – up through to actual concrete production. Maintenance and servicing plans drawn up in collaboration with the plant manufacturer would be extremely beneficial. The plant manufacturer and the operating personnel should collaborate in the carrying out of these maintenance tasks in the initial stages of concrete production, in accordance with the drawn up schedules. The entire plant should also undergo a thorough inspection, and serviced accordingly, by the manufacturing company at predetermined time intervals.

High concrete production output of consistently high quality concrete over an extended time scale can only be achieved with the utilisation of qualitative, high – grade plant components and professionally trained operating and maintenance personnel.

6 Investment and operating costs

Further important points that have to be considered during planning are the investment and operating costs.

Not only the maintenance costs have a huge influence on the operating costs but also the right planning of the aggregate feeding as well as the choice of the most suitable cooling system for the concrete production. For example with the right cooling system the necessary energy costs for the concrete cooling can be significantly reduced.

To decrease the investment costs the batching plants should be used for further projects. Therefore modular and semi – mobile systems have been proven. In this case the initial costs are spread to several projects.

As examples the projects Se San III and Son La can be named. Both batching plants have been used for two further projects.

7 Conclusion

State – of – the – art, qualitative first – class plant technology allows the realisation of extremely high standards in quality and production output for RCC – concrete. Modular plant systems which have been adapted to the respective requirements, and which can be reused for subsequent projects, are of utmost significance. Additional advantages of these plants are low foundation costs, minimal set – up times and optimum availability of wear – parts and spare – parts due to the incorporation of standard components.

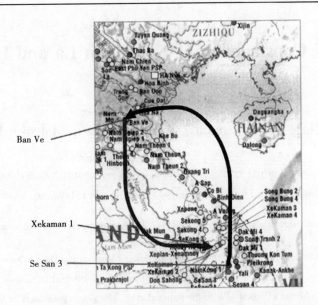

Fig. 9 Projects done with the same mixing plants

If operational safety for the entire plant has to be guaranteed, it is essential that the schedule from the storage of aggregates or binders up to the installation of the concrete itself, is thoroughly observed. Meticulous overall planning also guarantees optimum interaction (aggregate storage, cooling, sedimentation etc.), as well as preventing interface problems for the plant components. Maximum utilisation of the plant components or plant performance can thus be achieved, and fewer problems are presented during start – up and subsequent concrete production. If these benefits are to be maintained over lengthy periods of time, professional training of the maintenance and service personnel should also be incorporated in the overall planning.

Speed of Construction of Evn's Son La and Lai Chau RCC Dams

Nguyen Hong Ha[1], Marco Conrad[2], David Morris[3], Malcolm R. H. Dunstan[4]

(1. Son La Management Board (EVN), Vietnam; 2. AF – Consult Switzerland Ltd, Switzerland; 3. AF-Consult SwitOzerland Ltd, Vietnam; 4. Malcolm Dunstan and Associates, U. K.)

Abstract: The uppermost two dams in the cascade of power plants on the Da River in North West Vietnam have both been constructed ahead of schedule ensuring overall completion of both projects and providing the timely addition of 3 600 MW to the national grid. The power generated by the Son La project in the first year of full operation was some 8.5 billion kWh providing some 400 million USD of income annually compared to the investment cost of the project of some 2.8 billion USD. The completion of the Son La Project on time, the first large scale Government Project to do so, shows one of the advantages of RCC construction over other forms of dams, especially when generation income and reduction in contractor's establishment costs are taken into account. The paper describes the projects, the construction planning, and actual construction progress of the two dams and highlights some of the factors having contributed to the benefits gained.

Key words: RCC gravity dams, construction planning, construction methodology, project layout

1 Introduction

The River Da, a major tributary of the Red River in Nothern Viet Nam is an important source of power in a contry where annual demand is continuously increasing. Electricity of Vietnam (EVN) has constructed or is constructing a total of five projects with a total installed capacity of 6 250 MW.

Table 1 Hydro plants in the Da river catchment owned by EVN

Project	Start	Completion	Dam type	Installed Capacity (MW)
HoaBinh	1979	1994	Central Clay Core Rockfill	1 920
Son La	2005	2012	RCC gravity	2 400
Lai Chau	2011	2016	RCC Gravity	1 200
Huoi Quang	2009	2016	CVC Gravity	520
Ban Chat	2005	2013	RCC Gravity	210

The Da rises in China and has a total basin area of 52 600 km^2 and a mean annual flow of 854 m^3/s from a catchment of 26 000 km^2 at Lai Chau. The annual run-off at the Lai Chau site is 26 900 million m^3. The PMF is some 27 823 m^3/s into a reservoir with a live storage of some

710.9 million m³ and a surface area of 39.63 km² at full supply level. At Son La the catchment area is 43 760 km², the mean annual flow is 1 532 m³/s, and the annual run-off is 48 320 million m³ into a reservoir with a live storage of 5 970 million m³ and a surface area at full supply level of 224.3 km².

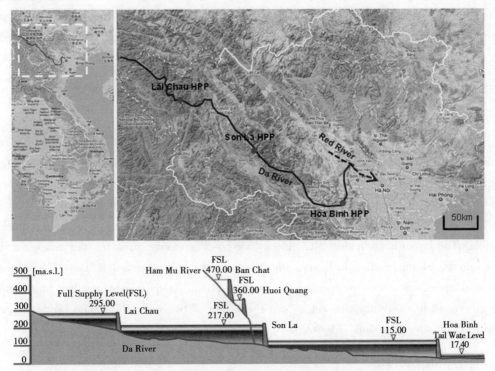

Fig. 1 **Da River catchment and cascade of EVN owned hydropower plants**

Because of the need to increase installed capacity as quickly as possible an RCC dam was selected for Son La as this saved some 3 years construction time compared to a conventional concrete gravity dam. In practice the RCC dam was completed ahead of programme and the units came on line as scheduled with the final unit commissioned in 2012.

Son La was the first major Vietnam Government project to be completed on time and the Government awarded the design of the Lai Chau Project to the same Designers and Contractors to ensure that it too would be completed on time. The essential parts of the project time schedule are, as always, river diversion, and diversion closure with construction of the dam to a suitable level sandwiched in between. In the cases of Son La and Lai Chau the diversion arrangements comprised an open channel on the right bank with culverts to carry dry season flow adjacent to the channel and forming part of the system of cofferdams to allow excavation and construction of the dam and powerhouse (located on the left bank). In order to meet the overall schedule it was essential to meet the critical dates for diversion closure and thus efficient and speedy construction was required.

So effective was the scheduling at Son La that the power generated by the Son La HPP in the first year of full operation was some 8.5 billion kWh, providing some 400 million USD of income annually compared to the investment cost of the project of some 2.8 billion USD. In the case of Lai Chau the RCC dam was completed ahead of schedule andthe diversion was closed June 20th, 2015. Unit 1 generation is scheduled for November 2015 and the third and final unit by the end of 2016. The completion of the Son La HPP on time, the first large scale Government project in Vietnam to do so, and the completion of the Lai Chau HPP at worst on time show one of the advantages of RCC construction over other forms of dams, especially when generation income and reduction in contractor's establishment costs are taken into account.

2 Son La RCC Dam

2.1 Layout of RCC dam and programme

The project comprises a 138 m high RCC gravity dam with a total RCC volume of 2.677 million m³. The RCC placement in the dam started on January 11th, 2008 and was complete by August 25th, 2010, after a total period of about 31.5 months. In order to provide an earliest construction start of the power intake structure integrated into the RCC dam the dam had to be divided into individual RCC blocks of limited volumes to reach the base of the power intakes as quick as possible. The layout of individual RCC blocks was also necessary to control the maximum RCC layer volumes within practicable limits, allowing fresh-on-fresh placement of consecutive layers, i.e. maintaining hot joints. Accounting for the corresponding requirements resulted in a complex arrangement of RCC construction stages and the layout of two main RCC conveyor branches to serve RCC placement on both sides of the CVC intake structure.

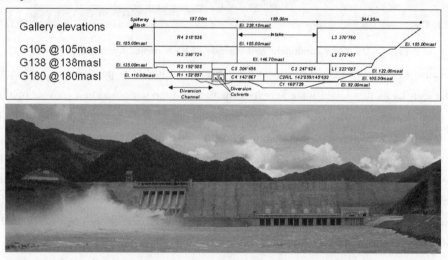

Fig. 2 Son La RCC block layout and downstream view of completed plant

The RCC programme was dictated by the layout of the project that had been approved by the Higher Authorities at an early stage and to change this to better suit RCC (a decision taken later) would have been time consuming and difficult. Thus, the block arrangement had to fit with the overall layout and the construction of the intake section needed to be a priority so that work on the CVC of the intake on top of blocks C3 and C5 could start as soon as possible. In fact, the RCC construction of the left and right abutment blocks overtook the CVC construction in the intakes by almost one year.

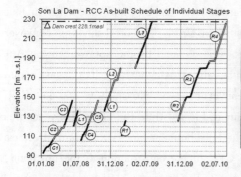

Fig. 3 Son La as – built RCC programme and completed RCC dam while intake CVC works are still ongoing

2.2 RCC speed of construction achieved at Son La

When the Son La RCC dam was complete it was the 7th fastest RCC dam in the world with an average placement rate of 84 995 m^3/month and a peak month of 200 075 m^3. The peak day during the construction period of 31.5 months was 9 980 m^3. A number of placement interruptions, some of them self – inflicted, occurred during construction (refer to [1]) and days lost due to non – self – inflicted interruptions (e.g. longer flood period than anticipated, resulting in a longer overtopping of block R1 in the diversion channel) totalled in 222 days. Deducting those from the overall RCC construction time, the RCC was essentially placed at a modified average rate of 110 632 m^3/month. It was seen that the climatic conditions, such as very hot temperatures and periods with high precipitation, had less impact on the overall average placement rate than the lost productivity due to coveyor and swinger relocations and reduced placement speed due to rather smaller layer areas and congestion. The main factors having contributed to the fast progress and the completion ahead of schedule were:

● RCC mix design and concept allowing the adoption of simple RCC placement methodologies, including avoiding segregation, placing hot joints, minimisation of lift joint preparation efforts and avoidance of bedding mixes, application of grout enriched RCC against interfaces, etc.;

● Capable RCC batching and transportation plant to support peak placement rates, as well as sufficient pre – cooling capacity to maintain the maximum allowable RCC placement temperature of 22 ℃ even during very hot periods and, thus, continuous placement;

● Dedicated and committed RCC materials supply including sufficient stockpiling of aggregate and cementitious materials;

● Intensive training of the contractors and supervision staff in three stages of full scale trial embankments.

3 Lai Chau RCC Dam

3.1 Layout of dam and programme

Construction of the Lai Chau RCC dam started in 2011 with excavation of the diversion channel and construction of the coffer dams. Foundation preparation and grouting started in 2012. The project comprises a 131 m high RCC gravity dam with a total RCC volume of 1.884 millon m³. RCC placement started on March 7th, 2013 in the narrow river channel and proceeded in four large sections, each comprising several blocks as depicted in Fig. 4. The need for three conveyor lines (Fig. 5) and the arrangement of individual blocks was dictated by the approval of the Authorities to the earlier project component layout and adaptations to better suit RCC construction would have been very time consuming. The centre conveyor served the spillway and left RCC blocks up to the intake; the left conveyor the left abutment blocks while the centre conveyor was repositioned to serve the right abutment blocks while the left abutment blocks were built. Because of other constraints on the site for access to serve the CVC intake construction at the upstream, there had to be a gap between the centre blocks and the start of the left side which slowed RCC construction.

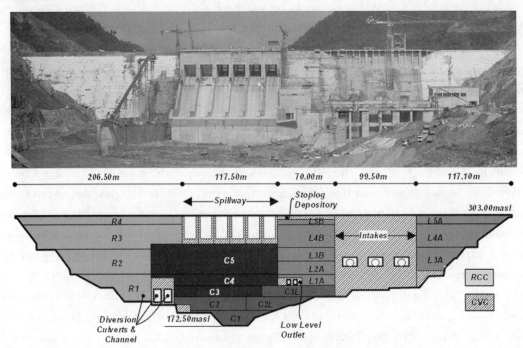

Fig. 4 **Lai Chau RCC block layout and downstream view of completed plant**

The first original programme allowed for two very long stops imposed by constraints from other sections of the project, namely the construction of the steel lined low level outlet structure in CVC and waiting for the wet season flow to reduce to allow construction of cofferdams in the river diversion channel and cleaning to start RCC at the right abutment. However, minor changes in the location of some of the smaller features on the dam, although not reducing the number of stopages, did allow sections of the dam to start a little earlier and be completed faster and, thus, a revised RCC schedule could be prepared as shown in Figure 6 and compared against the actual RCC progress.

Fig. 5　Lai Chau RCC conveyor routings (left) and RCC placement at right abutment in 01/2015 (right)

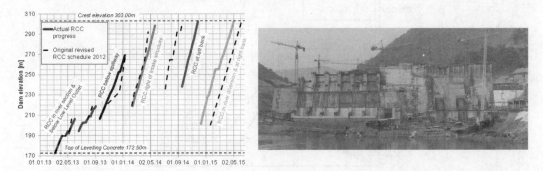

Fig. 6　Lai Chau as-built RCC programme and RCC dam under construction in 10/2014

3.2　RCC speed of construction achieved at Lai Chau

The RCC dam was complete by May 2nd, 2015, with the spillway CVC placed and hydromechanical equipment installed prior to diversion closure and start of reservoir impounding by June 20th, 2015, and only parts of the powerhouse to be constructed prior to the start of Unit 1 power generation scheduled by 11/2015. With a time span of 5.5 years from commencing the Technical Design until completion of the civil works Lai Chau can be considered an overall fast-track project.

The batch plant used was the one previously used at Son La and was rated at 720 m³/hour of compacted RCC. Generally a minimum of two layers were placed each day with three layers a day common and occasionally four. The maximum daily RCC placement averaged about 4 906 m³ in any 24 hr-period with a maximum day of 7 449 m³ of RCC placed. RCC placement was carried out on average for about 15 hr per day, resulting in an equivalent peak placement of about 10 000 m³ in a 24 hr-period. The total RCC construction time RCC was 786 days (25.8 months). The actual RCC production days accumulated to 505 days over the total construction time. The missing 281 days were the result of stoppages of 14 days or longer and including some 47 days lost during actual placing due to rain, breakdowns, etc.. Some of the stoppages, as also seen at Son La, can be considered self-inflicted (such as the longer gap due to required access to the dam's upstream side before continuing the RCC blocks on both sides of the power inake structure), those accumulating to about 125 days (i.e. 156 days of the total stoppages were not self-inflicted and may be accounted for as actual days lost). Depending on how the different figures are looked at, the commonly accepted measure of total RCC duration results in an overall average RCC placement rate of 72 907 m^3/month, while the modified RCC duration considering non-self-inflicted time lost results in a modified average RCC placement rate of 90 960 m^3/month. The peak month was recorded at 175 041 m^3/month. Looking at the actual RCC productive duration, the average placement rate is 113 475 m^3/month. Comparing these placement rates with all the RCC dams in the world and constructed so far (see[2], figure 7), the Lai Chau RCC dam will either be placed in the mid or at the top of the range.

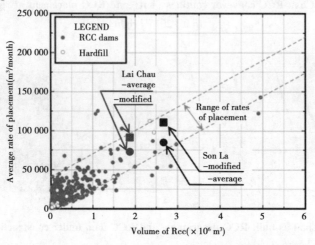

Fig. 7 World wide average rates of RCC placement versus RCC volume placed and rates of RCC placement achieved at Son La and Lai Chau

One of the main factors having contributed to the good RCC progress at Lai Chau, despite spatial conditions compared to Son La were more difficult, certainly was the experience the contractor gained at Son La. As at Son La, a number of full scale trial embankments were constructed at Lai Chau prior to start of RCC in the dam; the last and main trial embankment being completed about 3.5 months before start of the dam. This provided sufficient and refreshened training to the contractor in terms of the RCC placement methodology and in fact, slightly above 155 000 m^3 of RCC were placed in the first month. As another important aspect was the overhaul of plant which was moved from Son La and the replacement of used up equipment with new plant to ensure that downtime due to breakdowns were minimized.

4 Conclusions

The Son La and Lai Chau RCC dams have both been constructed at a highly creditable rate of placement and both RCC dams were completed ahead of schedule, allowing reservoir impounding to at least minimum operation level for the timely commissioning of power generation. Both projects proved that RCC was the right choice for the dam construction to save considerable time compared to other forms of dams and to provide early power for the national grid, resulting in very economic projects.

It has been seen in both projects that a number of factors must be got right to result in an efficient and successful RCC dam completed on time, on cost and with high quality. While most of these factors (see[2]) were got right at Son La and Lai Chau, some related to layout and construction planning weren't and resulted in significant lost time and, thus, in a reduced overall average RCC placement rate. It is quite common that conventional mass concrete construction as part of RCC dams progresses much slower than the RCC placement and it is thus generally aimed at minimizing CVC volume integrated into the RCC dam structure. The concept of external power intakes with the result of only having to construct the penstock penetration in CVC, compared to the complete intake structure being integrated into the RCC dam, has been successfully applied at the Yeywa and Upper Paunglaung RCC dams[3,4]. However, such proposal at Son La and Lai Chau would have affected the Higher Authorities'technical approval of the projects and would have resulted in a rather time consuming process and possible delays. Some optimisation at Lai Chau was possible by changing a large volume of CVC above the low level outlet to RCC and only encasing the steel lining in a limited volume of CVC. The previous proposal to locate the low level outlet on the right side of the spillway structure would have again affected the Authorities'earlier technical approval decision.

In order not to compromise the project programme the actual RCC placement rates must be much higher compared to the overall average and this must be addressed in the design of plant capacities and logistics, as well as in some items of the specification to allow for some flexibility to avoid interruptions to RCC placing (e. g. in regard to temperature control, adaptation of setting times). Although the Technical Specification at Lai Chau included a provision for adjusting the retarder dosage to control initial and final setting times depending on the layer surface area, RCC placement was limited to maximum three layers per day (occasionally four) to reduce loading on the formwork, although a larger number of RCC layers per day could have been placed.

The experience from Son La and Lai Chau showed that the approach to the RCC concept including low Vebe workability, appropriate setting time retardation to maintain hot joints without the application of bedding mixes and allowing the application of a simple RCC placement methodology was one very important success factor for the completion of the two dams ahead of schedule. Both projects also showed that training (preferably more rather than less) of contractors'and supervision staff prior to start of RCC placement in the dam, and leading to anadvanced progress on the learning curve before any RCC is placed in the dam, is a major contributor to rapid placement rates. Since the quality of the RCC structure is also linked to the speed of construction, the training together with the according quality control basically ensures that the RCC quality as designed is met.

Acknowledgements

The authors would like to thank Viet Nam Electricity, the Son La Management Board for Lai Chau, and PECC1 for permission to publish this paper. Regrettably one of the authors Mr. Nguyen Hong Ha, Vice Director of EVN, passed away during the preparation of this paper and it is thanks to his untiring leadership and drive that both Son La and Lai Chau RCC dams were completed ahead of schedule and to a high standard.

References

[1] Ha, N. H., Conrad, M., Morris, D. Review of the construction of the Son La RCC Dam. Proceedings of the 6th International Symposium On Roller Compacted Concrete (RCC) Dams, Zaragoza, 2012.

[2] Dunstan, M. R. H. How fast should an RCC dam be constructed? Proceedings of the 7th International Symposium on Roller Compacted Concrete (RCC) Dams, Chengdu, 2015.

[3] Kyaw, U. W., Zaw, U. M., Dredge, A., et al. Yeywa Hydropower Project, an Overview. Proceedings of the International Symposium Asia 2006, Bangkok, 2006.

[4] Zaw, U. M., Thar Htwe, U. M., Min San, U. Z., et al, Design and Construction Aspects of the Upper Paunglaung RCC Dam. Proceedings of the International Symposium Asia 2012, Chiang Mai, 2012.

Lessons Learnt from Operations of Some RCC Dams

François Delorme

(ELECTRICITE DE FRANCE (EDF) Hydro Engineering Center, France)

Abstract: This article presents the return of RCC experience works designed, constructed and operated by EDF in France and abroad (Riou dam, Petit Saut dam in French Guyana, Nakai dam in Laos, Rizzanèse dam) as a critical analysis of the various problems encountered both during first impounding and after 3 to 24 years of operation. Lessons learned from these examples are presented.

It is proposed that all professionals working in the field of RCC dam, inclueling Owners, Engineers, Consultants, Contractors and Operators, communicate more on this subject of RCC dams performances in order to benefit from these problems with useful information to all.

Key words: Performance, watertightness, leakages

1 Introduction

Realization of RCC dams has grown considerably since the early 1980s. Almost 700 large RCC dams as considered by ICOLD ($H > 15$ m) have been built or are under construction. A large variety of materials used, design and construction methods have led to a wide variety of RCC structures. The performance of most of these RCC dams during the first impounding, or even after several years of operation has not necessarily resulted in a lot of technical publications, in particular for the central issue of leaks and/or water collected within drainage systems of these structures in order to qualify the robustness of these different approaches. It is still further for the remedial works that could or should have been done to improve locally or globally the watertightness of these structures after end of construction.

Hereafter are pesented the performances on 4 RCC dams designed, owned and operated by EDF. Data coming from monitoring during first impounding and after 3 to 24 years as well as remedial works which had to be performed gave a new perspective on the performance ofeach selected RCC design.

2 Riou Dam (France)

2.1 Introduction

The Riou dam was the firstRCC dam built by EDF and the second one in France. It was also the first RCC dam performed with an exposed geomembrane. It was commissionned in 1991. The dam is 26 m high and 322 m long in crest. Total concrete volume is 46 000 m³ (89 % of RCC with only one size aggregate class, MSA 63 mm and 120 kg/m³ of special road cementitious material BARLAC;85% to 60 % Slag + 20 to 35 % Lignite Gardane Classe C Fly Ash + 0 to 5 % filler). RCC ensures only dam stability. Watertightness is provided by an exposed PVC – P geomembrane from CARPI CNT2800 (Thickness 2 mm) associated to a non woven geotextile (200 g/m²). No construction joints were performed within RCC body. The dam is for hydropower, irrigation and recreational pupose. Its normal water level (NWL) 638.4 a.s.l. is almost constant with only small variations (about 20 cm). The dam was submitted to 2 normal adminstrative dam safety assessment procedures, with full emptying of the resevoir 5 years after end of first impounding (1996) and again 11 years later (2007).

Fig. 1　General d/s view

Fig. 2　Upstream face before geomembrane placing

Fig. 3　Details of steel formwork attached on CARPI profiles and drainage facilities

Fig. 4　View of geomembrane in 2007

The lessons learnt on the behavior of this dam after 6 years were already presented in Chengdu RCC'99 Conference[1]. Concerning the hydraulic behaviour, the watertightness and drainage facilities always fulfill their role correctly. Piezometric levels are well within the dam stability calculation hypotheses. Drainage from the geomembrane is collected in 4 shafts in the central part

(Puits 1 to 4) and in 5 pipes on the downstream face (M1 – M5) in right bank and 6 in left bank (M6 – M11). 34 foundation drains were drilled (23 from the inspection gallery and 11 from downstream toe).

Overall water collected through the drainage facilities reached a total value of 700 L/min during first impounding with NWL——2.9 m, and have gradually decreased to 210 L/min after 8 months due sedimentation (mainly silt) within the reservoir. It is then almost constant and has reached now 45 L/min for the NWL. The water collected from the foundation drains is 10 L/min (and essentially from 1 drain DR14 with 7 L/min). Flow coming from the geomembrane facing is 3 L/min from the right abutment, about 0 L/min from the left abutment and 32 L/min (Puits 1 = 2 L/min, Puits 2 = 10 L/min, Puits 3 = 15 L/min, Puits 4 = 5 L/min) in the central part. This is mainly due to the fact that there is no complete watertightness closing between the upstream plinth, where is fixed the geomembrane, and the grouting curtain within the foundation which was performed from the control gallery in this part of the dam. As expected, the dam cracked (for thermal or foundation reasons) but it was without any consequence in term of leakage or dam safety[2].

4 Samples of geomembrane were taken during the 2007 safety review. The uptream face observations and test results show that the geomembrane presents always good performances. Only a few repairs were needed in the upper part between 638.2 and 638.3 and closed to each dam abutment (5 tears: 1 to 3 cm + 22 unsticking welds: 5 to 15 cm + 1 bullet hole 10 cm^2).

2.2 Lessons learnt

After 24 years of operation, the watertightness with exposed geomembrane of the Riou dam presents always good performance. Very few maintenance works were needed in 2007 (then after 16 years of operation). Leakages from usptream facing are limited to 35 L/min for 4,000 m^2 upstream facing and 22 m maximum water head. They could have been reduced with a more accurate design ensuring a full continuity between the plinth and the grouting curtain, which is now the recommended practice for such design.

Such design has beensuccessfully used on 15 RCC dams, among them Miel 1 in Colombia (which was in 2002 the highest RCC dam built in the world at that time H = 188 m) and with very good performances[3] with maximum flow recorded during first impounding at 120 L/min for 31 500 m^2 of geomembrane.

3 Petit Saut Dam (French Guyana)

3.1 Introduction

The Petit Saut damwas the second RCC dam built by EDF and the fourth one in France, but the first in French Guyana in a specific wet tropical environment. It is still the highest RCC dam with the largest RCC volume in France, and with the largest reservoir volume (3.5 km^3). It was commisionned in Jan. 1994 and it reached the NWL (35 a.s.l.) 18 months later due to problems encountered on earthfill saddle dams. The dam is 51 m high and 740 m long in crest. Total concrete volume is 410 000 m^3 (61 % of RCC with 4 size aggregate classes, MSA 50 mm and 120 kg/m³ of special road cementitious material LRCC:86 % Slag +1% to 2% CaO + 12% to 13% $CaSO_4$). RCC ensures only dam stability. Watertightness is provided by an upstream facing in conventional reinforced concrete (CRC) built in advance on the RCC. It was the first construction with such design in an RCC dam. This 1.2 m thick wall is used as formwork and in some part is more than 40 m high. The wall is linked to a CRC inspection gallery. Single waterstop joints ensure watertightness between each wall block (37) with spacing varying between 12 ~ 27 m. 20 construction joints were performed within RCC body. The upstream wall is equipped with 266 vertical drains at the interface CRC/RCC and connected with the inspection gallery. 292 drains inclined at 30°were drilled within the rock foundation from the inspection gallery.

The dam is for hydroelectric purpose and the reservoir level varies annually by 5 m. The dam was submitted to 2 normal administrative dam safety assessment procedures: with subaquatic ROV inspection 7 years after end of first umpounding (2002) and again 10 years later (2012).

Before first impounding, it was necessary to repair a lot of cracks on the upstream wall (2.6 km) which were the consequence of CVC placement with reduced MSA (25 mm instead of 50 mm) and water in excess as this concrete had to be placed using of concrete pump instead of with a crater crane (destroyed by activists during aworkers strike on site). Elements of Hypalon geomembrane (Chloro-Sulfonated polyethylene CSM) were stuck with an epoxy resin over the wall cracks. These remedial works have been efficient as no leakage were observed on the upstream wall in these treated areas (except at dam ends where the face was already covered by connection clay fills). Furthermore, no damage of these treatments has been observed during both sub-aquatic wall inspections. In 2012, 50% of the bands appeared blighted, but none showed tear or of separation.

Fig. 5 General view from downstream left bank during contruction

Fig. 6 Hypalon repairs on upstream face

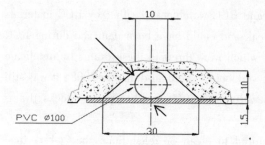 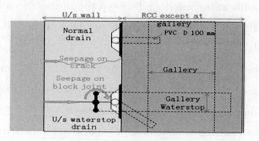

Fig. 7 **U/s wall drain details**   Fig. 8 **U/s wall and gallery seepage way and drainage**

The lessons learnt on the behavior of this dam, 2 years after the end of first impounding, were already presented in Chengdu RCC'99 Conference[1]. At that time, the total water collected was 750 L/min at NWL in July 1995 of which 25% was in relation with the upstream facing.

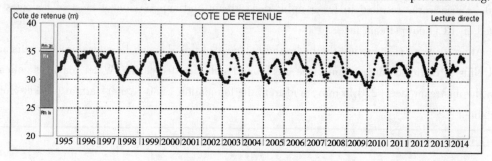

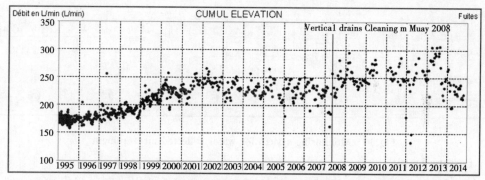

Fig. 9 **Evolution of reservoir water level (top) and total leakages for the upstream facing (bottom)**

Between 1995 and 2002 (see Fig. 9), the total leakage from the upstream face increased with a rate of 15 L/min/year. Several water leakages within the RCC dam body were also observed downstream (see Fig. 10). Most of the leakages on the downstream face were located at or close to the RCC transverse construction joints. These leakages can be put in relation with some waterstops defects or damages from the upstream wall. But this situation was not considered acceptable in the long term because of the water acidity (pH = 5) coming from the reservoir which could badly weathered RCC.

Another important subject of attention for the dam operator was that a large number of vertical drains at the interface between the upstream wall and RCC wereblocked either by RCC materials (the protective asbestos cement plates were too weak and could have been damaged during RCC placement) or by various kinds of waste materials which were threw in these drains by indelicate operators during RCC construction. After cleaning works of the vertical drains in 2008, it was still observed that 25% of the wall drains were inoperative and 60% for the inter – blocks drains downstream of the waterstop.

Extensive maintenance works were then required to clean or rehabilitate the 264 vertical drains (2010, 2011 & 2012) and it was also necessary to drill 4 news drains (2 in 2010/2011, 2 in 2012), to re – drill the exit of some drains for better possibility of maintenance (42 in 2010/2011, 26 in 2012). The boring of certain waterstop joints surrounding the gallery was revealed also effective (towards the roof downstream for the infiltrations insufficiently collected by the vertical drains, like those at the downstream toe for the percolations walking on in pressure under the gallery). These works resulted in lower uplift downstream and suppression of most of leakages observed on RCC downstream face.

After 20 years of operation the hydraulic behavior of this damcan be considered as satisfactory: Uplifts at bedrock/concrete interface are lowand stable or decreasing; Water leakages collected from the upstream face (220 L/min) and from the foundation (280 L/mm) are at present globally stable.

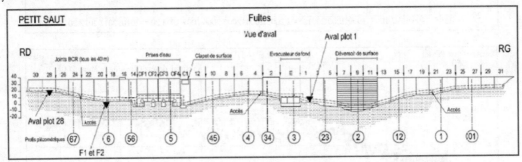

Fig. 10 Monitoring system for water collected measurement

3.2 Lessons learnt

After 20 years of operation the watertightness with upstream CRC facing built in advance on RCC presents acceptable performance. A lot of maintenance works were however needed since end of dam construction mainly due to defects and damages at the waterstop joints. Leakages from usptream facing are now limited to 220 L/min for about 19 000 m^2 upstream facing and 49 m maximum water head. It also appears necessary to avoid bends in dam elevation drains, which accentuate their blockage and to equip all RCC construction joints with elevation drains.

3 others dams have been built with suchRCC dam design Cantache dam (old Villaumur) and Sep dam in France[1], with better construction method for the upstream wall and the drainage facilities and with good results. It was also used for Ziga Spillway in Burkina Faso (2000 year) but no record in operation on this dam has been yet published.

4 Nakai Dam (Laos)

4.1 Introduction

TheNakai dam was the third RCC dam built by EDF and the first one in Laos. It is part of the Nam Theun 2 HPP for NTPC and has also a large reservoir volume (3.53 km³). It was commissioned in May 2008 and it reaches the NWL (538 a.s.l.) 17 months later. The dam is 39 m high and 415 m long in crest. Total concrete volume is 210 000 m³ (71% of RCC with 4 size aggregate classes, MSA 50 mm and 100 kg/m³ of cement and 100 kg/m³ of class C Fly Ash). Watertightness is provided by an upstream facing in conventional vibrated concrete (CVC) built in the same time as RCC. This 0.9 m wide CVC upstream face is enlarged to 1.3 m at double waterstop joints each 22~24 m (16 Monolith Joints (MJ) with 2 drains associated at each waterstop: 1 +3 U/s and 2 +4 D/s). The face was built with V – notch crack inducers with 6 m spacing in order to organize thermal cracking and 34 vertical drains were drilled in RCC (RD). 56 drains were drilled in foundation from the gallery. Details of design and construction have been given in[4,5]. The dam is for hydroelectric purpose and was designed and built in the frame of an EPC contract. The annual water level variation is between 6~10 m.

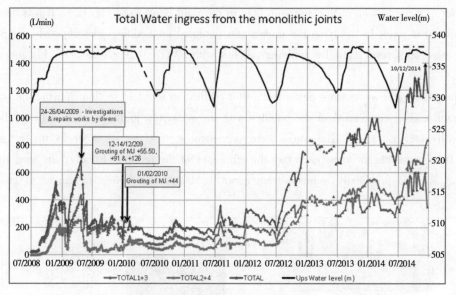

Fig. 11 Evolution of reservoir water level and total leakages for
MJ drains between (2008~2014)

During first impoundment, leakages appeared in the gallery. The leakages coming from uncontrolled cracks crossing the dam upstream face through RD were small (total less than 60 L/min and individually less than 2 L/min), confirming the efficiency of the crack inducers. All the thermal cracks on the concrete facing were localized at the bottom of the crack inducers, and they have been properly sealed[5]. Drainage flows from the foundation was limited to 20 L/min except during floods where drains are influenced by the tailwater level (maximum flow of 200 L/min). Associated uplifts at bedrock/concrete interface were low and well within design hypotheses.

Localized seepages were observed at some MJ (see figure 3). The leakages were mainly due to defects around the vertical PVC waterstops and at the junction with the plinth. The flow rate coming from MJ drains reached 680 L/min at NWL − 0.75 m in May 2009 and the total flow rate was around 820 L/min. Corrective measures were implemented between Nov. 2008 and Mar. 2010 after investigations by divers: soil dumping from the dam crest, application of underwater sealants using various products (SIKA & SCUBAPOX), spreading of rice powder with divers and finally grouting of 4 upstream drains of MJ in right bank after series of water tests. A flood which occurred in July 2009, with a large income of fine materials, helped to decrease lightly but temporally the leakages. The MJ flows were finally reduced to less than 150 L/min in mid 2010. The total flows increased after irreversibly at an annual rate of 100 L/min up to mid 2012.

Unfortunately, these previous repairs were not sufficient or durable and since mid 2012, the leakages have increased again irreversibly with an annual rate of 550 L/min up to a total value of 2 500 L/min at the end of 2014 (1 400 L/min from MJ drains and 1 100 L/min from cracks or parasite infiltrations).

New diver investigations in Feb. 2014 permitted to identify the most critical points causing these leakages. It was mainly due to 5 of the 7 MJ of the CVC spillway structures and 2 construction defects at the bottom of the upstream face in the right bank (see Fig. 12). It was decided to repair all these zones by applying the underwater patented methodology developed by CARPI and already used on several dams by applying patches with PVC − P geomembrane (Platanovryssi in 2002).

The underwater works were performed, after realization of a test section showing the possibility to reach a good connection of the patches with the waterstop joint via an underwater borehole filled with watertight resin. The works were successfully performed in 3 weeks up to mid June 2015. It was possible to follow each day the efficiency of the repairs. Finally, the total flow rate from the MJ was drastically reduced from about 1 500 L/min before treatment to about 100 L/min.

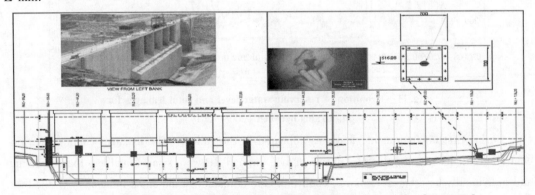

Fig. 12 Localization of the 7 underwater CARPI patches and example of one patch

4.2 Lessons learnt

The CVC upstream facing placed in the same time as RCC provides good watertightness, due to good design and construction of the V-notch crack inducers. However, the problems, as often in such works and structures, came on defects or damages for waterstops at monolithic construction joints. The initial design proposed to EPC candidates during tender process was based on exposed geomembrane watertightness. The EPC Contractor indicated at that time that such design was too much expensive compare to the alternative finally built. When considering such final design, permanence vigilance is required on waterstop quality control during all construction phases in order to avoid additional expenses through remedial works.

5 Rizzanèse Dam (France)

5.1 Introduction

The Rizzanèse dam was the fourth RCC dam built by EDF and the first one in Corsica Island. It is also the first high Hardfill dam built in France. Its design was in fact finalized in 1997, but construction was unfortunately delayed during 10 years for administrative reasons [6]. It was commissioned in May 2012 and it reaches the NWL (541 a.s.l.) 5 months later. The dam is mainly for peak power hydroelectric purpose with a small reservoir (1.3 hm^3) and high daily water variation level (12 m). The dam is 41 m high and 140 m long in a curved crest. Total concrete volume is 72 000 m^3 (89% of Hardfill material with 2 size aggregate classes, MSA 63 mm and 80 kg/m^3 of cement). Hardfill ensures only dam stability. Watertightness is provided by a covered PVC – P geomembrane from CARPI CNT3750 (thickness 2.5 mm) associated to a non woven geotextile (500 g/m^2) in the lower part, below El. 520, inclined at 1H/1V, the protection being with a 1 000 g/m^2 geotextile and backfill. In the upper part, above EL. 520, the upstream face is vertical and use the CARPI patented system with the geocomposite CNT2800 attached to concrete precast panels during their fabrication. Details of design and construction have been given in [6,7]. Directly downstream of the panels, drainage is ensured by a 1 000 g/m^2 geotextile, which also acts as anti – puncture layer for the geomembrane.

The drainagecapacity was reinforced by the fact that the contractor placed 60 cm of CVC downstream for a conservative stability of the precast elements during construction and to distance as far as possible construction equipments from the precast panels for work safety. The flows collected by the draining geotextile were harnessed by vertical 300 mm half – pipes located behind each column of panels, so every 3 m. These half – pipes are connected to pipes linked to the drainage gallery, either individually along it, or grouped in the abutments. Such safe idea "to reinforce the drainage collection" was not adopted in the previous projects built of this type. It was, in fact, an unsound design and the cause of problems encountered during the first impounding of the reservoir.

The half – pipe gives limited areas for the geomembrane with no support. The water pressure applies on the geomembrane which ensures the link between the two precast panels with the geocomposite. Under the upstream pressure, the geomembrane tends to move in the void of the half – pipe (see Fig. 13, Fig. 14) and in some cases (but not all) tends to snatch the geocomposite from the precast panel. In some cases, the geocomposite occupied all the voids of the drainage collec-

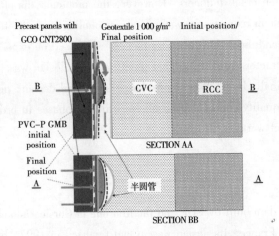

Fig. 13 Model of geomembrane deformation within half-pipe under water pressure

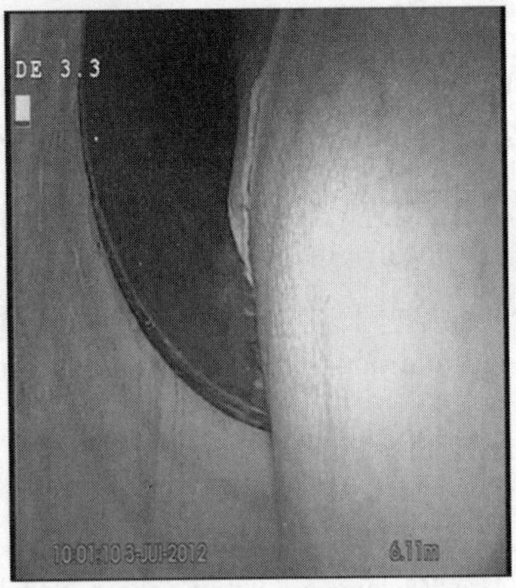

Fig. 14 Endoscopic view in one half-pipe affected by water pressure

tor. In some cases, the horizontal weld between the geocomposite and the geomembrane which acted in tensile peel was not sufficiently strong and the weld opened with leakages occurring within the drainage collector. With pressure rising, opening is widening as a "zipper" liberating easily huge flows. Although there was no safety risk for the dam itself, it appeared that it was a generic issue, and so it was decided to stop the first impounding at El. 529.5 (see Fig. 15, total flow from geomembrane 1 650 L/min). After lowering the reservoir, a first quick repair was initiated in 07/2012 (one month). The works consisted in filling the half-pipe collectors either with cement grout in areas where defects had been observed, or with gravels where there was no leak. Thus the geomembrane would recover a rigid support. The works were easy on the banks where the collector is accessible directly but on the spillway, it was necessary to bore the top of the precast panel to cut the geomembrane for being able to fill the collector and then to repair it. The reservoir was filled again. Some new defects appeared above the first level reached (530.3). They have been underwater treated by clogging the gaps between panels with epoxy resin and by grouting with polyurethane acqua-reactive resin. Then the total flow from the geomembrane decreased from 1 200 L/min to less than 200 L/min. In order to ensure the durability of the watertightness a direct and dry repair was decided at the 12 observed defects. It was done by boring the precast panel then after verification of the geomembrane integrity or damage and support presence, repairing, if necessary, the damaged weld, after cleaning the geomembrane and welding a first layer of geomembrane and a second one (see Fig. 16). The panel hole was finally filled with non-shrink SELTEX mortar. These works were performed in 3 weeks in sept. 2013. After new impounding of the reservoir, the total flow from geomembrane has been thus stabilized at about 50 L/min.

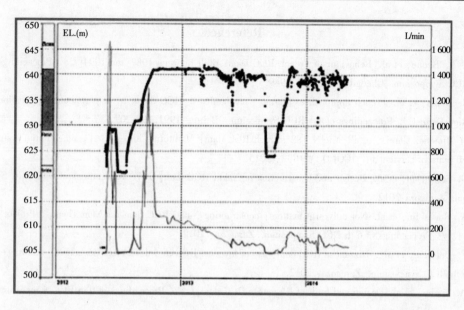

Fig. 15　Evolution of reservoir water level and total leakages for the geomembrane

5.2　Lesson learnt

The reinforcement of the drainage which was designed has finally created weakness areas for this watertightness type. Such drainage design shall be avoided in future projects. These events show however that once a defect is localized, it is easy to repair such covered watertightness and finally to reach acceptable flow. These issues were not encountered in the past on most of the dams built with such design, because drainage was ensured only by the geotextile, except may be at Urugua – I dam where similar large leakages were noted and could have similar causes [8].

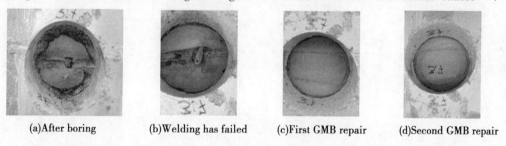

(a)After boring　　(b)Welding has failed　　(c)First GMB repair　　(d)Second GMB repair

Fig. 16　Example of final repair for damaged geomembrane attached to precast panel

6　Conclusion

The fourvarious RCC dams presented show that we have to learn not only from method of RCC mix design, RCC design and construction methods, where large number of technical papers have generally been published in the past and still today but also on the actual performance of these proposed designed.

It is proposed that all professionals working in the field of RCC dams, induding Owners, Engineers, Consultants, Contractors and Operators, communicate more on this subject of RCC dams performances in order to benefit from these problems with useful information to all.

References

[1] J. P. Becue, et al. Behaviour of French RCC Dams Built between 1987 and 1994[C]. Proceedings of 3rd RCC Symposium. Chengdu: 1999:933-949.

[2] B. Denis, et al. French approach to thermo – mechanics in RCC dams – Experimental application to the Riou Dam[C]. Proceedings of 1st RCC Symposium. Beijing:1991. 198-207.

[3] M. Jiménez Garcia, et al. Miel 1 188 m high RCC dam: Monitoring an exposed geomembrane system after 11 years of service[J]. ICOLD Q99R32, 2015.

[4] G. Stevenson, et al. Design and construction of Nakai RCC dam[C] // Proceedings of 6 th RCC Symposium. Zaragoza:2012.

[5] A. Rousselin, et al. Not only size matters: constructing Nakai RCC dam for Nam Theun 2 Project in Laos [C] // Proceedings of 6 th RCC Symposium. Zaragoza: 2012.

[6] F. Delorme, et al. . Rizzanèse RCC dam: use of low cost RCC in a thick cross section[C] // Proceedings of 6 th RCC Symposium. Zaragoza: 2012.

[7] A. Lochu, et al. Design and built of Rizzanèse RCC dam[C] // Proceedings of 6 th RCC Symposium, Zaragoza: 2012.

[8] A. C. Lorenzo, et al. Behavior of Urugua – I dam[C] // Proceedings of 1st RCC Symposium. Beijing:1991: 470-479.

Research and Application of Cinder and Ash from Coal – fired Power Plant Used as Admixture in RCC of Guanyin yan Hydropower Station

Xu Xu[1], Yi Junxin[1], Li Xiaoqun[2], Liu Yingqiang[2]

(1. Kunming Engineering Corporation Limited, Kunming, 650051, China;
2. Datang Guanyinyan Hydropower Development Co, Ltd, Panzhihua, 617000, China)

Abstract: The concrete gravity dam of Guanyinyan hydropower station has a height of 159 m. The concrete quantity of the dam is about 875×10^4 m^3 and among them RCC quantities is about 458×10^4 m^3. The admixture quantity which the dam need totally is about 70×10^4 t. During construction the feasibility research of cinder and ash from Coal – fired Power Plant number 504 used as admixture in concrete was carried out in order to enhance assurance of admixture supply and save investment. By carrying out the following researches : the judgment of the quality uniformity of cinder and ash, ash and cinder's optimal fineness research, the test of mix proportion of concrete using cinder and ash, and the research of temperature control standards and measures, it shows that, after being dried and ground and meeting with some certain requirements, the cinder and ash from Coal – fired Power Plant number 504 can be used as RCC's admixture in Guanyinyan hydropower station. Comparing with the concrete using II – level fly ash, the concrete using the cinder and ash from Coal – fired Power Plant number 504 has almost the same usage of water and performance, and its standard and measures of temperature control seem like no different from the concrete using II – level fly ash. Now the concrete gravity dam of Guanyinyan hydropower station has been completed. The total amount of the cinder and ash from Coal – fired Power Plant number 504 used in the dam reached about 60×10^4 t. The research and application has great meaning of technology, environmental protection and economy.

Key words: Cinder and ash from Coal – fired power plant, admixture, RCC, the mix proportion of concrete, Temperature control, Guanyinyan hydropower station

1 Overview

Guanyinyan hydropower station lies the middle reaches of the Jinsha River between Huaping County, Lijiang City, Yunnan Province and Panzhihua city, Sichuan province. The dam is blend one which consists of RCC dam and core – wall rockfill dam. RCC dam has a height of 159m and a bottom width of 150.5 m. The concrete quantity of the dam is about 875×10^4 m^3 and among them RCC quantities is about 458×10^4 m^3. The admixture quantity which the dam need totally is about 70×10^4 t. At the beginning of construction II – level fly ash was used as admixture in dam concrete. Considering the shortage of supply of II – level fly ash from Coal – fired Power Plant number 504, the feasibility research of cinder and ash from Coal – fired Power Plant number 504 used in concrete was carried out in order to enhance assurance of admixture supply and save investment.

Coal – fired Power Plant number 504 which produces 20×10^4 t II – level fly ash per year lies the west area of Panzhihua city, Sichuan province. Madiwan ashery of Coal – fired Power Plant number 504 which has come into use since 1994 is 30 kilometers away from Guanyinyan hydropower station. Its design storage is 912×10^4 t and 600 ~ 800 t cinder and ash are gotten in every day. The cinder and ash in it were composed of two parts: one part is fly ash below grade, the other part is cinder after falling to the bottom of furnace, water quenching and broken.

2 The technical route of the research

Thetechnical route of the research: ①take samples from Madiwan ashery and test, judge the quality uniformity of cinder and ash, ②research ash and cinder's optimal fineness according to activity index and ratio of water demand after drying and grinding it into different fineness, ③research whether it is feasible for cinder and ash to be used as admixture in concrete by carring out the mix proportion of concrete test according to its major performance index, ④carry out the research of temperature control on the concrete using cinder and ash, propose temperature control standards and measures and apply them to the project construction.

3 Research on the quality uniformity of cinder and ash

Twenty drilling holes were arranged and 120 samples which weigh 4 200 kg were taken. Test results of these samples are listed in the Table 1 and Table 2.

Table 1 Statistical result of Chemical analysis of the cinder and ash

Items	Testing items(%)									
	SiO_2	Fe_2O_3	Al_2O_3	CaO	MgO	K_2O	Na_2O	Alkali content	SO_3	Loss on ignition
Number oftesting classes	24	24	24	24	24	24	24	24	24	48
Average	52.47	3.92	28.80	2.72	3.63	1.02	0.38	1.05	0.17	3.48
Maximum	53.67	7.46	29.84	3.45	3.91	1.08	0.43	1.11	0.25	4.86
Minimum	51.54	2.43	26.33	2.34	3.40	0.96	0.32	1.00	0.05	2.88
Standard deviation	0.49	1.64	1.07	0.29	0.14	0.05	0.03	0.04	0.05	0.51

Table 2 The testing results of fineness, ratio of water demand and activity index of cinder and ash

Items	Ratio of water demand (%)	Density (g/cm³)	Bending strength of mortar (MPa)		Compressive strength of mortar (MPa)		Activity index of mortar (%)	Fineness (%)	
			7 d	28 d	7 d	28 d	28 d	residue on 0.08 mm sieve	residue on 0.045 mm sieve
Number of testing classes	60	24	60	58	60	58	58	36	36
Average	100.4	2.4	4.5	7.2	20.0	36.1	72.1	1.0	14.8
Maximum	102.4	2.43	4.9	7.8	21.9	38.4	76.8	3.2	22.7
Minimum	99.2	2.38	4	6.6	17.8	33.5	67.0	0.5	6.9
Standard deviation	0.7	—	0.2	0.3	0.8	1.2	2.4	0.6	3.5

(1) It shows that the carbon content is steady for the maximum of samples 'loss on ignition is 4.86%, the minimum is 2.88%, the average is 3.48% and all of them are below 8% which is stipulated in the technical specification of fly ash.

(2) It shows that the quality uniformity of cinder and ash is good according to the testing result of the 24 samples' density whose maximum is 2.43 g/cm³, minimum is 2.38 g/cm³ and average is 2.4 g/cm³.

(3) 50% of the total samples, i.e. 60 samples were taken and were ground and were tested. Its activity indexat the age of 28 days is between 67% and 76.8%, its average is 72.1%. The ratio of water demand is between 99.2% and 102.4%, its average is 100.4%. The test results listed in the table 2 shows that the quality uniformity of cinder and ash is good.

(4) Radioactivity of the samples is qualified according to testing result.

A conclusion can be made that thequality uniformity of the cinder and ash from Coal – fired Power Plant number 504 is good according to the testing results of Chemical analysis, loss on ignition, density, radioactivity, ratio of water demand and activity index. The next research work can be carried on.

4 Research of optimal fineness

The cinder and ash were ground into five different fineness and were tested. Results of fineness, specific surface area, content of vitreous body, ratio of water demand and activity index are listed in the Table 3.

Table 3 The testing results of the cinder and ash in different fineness

Number	Fineness residue on 0.045 mm sieve (%)	Specific surface area (m²/kg)	Content of vitreous body(%)	Ratio of water demand (%)	Compressive strength (MPa)				Activity index(%)			
					7 d	28 d	90 d	180 d	7 d	28 d	90 d	180 d
Number 5	3.7	538	72	100.4	19.7	36.5	58.3	70.7	63.5	74.5	93.4	103.8
Number 4	6.3	510	70	100.4	19.2	34.4	56.4	65.8	61.9	70.2	90.4	96.6
Number 3	10.4	465	67	99.6	18.5	33.3	55.3	65.8	59.7	68.0	88.6	96.6
Number 2	17.8	395	69	98.0	18.2	32.4	52.8	64.3	58.7	66.1	84.6	94.4
Number 1	23.2	382	72	100.4	17.4	30.4	50.7	62.6	56.1	62.0	81.3	91.9

Testing the samples by SEM showed that the shapes of 5 samples with different fineness which consist of spherical particles and irregular particles is basically the same. The only difference is the content of vitreous body and the size of particles. What the Fig. 1 shows is the SEM photo of No. 4.

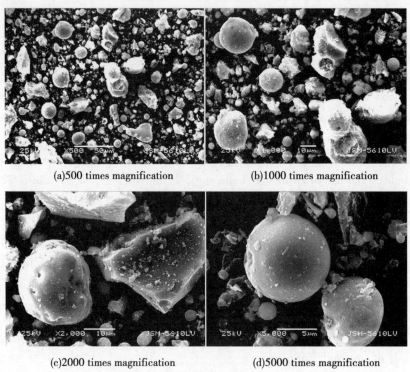

(a)500 times magnification (b)1000 times magnification

(c)2000 times magnification (d)5000 times magnification

Fig. 1 the SEM photo of No. 4 (in the fineness of 5% to 10%)

The testing result shows that the smaller the fineness of the cinder and ash is, the higher the activity index it has, and the ratio of water demand of different fineness is between 98.0% and 100.4%. The activity index at the age of 28 d is greater than 70% and the one at the age of 90 d is greater than 90% when the specific surface area is above 500 m²/kg.

Mineral Admixtures For High Strength And High Performance Concrete (GB/T 18736—2002) has made rules about activity index: the 7 d's activity index of level-Ⅱ ground fly ash is no less than 75%, and the one of 28 d is no less than 85%. The cinder and ash in this article, which is fly ash below grade and cinder after falling to the bottom of furnace, water quenching and broken, doesn't exactly equal the ground fly ash. Meanwhile, considering that the design age of RCC of this project is mostly 90 d or 180 d, long-age activity should be taken full advantage of and the cinder and ash should not be ground too finely. Therefore the optimal ash and cinder is the one with specific surface area between 500 m^2/kg and 600 m^2/kg, the 28 d's activity index no less than 65% and the 90 d's activity index no less than 85%. Moreover, the following indexes are required: $SO_3 \leqslant 3\%$, loss on ignition $\leqslant 8\%$, $Cl^- \leqslant 0.02\%$, $f-CaO \leqslant 1\%$, water content $\leqslant 1\%$, the ratio of water demand $\leqslant 105\%$.

5 The test of mix proportion of concrete

Testing conditions: 42.5 – grade P · MH + the cinder and ash with optimal fineness (in contrast with Ⅱ – level fly ash) + limestone aggregate from Longdong quarry. The mix proportion of concrete is listed in the Table 4.

Contrasting testing results of two kinds of admixtures.

(1) Water contents of concretes using two kinds of admixture are the same that the one of three – graded RCC is 80 kg/m^3 and the one of three – graded RCC is 92 kg/m^3.

(2) The compressive strength, impermeability grade, frost resistant grade of all concretes meet the design requirements.

(3) Despite the other indexes, only ultimate tensile strain values of the concrete using cinder and fly ash is slightly less than the concrete using Ⅱ – level fly ash. In general, the performance of the concrete using cinder and fly ash seems like slightly inferior to that of Ⅱ – level fly ash, but there is not great difference between them.

Table 4 The mix proportion of concrete and its performances

Strength grade	Admixture		Dosage (kg/m^3)			Performances (at the age of 90 d)						
	name	Content (%)	W	C	F	Compressive strength (MPa)	Tensile strength (MPa)	Elasticity modulus of static pressure ($\times 10^4$ MPa)	Impermeability grade	Frost resistant grade	Ultimate tensile strain values ($\times 10^{-6}$)	Coefficient of linear expansion (10^{-6}/°C)
Three – graded RCC $C_{90}15$	Ⅱ – level fly ash	60	80	64	96	34.7	2.68	4.397	W6	>F100	76.1	4.824
	Cinder and ash		80	64	96	35.7	2.96	4.346	>W6	F100	75.6	5.140
Three – graded RCC $C_{90}20$	Ⅱ – level fly ash	50	80	80	80	34.9	3.38	4.555	>W6	>F100	85.8	5.318
	Cinder and ash		80	80	80	39.0	3.24	4.594	>W6	>F100	81.1	5.113
Two – graded RCC $C_{90}20$	Ⅱ – level fly ash	50	92	92	92	36.6	3.61	4.513	W8	>F100	88.2	5.236
	Cinder and ash		92	92	92	38.2	3.86	4.544	>W8	>F100	82.9	4.888

6 Temperature control standards and measures of RCC

Temperature control standards of RCC using two kinds of admixture are listed in the Table 5.

Table 5 Temperature control standards of RCC(℃)

Concrete	Admixture	Allowable maximum temperature of strong constraint zones	Allowable maximum temperature of weak constraint zones	Allowable maximum temperature of non-restraint zones
Three-graded RCC $C_{90}20$	II-level fly ash	27.5	29	31
	cinder and fly ash	27	28.5	31

Contrasting temperature control standards and measures of two kinds of admixture, there is no essential difference between them. Allowable maximum temperature of cinder and fly ash is less 0.5℃ than that of II - level fly ash. Precooling of aggregates and water cooling are all required by two kinds of concretes, however, the distance of pipes in the concrete using cinder and fly ash is closer. And the surface protection standards are slightly different: the surface protection standard for the concrete using II - level fly ash is 10 kJ/(m² · h · ℃) and the protection last not less than 90 days, the surface protection standard for the concrete using cinder and fly ash is 8 kJ/(m² · h · ℃) and the protection last not less than 180 days.

7 Epilogue

The research shows that the cinder and ash from Coal - fired Power Plant number 504 can be used as RCC's admixture in Guanyinyan hydropower station after being dried and ground and meeting with some certain requirements. Now the concrete gravity dam of Guanyinyan hydropower station has been completed. The total amount of the cinder and ash from Coal - fired Power Plant number 504 used in the dam reached about 60×10^4 t。

Fly ash has been widely used inhydroelectric project and civil engineering projects, And ground fly ash are used in hydroelectric project to an extent. But the cinder and ash from Coal - fired Power Plant has not yet used as admixture in large hydroelectric project. The use of the cinder and ash from Coal - fired Power Plant number 504, which expands source of admixture and enhance use ratio of waste and reduce investment, has great meaning of technology, environmental protection and economy.

References

[1] XuXu, etc. The feasibility research of cinder and ash from Coal - fired Power Plant number 504 used as admixture in concrete: the judgment of the quality uniformity of cinder and ash[R]. Kunming: Kunming Hydropower investigation, design, research institute, CHECC, 2009.

[2] Hong Zhang, etc. The feasibility research of cinder and ash from Coal - fired Power Plant number 504 used as admixture in concrete: the Reports of admixture and its Concrete[R]. Kunming: Kunming Hydropower inves-

tigation, design, research institute, CHECC, 2010.
[3] XuXu,etc. The feasibility research of cinder and ash from Coal – fired Power Plant number 504 used as admixture in concrete: Reports of research of temperature control and economic comparison[R]. Kunming: Kunming Hydropower investigation, design, research institute, CHECC, 2010.
[4] Bofang Zhu. Thermal stresses and temperature control of mass concrete[M]. Beijing:China electric power press,2003.
[5] A. M. Nevil(author), Shuhua Liu, Faguang Leng, Xinyu Li, Xia Chen(translator). Properties of concrete [M]. Beijing:China Architecture & Building Press. Beijing. 2011.
[6] Kunhe Fang. Material, structure and Properties of RCC[M]. Wuhan :Wuhan university Press. Wuhan. 2004. 2.

The Relationships between the In – situ Tensile Strength Across Joints of an RCC Dam and the Maturity Factor and Age of Test

Malcolm Dunstan[1], Marco Conrad[2]

(1. Malcolm Dunstan and Associates, U. K. ;2. AF-Consult Switzerland Ltd. , Switzerland)

Abstract: Cores are usually extracted from dams at various ages, due to the limitation of access and due to the progress of the placement of the concrete. Thus the specimens from the cores are frequently tested at ages that vary considerably. Therefore when considering the tensile strength of cores containing joints between layers, there are two variables; first the Modified Maturity Factor (that defines the exposure time between the layers and the ambient temperatures at the time, both these factors influencing the strength at the joints) and secondly the age of test. With RCC, containing high proportions of pozzolan, there can be a considerable increase in strength with age. Thus a three – dimensional model has been developed that relates the direct tensile strength across joints with the Modified Maturity Factor and the age of test. These models have been used for four dams in South – East Asia, for which several hundred results were available. The shapes of the four models were all different, and in two of them the tensile strength did not reduce as much as was expected with increasing maturity at the joints. It is considered to be important to be able to determine the shape of such 3D models for each dam so that the most economic solution can be developed for that particular dam.

Key words: RCC, dams, cores, Maturity Factor, age of test, tensile strength

1 Introduction

Cores can be taken from an RCC dam at any age from circa 90 days to several years (for example at Upper Stillwater dam in the USA, cores were taken at an age of 14 years[1]). Nevertheless the majority of tests are undertaken when the RCC is between 100 and 1000 days old. However between these two ages the in – situ strength (both compressive and to a lesser extent tensile) can increase by 50% to 100% and in extreme cases (e. g. New Victoria dam in Australia) by a factor of three or four.

In addition, at the same time there is the maturity of the joint. As the maturity increases (both in terms of time and temperature), there is usually a decrease in the tensile strength across the joint. Finally there is the actual joint treatment (including the use, or non – use, of bedding mixes. The latter can result in a completely different pattern of development of vertical in – situ tensile strength[2]).

Thus there are a number of influences on the performance of horizontal joints in RCC dams. In order to study these influences, MD&A has developed a three-dimensional model that relates the in-situ vertical direct tensile strength across the joints to the age of the test and the maturity of the joint.

2 Modified Maturity Factor

In order to be able to specify joint treatments in RCC dams, practitioners have been using Maturity Factors that are a product of temperature and exposure time (i.e. the time gap between the layers). Unfortunately there are a number of problems with the use of these Factors. Firstly both deg C. hr and deg F. hr have been used. These two units are incompatible. For example if the temperature is 10 ℃ (50 °F) and a limiting exposure time of 24 hours has been defined, the Maturity Factor would be 240 deg C. hr or 1200 deg F. hr. If the temperature increases to 20 ℃ (68 °F), with a Maturity Factor of 240 deg C. hr, the limiting exposure time would be 12 hours, while with a Maturity Factor of 1200 deg F. hr the limiting exposure time becomes 17.6 hours. Thus both systems cannot be correct. In addition, RCC can be placed in air temperatures below freezing. For example at Platanovryssi the minimum temperature at which RCC was placed was -8 ℃ (18 °F). This RCC probably has the best in-situ properties so far found in any concrete dam[3]. With such a temperature there would be a negative Maturity Factor when deg C. hr units are used.

These problems with Maturity Factors can be addressed in the following manner. Below -12 ℃ (10.5 °F), concrete does not appear to gain strength with time[4]. Consequently this temperature is used as an origin. Thus to calculate the Modified Maturity Factor (MMF) of a concrete 12 deg C is added to the temperature (or similarly 10.5 deg F is subtracted).

Therefore if an RCC has an exposure time of 24 hours at a temperature of 10 ℃ (50 °F), it will have a Modified Maturity Factor of $[24 \times (10+12)]$ = 528 mod. deg C. hr or $[24 \times (50-10.5)]$ = 948 mod. deg F. hr. Assuming the temperature increases to 20 ℃ (68 °F), with the same Modified Maturity Factors the limiting exposure time becomes $[528/(20+12)]$ = 16.5 hours or $[948/(68-10.5)]$ = 16.5 hours.

3 Shape of models

3.1 **Hot joints**

The great majority of joints in an RCC dam are (or definitely should be) "hot" joints, for which there is negligible treatment other than to keep the surface clean and damp. No bedding mixes should be used as they are expensive and more importantly slow the rate of placement, or at least have the potential to do so.

Thus the base 3D model for an RCC dam is developed from the results of tests on hot joints. The performance of other forms of joint treatment can then be compared to the base model.

3.2 Yeywa

The RCC at Yeywa dam, which is located in Myanmar, was placed between February 2006 and December 2008. The volume of RCC was 2.473 million m³ and the mixture proportions were 75 kg/m³ of Portland cement and 145 kg/m³ of a local natural pozzolan[5]. A retarder was used throughout to delay the Initial Setting Time to (21 ±3) hours.

The 3D model of the results of tensile tests on cores containing hot joints is shown in Fig.1.

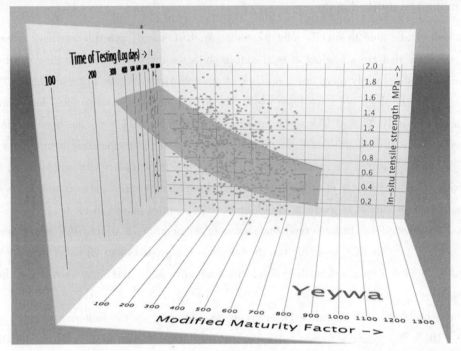

Fig. 1 3D model relating the in–situ vertical direct tensile strength of hot joints at Yeywa to the age of test and Modified Maturity Factor

The great majority of tests on the cores from Yeywa were undertaken between the ages of 400 and 1000 days, i.e. after the design age of 365 days. It can be seen that there is little increase in strength after an age of 400 days and this could be a function of the use of a natural pozzolan in the cementitious material. The range of MMFs was from zero (i.e. the parent RCC with no joints) to 750 mod. deg C. hr (an exposure time of circa 18 hours at an average air temperature of 30 ℃).

What is perhaps surprising regarding the shape of the model in Fig. 1 is that the rate of decrease in the vertical in–situ direct tensile strength across the joints decreases as the MMF increases, rather than increasing which is what is traditionally thought to happen (see Section 3.4).

3.3 Son La

The RCC at Son La dam, which is located in Viet Nam, was placed between January 2008 and August 2010. The volume of RCC was 2.677 million m³ and the mixture proportions were 60 kg/m³ of Portland cement and 160 kg/m³ of a low–lime flyash, a significant proportion of which was retrieved from ash lagoons[6,7]. A retarder was used throughout to delay the Initial Setting Time to (21 ±3) hours.

The 3D model of the results of tensile tests on cores containing hot joints is shown in Fig. 2.

The great majority of tests on the cores from Son La were undertaken between the ages of 100 and 500 days, i.e. bracketing the design age of 365 days. As can be seen there is significant increase in strength during this period. The range of MMFs was from zero to 750 mod. deg C. hr.

The shape of the model in Fig. 2 has the same general shape as that in Fig. 1 although it is flatter and at a higher level.

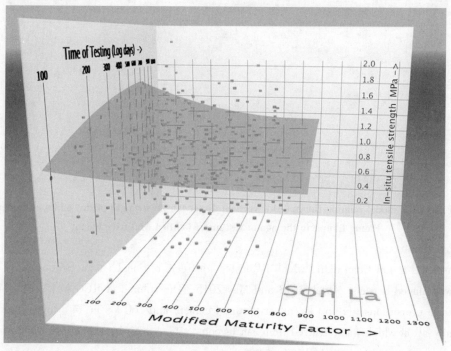

Fig. 2 3D model relating the in-situ vertical direct tensile strength of hot joints at Son La to the age of test and Modified Maturity Factor

3.4 Upper Paung Laung

The RCC at Upper Paung Laung dam, which is located in Myanmar, was placed between January 2011 and December 2013. The volume of RCC was 0.936 million m³ and the mixture proportions were 90 kg/m³ of Portland cement and 140 kg/m³ of local natural pozzolan[6]. A retarder was used throughout to delay the Initial Setting Time to (21 ±3) hours.

The great majority of tests on the cores fromUpper Paung Laung were undertaken between the ages of 300 and 900 days. As with Yeywa (see Fig. 1), at which the same natural pozzolan was used, there is little increase in strength after an age of 300 days. The range of MMFs was from zero to 600 mod. deg C. hr.

The shape of the model in Fig. 3 is closer to the normal expectation with the rate of decrease in the vertical in-situ direct tensile strength increasing as the MMF increases. With this shape, it is reasonably clear at which MMF the average tensile strength equates to the design strength.

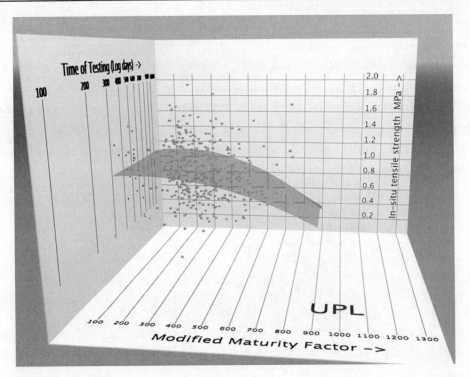

Fig. 3 3D model relating the in-situ vertical direct tensile strength of hot joints at Upper Paung Laung to the age of test and Modified Maturity Factor

3.5 Lai Chau

The RCC at Lai Chau dam, which is located upstream of Son La (see Section 3.3) in Viet Nam, was placed between March 2013 and May 2015. The volume of RCC was 1.884 Mm^3 and the mixture proportions were 60 kg/m^3 of Portland cement and 160 kg/m^3 of a low-lime flyash, practically all of which was retrieved from ash lagoons[7,9]. A retarder was used throughout to delay the Initial Setting Time to (21 ±3) hours.

The 3D model of the results of tensile tests on cores containing hot joints is shown in Fig. 4.

The great majority of tests on the cores from Lai Chau were undertaken between the ages of 100 and 800 days, i.e. nicely bracketing the design age of 365 days. It can be seen that the increase in strength after an age of 100 days is not significant, which is a little different when compared to the Son La results (see Fig. 2). The range of MMFs was from zero to a little over 800 mod. deg C. hr.

The shape of the model in Fig. 4 shows little decrease in strength as the MMF increases, indeed the whole model is nearly horizontal. The average vertical in-situ tensile strength across the horizontal joints varied by less than 10% from the tensile strength of the parent, unjointed, RCC independent of the joint treatment[8]. This is an ideal scenario for an RCC dam.

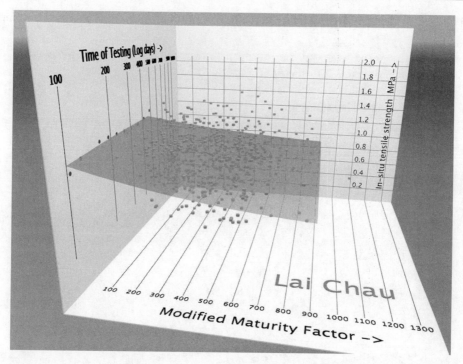

Fig. 4   3D model relating the in-situ vertical direct tensile strength of hot joints at Lai Chau to the age of test and Modified Maturity Factor

4   Other joints

4.1   Introduction

Warm joints, which are the transition from a hot joint to a cold joint, are the most difficult of the joints treatmentsin RCC dams to prepare. Preparation of cold joints is relatively easy although they can be very expensive and take time. High-pressure water is usually used to create an exposed-aggregate finish to the surface of the joint. As long as a cold joint is planned, it should not impact on the programme but if not planned, it can have a significant effect on the rate of progress. Super-cold joints are cold joints in which a bedding mix of some form is used. None of other joints discussed in this paper, i.e. warm and cold joints, had a bedding mix.

4.2   Warm joints

There are various treatments for warm joints, although for the four dams discussed in this paper, a road brush was used to scarify the surface. On the joints just after a hot joint a brush with plastic tines was used and for joints just prior to a cold joint, a brush with steel tines was used.

Fig. 5 shows the results of tests on cores containing warm joints at Yeywa and Fig. 6 at SonLa (neither Upper Paung Laung nor Lai Chau had sufficient warm joints tested for an analysis). In both these Figures the design strength is shown as a horizontal plane through the model. In both Figures the same pattern can be seen with the early-age warm joints (i.e. just after the end of a hot joint) having relatively low strength, indeed at Yeywa somewhat below the design strength – albeit for very few results – while the longer-age warm joints have very much higher strengths.

In both cases, as the MMF increases the vertical in‐situ tensile strength across the warm joints also increased.

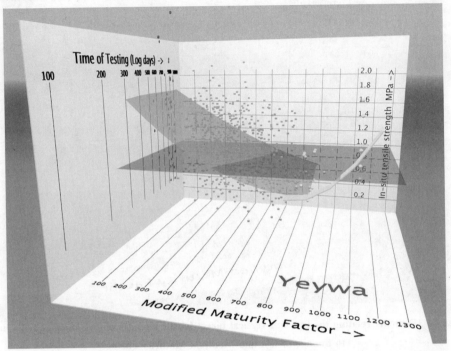

Fig. 5 The in‐situ vertical direct tensile strength of warm joints at Yeywa compared to the 3D model created from the results of tests on hot joints

4.3 Cold joints

The very few cold (and super‐cold) joints, for which tests have been undertaken on these four RCC dams, have been found to have vertical in‐situ tensile strengths across the joints approaching that of the parent (unjointed) RCC.

5 Discussion of results

3D models correlating the vertical in‐situ direct tensile strength of jointed and unjointed cores, the Modified Maturity Factor (MMF) and the specimen age at time of testing have been created from comprehensive core testing data sets obtained from four recently‐completed RCC dams in Southeast Asia, of which two used a natural pozzolan and two a treated lagoon ash (with some reduced‐carbon flyash) in the cementitious materials. These 3D models allow a more complete view of the in‐situ direct tensile strength across the joints compared to the more usual 2D correlation between in‐situ vertical direct tensile strength across joints and the MMF.

The 3D models presented in this paper show three different general shapes for the performance of hot joints related to the age of testing and the MMF:

(1) Little increase of in‐situ direct tensile strength with testing ages beyond 365 days and decreasing drop of in‐situ direct tensile strength across joints with increasing MMF.

(2) Significant increase of in‐situ direct tensile strength with testing ages beyond 365 days

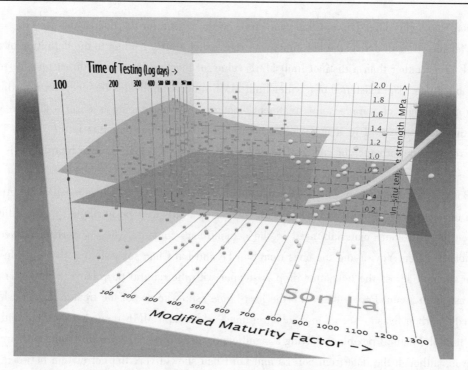

Fig. 6 The in – situ vertical direct tensile strength of warm joints at Son La compared to the 3D model created from the results of tests on hot joints

and decreasing drop of in – situ direct tensile strength across joints with increasing MMF.

(3) Little increase of in – situ direct tensile strength with testing ages beyond 365 days and increasing drop of in – situ direct tensile strength across joints with increasing MMF.

Although the three different shapes represent what can be considered the reality of lift – joint performance in RCC dams, caution is advised with the actual interpretation of the core – testing data, because the quality of core drilling, the core storage, the test specimen preparation and the size and rigidity of the tensile strength testing equipment can have a considerable influence on the direct tensile strength test results. However, there are two major influences, which could explain the three different general shapes for the performance of hot joints; first the retarder performance (i. e. delay of initial and final setting times) and secondly the type of pozzolanic material in the RCC mix (in this case natural pozzolan and low – lime ash).

The effectiveness of a setting time retarder inthe RCC to extend the MMF limit for meeting the design direct tensile strength across lift joints can decrease, if, for example deleterious material such as clay is present in the RCC mix. In the case of Upper Paung Laung, some of the core data represented part of the dam where problems occurred due to contamination of the aggregates with silty – clay fines. These fines also caused some problems with shorter than specified initial setting times. This, together with the hydration characteristics of the Portland cement and natural pozzolan, could well have contributed to the increasing drop of the vertical in – situ direct tensile strength across the hot joints with increasing MMFs above 400 to 500 mod. C. hr (up to where the retardation apparently was effective). In the cases of the other three data sets, the setting time retardation was highly effective and this could well be the reason for the decreasing drop of in – situ direct tensile strength across the hot joints with increasing MMF, leading to the definition of a limit MMF range rather than a distinct limit MMF value for meeting the design direct tensile strength across lift joints. The existence of a wider limit MMF range complicates the decision regarding the timing and method of warm – joint treatment, potentially commencing a warm – joint treatment when the lift joint is still in a hot – joint condition. It is possible that this is the reason for the rather reduced performance of very early – age warm joints when high – range retardation is effective, as seen in the few data from Yeywa and Son La.

Both, the strength gain with RCC age and the rate of in – situ direct tensile strength drop across the hot joints before it decreases with increasing MMF may be attributed to the difference between the pozzolanic materials used in the cementitious content. The comparison between the data obtained from Yeywa and the ones from Son La and Lai Chau suggests that the in – situ direct tensile strength across the hot joints drop is steeper with the natural pozzolan (in this case a volcanic ash from Mount Popa) than with the low – lime ash. The use of the treated lagoon ash (and reduced – carbon flyash) seems to result in a much wider limit MMF range between hot and warm joints than the natural pozzolan, having a significantly beneficial effect on lift – joint performance. However, although the data from Son La and Lai Chau show a very flat correlation between the in – situ direct tensile strength across the joints and the MMF around the MMF specified limit for a hot joint, it should be expected that there would be a considerable strength drop once the MMF is clearly beyond the limit MMF range for hot joints.

6 Conclusions

The following general conclusionscan be drawn with a focus on rapid RCC placement and maximizing the proportion of hot joints in the RCC dam:

(1) The application of an effective high – range setting time retarder is highly beneficial;

(2) Ways should be found during the identification of RCC materials and during the RCC Trial Mix Programme to make high – range setting time retardation work, since this reduces the probability of unplanned warm joints (potentially becoming unplanned cold joints) and thus reduces time and the cost for lift – joint treatment during dam construction;

(3) Obtain an understanding of thebehaviour of the in – situ direct tensile strength across the hot joints with increasing MMF through the evaluation of direct tensile strength tests across jointed cores from the Full – Scale Trial Embankment to understand the implications of timing and method of early – warm joint treatment on lift – joint performance;

(4) Generally achieve RCC placement within the MMF hot – joint limit and/or limit the MMF range by specifying a simple RCC placement methodology and by considering that the Initial Setting Time, the MMF, the RCC placement area and the plant capacities are all interrelated in a hot – joint scenario.

Acknowledgements

The authors would like to thank Electricity of Viet Nam, the SonLa Management Board, PECC1 and the Department of Hydropower Implementation of the Ministry of Electric Power in Myanmar for permission to publish this paper. The paper is dedicated to Mr Nguyen Hong Ha, Vice Director of EVN, who passed away during the preparation of this paper and it is thanks to his untiring leadership and drive that both Son La and Lai Chau RCC dams were completed ahead of schedule and to a high standard. Similarly it was due to the energy and perseverance of H. E. U. M. Myint, at the time Deputy Minister, that RCC dams were introduced into Myanmar and that the RCC placement at Yeywa was completed well ahead of schedule.

References

[1] T. P. Dolen. Long – term performance of Roller – Compacted Concrete at Upper Stillwater dam, Utah, USA, Proceedings of 4th International Symposium on Roller – Compacted Concrete (RCC) Dams, Madrid, Spain, November 2003.

[2] M. R. H. Dunstan and R. Ibáñez – de – Aldecoa, Quality Control in RCC dams using the direct tensile test on jointed cores, Proceedings of 4th International Symposium on Roller – Compacted Concrete (RCC) Dams, Madrid, Spain, November 2003.

[3] J. Stefanakos, M. R. H. Dunstan. Performance of Platanovryssi dam on first filling[J]. Proc Hydropower and Dams, London, 1999.6, (4).

[4] A. M. Neville. Properties of Concrete (p.274), 2nd Edition, Pitman, London, 1977.

[5] H. E. U. M. Myint, U. M. Zaw, A. Dredge, et al. Yeywa Hydropower Project, Myanmar – Current developments in RCC dams, Q. 88 – R. 3, XXIIIth ICOLD Congress, Vol. 1, Brasilia, 2009

[6] U. M. Zaw, O. Voborny, N. H Ha, D. Morris and M. R. H. Dunstan, Choice of pozzolans for use in the cementitious content of large RCC dams, Proceedings of 6th International Symposium on Roller – Compacted Concrete (RCC) Dams, Zaragoza, Spain, October 2012.

[7] M. R. H. Dunstan, N. H. Ha, D. Morris. The major use of lagoon ash in RCC dams, Proceedings of 7th International Symposium on Roller – Compacted Concrete (RCC) Dams, Chengdu, China, September 2015.

[8] M. Conrad, D. Morris, M. R. H. Dunstan. A review into the tensile strength across RCC lift joints – Case studies of some RCC dams in Southeast Asia[J]. International Journal on Hydropower and Dams, 2014, 21 (3)

[9] N. H. Ha, N. P. Hung, D. Morris and M. R. H. Dunstan, The in – situ properties of the RCC at Lai Chau, Proceedings of 7th International Symposium on Roller – Compacted Concrete (RCC) Dams, Chengdu, China, September 2015.

Investigation on Pulse Velocity Changes in RCC with Different Cement Content and Different Types of Admixtures

Shabani, N., Araghian H. R.

Abstract: RCC dams are going to be extended in water resource projects, so the new methodologies are also needed to evaluate this type of concrete. One of these methods is pulse velocity measurement for assessment of the quality and evaluation of uniformity of RCC. In this research, pulse velocity through the RCC (as a non-destructive test has been measured and also its changes have been detected. For this purpose 20 types of Mix designs with the same aggregates but different cementitous contents (100, 125, 150, 175 & 200 kg) and three types of chemical admixtures (retarder, plasticizer and super plasticizers) have been made in the laboratory and water cured up to 90 days. After measurement of pulse velocity in cylindrical 15 × 30 samples, other mechanical tests such as the compressive strength, Indirect tensile strength, Static & dynamic elasticity modulus and permeability of the samples have been tested and the results were evaluated.

Key words: Pulse Velocity measurement, Roller Compacted Concrete, admixtures

1 Introduction

Non-destructive testing is one of the measures required for technical inspection and assessment of the structures. As the name implies, non-destructive testing methods without any damaging to the structure assess the situation. In general, by using non-destructive tests on concrete structures important characteristics such as uniformity, homogeneity, depth of surface cracks & modulus of elasticity Will be evaluated.

One of the non-destructive test methods is ultrasound. In this way, by measuring the pulse rate in various parts of the concrete structure valuable information could be found. Considering the similarity of Roller compacted concrete with conventional concrete in hardened state & also excessive use in water resources projects as well as the other advantages mentioned new methods such as the use of ultrasound speed to assess this type of concrete is also taken into consideration.

With regard to specific circumstances and dry state and zero slump of Roller compacted concrete, in this research it is tried to examine the quality and uniformity of this type of concrete by using pulse rate method In this study 20 different concrete mixture with fixed amount aggregates and different cement content of 100, 125, 150, 175 and 200 kg per cubic meter and three types of chemical admixtures such as plasticizer, super-plasticizer, retarder and also without admixtures have been made and cured in the laboratory up to 90 days. After definition of pulse velocity in cylindrical samples (300 mm × 150 mm), compressive strength, splitting tensile strength (Brazilian), elasticity modulus (static and dynamic) and permeability have been determined.

2 Theory of research

Ultrasonic pulse velocity method is based on measuring the speed of pulses which is sent by a transmitter to a receiver on the other side of the. The speed of sound in is related to concrete modulus of elasticity and density, which are influenced by numerous factors, including the type of cement, concrete age, water - cement ratio, the ratio of aggregate used in concrete and concrete curing conditions[3]. So Pulse rate can be set for determination of uniformity of concrete, estimated concrete strength, modulus of elasticity, measured properties of concrete with the passage of time, the degree of hydration of cement, concrete durability and can be used to estimate the depth of the cracks[4,5]. This method has been developed in recent years and one of its advantages is that the equipment is portable.

Mechanical pulses concrete wave that includes three types of waves including longitudinal waves (compression), shear wave (transverse) and surface wave. Longitudinal waves are the fastest waves are suitable for testing. The relationship between velocity, modulus of elasticity and density of concrete is as follows:

$$V = \sqrt{\frac{(1-\mu) E_d}{\rho(1+\mu)(1-2\mu)}} \qquad (1)$$

Which: V = longitudinal wave speed, E_d = dynamic elasticity modulus, ρ = density and μ = concrete poisson's ratio. [6]

Since the range of Poisson's ratio in concrete (between 0.15 to 0.25) and variety of density for different concrete types, can be said that the pulse velocity (V) and dynamic modulus of elasticity (E_d) are related together and associated with changes in modulus of elasticity pulse rate also changes.

3 Materials & Methods

Method of the research is based on Laboratory tests and the materials used to make the roller compacted concrete mixtures was aggregates, cementitious material, water and the admixtures. Aggregates consisted of three categories such 0 ~ 5 mm crushed sand, rounded 0 ~ 20 mm and crushed 20 ~ 50 mm sources on which have been provided from CHAMSHIR dam borrow areas. Quality control tests such as determination of specific gravity (bulk, real saturated surface dry), determination of water absorption have been conducted on the aggregates. The ratio of aggregates in the mixtures was 30% for 0 ~ 5 mm sand, 45% for 0 ~ 20 mm, 25% for 20 ~ 50 mm. Mixed aggregate size distributioncurve is shown in Fig. 1.

Cement used was type 2 and physical and chemical tests have been conducted on cement showed its compliance with ASTM C150. Water for RCC mixtures were obtained from drinking water supply. Determination of density and dry residue tests were conducted on admixtures used in this research. The results of the physical and chemical tests, were presented respectively in Tables 1 and 2 respectively. The results of admixtures tests are shown in Table 3.

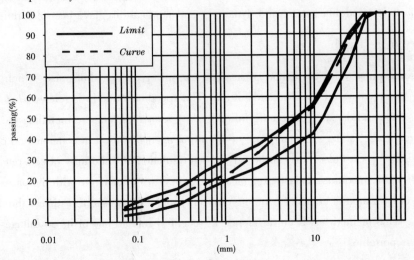

Fig. 1　Combined particle grading (Dash red line shows total grading of mixes)

Table 1　Physical test results of cement

Test	Normal consistency	Autoclave Expantion	Density	Blaine	Setting time (Vicat)		Compressive Strength (28 days)
					Final	Initial	
	(%)	(%)	(g/cm^3)	(cm^2/g)	(min)	(min)	(psi)
Result	21.2	+0.08	3.17	3 100	170	230	3 200

Table 2　Chemical test results of cement

Combined	SiO_2	Al_2O_3	Fe_2O_3	CaO	MgO	Na_2O	K_2O	SO_3	C_3S	C_2S	C_3A	C_4AF
%	21.36	5.30	4.70	62.30	1.75	0.27	0.61	2.12	42.90	28.90	6.09	14.3

Table 3　Test results of chemical admixture

Admixture	Super-plasticizer		Plasticizer		Retarder	
Test	Density (g/cm^3)	Dry Remaining (%)	Density (g/cm^3)	Dry Remaining (%)	Density (g/cm^3)	Dry Remaining (%)
Result	1.20	49.05	1.14	29.8	1.12	27.9

Five different amount of cement have been used for making the RCC mixtures including 100, 125, 150, 175 and 200 kg. Three different type of admixtures including plasitciser, super plasticiser and retarder were also used in the samples. Total number of RCC mixtures was 20 different mixes including control mixtures. Mixtures components used in this research for one cubic meteris shown in Table 4.

Table 4 Mix designs with different cement content and different admixtures

ID	RCC-100				RCC-125				RCC-150				RCC-175				RCC-200			
	Tes	Ret	Pls	Spl	Tes	Ret	Pls	Spl	Tes	Ret	Pls	Spl	Tes	Ret	Pls	Spl	Tes	Ret	Pls	Spl
Agg (kg)	2 150				2 150				2 150				2 150				2 150			
C (kg)	100				125				150				175				200			
W (kg)	110,100,93.5,88				113,105,97,90				115,108,100,93				127,118,110,104				135,127,120,113			
W/C	1.10,1.00,0.94,0.88				0.90,0.84,0.78,0.72				0.77,0.72,0.67,0.62				0.73,0.67,0.63,0.59				0.68,0.64,0.60,0.57			

Tes = Testifier, Ret = Retarder, Pls = Plasticizer, Spl = Superplasticizer

After preparation of the final mixes, aggregates and other constituents to produce 0.1 cubic meters of concrete was carefully weighted. Aggregates were poured first into the mixer and allowed to mix with each other. Then cement was added to the mixed aggregate and re-mixing took place. Finally, the water admixture were added to the mixture and mixing continued for 3 to 4 min. In order to determine the water-cement ratio (W/C) for each mix, based on Vebe time and with a trial and error tried to achieve the optimal water content for a optimum Vebe time of 10 to 14 s.

Specific gravity and vebe time of fresh concrete according to ASTM C 1 170 by using a 12.5 kg surcharge have been determined. To make a 300 mm × 150 mm cylindrical samples according to standard ASTM C1176 a 9.1 kg surcharge and a vibrating table were used. The samples removed from molds after 24 hours and were kept in curing pool up to 90 days. 90-day compressive strength test samples according to ASTM C-39, Brazilian tensile strength according to ASTM C-496, the modulus of elasticity according to ASTM C-469, permeability according to EN 12390-08 and pulse velocity according to ASTM C-597 were performed. By using pulse velocity, poissons' ratio and density of each sample, dynamic modulus of elasticity was obtained by using equation 1.

For determination of ultrasonic pulse velocity the device shown in Fig. 2 is used. The apparatus consists of a electrical pulses generator, amplifiers and electronic instrument for measuring the pulse transit time between the transmitter and receiver. Usually to carry out tests on concrete generators with a frequency between 25 and 100 kHz is used. High-frequency generators for short distance transport of pulses or thin concrete sections and generators with low frequency pulse are used for a relatively long path and thick concrete sections. But in general, in most cases, 50 to 60 kHz frequency generators are appropriate. In this study, the frequency is 54 kHz.

4 Analysis of laboratory results

Fig. 3 (a) shows that by increasing the amount of cement and with a fixed amounts of other

Fig. 2 Pulse velocity apparatus used in this study

components ultrasonic pulse velocity increased in all of the mixtures. The use of plasticizer admixtures in comparison with retarders is more effective on the speed of ultrasonic pulses through the roller compacted concrete. Super – plasticizer admixtures is more effective on pulse velocity in concrete compare with two other type of admixtures in fixed cementitious materials. The reason can be assumed that better dispersion of the cement particles and also decreasing the water cement ratio produces a denser concrete and the pulse velocity is increased.

It is interesting that pulse velocity in mixtures with plasticizers is slightly more than retarded mixtures with fixed cementitious materials. All of the mixtures with admixtures have a higher pulse velocity than control mixture. This is considered due to the lower water cement ratio of mixes with admixtures and better dispersion of cement particles. In lower compressive strengths, between 10 to 20 MPa retarders are the best admixtures for the studied mixtures. Between 17 ~ 25 MPa compressive strengths, plasticizers and super – plasticizers have equal effect on pulse velocity. But beyond the 25 MPa (150 kg cementitious materials) compressive strength, super – plasticizers are the most effective in Roller compacted concrete mix designs. This is probably due to the higher cement in the RCC.

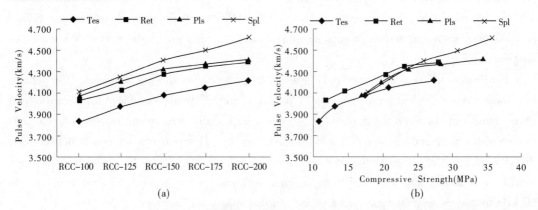

(a) Pulse velocity changes with different cement content and different types of admixtures, (b) Pulse velocity versus compressive strength relationship

Fig. 3

Longitudinal wave velocity depends on the elastic properties of roller compacted concrete and it is a function of the modulus of elasticity and consequently it is associated with compressive strength. Therefore, as in Fig. 3 (b), 4 (a) and 4(b) observed, By increasing the compressive strength, tensile strength and modulus of elasticity ultrasound pulse velocity is also increases.

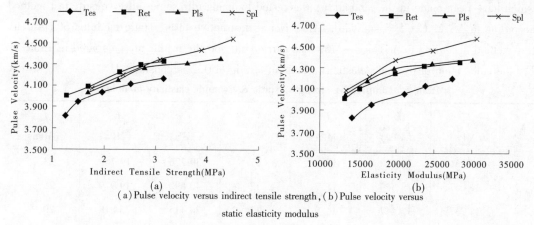

(a) Pulse velocity versus indirect tensile strength, (b) Pulse velocity versus static elasticity modulus

Fig. 4

Fig. 5 shows the relationship between ultrasonic pulse velocity and permeability and also porosity. Increasing the velocity of pulse represents the low porosity and consequently the low permeability of roller compacted concrete. Using the plasticizers and super – plasticizers increases the density of RCC and then increases the pulse velocity in roller compacted concrete. Retarders also causes the cement particles hydrated slower and the crystals be smaller, more stable and closer together, that this reduces the porosity, permeability and thus increase the pulse velocity in roller compacted concrete.

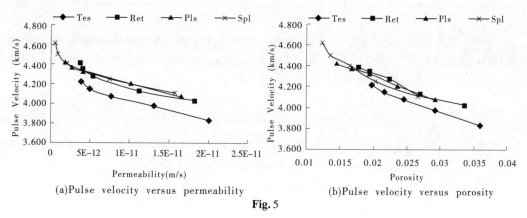

(a) Pulse velocity versus permeability (b) Pulse velocity versus porosity

Fig. 5

5 Dynamic Elasticity Modulus Calculation

Dynamic modulus of elasticity of concrete is related to a strain of very small spot. Ratio of dynamic elasticity modulus (E_d) to static elasticity modulus (E_c) is not fixed. BS 8110: part 2 for normal concrete represents the relationship between these fore – mentioned elasticity modulus as follow:

$$E_c = 1.25E_d - 19 \qquad (2)$$

For Stress analysis in the dams, which may be affected by the earthquake load, dynamic modulus of elasticity should be used which may be determined by equation 2 or by ultrasonic test. In this study, dynamic modulus of elasticity of cylindrical roller compacted concrete samples which have been made in the laboratory, was tested by using ultrasonic pulse velocity test method according to ASTM C – 597 standard. After Non – destructive tests, static modulus of elasticity test method according to ASTM C – 469 also carried out and the results are compared in Table 5. For the calculation of dynamic elasticity, Poisson's ratio of 0.2 is considered.

Table 5 compressive strength, static & dynamic elasticity modulus

No.	E_d (MPa)	E_c (MPa)	f_c (MPa)	No.	E_d (MPa)	E_c (MPa)	f_c (MPa)
1	30 761	14 310	10.7	11	39 226	19 763	23.7
2	34 057	13 340	11.7	12	40 690	19 979	25.9
3	34 737	13 447	16.9	13	36 115	23 818	20.9
4	35 423	13 525	17.9	14	39 589	23 815	23.1
5	33 051	16 680	13.1	15	40 046	24 768	28.4
6	35 682	15 178	14.7	16	43 464	24 890	30.7
7	37 079	16 325	19.8	17	37 308	27 568	27.4
8	37 877	16 563	21.3	18	40 414	28 580	28.1
9	34 822	21 068	17.3	19	40 968	30 282	34.5
10	38 324	20 063	20.4	20	44 759	30 628	35.8

According to the Table 5 values in the range of 1.35 to 2.62 for ratio of dynamic to static elasticity is achieved. It is evident that with increasing the compressive strength of samples, the ratio of dynamic to static elasticity modulus is decreased.

6 Conclusion

• By increasing the amount of cement and fixed amounts of other components, the pulse velocity in roller compacted concrete increases in all of the cement content.

• Use of a plasticizer admixtures compared to retarder is more effective on pulse velocity through roller compacted concrete and super – plasticizers have more effect compared with other plasticizers and retarders.

• The use of plasticizer and super – plasticizer admixtures in the roller compacted concrete mix design makes better the compaction and the use of retarders also causes the cement gently hydrate and crystals smaller, more stable and closer together so that the porosity and permeability of RCC reduces and pulse velocity increases.

• The ratio of Dynamic Elasticity Modulus to Static Elasticity Modulus will be decreased by increasing the cement content and thus by increasing the compressive strength.

• Ultrasonic method can be used as a suitable test for checking the roller compacted properties such as density (& compaction), tensile strength, elastcity modulus and compressive strength.

References

[1] ACI (2011). Roller-Compacted Mass Concrete. Reported by ACI Committee 207, ACI 207.5R-11.
[2] U. S. Army Corps of Engineers. Engineering and design manual, EM 1110-2-2006, 15 Januray 2000.
[3] Bungey. J. H., Millard. S. G. 1996. Testing of concrete in structures. Third Ed. Blackie Academic & Professional, an imprint of Chapman & Hall.
[4] Krautkramer J., Krautkramer M. Ultrasonic testing of materials[M]. Berlin: Springer,1990.
[5] Blitz J., Simpson G. Ultrasonic methods of non-destructive testing,London, Chapman & Hall, 1996.
[6] ASTM C597: Standard Test Method for Pulse Velocity through Concrete (2009).
[7] K. B. Sanish, Manu SANTHANAM. Characterization of Strength Development of oncrete Using Ultrasonic Method. 18th World Conference on Non-destructive Testing, 16-20 April 2012, Durban, South Africa.

Conclusion of and Suggestions for Engineering Materials and Test, Application and Practice for Longtan Roller Compacted Concrete Dam

Ning Zhong

(Sichuan Ertan International Engineering Consulting Co., Ltd., Chengdu, 610000, China)

Abstract: Longtan Roller Compacted Concrete Dam features highengineering technical difficulty and complex construction, and is beyond previous understanding in terms of engineering materials and test research. Beneficial practice and exploration were conducted in these fields during construction. Valuable experience was also accumulated for construction of similar projects in the future through refining the standards on use of raw materials, strengthening material management, and conducting test, research and practice on mix proportion.

Key words: Longtan, roller compacted concrete, engineering materials, mix proportion test

1 Foreword

Longtan Roller Compacted Concrete (RCC) Dam, the highest roller compacted concrete gravity dam so far, features maximum dam height of 216.5 m and crest length of 849.44 m. Therefore, the construction of it is divided into two phases. The dam is characterized with high construction difficulty, large technical content, and no reference of similar project experience can be found. What's more, the project site is in north Guangxi where is dominated by high-temperature, rainy climate and harsh meteorological conditions. Through great efforts in making technical breakthrough and tests, new technologies and management mode were adopted in Longtan Project. Construction in the weather conditions all year round was realized. Monthly average placing intensity reached 202 000 m^3, and maximum monthly placing intensity reached 316 000 m^3. There are many successes in terms of engineering materials and test in this project. The RCC Construction Quality Control Standard for Longtan Project was then prepared to put forward accurate requirements for engineering materials, test, quality control standard, etc., and all the aspects as above are worth summarizing and using for reference.

2 Raw material selection and application

Forachieving the purpose of optimizing the concrete mix proportion to mitigate adiabatic temperature rise of concrete, improve concrete performance and enhance concrete crack resistance, raw materials were effectively selected and applied.

2.1 Cement

Cement selection for roller compacted concrete has become relatively mature at present, and moderate heat Portland cement which meets national specification (GB200 – 2003) is mainly selected. In regard of roller compacted concrete for Longtan Project, national specifications shall be met. Additionally, certain detailed requirements and improvement for cement products are required for manufacturer according to actual conditions of the Project.

1) Hydration heat and MgO content

Hydration heat is an important technical control indicator for cementused for dam concrete. Currently, the hydration heat of cement produced by domestic moderate heat cement has fully met or been below hydration heat indicator for moderate heat cement. Through on – site sampling inspection, the 7 d hydration heat of moderate heat cement from Guangxi Liuzhou Cement Plant and Sanxia Cement Plant as suppliers of Longtan Hydropower Plant Project is respectively between 268 kJ/kg and 271 kJ/kg, below the standard of 293 kJ/kg for moderate heat cement. In regard of cement supplier selection, priority shall be given to the manufacturer with relatively lower hydration heat indicator for the purpose of minimizing difficulty in temperature control due to adiabatic temperature rise.

Major crack is in particularthe major concern in dam concrete since it may damage integrated safety of the whole dam While autogenous volume deformation of hydraulic concrete is generally between $(-100 \sim 100) \times 10^{-6}$, which is actually equivalent to the temperature deformation cause by temperature change of 10 ℃. Thus, it indicates that autogenous volume deformation greatly affects crack resistance of concrete. MgO in cement is characterized by unique delayed expansion deformation which meets dam concrete characteristics of slow heat dissipation and slow shrinkage deformation by temperature drop. This kind of delayed expansion deformation of MgO can offset part of concrete shrinkage deformation by temperature drop, thus crack of dam due to temperature stress is prevent. The hydration – heat 42.5 cement of Guangxi Yufeng Group Ltd. adopted for Longtan Project has MgO content between 4.09% and 4.22%, meeting specification requirements. However, it is necessary to measure concrete deformation to check whether cement is in expansion deformation before confirming whether MgO content is proper. If MgO content in cement cannot meet requirements, additional MgO can be mixed to solve the problem; some beneficial explorations were conducted in this aspect in Longtan Project. During roller compacted concrete construction for downstream cofferdam, 14 kg/m³ characteristic MgO (with MgO content of 60%) was mixed. The application of mixing MgO was under exploratory research and application in Longtan Project, but was not applied to main works of the dam in consideration of the importance and risk degree of Longtan Dam Project. Further research and application can be conducted in this aspect for other projects with conditions met.

2) Cement fineness

The finer cement granularity is, the betterbleeding effect of roller compacted concrete will be. The faster hydration reaction will be, the quicker strength improvement will be. However, due to the inherent characteristics of roller compacted concrete, the concrete is in low early strength, and crack is easy to generate under temperature influence. Therefore, cement fineness indicator shall be put under control to prevent heat release in advance due to too fast hydration reaction. Upper limit of specific surface area of moderate heat cement was specified in Longtan Dam Project to control specific surface area around 300 m^2/kg. It is to not only facilitate temperature control, but also avoid lowering early strength of concrete too much.

2.2 Flyash

Flyash can be mixed into concrete to cause water – reducing potential energy by its granule shape. Its micro aggregate effect will cause compact potential energy, and pozzolanic effect will cause activation potential energy. As a result of these characteristics, the water and cement consumptions can be reduced, compactness can be improved, bleeding can be reduced, internal temperature rise can be mitigated, durability can be improved, and alkali – aggregate reaction can be restricted.

Flyash is one of the main materials of roller compacted concrete. The grade – I ash was adopted for flyash of roller compacted concrete for Longtan Project, with maximum dosage of 66%. Flyash is an auxiliary product from power plant production and is not produced as a particular product. Quality of flyash is affected by coal source, coal ash collection mode and storage mode and so on. While flyash has become an important raw material for dam construction, and its quality and supply conditions will impose huge influence on project construction quality and schedule.

According to experience of flyash application in Longtan Project, the following issues need to be paid attention to in terms of flyash application for roller compacted concrete:

1) Fineness, ignition loss and water requirement ratio

As flyash mainly serves for causing micro aggregate effect in concrete to improve concrete performance, the flyash fineness indicator is an important reference for the effect. Generally, the finer flyash is, the larger specific surface area will be, and the easier flyash activity can be activated. Therefore, the finer flyash is adopted, the higher early strength and better durability of concrete will be. In addition, since roller compacted concrete belongs to hard concrete, construction and workability are the key to concrete control for the reason that only assurance of easy concrete bleeding during rolling can the concrete interlayer bonding effect be ensured. Pozzolanic activity of flyash cannot bring into full play unless flyash is fine enough. Flyash not only serves as filling material, the glass beads in flyash can play the role of lubrication in fine particles, which can improve concrete bleeding capacity and performance of concrete of being roller compacted. During flyash selection and quality control, it is absolutely prohibited to adopt the flyash with fineness out of standard, and fineness indicator is better to be as small as possible.

Ignition loss indicator of flyash is essentially the content of substance not burnt up in flyash. Main content of this part of substance is carbon. Flyash ignition loss affects water requirements greatly and is in direct proportion to water requirement. The ignition loss is better to be not more than 5%. When ignition loss is above 10%, there will be no beneficial effect to flow expansion degree by flyash. Ignition loss becomes higher with the increase of carbon content in flyash, and flyash is easier to float on surface during concrete mixing, transportation and forming. The appearance and inherent quality of concrete thus will be affected. In addition, larger ignition loss will weaken use effects of water reducing agent and increase concrete free water consumption, which will be very easy to cause bleeding during rolling of RCC.

As for important roller compacted concrete project, Longtan Project can be used for reference to give priority to grade – I flyash. In case if difficulty in supply, or in consideration of cost saving, it is considered that quasi – grade – I flyash can also meet design indicator requirements. Since the indicators of quasi – grade – I flyash can meet requirements for grade – I flyash except that water requirement ratio is extended to not greater than 100%.

2) Suggestions for flyash quality management and storage

Flyash is taken as an auxiliary product of power plant, and is not under strict quality management by power plant. Many power plants only conduct simple testing, and testing method, equipment and frequency cannot meet specification requirements, causing failure of much flyash to meet required indicators in incoming testing. Therefore, it is suggested to perform detailed survey on supplier before tendering for flyash for similar projects. Special testing station can be set or personnel can be assigned to perform on – site patrol inspection and spot check in supply period, so as to ensure quality of flyash transported to the construction site and avoid unqualified flyash and further avoid waste of manpower and time for handling quality problems.

Because of hugequantities and high construction strength of Longtan Project, 15 flyash suppliers were selected at the most. However, the flyash from different suppliers always differs greatly in quality, chemical composition, color, etc. , and differs slightly in mix proportion compatibility, bringing about high difficulty to concrete quality control. Additionally, use of the whole storage tank of flyash would always be affected by product quality problems of a certain manufacturer (batch). To better perform quality control and facilitate finding out possible quality problems from flyash sources, fine adjustment can be conducted to mix proportion of flyash from different suppliers during construction. It is also suggested to reduce supplier number during purchasing and avoid tank storage of mixed flyash to the greatest extent.

2.3 Concrete additive

Additive is another indispensable functional material for modern hydraulic concrete. Two kinds ofadditive will be mainly used in concrete for roller compacted concrete dam project. One is efficient set retarding and water reducing agent which integrates water reducing and set retarding functions, the other is air entraining agent with main function of air entraining when being mixed into concrete. The latter is to improve permeation resistance and frost resistance of concrete, and with functions of slight water reducing effect and improvement of workability of concrete mixture. The current efficient set retarding and water reducing agent includes two types: naphthalene type and polycarboxylic acid type. The latter type features sound water reducing effects, generally with water reducing rate above 30%, and is usually applied to high – grade normal concrete rather than roller compacted concrete.

The quality of efficient set retarding and water reducing agent shall not only meet indicators of technical standards, but also meet construction requirements. Concrete setting time is very important during roller compacted concrete construction. Setting time in technical standards for efficient set retarding and water reducing agent is expressed by time difference in the test of mixing the agent into neat cement paste, resulting in certain difference from concrete setting time. What's more, the setting time of the same concrete varies with weather and air temperature. In roller compacted concrete construction of Longtan dam in early summer of 2005, it was usually hard to make concrete setting time meet on-site construction requirements due to lack of understanding of efficient set retarding and water reducing agent and on-site construction. In the construction in early summer, additive formula was still that for winter construction, and concrete setting time was hard to meet requirements in summer. The problem was not solved until manufacturer provided winter-type and summer-type additive on the basis of requirements for concrete setting time in different seasons upon research with manufacturer. The case provides reference for similar projects, and puts forward requirement for manufacturer to provide efficient set retarding and water reducing agent with different setting times according to different seasons.

Initial setting time of concrete is the main basis of determination of interlayerinterval for roller compacted concrete construction. However, the initial setting time obtained on the basis of different data in accordance with the Test Code for Hydraulic Roller Compacted Concrete varies significantly. Therefore, research and test were conducted on construction site for Longtan Project and test indicators for initial setting time of roller compacted concrete meeting on-site and actual conditions were obtained. Penetration resistance between 5 MPa and 6.5 MPa was considered to be proper and reflected real initial setting characteristics of roller compacted concrete. To guide on-site construction and formula selection of additive manufacturer, and to further facilitate determination of proper efficient set retarding and water reducing agent, targeted test and research in this aspect should be conducted for different projects.

In consideration of absorption effect of flyash in roller compacted concrete, high-quality air entraining agent shall be mixed to improve permeation resistance, etc. of roller compacted concrete. After test for roller compacted concrete for Longtan Dam, ZB-1RCC15 efficient set retarding and water reducing agent and ZB-1G air retraining agent from Zhejiang Longyou Additive Plant and JM-II efficient set retarding and water reducing agent from Jiangsu Research Institute of Building Science (Bote New Materials Co., Ltd.) were adopted.

2.4 Concrete aggregate

1) Coarse aggregate

Coarse aggregate takes up about 60% in concrete, and is an important factor related to concrete performance. Aggregate is an important factor that affects concrete strength. Generally, weak link of low-grade concrete (below 40 MPa) is bonding surface among aggregates. Aggregate is generally free of great damage, but aggregate shall be selected in terms of concrete for crest overflow surface. Though concrete for overflow surface of the Project is in grade C35 and not high in strength, certain shock resistance and abrasion resistance are also required. It is concluded from research on mix proportion in Longtan, Pubugou and other projects that large proportion of aggregate in concrete is the main quality loss factor in concrete shock resistance and abrasion resistance. It is suggested to carry out certain in-depth test for aggregates before preparing concrete for overflow surface, since only proper aggregate selection can ensure sound shock and abrasion resistance for concrete on the premise of meeting strength indicators.

The surface of coarse aggregate in manual processing is easy to cause powder coating by aggregate abrasion during transportation (in particular by belt conveyor) and storage. There was powder coating during roller compacted concrete construction for Longtan Project and Baise Hydropower Plant Project, and bonding strength between aggregate and mortar would be lowered. Roller compacted concrete is characterized by low bonding material content and high hardness, and shows low mortar coating degree in aggregate. So roller compacted concrete is under greater influence of powder coating than ordinary concrete. The crusher dust coated on the surface of coarse aggregate will lower compressive strength, splitting tensile strength and axial tensile strength of roller compacted concrete, and will lower concrete frost resistance to certain degree from research on powder coating in Baise Hydropower Plant Project and Longtan Project. Bonding surface will be broken under continuous freezing and thawing cycles, which will prevent concrete from meeting corresponding anti-freezing indicators. Secondary screening and washing process was added for concrete mixing system of Longtan Dam Project, and coated powder content in aggregate was lowered from maximum 2.02% to maximum 0.34% upon washing through test and statistics. The aggregate quality indicator was greatly improved, and aggregate quality in mixing system was ensured.

2) Fine aggregate

Roller compacted concrete has high quality requirements for sand. Main quality indicators of sand include fineness modulus, crusher dust (with granularity not higher than 0.16 mm) content, and particle (with granularity below 0.08 mm) content. Fineness modulus of sand for Longtan Project, Three Gorges Project, etc. is generally required to be between 2.4 and 2.8. The artificial sand adopted for roller compacted concrete has higher and stricter requirements for crusher dust content than normal concrete. Limestone powder (below 0.08 mm) features certain activity, and may engage in hydration reaction of cementing material. In addition, crusher dust particles have micro aggregate effects to fill the gap among cement particles and optimize gradation of fine powder (including cement and active mixture). Crusher dust ensures sound binding force among slurry and high mechanical interlocking capacity by its shape of multi-edge body. Therefore, crusher dust can improve workability and compactability of roller compacted concrete, and is good for improving concrete strength. The required quality standard for sand of roller compacted concrete in Longtan Project is "crusher dust content shall be between 18% and 22%, and the content of particle with $d \leqslant 0.08$ mm shall be above 50% crusher dust content". During construction of Longtan Project, there was once the problem of inadequate crusher dust content in sand for roller compacted sand due to equipment problem, thus concrete compactability was seriously deteriorated, bleeding effect was unsatisfactory, and layer treatment became difficult. As for sand for

RCC in most of projects, quality requirements for sand for RCC can be developed with reference to the sand for RCC in Longtan Project, since the requirement in Longtan Project refers to a relatively mature quality standard for sand for RCC, and will facilitate project construction.

3) Water content in aggregate

Water content in aggregate is not a problem for many projects, buthigh attention shall be paid to water content in aggregate for RCC dam under high-strength production. Water content in aggregate was out of standard for many times during RCC placing of Longtan Project. Due to low water consumption for RCC itself, (such as water consumption of about 80 kg/m^3 for Longtan Project, in comparison with aggregate consumption of 2 200 kg/m^3), 1% change of average water content in aggregate will result in mixing water reduction by 22 kg/m^3. With water content in additive solution counted in, there will be much less free water in concrete mixing, bad for concrete quality control and stable production. In particular for ice-mixing concrete in high-temperature season, ice may not be mixed or only a little amount of ice can be mixed. If small-size aggregate with high water content is led in air cooling silo, silo is easy to be frozen by aggregate. Therefore, aggregate demand in peak hours shall be fully estimated during aggregate production, screening and washing. Enough adjustable silo capacity shall be provided for aggregate; fine aggregate shall be provided with dewatering time of at least a week, and coarse aggregate shall be provided with dewatering time of at least 3 days. In this way, aggregate in stable quality with low water content can be kept, and this is an important premise for stable concrete production.

3 Mix proportion design of Longtan Project RCC

Fine concrete mix proportion is a prerequisite for smooth development of dam concrete construction. A proper construction mix proportion shall meet design indicator requirements, ensure sound workability and quality control, and shall result in sound economic control indicators.

Since it takes long age for concrete to reach corresponding performance indicators, the design institute had provided mix proportion through referring upon in-depth research. As it will be a long process of conducting design of construction mix proportion on construction site, it is necessary to prepare a thorough mix proportion test plan in advance to prevent failure of mix proportion to meet construction requirements.

According to the application of mix proportion design in Longtan Roller Compacted Concrete Dam Project, the following views are put forward:

1) Designed age in regard of RCC strength grade

Control indicators for Longtan Roller Compacted Concrete Dam adopted 28 d and 90 d age. Since later strength of concrete is greatly improved by high flyash dosage, 90 d concrete strength indicator is more than needed. Dam concrete placing features long period, and dam cannot be put into operation until at least half a year after completion of concrete placing. Therefore, it is considered that designed age for RCC is better to be 180 d, and 90 d can also be taken as an intermediate control standard for the purpose of facilitating concrete quality control. Such long age design facilitates both temperature control and cost saving.

2) VC indicator

VC refers to an indicator for reflecting plasticity during roller compacted concrete construction, and is an important parameter of roller compacted concrete workability and one of the control objectives in concrete mix proportion design. Roller compacted concrete is mainly in low VC design at present, and VC is generally controlled between 5 s and 7 s for the reason that roller compacted concrete with this indicator features best workability. Since VC is sensitive to the change of water consumption of RCC, it is generally better to be designed in a wide range in mix proportion design to ensure wide adjustment range of concrete production control and enable the concrete to adapt to construction in different conditions (high temperature, light rain, etc.). VC control range for roller compacted concrete for Longtan Project is: 5~7 s ± 2 s for mixer outlet, and 5~7 s ± 3 s for placing surface.

3) Cementing material dosage

Take RIII (C9015W4F50) RCC of Longtan Dam for example, cementing material dosage for RCC in this grade for Longtan Project is: 56 kg/m^3 for cement and 109 kg/m^3 for flyash, and total dosage of cementing material is 165 kg/m^3. It can be seen from statistics of 113 domestic roller compacted concrete dams, the dosage of cementing material between 160 kg/m^3 and 189 kg/m^3 takes up 46.9%. Cementing material dosage for Gra. 3 roller compacted concrete of Longtan Project is between 165 kg/m^3 and 195 kg/m^3, and that for Gra. 2 roller compacted concrete is 220 kg/m^3. It can be reflected from rolling test and actual construction in Longtan Project that bleeding is difficult and workability is poor in case of low cementing material dosage for roller compacted concrete and high VC control. There are many influence factors in actual roller compacted concrete construction, the consequence of poor interlayer bonding is directly reflected. Therefore, it is suggested that cementing material dosage for common roller compacted concrete dam shall be at least controlled at 160 kg/m^3 or properly higher; and flyash dosage can be properly higher to serve as one of the most effective actions for ensuring sound interlayer bonding. Maximum flyash dosage for roller compacted concrete of Longtan Project is 66%. On the premise of meeting concrete design grade, it is necessary to ensure sound bleeding capacity for concrete to guarantee interlayer bonding quality of roller compacted concrete. Specific optimal cementing material dosage shall be determined through indoor mix proportion test and on-site rolling test.

4) Mortar mix ratio indicator of RCC

Mortar mix ratio indicator refers to the ratio of slurry and mortar in RCC, and serves as an empirical datum for reflecting actual inherent bleeding capacity and compactability of roller compacted concrete. The mortar mix ratio for most of domestic roller compacted concrete ranges from 0.3 to 0.39. Through expert consultation, the mortar mix ratio for RCC of Longtan Project was recommended to be 0.42. The mortar mix ratio for RCC of Longtan Project can be calculated by two methods: I. mortar mix ratio upon pure cementing material calculation is below 0.42; II. Micro powder with grain size of 0.08 mm in artificial sand is taken as cementing material for calculation, and mortar mix ratio will exceed 0.42. It was considered in expert consultation and discussion on mix proportion that the second calculation method was proper, for the reason that the particles with grain size of 0.08 mm has certain activity by surface effect, and are the substance forming concrete bleeding. It can be concluded from practice in Longtan Project that mortar mix ratio indicator for roller compacted concrete is better to be 0.42, and the indicator can be used for reference for similar projects.

5) In-situ shear test for RCC is recommended

In-situ shear test refers toreal shearing resistance test between layers of concrete dam monolith for RCC upon simulation of construction area and in actual rolling conditions. The in-situ shear test for Longtan Project was carried out in RCC process test area, and interlayer shearing strength by different rolling times, different interlayer intervals, different interlayer treatment (mortar paving after surface roughening, Gra. 1 concrete paving after surface roughening) and other different combinations can be tested. The f' and c' in test results and shearing strength indicators play an important role in guiding roller compacted concrete construction and ensuring shear resistance and skid resistance of dam body.

4 Conclusion

(1) Longtan Dam is the first domestic 200 m-level roller compacted concrete dam, and has created many domestic precedents in terms of roller compacted concrete technology and management. Much attention was paid to material and test research, and the standard suitable for Longtan Dam was prepared.

(2) As performance of roller compacted concrete is highly sensitive to quality conditions and composition of materials adopted, effective actions shall be taken to ensure material quality, and targeted refining shall be performed for some indicators.

(3) During roller compacted concrete construction and mix proportion design, it is required to take into consideration balance among performance, temperature control and construction (ensuring interlayer bonding quality), and take mortar mix ratio as inherent actual indicator to verify mix proportion design.

The In-situ Properties of the RCC at Lai Chau

Nguyen Hong Ha[1], Nguyen Pham Hung[2], David Morris[3], Malcolm Dunstan[4]

(1. Son La Management Board (EVN), Vietnam; 2. PECC1 (EVN), Vietnam;
3. AF - Consult Switzerland Ltd., Switzerland; 4. Malcolm Dunstan and Associates, U.K.)

Abstract: Lai Chau, a 130-m high RCC dam on the Da River in North-West Viet Nam, has been very comprehensively cored. During the first three Stages of coring - there are two further Stages still to come - a fraction under 450 m of core was extracted from the dam body. In all, 1 343 samples of core have been tested, 307 in compression, 340 in direct tension of the parent (unjointed) RCC, 677 in direct tension of cores containing joints and 164 for modulus in compression and Poisson's ratio. The paper describes the analysis of all these results and the relationships between the compressive and tensile strengths of manufactured cylinders and the same strengths of cores.

All the strengths conformed to the Specified requirements, indeed, apart from the direct tensile strengths of cylinders in the laboratory, they exceeded the requirements by some margin.

Further cores are to be extracted from the last two sections of Lai Chau as soon as the RCC has reached a suitable age and access is possible. Such comprehensive testing of the in-situ properties of Lai Chau will enable the Higher Authorities in Vietnam to be confident of the performance of the RCC in the dam body.

Key words: RCC, dams, construction, cores, in-situ properties

1 Lai Chau project

The Lai Chau Hydropower Project is located on the River Da in North West Viet Nam. It is the uppermost Project in a cascade of three dams (Lai Chau, Son La and Hoa Binh), all of which are owned by EVN (Electricity of Viet Nam). Together the three Projects have a combined installed capacity of 5 520 MW. The Lai Chau RCC dam is a 131-m high gravity structure with a total volume of 2.5 million m³ of which 1.884 million m³ is RCC. The installed capacity is 1 200 MW[1].

2 Rate of placement of the RCC at Lai Chau

2.1 Rates of placement

The RCC placement started at Lai Chau on 7 March 2013 and was completed on 2 May 2015, 1.884 million m³ of RCC having been placed in that time. However the placement was complicated by decisions made at the Feasibility Study stage and which would have been difficult and time consuming to change and there were a number of breaks in the RCC placement[2]. The overall average rate of placement was 72 900 m³/month, whereas with more appropriate planning for an RCC dam and with a reduction in the number of breaks in the placement, the average rate placement could have been over 90 000 m³/month. These rates of placement have been compared to the rates of some 400 RCC dams in Fig. 1. It can be seen that the overall average rate of placement is approximately halfway between the upper and lower rates of placement for the volume of RCC in Lai Chau. Nevertheless, the rate of placement when the unnecessary breaks are neglected, would have been right at the top of the range.

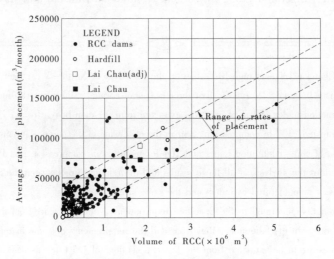

Fig. 1 Rates of placement of the RCC at Lai Chau

The actual placement can be seen in Fig. 2 which shows the dam on 22 August 2013 (some five and a half months after the start of a placement) compared to the completed dam on 5 May 2015, some four to six weeks ahead of schedule and less than two years later.

2.2 Treatment of the horizontal joints

Within the placement of the RCC there was very good planning and the number of hot joints was just under 98% of the total number of joints (Table 1). Any figure above 90% for the hot joints in an RCC dam is excellent and a figure approaching 98% is exceptional. By maintaining practically all the joins as hot joints, not only is the time lost due to joint treatment kept to a minimum, the quality of the joints is excellent as fresh RCC is placed on fresh RCC.

Table 1 Forms of joint treatment used in Lai Chau (from cores tested to date)

Items	Hot	Warm	Cold	Supper – cold	Total
No. of joints	1 053	11	4	9	1 077
Percentage	97.8	1.0	0.4	0.8	100.0

3 Location of cores

Details of the eight cores taken to date in the Central and Left – central Sections are shown in

(a)Placement of RCC at Lai Chau on 22 August 2013

(b)Completed Lai Chau dam on 5 May 2015

Fig. 2

Table 2. The cores can be located relative to the joints shown in Fig. 3.

Table 2　Locations and Elevations of the cores taken to date from the dam body at Lai Chau

Core	Distance left of Joint(m)		Distance from u/s face(m)	Elevation (m ASL)	Depth (m)
C2.1	0.8	14/15	23.75	206.1	36.0
C2.2	5.2	14/15	65.75	206.1	41.7
C2.3	6.5	14/15	105.75	206.1	38.4
C4.1	4.0	11/12	8.75	233.0	38.3
C5.1	2.0	13/14	8.75	264.6	84.5
C5.1 – Stage 1				[264.6]	[27.6]
C5.1 – Stage 2				[233.0]	[39.7]
C5.1 – Stage 3				[194.2]	[17.2]
C5.2	2.0	16/17	8/.75	264.6	72.0
C5.2 – Stage 1				[264.6]	[29.5]
C5.2 – Stage 2				[233.0]	[42.5]
L5B.1	10.0	10/11	8.75	301.8	83.0
L5B.2	4.0	9/10	8.75	298.2	58.0
TOTAL length of core					451.9

451.9 m of cores have been extracted to date. This is equivalent to some 1 506 layers. 1 343 specimens have been tested, i.e. 89.2% of the total length of cores. This is an extremely high percentage, given that the core barrels were only 1.5 m long and the frequent breaks that were therefore needed to extract the cores.

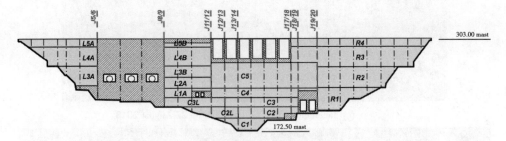

Fig. 3　Upstream Elevation of Lai Chau dam showing the location of the significant joints

4　Strength development of cylinders from Lai Chau

4.1　Introduction

The RCC at Lai Chau had mixture proportions of 60 kg/m^3 of Portland cement and 160 kg/m^3 of low-lime flyash, practically all sourced from ash ponds[3]. A retarder was used throughout to delay the Initial Setting Time to (21 ±3) hours.

Samples were taken from the RCC every 2 000 m^3 for testing in compression and every 10 000 m^3 for testing in direct tension. There are now 530 sets of manufactured specimens for which strengths up to an age of one year have been obtained and 778 for which 28-day results are available.

4.2　Development of cylinder compressive strength with age

The development of cylinder compressive strength with age at Lai Chau is shown in Fig. 4. When compared to the development of strength at Son La at which the RCC had similar mixture proportions (after making an allowance for the different shaped specimens, i.e. cubes at Son La and cylinders at Lai Chau), the strengths at Lai Chau are slightly lower than those at Son La at early ages (up to 56 days), are very similar between 91 and 365 days, but there are indications in the long term that Lai Chau could have slightly higher strengths. Essentially there is not a great difference.

4.3　Development of direct tensile strength with age

The development of cylinder direct tensile strength with age is shown in Fig. 5. The strengths are somewhat disappointing and lower than might have been expected (Section 6.2); nevertheless it is the vertical in-situ direct tensile strength across joints that is the most important criterion for the assessment of the performance of the dam (Section 5.5).

5　In-situ properties of the RCC at Lai Chau

5.1　Introduction

1 343 sections of core of RCC from Lai Chau were tested for strength (or modulus). Given

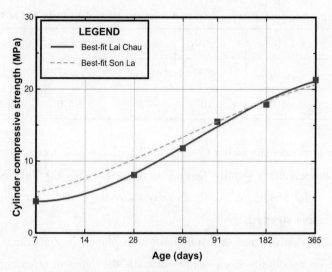

Fig. 4 Development of cylinder compressive strength with age of the RCC placed in the dam body at Lai Chau compared to that at Son La

Fig. 5 Development of direct tensile strength with age of the RCC placed in the dam body at Lai Chau

that 451.9 m of core (Table 2) have been extracted to date from the dam – equivalent to 1 506 300 – mm long specimens – this is a very high percentage. Effectively practically every length of core that could have been tested has been tested with minimal losses due to poor – quality RCC (or poor – quality coring). 340 of the specimens were tested in direct tension of the parent (un-jointed) RCC and 677 in direct tension of cores containing joints.

5.2 Density

The densities of the 1 343 sections of core have been tested for density and the results are summarized in Table 3.

Table 3 Densities of cores extracted from the RCC dam body at Lai Chau

No.	Average (kg/m³)	S. D. (kg/m³)	C. o. V (%)	Maximum (kg/m³)	Minimum (kg/m³)	Characteristic (kg/m³)	ICOLD rating
1 343	2 576	26.0	1.01	2 660	2 510	2 554	Good

The variability of the density with a Coefficient of Variation of 1.01%, is right on the boundary between good and excellent Quality Control as defined in ICOLD Bulletin No. 126[4]. The characteristic density of 2 554 kg/m³ is significantly above the 2500 kg/m³ required.

5.3 Core compressive strength

All 307 core compressive strength results have been plotted in Fig. 6 and the best – fit relationship between those results has been compared to the development of strength of the cylinders manufactured on site (Fig. 4). As can be seen the two relationships are parallel but the cores have a lower strength than the cylinders, with a difference of circa 4 MPa. This is somewhat different from the situation at Son La where the core strength was fractionally higher than the equivalent cylinder strength.

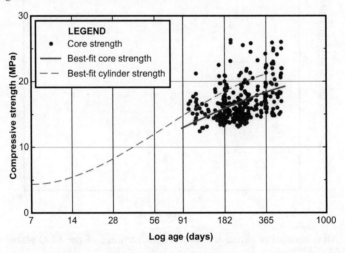

Fig. 6 Development of core compressive strength with age of the RCC placed in the dam body at Lai Chau compared to the cylinder compressive strength

5.4 Vertical direct tensile strength of cores containing unjointed RCC

All 340 direct tensile strength results of unjointed cores have been plotted in Fig. 7 and the best – fit relationship between those results has been compared to the development of tensile strength of the cylinders manufactured on site (Fig. 5). The two relationships are somewhat different with the tensile strengths of the cylinders increasing at a reasonable rate with age, whereas the tensile strengths of the unjointed cores seems to be approaching a maximum value at an age of circa 365 days. At that age the cylinders (which probably have somewhat lower strengths than the actual values due to some difficulties with the direct tensile testing in the site laboratory (Section 6.2) - the cores were tested in PECC1's main laboratory in Hanoi) and cores seem to have similar direct tensile strengths.

5.5 Vertical direct tensile strength of cores containing horizontal joints between layers

The review of the in-situ vertical direct tensile strength of cores containing joints is not straightforward because there are two variables, first the age of test and secondly the MMF (Modified Majority Factor). MD&A have therefore developed the concept of three-dimensional models relating the direct tensile strength to the age of test (on a log scale) and the MMF[5].

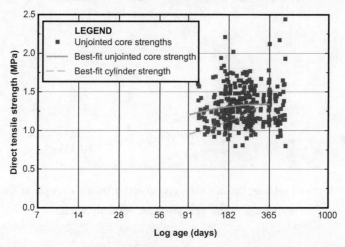

Fig. 7 Development of direct tensile strength of unjointed cores with age of the RCC placed in the dam body at Lai Chau compared to the direct tensile strength of manufactured specimens

There is a preponderance of hot joints in the dam body at Lai Chau (Section 2.2); this is very beneficial for good performance at the horizontal joints. Fig. 8 and Fig. 9 show the 3-dimensional model created from the results of the direct tensile testing of cores containing hot joints. The majority of the results are at ages less than the 365-day design age. It is clear that the model is very flat', with little increase in strength with age (c.f. the direct tensile strength of the unjointed specimens - see Fig. 7) and surprisingly little decrease in strength as the MMF increases. At the limiting MMF (for hot joints) of 800 mod. deg C. hr, the model is still at a level in excess of the design strength of 0.94 MPa (Section 6.1).

There are very few joints in Lai Chau other than hot joints. Those that have been tested are shown in Table 4.

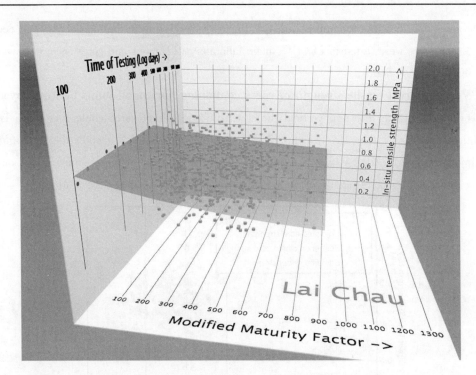

Fig. 8　3D model relating the in-situ vertical direct tensile strength at Lai Chau to the age of test and Modified Maturity Factor

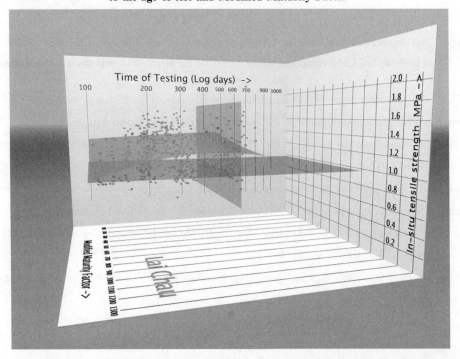

Fig. 9　3D model relating the in-situ vertical direct tensile strength at Lai Chau to the age of test and Modified Maturity Factor (with design age and design strength shown)

Table 4 In – situ vertical direct tensile strength of super – cold joints at Lai Chau

Joint	Block	Core	MMF (mod. deg C. hr)	Age of test (days)	Log age (log days)	Tensile strength (MPa)
Super – cold joints						
67/66	C5	C5.2	4 390	158	2.198 66	1.19
1/0	C1A	C2.2	5 317	145	2.161 37	1.07
1/0	C1A	C2.3	5 317	139	2.143 01	1.61
34/33	C3L	L5B.1	5 439	438	2.641 47	1.40
28/27	C2L	C4.1	13 972	338	2.528 92	1.39
1/0	C3	C5.2	66 958	289	2.460 90	1.24
Average			16 899	251		1.32
S.D.						0.190
C.o.V						14.5%
Characteristic						1.15

There are six results for super – cold joints in the Table, i.e. two thirds of all the super – cold joints in the dam to date (Table 2). The average strength of the six results is 1.32 MPa; although the average age of 251 days is less than the design age, the strength at the design age might not be that much greater, nevertheless it is a more than satisfactory strength and essentially equal to the strength of the unjointed cores.

During the coring of the Left and Right Sections of the dam, efforts are be made to try to retrieve cores containing the very few warm joints, so that the strengths of this form of joint can also be reviewed.

6 Comparison between the actual results achieved at Lai Chau with those that were specified

6.1 Design and characteristic strengths

For mass concrete, such as the RCC at Lai Chau, an allowable failure rate of 20% has been defined in accordance with normal practice. The margin (i.e. the difference between the characteristic strength and the design strength) can be calculated from the Standard Deviation (and thus from the Coefficient of Variation) and the allowable failure rate. The design strengths (i.e. the strength that has to be exceeded by the average of all the results) and the characteristic strengths for each of the properties required at Lai Chau are shown in Table 5.

Table 5 The characteristic and design strengths for the various strength properties at Lai Chau

No.	Property	Characteristic strength (MPa)	Coefficient of Variation (%)	Standard Deviation (MPa)	Design strength (MPa)
1	Cylinder compressive strength	16.5	10	1.65	17.9
2	Direct tensile strength of cylinders	1.30	15	0.195	1.47
3	Core compressive strength	14.7	12.5	1.84	16.3
4	Direct tensile strength of unjointed cores	1.10	15	0.165	1.24
5	Direct tensile strength of jointed cores	0.80	20	0.160	0.94

(N. B. the direct tensile strength of jointed cores is the only real design criterion for Lai Chau (and practically all RCC dams); all the other properties are effectively 'equivalent' strengths).

6.2 Comparison between the actual results and the design strengths

The average results of the in-situ tests on the RCC taken from the dam body at Lai Chau at the design age of 365 days have been compared to the design strengths at the same age in Table 6.

Table 6 Comparison between the actual results and the design strengths

	Property	Design strength (MPa)	see	Actual average strengths (MPa)	Exceedance
1.	Cylinder compressive strength	17.9	Fig. 4	21.3	19%
2.	Direct tensile strength of cylinders	1.47	Fig. 5	1.39	(see below)
3.	Core compressive strength	16.3	Fig. 6	18.2	12%
4.	Direct tensile strength of unjointed cores	1.24	Fig. 7	1.35	9%
5.	Direct tensile strength of jointed cores				
	Hot joints (@ MMF of 600)	0.94	Fig. 8 & 9	1.28	36%
	Hot joints (@ MMF of 800)	0.94	Fig. 8 & 9	1.23	31%
	Super-cold joints (@ 250 days)	0.94	Table 4	1.32	40%

There were a number of concerns expressed regarding the testing of cylinders in direct tension in the Lai Chau site laboratory (Section 4.3). On the basis of the cylinder compressive strengths and the Trial Mix Programmes undertaken for the dam at the main PECC1 laboratory in Hanoi (where the cores were tested in direct tension) it might be expected that the average direct tensile strength of the cylinders at an age of 365 days should be 1.64 MPa, an exceedance of 12%.

It can be seen in Table 6 that (apart from the direct tensile strength of the cylinders), the actual average strengths exceed the design strengths by some margin. In particular for the most important property, that is the direct tensile strength across the horizontal joints, the actual values exceed the design strength by some 30 to 40%.

7 Conclusions

A very economic and well – performing RCC mixture has been designed for the Lai Chau RCC dam, which comfortably met all the design strength criteria established for the dam. A reduction of the Portland cement content below the selected 60 kg/m^3 would have been possible, but this would have resulted in the lowest cement content used in any concrete dam so far constructed in Viet Nam. The Higher Authorities therefore decreed that the cement content should be 60 kg/m^3. The testing data presented include the results from RCC specimens containing two different cements, But Son PC40 and Yen Binh PC40, which both had extensively been tested during the RCC Trial Mix Programmes as well as in the Full Scale Trial Embankments. Due to this testing, the change of cement source for the remaining about 40% of the RCC volume could have comfortably been executed without any other change to the RCC mixture proportions, apart from a minimal adjustment of the retarder dosage, and more importantly without any programme delay. This shows that the efforts and cost associated with an expanded RCC Trial Mix Programme, and with the anticipation that a change in materials may be executed during the RCC dam construction, are justified, since the acquisition of knowledge in advance avoids possible delays or quality problems during construction.

The results presented in this paper clearly show that a well – designed and workable RCC mix, along with the specification of simple RCC placement methodologies, contribute to a rapid RCC placement rate, thus maximizing the proportion of hot joints in the dam body and particularly leading to very good in – situ parent and lift – joint properties. It also enables the Owner of the dam, in this particular case EVN, to have an early return on their investment.

Acknowledgements

The authors would like to thank Electricity of Viet Nam, the Son La Management Board for Lai Chau and PECC1 for permission to publish this paper. Regrettably one of the authors Mr Nguyen Hong Ha, Vice Director of EVN, passed away during the preparation of this paper and it is thanks to his untiring leadership and drive that both Son La and Lai Chau RCC dams were completed ahead of schedule and to a high standard.

References

[1] N. H. Ha, M. Conrad, D. Morris, P. T. Hoai. The construction and Quality Control of Lai Chau Dam[C] // Proceedings of 7th International Symposium on Roller – Compacted Concrete (RCC) Dams. Chengdu, China, September 2015.

[2] N. H. Ha, M. Conrad, D. Morris, M. R. H. Dunstan. Speed of construction of EVN's Son La and Lai Chau RCC dams[C] // Proceedings of 7th International Symposium on Roller – Compacted Concrete (RCC) Dams. Chengdu, China, September 2015.

[3] M. R. H. Dunstan, N. H. Ha and D. Morris. The major use of lagoon ash in RCC dams[C] // Proceedings of 7th International Symposium on Roller – Compacted Concrete (RCC) Dams. Chengdu, China, September 2015.

[4] ICOLD. Roller – compacted concrete dams – State of the art and case histories, Bulletin No. 126. ICOLD, Paris, 2003.

[5] M. R. H. Dunstan, M. Conrad. The relationships between the in – situ tensile strength across the joints of an RCC dam and the Modified Maturity Factor and age of test[C] // Proceedings of 7th International Symposium on Roller – Compacted Concrete (RCC) Dams, Chengdu, China, September 2015.

Stress Meter and Strain Gauge Measurements at Upper Paunglaung RCC Dam

U Maw Thar Htwe[1], U San Wai[1], Christof Rohrer[2], Stuart L. L. Cowie[2]

(1. Ministry of Electric Power, DHPI, Myanmar; 2. AF – Consult Switzerland Ltd., Switzerland)

Abstract: The paper gives a brief overview of the layout of the Project and the RCC Dam and concentrates on the installation, recording and study of the results of the strain gauges and stress meters which had been installed during the construction of the RCC dam.

First the idea of introducing of strain gauge and stress meter instruments will be explained. The data are very useful to get a better picture of early temperatures, deformations and stresses of the RCC. The experience gained is very important for the understanding of the thermal stresses and strain capacity of the concrete. These recorded data help to improve the thermal analysis of RCC dams and to take early measures at the construction site to prevent thermal surface cracking.

For mid and long term, the recorded data show tendencies over time for temperature, deformation and stress changes. Effects due to reservoir impounding and cold winter or hot summer periods could be observed as well as effects of additional heat sources such as construction of the CVC Spillway Chute, which was built on top of the RCC downstream face of the Upper Paunglaung Dam.

The paper presents all relevant results and interpretation and try to give an preliminary conclusion for the thermal behaviour of the RCC in Upper Paunglaung.

Key words: RCC Dam, Strain Gauges, Stress Meters

1 Project overview

The Upper Paunglaung Hydropower Project is located on the Paunglaung River, about 50km east of the Capital NayPyi Taw, in Mandalay Division, Myanmar. The Project is owned by the Ministry of Electric Power with the main purpose of power production for the national grid. The 140 MW Upper Paunglaung Hydropower Project comprises a RCC dam of gravity type with a height of 103 m and with a crest length of 530 m. The Gravity dam is divided into 27 concrete blocks of maximum 20 m width, and contains a total concrete volume of 950 000 m^3. The Upper Paunglaung dam is the second RCC dam project in Myanmar and the first built by local contractors. The RCC dam construction has been completed end of 2013, and the plant installation has been completed by end of 2014.

2 Introduction

InMyanmar no fly ash is available. During the planning of the first RCC dam in Myanmar, the Yeywa HPP, an alternative source for cementitious material – a natural pozzolan – had been found at Mount Popa, a volcano in central Myanmar. The Upper Paunglaung RCC is also using the same natural pozzolan in the RCC mix and as natural pozzolan is of almost unlimited amount available, future dam projects and other concrete works will utilise this local and relatively cheap material as well.

A RCC mix using a local and unique naturalpozzolan as cementitious material may behave different from a RCC mix with fly ash and not so many RCC dams worldwide were built with natural pozzolan so far.

Therefore, it is of upmost interest to gain knowledge about the effective RCC properties such as E – modulus, strain capacity and thermal expansion coefficient, as well as the maximum tensile stress occurred at different locations in the dam body and surfaces of the Upper Paunglaung RCC dam.

During early stage ofRCC dam construction it has been decided to install additional instruments namely Stress Meters and Strain Gauges in one of the RCC stages. The detailed information of the RCC short and long – term behaviour gained in this way should help to understand the RCC behaviour using natural pozzolan in the mix.

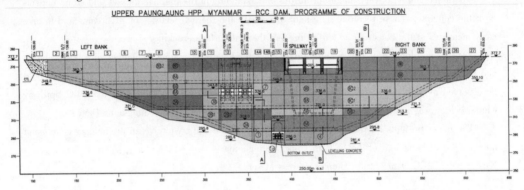

Fig. 1 Upper Paunglaung RCC dam with stages and location of Instruments

It is furthermore the intention to compare the measured effective temperatures, deformations and stress values of an actual RCC dam project under construction with the performed FE thermal analysis and to use the measurement for the finite element model calibration. The gained measurement data shall help for future RCC dam designs and especially FE thermal analyses to consider the specific behaviour and properties of a RCC with natural pozzolan from Mount Popa.

2 Instrumentation Type and Installation Arrangement

The stress meters installed at Upper Paunglaung RCC dam are the Geokon concrete stress meter Model 4370, which are designed to measure tensile and compressive stresses in mass concrete. The concrete stress meter comprises a short vibrating wire load cell in series with a concrete cylinder. The vibrating wire load cell measures the load imposed on the inner concrete cylinder by stresses in the surrounding concrete.

The strain gauges installed at Upper Paunglaung RCC dam are the Geokon deformation meter

Model 4430, which is designed to measure axial strains or deformations in concrete.

Strain gauges have been installed at RCC Stage 9A at two Elevations (EL. 322.50m and EL. 331.50m) of two RCC Blocks (Block 16 and Block 18), and at RCC Stage 9C at two Elevations (EL. 322.50m and EL. 331.50m) of Block 20.

For each Block and elevation, the strain gauges have been installed near the upstream face in both horizontal directions, near the downstream face in both horizontal directions, in the centre of the block in both horizontal directions and across one block joint. The instrumentation layout follows a proposal made in [1].

Stress Meters have only been installed in Block 20, in pairs, but with similar arrangement as for the Strain Gauges.

In Block 16, Block 18 and Block 20 totally 44 strain gauges and 24 stress meters have been installed (see Fig. 2). All strain gauges measure the deformation of the RCC within the instrument length and the RCC Temperature. All stress meters measure stress of the RCC in the load cell and also the RCC Temperature.

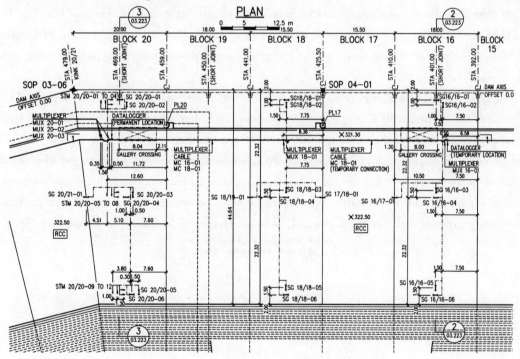

Fig. 2 The layout of the stress meter and strain gauge instruments at elevation 322.5 m

3 Installation of instruments

All instruments have been installed after placing and compacting of the designated RCC layer, by digging trenches for the cable routing and embedding of the instruments, or by using formwork box out templates. The strain gauges just can be laid down in the trench and backfilled with RCC. The Stress Meter steel pipes have to be filled first with RCC and compacted, then laid down into the trench and properly connected with the surrounding RCC, by backfilling with RCC.

As the strain gauges are long and thin, it has tobe made sure that the strain gauges are not bending during placing and compacting of the upper RCC layers. Therefore, it shall not be allowed to cross these areas with heavy equipment for the next two to three layers. Compaction of the RCC should preferable be done only with small roller or vibrating plate.

4 Time elapsed Results

The time elapsed measurements are presented in the following subchapters. The measurement time steps of all instruments were 20 minutes for the first 6 months, the 1 hour for the next 12 months and then fixed to 2 hours.

4.1 Temperature development

The quite detailed and accurate temperature readings allowed plotting early concrete temperature development curves of the first days after placing. Fig. 3 below shows the actual RCC temperatures at different locations of Block 16 EL. 322.50, influenced by heat of hydration, day or night time as ambient temperatures and solar radiation effects, as well as of effects of covering the RCC by the following layer with placing temperature of 20 ℃. It can be seen that the RCC layer 5 actually cools down during early heat of hydration period due to night time and covering with cold concrete.

Fig. 4 shows the long term temperature development of several instruments at Block 18 EL. 331.5m. The maximum RCC temperature in the core of the dam was recorded as well as the temperature developments at the faces and influences of construction of the Spillway or cold ambient temperatures could be caught. The maximum RCC temperature recorded gives a good idea about adiabatic temperature rise, but not solely because the recorded temperature includes also temperature effects of ambient temperatures and solar radiation. The cooling rate of the RCC in the dam body after maximum temperature is reached, is in average 0.35 ℃/month as recorded in Fig. 4.

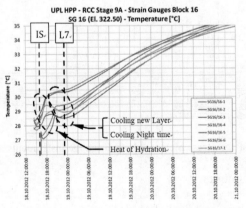

Fig. 3　Early age RCC temp. development

The recorded temperature increase due tospillway construction at the downstream face is ap-

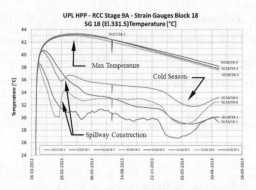

Fig. 4　Long-term RCC temp. development

prox. 6 ℃ for location 2 m away from face and 2.5 ℃ for location 4.5 m away from the surface. This resulted in a recorded expansion of 0.2 mm and 0.1mm respectively. This information establishes useful input for future FE Analyses to study the thermal and structural influence depth into the RCC.

4.2 Deformation development

The deformation developmentmeasured with the strain gauges in shown in Fig. 5 and Fig. 6. Fig. 5 presents the deformations (in mm) at Block 18 EL. 322.5, at several locations of the horizontal section. Fig. 6 shows the deformation at Block 20, EL. 322.5, at similar location of the horizontal section.

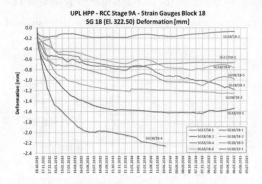
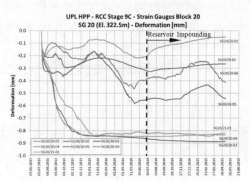

Fig. 5　Deformation development Block 18　　**Fig. 6　Deformation development Block 20**

The comparison of deformations at block 18 and block 20 shows that the deformations at block 18 are much higher than at block 20. By studying the deformation data, it has been found out that various strain gauges had been overstressed after installation and covering with RCC, or bent due to heavy equipment passing over it. Therefore, the strain gauges of block 18 can be used only indicatively. The deformation at block 20 show very nicely the behaviour of the RCC at upstream face with SG20/20 – 01 and SG20/20 – 02, at the downstream face with SG20/20 – 05 and SG20/20 – 06, influenced by the ambient conditions, and at the core with SG20/20 – 03 and SG20/20 – 04 and at core block joint SG20/21 – 01.

The interesting result at Block 20 is to see that in the core the deformation do not increase (expansion or release of contraction) again although the concrete already started to cool down, similar as shown in Fig. 4. However, the compressive stresses in the core decreased.

The impact on the RCC deformation of the reservoir impoundingwas recorded at the instruments of the upstream face in both directions, parallel and perpendicular to the dam axis.

The recorded deformationswere compared with the design deformations of the FE Analysis. The range of deformations was the same, but the FE model could not include all the detailed deformation developments as presented in Figure 6, due to the application of averaged inputs in the design analyses.

4.3 Stress development

The stress development measured with the stress metersis shown in Fig. 8, and corresponding RCC Temperatures are given in Fig. 7. Fig. 8 presents the stresses (in MPa) over time at various locations of Block 20 at EL. 331.5m. A remarkable record occurred one month after RCC placing as the temperatures and the stresses (SM20/20 – 05 to SM20/20 – 08) dropped in the core of the dam due to placement stop after RCC Stage 9C1. The impact could be recorded so clearly because the concrete cover from the instruments to the top of stage is only 4.5 m.

The similarity of the temperature development and stress development can be observed by comparing both Figures. The temperature curves close to the concrete face are similar to the stress curves (SM20/20 – 09 to SM20/20 – 12). However, the stress readings do not confirm the deformation readings.

The compressive stress increase in the core, and decrease at the surface due to reservoir impounding can be observed. The stress increase in the core due to impounding is in the range of 0.4 MPa in flow direction.

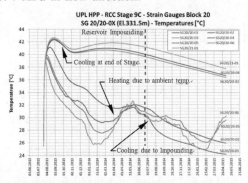

Fig. 7 Stress Meter Temperatures at Block 20

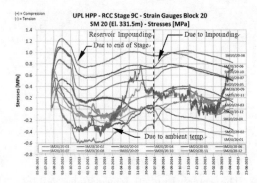

Fig. 8 Stress development at Block 20

5 Temperature elapsed results

The gained temperature, deformation and stress data are further analysed by plotting them in other relations, for example in dependence of temperatures.

5.1 Temperature vs. deformation

Deformation developmentis plotted against temperature development to find relations such as linearity.

Fig. 9 shows the deformation – temperature hysteresis of the RCC core of block 20,

EL. 322. 5m and EL. 331. 5m. It can be observed that during initial heating, the RCC deforms linear and of same rate for all measurement locations in the core (phase 1). During cooling (phase 3), the rates are of similar range as well. In between these two phases of high heating and cooling rates, a non-thermal deformation component can be observed (phase 2), which is formed due to static loads, creep or shrinkage and restrain effects. But also the thermal deformation component can be still recognised, for example the impact of the earlier described cooling effect on top of the Stage 9C.

Fig. 10 shows the deformation – temperature hysteresis of the RCC near the surfaces of block 20, EL. 322. 5m and EL. 331. 5 m.

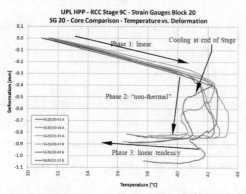 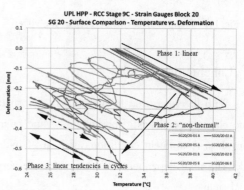

Fig. 9   Temp – Def hysteresis in the core   Fig. 10   Temp – Def hysteresis at the surface

Similar cooling and heating rates as in the core can still be recognised (phase 1), but the cycles of cooling and heating at the surface are faster and of many more numbers. This is mainly influenced by the daily and seasonal ambient conditions (phase 3). Similar non-thermal deformations in between these cycles are observed again near the surface of the RCC (phase 2), and interestingly of similar extend. The figures also show that at the core, the RCC has not yet cooled down much and the deformation are in contraction, whereas at the surface, the RCC has cooled down completely and the deformation already changed from contraction to expansion.

5.2   Temperature vs. stresses

Stress development is plotted against temperature development to identify if a clear relation can be observed.

Fig. 11 presents the stress – temperature hysteresis of the RCC in the core and Fig. 12 the stress – temperature hysteresis at the surface. Similar as for the deformation hysteresis, during the initial heating of the RCC, the stresses in the core and near surface increase according to a distinct relation (phase 1). In the core (Fig. 9) a defined relation during the following cooling phase was recorded at almost all measurement points (phase 3). At the surface the tendency and shape of the following cycles can be interpreted as similar, but many cycles of heating and cooling, especially daily ambient temperature cycles complicates the figure of hysteresis.

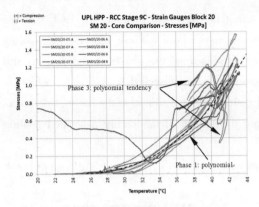 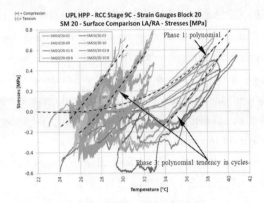

Fig. 11 Temp – Stress hysteresis in the core   Fig. 12 Temp – Def hysteresis at the surface

5 Conclusion

With the measured temperatures, deformations and stresses at Upper Paunglaung RCC dam time elapsed graphscould be generated and compared with each other. Influences due to construction phases, ambient conditions or heat of hydration of the RCC could be identified and presented. The time dependent influences can be very well used for thermal design of future RCC dams in Myanmar.

Fig. 13 Completed RCC Dam at Upper Paunglaung, Myanmar

The deformation and stress hysteresis at core and surface give a good picture of thermal and non – thermal behaviour of the RCC at Upper Paunglaung. The linear rates and trend lines of the thermal deformation as well as the quantity of non – thermal components give good picture for future dam Design, as well as the polynomial rates of the thermal stresses.

The next important steps will be the comparison with other RCC dams using natural pozzolan and fly ash to find similarities and differences, and the implementation of the above presented data in the thermal FE analysis. But those steps cannot be covered by the scope of this paper and further analysis of the data is required due to the complexity of influences.

It can be concluded that the installation of the additional instruments gave very useful and detailed results and a lot of knowledge could already be gained, even the analysis of the results is not yet completed.

Acknowledgements

Gratitude must be accorded to MOEP (Ministry of Electric Power) and AF – Consult Switzerland Ltd. for permission to publish this paper, but the views expressed in this paperare purely those of the authors. Many thanks are also due to the DHPI (Department of Hydropower Implementation) engineers on the construction site who made this project to a success and to our RCC Expert M. Dunstan.

References

[1] M. Conrad, Q. H. W. Shaw, M. H. R. Dunstan. Proposing of standardized approach to stress – strain instrumentation for RCC dams[C] // Proceedings of the 6 th International Symposium on Roller Compacted Concrete (RC) Dams, Zaragoza, 2012.

[2] Ch. Rohrer, et al.. Design and Construction Aspects of the Upper Paunglaung RCC Dam, Paper of ASIA 2012 Conference – Water Resources and Renewable Energy Development in Asia, of Hydropower and Dams, 2012.

[3] Ch. Rohrer, et al.. Experience made with retarder admixtures at two RCC dams in Myanmar, Paper of 6 th International Symposium on Roller Compacted Concrete (RCC) Dams, Zaragoza, 2012.

[4] Ch. Rohrer, et al.. Completion of the Upper Paunglaung RCC Dam, Paper of ASIA 2014 Conference – Water Resources and Hydropower Development in Asia, of Hydropower and Dams, 2014.

The Revival of Dam Building in Afghanistan: Ministry of Energy and Water Islamic Republic of Afghanistan

Sayed Karim Qarloq

(Ministry of Energy and Water, Afghanistan)

Abstract: Main objective:

The main objective of this presentation is to describe dam history and the role of dam construction in the water resources and economy of Afghanistan.

Sub objectives:

● Describe history of dam construction in Afghanistan.

● Brief description about Sha Wa Arus RCC Dam project.

1 Introduction

● Afghanistan Isa Landlocked country. Located in central Asia, covering an area of 647 500 square kilometers (251 772 square miles).

● It has a total population of about 30 million.

● It has Only 12% of its land arable.

● It has 17% of itsland occupied by river valleys such as Harirud, Helmand and Kabul.

● Its bulk of water flows into neighboring country.

● Dams are critical in preserving and developing the water resources of the country.

2 Water resources in Afghanistan

● Annual water resource volume at 95 billion m^3, consisting of approximately 88% (84 billion m^3) surface water and 12% (11 billion m^3) groundwater.

● Ministry of Energy and Water leads the development and management of water resources in Afghanistan.

● Dams are critical to the development of Afghanistan.

3 Existing dams

● Dams has been an integral part of the water resources infrastructure in Afghanistan since 1918. As shown in the following slide, many dams were built in the 20 th century.

● Due to the population it serves, the Kabul River is one of the most important rivers in the country and a number of dams have been constructed on it.

- Naghlo concrete gravity dam, 110 m high, is the highest dam and generates 97 MW of electricity.
- Ninety-one meter high Kajaki is the highest embankment dam with a command area of 176 000 hectares.

4 Dams under construction (table 1)

Table 1 Dams under Consfracfion

Project	Sha-Wa-Arus	Pashdan	Machalgho	Almar	Kamal Khan	Salma
River	Shakardara	Karukh	Machalgho	Almar	Hilmand	Harirud
Dam Type	RCC Dam	Rock Fill	Rock Fill	Concrete gravity	Clay Core	Rockfill
Height (m)	77.5	49	55	68	17.5	107
Power (MW)	1	2	0.4	0	9	42
Contract Value (US $)	$ 48 M	$ 117 M	$ 31 M	$ 51 M	$ 85 M	$ 180 M

5 Dams under design and feasibility study (table 2)

Table 2 Dams under design and feasibility study

Dam Project	Purpose	Irrigated Area (ha)	Power (MW)	River Basin	Status
Kelagai	Irrig + Power	92 000	60	Amu	Preliminary Design
Upper Kokcha	Irrig + Power	132 000	45	Amu	Feasibility/Preliminary Design
Lower Kokcha	Irrig + Power	6 000	445	Amu	Feasibility/Preliminary Design
Warsaj	Irrig + Power	40 000	40	Amu	Feasibility/Preliminary Design
Qala-i-Momai	Hydropower	N/A	420	Amu	Feasibility/Preliminary Design
Hasan Tal	Irrigation	8 970	N/A		Feasibility/Preliminary Design
Dagh Dara	Hydropower	N/A	120	Kabul	Feasibility/Preliminary Design
Gamberi	Irrig + Power	35 000	50	Kabul	Detail Design
Kama	Irrig + Power	18 000	43	Kabul	Detail Design
Gul Bahar/Panjsher	Irrig + Power + Drinking	70 000	120	Kabul	Detail Design
Shahtoot	Irrig + Drinking	1 966	1.2	Kabul	Detail Design
Khesht Pul	Irrigation	2 860	N/A	Kabul	Feasibility/Preliminary Design
Sultan Ibrahim	Irrigation	7 800	N/A	Kabul	Feasibility/Preliminary Design

Table 2

Dam Project	Purpose	Irrigated Area(ha)	Power(MW)	River Basin	Status
Cheshmah – e – Shefa	Irrig + Power	200 000	180	Northern	Feasibility/Preliminary Design
Upper Amu	Irrig + Power	605 300	1000	Amu	Feasibility/Preliminary Design
Bakhshabad	Irrig + Power	138 605	27	Helmand	Detail Design
Kuner Shall & Shegai	Hydropower	N/A	1098	Kabul	Feasibility/Preliminary Design
Agha Jan (Uruzgan)	Irrig + Power			Helmand	Feasibility/Preliminary Design
Manogayee	Irrig + Power			Kabul	Feasibility/Preliminary Design

6　Sha Wa Arus RCC Dam

This project situated at the north side about 22 km far from Kabul.

6.1　General information

- Contract signed: May 8, 2010.
- Price: 48 149 262 $.
- Duration: 55 Month.

6.2　Purpose

- Irrigation: 2 700 hac.
- Power generation: 1.5 MW – turbin 2unit.
- Potable water: 5 milion cum/year.
- Flood Control.

6.3　Specification

- Type of the dam: Roller Compacted Concrete.
- Height of the dam: 77.5 m.
- Lent of the dam crest: 303 m.
- Wide of the dam foundation: 69 m.
- Crest Wide: 5 m.
- Catchment area: 97 km^2.

- Catchment lent: 15 km.
- Concrete volume: 330 000 cum.
- Storage capacity: 9.5 milion cum.
- Foundation excavation volume: 80 000 cum.
- Diversion system: precast box with inlet and outlet structure. $L = 160$ m.
- Foundation consolidation grouting: 16 500 m.
- Curtain grouting: 10 000 m until now 34% done.
- Concrete until now: 171 144 cum.
- Road around the Dam: 5.29 km.
- Permanent building area: 800 m^2.

6.4 Total prospect concrete volume (table 3)

Table 3 Total prospect concrete volume

Structure name	Concrete quantity (cum)
Diversion culvert	5 000
Spillway tailrace canal and other minor canals	2 000
Dam body	330 000
Total concrete quantity	337 000

All of the used materials aggregate, cement water and additional requirement determined based on the test and standard code recommendation!

Aggregate – the available rockquarry and river bed aggregate was suitable for concreting works.

Water – the quality of water in the river and spring in upstreamof the dam axis was acceptable for concrete mixing, aggregate washing and curing.

Cement – cement with low hydration heat (according ASTM C 150 type 2 or pozolanic material) used.

The table 4 shows concretes mix design that used to the dam body.

6.5 Problem

Unfortunately some cracks appeared on the conventional concrete at the both u/s and d/s dam face and around all openings, after a few days of the concreteplacement, the investigation performed by local expert but not got satisfactory result final Mr. Timothy P. Dolen, P. E. Civil Engineer Dolen and Associated from American invited to more investigation and professionally recommendation so the following is his inspection and recommendation:

Table 4 Concretes mix design that used to the dam body

Class	Description	Age	Cement content	W/C	Aggregate					Admixture			Air content
					0–5mm	5–50mm	0–50N·mm	5–19mm	19–38mm	Fluid AS105	Air G100		
		day		—	(kg/m³)	(kg/m³)	(kg/m³)	(kg/m³)	(kg/m³)	(%)	(%)		(%)
50/12	RCC	180	150	0.8	1085	1090	1328	437	437	—	—		1–2
50/30	Dam face CVC	180	350	0.45	917	930	—	—	—	0.7	0.1		3–5
50/25	Openings CVC	180	325	0.45	940	960	—	—	—	0.7	0.1		3–5
50/20	Abutments CVC	180	280	0.45	983	998	—	—	—	0.7	0.1		3–5
50/15	Filling CVC	90	230	0.5	1 020	1 034	—	—	—	0.7	—		1–2
19/25	Precast CVC	180	350	0.45	1 003	—	—	833	—	0.7	0.1		3–5
19/30	Spillway	180	400	0.4	965	—	—	800	—	0.7	0.1		3–5

Cracking was observed in conventional vibrated concrete (CVC) on the upstream (U/S) and downstream (D/S) dam are spaced from 3 to 10 m intervals between induced contraction joints.

Several continuous, vertically orientated cracks were visible on the U/S face of the dam and one continuous crack from bottom to top of the D/S face. Similar cracking is also observed in the walls of the gallery.

a. Likely causes of cracking include (1) early age drying shrinkage due to insufficient curing combined with (2) thermal induced shrinkage from high interior – exteriorconcrete temperature gradients in the CVC.

b. Reported lack of curing can increase amount of and rate of surface cracking of CVC facing. High cement content of CVC increases thermal cracking potential and couldpropagate into CVC facing and RCC.

c. The 2 m zone of upstream CVC is wider than typical RCC dams, and increases cracking potential, especially with the apparent lack of continuous curing andrelatively high (375 kg/m^3) cement content of the mixture.

d. Observed cracks were generally about 1 to 3 mm wide at the surface and many propagate through at least the CVC facing. Widening of cracks over time isconsistent with ongoing thermal contraction of the dam.

e. It is not apparent if these cracks have propagated completely through the dam cross section from either the U/S to the D/S face at this time. A detailed crack mapshould be developed to note specific locations of cracks focusing on possible continuity between the upstream face, gallery, and downstream face of the dam.

f. Quality control (QC) tests for CVC indicate the compressive strength is near the specified (design strength plus 5 MPa) strength for Class 35 MPa concrete. Thecement content of the CVC cannot be reduced without affecting the strength necessary for freezing and thawing durability.

g. The cement content of CVC facing concrete could be reduced if measures are taken to decrease the mixture water content and proportionately decrease the cement content according to the specified maximum 0.45 W/C ratio; such as by using a super plasticizer and/or decreasing the sand content of the mixtures.

h. Several vertically oriented cracks are also noted in the walls of the gallery, many of which have seepage. The source of water in the gallery cracks is unclear. Water could be entering the dam from water impounded upstream (1) through above noted vertical cracks, (2) through contraction joints (passing around water stop), (3) through horizontal cracks or un – bonded horizontal lift lines originating from the upstream face of dam, (4) from top of dam surface ponding or snowmelt, and (5) from foundation seepage or the foundation drilling and grouting operations.

i. The path of water could be through both discrete cracks or through a combination of discrete cracks in CVC entering into cold joints or zones of poorly compacted RCC.

Induced Contraction Joints

a. Cracks are also opening in the purposely induced, vertical contraction joints in both the CVC facing and beginning to follow into the RCC as designed.

b. Construction procedures for embedded crack inducers, water stop, and drains are not of satisfactory quality due to movement of vertical PVC sheeting used to induce vertical cracks and misalignment of the plastic sheeting, water stops, and/or drains.

c. Some induced cracks observed in the U/S, CVC facing appear to branch around the waterstop. This defeats the original intent of centering induced cracks between water stop and drains, increasing the potential higher flows of crack leakage with higher head of water in the reservoir.

d. Improving the design and construction practices for crack inducing features of contraction joints is recommended. Use of stiff plastic sheeting or galvanize sheet steel, with an external support system has been found successful in other RCC dams.

QC Testing and Reporting

a. There is a lack of consistent documentation of QC testing in the Monthly Progress Reports. Better record keeping is necessary to document all QC testing and follow the placement of CVC and RCC in the dam on a daily, weekly, and monthly basis.

b. Record of lift by lift placement progress is inconsistent and/or not available.

Analysis of The Running State of Guizhou Guangzhao Roller Compacted Concrete Gravity Dam

Yang Ning'an

(Guizhou Qianyuan power Limited by Share Ltd, Guiyang, 550002, China)

Abstract: The dam of the Guangzhou hydropower station is a full section RCC gravity dam, The maximum dam height of 200.5 m, crest length of 410 m, is currently the world has been built, in the construction of the highest RCC gravity dam. In order to understand and grasp the Guangzhao hydropower station RCC gravity dam running state, in the buried at different positions of the dam deformation monitoring, stress and strain monitoring, seepage pressure monitoring, such as monitoring instruments, monitoring the running state of the dam. This article through the Guangzhao of RCC gravity dam deformation, stress and strain, seepage in the early monitoring data analysis of osmotic pressure and joint operation, the level of the dam deformation and subsidence deformation is very small, much smaller than design value, the stress and strain monitoring show that the stress and strain measurement value is small, the stability of the dam. Dam section between the transverse crack in open state as a whole, have the effect of the induced slit dam design, the effect is better. Dam and on both sides of the slope rock and alveolar concrete and bedrock joint in good condition, contact state stability. In front of the dam before the main drain uplift pressure intensity coefficient between 0.042 ~ 0.134, the downstream the residual strength of uplift pressure coefficients between 0.031 ~ 0.214, either upstream or downstream anti – seepage curtain and drainage effect is good. Dam foundation uplift pressure, dam seepage flow and water relation is not obvious. Total water weir measured peak flow of 21.10 L/s, is far less than the dam seepage flow estimation. The dam running in good condition.

Key words: gravity dam, running, state analysis

1. Engineering Survey

Guizhou Beipanjiang Guangzhao of RCC gravity dam for the whole section, crest total length of 410 m, crest elevation of 750.50 m. Maximum height of 200.5 m, is currently built in the world, under construction highest roller compacted concrete gravity dam, is divided into 20 dams, namely four overflow section and bottom outlet dam, a dam and the elevator shaft 15 slope retaining dam. The dam concrete is about 2 800 000 m³, of which the roller compacted concrete is about 2 400 000 m³, the normal concrete is about 400 000 m³. Non overflow section of spillway dam crest crest width 12 m, platform width 33 m, the maximum bottom width of 159.05 m, crest length of 410 m. Composed of the left and right bank and non – overflow dam on the river bed, the left and right bank non – overfall dam respectively 163 m and 156 m. On the river bed and the

dam section of long 91 m. The typical section of the dam is shown in Fig. 1 and Fig. 2.

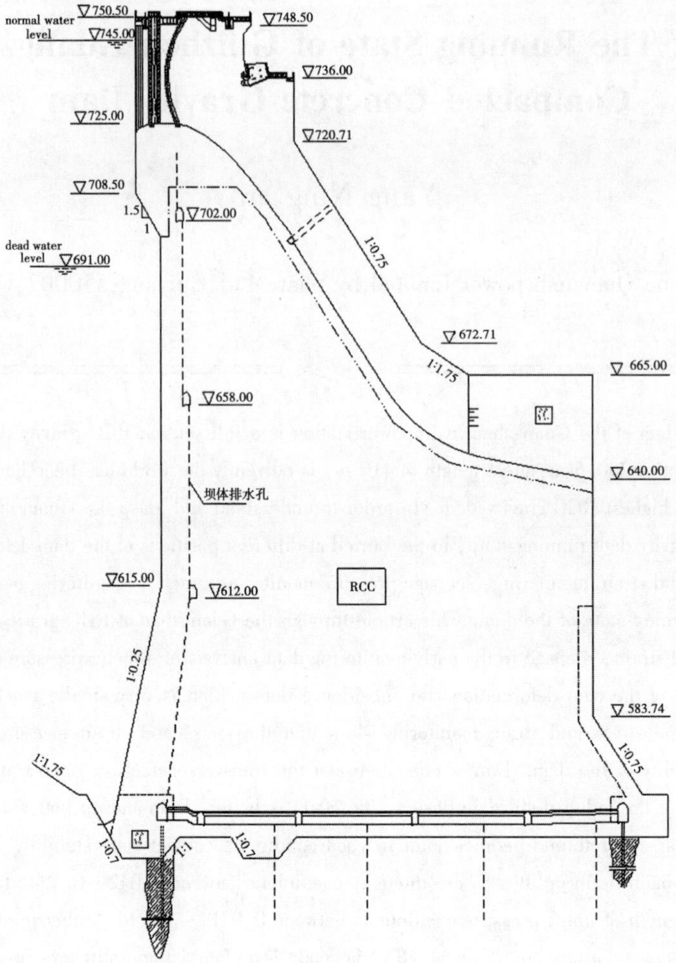

Fig. 1 Spillway section

2 Observation facility

2.1 Deformation monitoring

Including horizontal displacement, vertical displacement, tilt, contact joint and crack opening, the main means of monitoring with optic alignment method and forward intersection method, dam and dam line of level and vacuum laser collimation system, dam is inverted plumb line, hydrostatic leveling system, vacuum laser collimation system, induced joints and contact seam joint meter.

2.2 Seepage monitoring

Including the dam foundation seepage observation, the dam seepage pressure and seepage pressure monitoring, around dam seepage monitoring, uplift pressure monitoring etc., the main method for monitoring a weir, embedded type osmometer and uplift pressure meter etc..

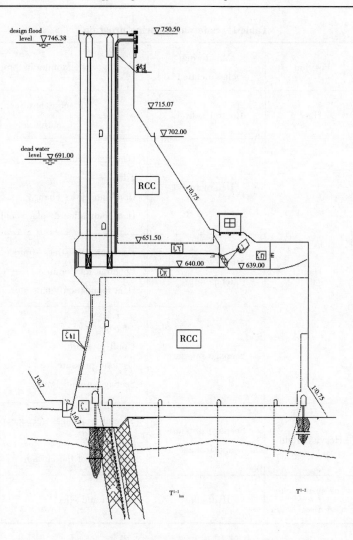

Fig. 2　Dam section profile

2.3　Stress and strain monitoring

Including the dam stress and strain, temperature monitoring, the main monitoring means are buried in the instrument, there is a temperature measurement cable, stress meter, no stress meter, thermometer, etc..

The dam monitoring program is shown in Table 1.

Table 1 Safety monitoring program

No.	Position	Monitoring classification	Monitoring project
1	Hub	Control network	Plane control net, level ontrol network
2	Dam	Deformation	Dam deformation Tilt Crack and joint change Dam foundation displacement
		Stress strain and temperature	Dam concrete stress strain Concrete temperature Dam foundation temperature
		Seepage pressure	Seepage flow Uplift pressure Osmotic pressure Seepage around Dam
3	Heavy curtain	Seepage pressure and leakage	Special geological seepage pressure after the curtain line End point of the curtain
4	The system can release flood waters	Hydraulics	Reason and effect

3 Running state analysis

3.1 Dam deformation

3.1.1 Horizontal deformation

3.1.1.1 Vertical monitoring

1# (Left abutment): IP1 vertical to the upstream and the deformation, but the displacement is less than 0.2 mm.

2# (Section 5) vertical: upstream and downstream direction displacement, mainly towards the downstream direction deformation, the crest and the maximum displacement is 6.13 mm; about shore direction displacement is mainly towards the right bank of the deformation, maximum displacement is 1.05 mm.

3# (Section 10) vertical: IP3 - 1 (EL560 m) upstream and downstream direction displacement mainly toward the upstream deformation, the maximum displacement 2.05 mm. About shore direction displacement between - 0.17 ~ 0.2 mm changes. IP3 - 2 (EL. 560 m) upstream and downstream of the direction of the displacement mainly toward the upstream deformation, the maximum displacement 0.84 mm, left and right shore direction displacement mainly towards the left bank of the deformation, the maximum displacement of 0.2 mm. PL3 -4 EL612 m upstream and downstream of the direction of the displacement, December 2008 before toward the downstream deformation, maximum displacement 0.53 mm, then mainly toward the upstream deformation, maximum displacement 1.23 mm; about shore direction displacement −0.2 ~ 0.70 mm. PL3 -3 (EL.658 m) on the downstream direction of the main downstream deformation, the maximum displacement of 2.39 mm, the direction of the left bank −0.3 ~0.78 mm. PL3 -2 (EL.702 m) upstream and downstream of the direction of the displacement, December 2008 before toward the downstream deformation, the maximum displacement 2.56 mm, then mainly toward the upstream deformation, the maximum displacement 5.04 mm. About 1.1 shore direction displacement to −1.1 ~0.24 mm. PL3 -1 (EL. 750 m) upstream and downstream of the direction of the displacement mainly toward the upstream deformation, maximum displacement 8.1 mm, left and right shore direction displacement mainly towards the right bank deformation. The maximum displacement 1.06 mm.

4# (Section 16) vertical: IP4 (EL. 560 m) upstream and downstream direction displacement mainly toward the upstream deformation, the maximum displacement 0.34 mm, left and right shore direction displacement −1.19 ~ 0.74 mm. 612 m to 750 m elevation on the downstream direction displacement mainly toward the downstream deformation of each measuring point, the maximum displacement of 1.6 ~ 6.81 mm. About shore direction displacement is mainly towards the left bank of deformation, the maximum displacement 6.39 mm.

5# vertical (right abutment): IP5 direction displacement is mainly to the upstream, but the maximum value is only 0.5 mm.

3.1.1.2 Vacuum laser alignment monitoring

1) 658 m height vacuum laser alignment

From the LA1 to LA8, the value of the displacement measurement is poor, and all the measured values are in the −1.5 ~0.5 mm, the deformation direction is lower.

2) the vacuum laser collimation

LA − D1 to LA − D11 on downstream direction displacement measurement value of difference, LA − D1 to LA − D11 after installation are biased downstream deformation. In March 2010 before the 10th of each measuring point towards the downstream of maximum displacement respectively 1.38 mm, 5.04 mm, 6.16 mm, 7.35 mm, 7.68 mm, 9.45 mm, 10.98 mm, 8.25 mm, 7.20 mm, 5.63 mm and 2.52 mm. From LA − D11 to LA − D1, the measured values can be seen, the closer to the center of the river bed, the greater the displacement, the more the slope, the smaller the displacement value.

3.1.2 Vertical displacement

3.1.2.1 Based dam displacement meter

1) No. 10 dam foundation of multi − point displacement meter

The displacement of the M10 − 1 is very small, and the variation range is ±0.3 mm. The displacement of the dam is slowly increasing, and the maximum settlement is 1.97 mm, which occurs at the point of L3.

M10 - 2 of L1 and L4 measuring point and the change law of M10 - 1 similar, before the water displacement is very small, almost to 0. After the impoundment of the reservoir gradually slow increase, the settlement and displacement are 1.2 mm; L2 before impoundment settlement displacement is very small, almost to 0, in July 2008 after the impoundment of the reservoir to Inter, periodic variation, luffing + 1.2 mm, the settlement and displacement is very small, almost to zero.

M10 - 3 of L3 bit storage after settlement of displacement increases slowly, at the end of 2008 increased to 2.53 mm after settlement displacement values decrease gradually, at present for 0.34 mm; the rest of the measuring point installation gradually slow uplift, the L2 measured the uplift displacement is the biggest, 1.59 mm.

The settlement of the M10 - 4 is very small, and the variation range is from - 0.5 to 0 mm. The displacement of the dam is slowly increasing, and the maximum settlement is 2.76 mm, which occurs at the point of L3.

2) No. 11 dam foundation of multi - point displacement meter

M11 - 1 of L1 measuring point of displacement is very small, after the installation of the slow uplift, at present uplift displacement is 0.5 mm; L2 measuring points water displacement front is almost zero, after the impoundment of the settlement displacement slow growth, by the end of 2008 reached a maximum of 2 mm, the displacement value of 1.5 mm; L3 measuring points have been bad; L4 measuring points installed slowly after the settlement, the settlement and displacement of 0.5 mm.

The measured value of - 0.9 is very small. The change range of the measurement points is from M11 - 2 to 0.4 mm, and the main points are the uplift deformation, and the uplift displacement of L1 is the maximum, which is 0.94 mm.

The measured values of the to - 0.7, the deformation state of the - 0.2 mm ~ M11 - 3, the deformation of the deformation, and the deformation of the current settlement, the maximum displacement of the L1 measurement point is 2 mm.

The settlement of the M11 - 4 is very small, almost zero, and the displacement of the dam is slowly increasing, and the maximum settlement is 3.85 mm, which occurs at the point of L4.

Overall, dam foundation deformation dominated by subsidence downstream sedimentation slightly larger displacement and upstream displacement, may and after the impoundment of the dam toe pressure increases, the dam heel stress decreases. The largest settlement displacement only 3.85mm, each measuring point of the measured amplitude are basically small and 4 mm, the measuring point displacement differential settlement is small, uneven settlement value is very small, indicating that the dam foundation consolidation grouting effect is very good, the compressive strength of the rock mass of dam foundation is very high, is conducive to the overall stability of the dam.

3.1.2.2 Static level

1) Static leveling of dam foundation corridor (560 m)

Dam longitudinal direction: due to static standard reference point of the bimetallic standard values have a bigger system error, the analysis, La1 as a reference point and other measuring point of the measured values are relative to the variation of the LA -1. LA -2 to LA -7 after installation are through the settlement mainly, in the settlement process changes with the seasons is certain cyclical changes. 5 months from December to next year's settlement value is gradually increased, from June to November settlement values decrease gradually. The maximum settlement displacement of LA -7 ~ 2.29 mm are 3.13 mm, 2.04 mm, 2.55 mm, 1.30 mm and 1.43 mm, and the subsidence displacement is 1.06 mm, 2.41 mm, 1.92 mm, 1.58 mm, 0.33 mm, LA -2 and 0.57 mm respectively. From the settlement of the settlement, the settlement value of the dam foundation is slightly larger than that of the curtain grouting tunnel.

In the horizontal direction of the dam, the displacement of the measured points was measured by the double metal SJ3 -1 as the base point. LSB -0 changes periodically, the measurement range is + 0.6 mm; LSB -1 in before May 2009 through the settlement, maximum sedimentation value was about 0.8 mm, then settlement displacement changes periodically, change range of 0.7 ~ 0.5 mm; LSB -2 after installation settlement displacement slowly increased, the settlement and displacement of 1.5 mm; LSB -3 to settlement based, settlement and displacement changes periodically, the maximum settlement value is about 0.6 mm; LSB -4 after installation settlement displacement slowly increased, the settlement displacement is 2.5 mm. From the settlement of the settlement, the settlement of the lower reaches is slightly larger than that of the upstream. Dam horizontal hydrostatic leveling system measured value distribution and the dam foundation of Multipoint Displacement Meter sedimentation value distribution are almost the same, the measured value of the difference is not large, two sets of system measurement values confirm that the dam foundation settlement, foundation conditions better, is conducive to the stability of the dam.

2) Static leveling system of 702m elevation dam

To the LS3 -1 as the reference point, the other points of the measured value are relative to the change of LS3 -1.

To LS3 -8 LS3 -2 values showed cyclical changes, the settlement deformation, variable range respectively for -1.11 mm to found and -1.86 ~ 2.34 mm, 0.94 ~ 2.58 mm, -0.56 to 3.31 mm, -0.40 to 3.01 mm, 0.00 ~ 2.83 mm and -0.03 to 2.48 mm. From the measured values, the closer to the river bed, the greater the settlement value.

3.1.2.3 Vacuum laser alignment

(1) 658m height vacuum laser alignment system

LA1 ~ LA8 on the downstream direction of displacement measurement value is poor, the observation results are for reference only, all the measured values are in the -0.7 ~ 1.2 mm, the vertical displacement is small.

(2) the vacuum laser collimation system

LA - D1 to LA - D11 on downstream direction displacement measurement value of difference and the observation results are for reference only. LA – D1 to LA – D11 installed after the gradual settlement of, March 2010 10 before the measured maximum displacement were resectively, 1.6 mm, 1.94 mm, 2.09 mm, levels in the crop, 2.14 mm, 2.24 mm, 2.01 mm and 0.75 mm. Seen from LA – D1 ~ LA – D11 measurements respectively, the riverbed site settlement than near the slope displacement settlement site.

Vertical and vacuum laser collimation system monitoring results show that after the impoundment of the reservoir, and the level of dam deformation is very small, maximum horizontal displacement of only 8 ~ 10 mm, far less than the typical dam 3D nonlinear calculation 93.1mm (normal water level conditions), the dam stability is very good.

3.1.3 Contact seam and structural fracture

3.1.3.1 Contact seam

Dam in the 574.80 m, 615.30 m, 645.00 m and 696.00 m elevation dam and cross – strait slope bedrock contact seam layout crack, contact joint opening and closing degree of monitoring.

From the crack measured value process line and its corresponding temperature curves of dam show that, crack opening and closing is mainly affected by the influence of dam body temperature change. Crack measured value and negatively correlation with concrete temperature change, initial dam body concrete temperature higher and expansive concrete, crack a pressure closed characteristics; post dam temperature decreased, concrete shrinkage, crack a tensional state.

The analysis of observation results shows that the gravity of the dam body is affected by the gravity of the dam body, and the contact state of the low height contact seam is better than that of the high contact seam. Low elevation contact joints were closed or compression; part of the high elevation contact joints was slightly open state, but with open values were small and open process occurs in the initial stage of the embedded instrument, now are in a stable state. That dam and bedrock slope on both sides and alveolar concrete and rock joint in good condition, contact state stability.

3.1.3.2 Structure seam

Joint measurement measuring value process line and its corresponding dam temperature curves show, contraction joint opening and closing is also mainly due to the influence of dam body temperature change. Measured values and negatively correlation with concrete temperature change, when the dam body concrete temperature was elevated, expansive concrete, crack a closed pressure; on the contrary, dam body temperature drops, concrete shrinkage, crack a tensional state.

The analysis on the observation results show that the dam transverse joints, on the whole, in an open state, played a role in the design of dam induced slit, effect is good, the transverse joints tends to be stable.

3.2 Stress strain

3.2.1 Unidirectional strain gauge

1. The strain gauge measured should stress and strain are in the pressure strain gauge has been deducted from concrete autogenous volume deformation), in line with the current dam condi-

tion, the maximum compressive strain for 268 Mu epsilon.

2. Strain gauges embedded in the early due to concrete hydration heat effect, concrete temperature rising stage strain meter is basically in the state of tension; to be concrete temperature gradually decreased, tensile strain decreases the compressive strain increases. Therefore, the maximum tensile strain was in early 2006 2.

3. From the feature value statistics can be found in March 2006 so far, maximum tensile strain was 37.4 Mu epsilon, instrument number SB2 - 6, elevation 555.75 m, dam longitudinal 0 + 142.05, right dam 0 + 010.25, appear date in 2006 on February 8. Maximum compressive strain is 279.40 Mu epsilon, instrument number SB2 - 7, elevation 559.50 m, vertical bar 0 - 012.00, right dam 0 + 010.25, appear date to January 2008 4.

4. The strain measured values and are related to the temperature change and the change of reservoir water level, specific for the strain gauge measurements, with the rise of temperature of concrete, the tensile strain is increased to a certain degree, reduce the temperature, compressive strain is increased to a certain degree; strain gauge measurements, with the increase of the reservoir water level, pressure should be increased, whereas the tensile strain increases.

3.2.2 Five strain gauge

Most of the measured value of the strain gauge group is the pressure strain. From the spatial distribution of point of view, the general rules for the elevation lower strain gauge group compression and tension in the smaller, the higher elevation on the contrary; after the impoundment of the reservoir, the impact of rising water level, tension of dam heel, toe pressure, in line with the general rule. From the perspective of time distribution, generally in the strain gauge embedded initial tension, after with the passage of time, tensile strain have different degree decreases and the compressive strain is increased to varying degrees, analysis the reason for the strain gauge buried after after concrete water of thermal effect of, the concrete temperature rise stage strain gauge is basically in a state of tension; to be concrete temperature gradually decreased, compressive strain increases; in addition under the late gate storage upstream reservoir water level was elevated to one of the reasons for this phenomenon.

3.3 Seepage pressure

3.3.1 The uplift pressure

From the process curve of the dam foundation seepage pressure gauge and the upstream water level, it can be seen:

(1) osmometers buried early, the value measured by construction influence, most measurement value decreases.

(2) of dam foundation uplift pressure measured value in the water did not change significantly before and after, most of the measuring point is slightly affected by the upstream water level and the measured values are smaller, indicating that the dam curtain and drainage effect is good.

By 10# 11#, the two typical dam section of the dam foundation uplift pressure distribution figure:

Upstream in the uplift pressure of the main drainage after upstream anti - seepage curtain and drainage holes after the sharp decline, upstream anti – seepage curtain and drainage effect is good; the uplift pressure at the downstream of the residual after downstream seepage curtain and drainage system also decreased significantly, indicating downstream of anti – seepage curtain and drainage water effect is good.

(3) in large, downstream of the impervious curtain lateral uplift pressure, the pressure difference between the two curtain is very small.

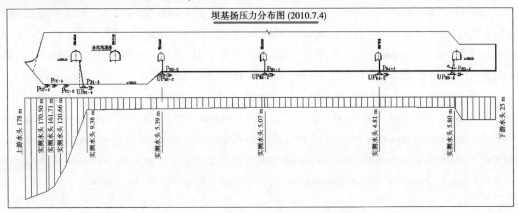

Fig. 3 The typical dam foundation uplift pressure distribution

The uplift pressure coefficient is an important factor to determine the effect of the curtain, the uplift pressure coefficient calculated by the following methods:

Upstream measuring point:

$$\text{Uplift pressure coefficient} = \frac{\text{Measured water level}}{\text{The upstream water level elevation} - \text{Instrumentation}}$$

Downstream point:

$$\text{Uplift pressure coefficient} = \frac{\text{Measured water level}}{\text{The downstream water level elevation} - \text{Instrumentation}}$$

After the upper reaches of the curtain, the measurement points are PB1 – 1 ~ PB1 – 10, and the reduction coefficient of the measuring points in the upper reaches of the upper reaches of the curtain wall is shown in Table 2.

Table 2 The test points in the upstream of the dam foundation

Point number	Osmotic coefficient (α_1)	Point number	Osmotic coefficient (α_1)
PB1 – 1	0.281	PB1 – 6	0.053
PB1 – 2	0.060	PB1 – 7	0.071
PB1 – 3	0.042	PB1 – 8	0.131
PB1 – 4	0.134	PB1 – 9	0.061
PB1 – 5	0.099	PB1 – 10	0.050

From Table 2 can be seen in front of the dam main drainage hole before the uplift pressure intensity coefficient alpha 1 except PB 1 – 10.281, rest between 0.042 ~ 0.134 between, main drainage hole contrast the hydraulic building load code for design of DL5077 – 1997 suggestions Yang pressure strength coefficient, that dam upstream anti – seepage curtain and drainage effect of good.

After the test point is PB3 – 1 ~ PB3 – 7, the reduction coefficient of the measuring points in the downstream of the curtain wall is shown in table 3.

Table 3 The seepage pressure coefficient of each point in the lower reaches of the dam foundation

Point number	Osmotic coefficient (α_2)	Point number	Osmotic coefficient (α_2)
PB3 – 1	0.053	PB3 – 5	0.167
PB3 – 2	0.214	PB3 – 6	0.031
PB3 – 3	0.170	PB3 – 7	0.142
PB3 – 4	0.187		

From table 3 can be seen between the lower residual uplift pressure strength coefficient between 2 0.031 ~ 0.214; contrast the hydraulic building load code for design of DL5077 – 1997 suggestions residual uplift pressure intensity coefficient, indicating that the downstream of the dam anti – seepage curtain and drainage effect is excellent.

3.3.2 Dam seepage

Dam seepage flow were monitored by measuring weir, in 559.50 m corridor of the dam left, on the right bank of the layout of a amount weir observation of left and right bank foundation leakage amount, in 559.50 m corridor foundation set well before the design of a weir, observation of dam foundation seepage volume.

Before the impoundment of the reservoir, left, on the right bank of the weir measured value is small, after the impoundment of the reservoir, left, on the right bank of the weir measured value increased, but the late changes to the water level in the reservoir is not consistent; right values have obvious seasonal variations, on the left bank of the measured values had no obvious regularity, measured value is relatively stable; total weir measured values also existing seasonal variations.

The monitoring results show that the leakage of dam foundation is influenced by the change of reservoir water level, but the water level changes.

3.3.3 Around dam seepage

The seepage flow of the dam is mainly influenced by rainfall and mountain water seepage. The measured water head of the seepage pressure meter increases with the increase of rainfall, and decreases with the increase of the rainfall.

4 Conclusion

The characteristics of dam operation can be obtained by analyzing the monitoring data of dam deformation, stress strain, seepage pressure and so on:

(1) The riverbed dam section level deformation slightly larger than the dam slope on both sides of the Taiwan Strait, the upper deformation is larger than the lower deformation; the riverbed site settlement slightly larger than the slope of dam displacement, the downstream displacement slightly larger than the upstream displacement. Dam deformation is consistent with the deformation of gravity dam. The maximum horizontal displacement of the dam is between 8 ~ 10 mm, far less than the design calculation value; the foundation of the maximum settlement is about 3 mm, the uneven settlement is very small, the deformation value of the dam is within the range of design.

(2) The deformation of dam is mainly influenced by age and temperature, and the water level is small. The current limitation is the main factors to influence the deformation of the dam body; crest measuring point (near the downstream surface level deformation by dam thermal expansion effect, temperature rise dam's horizontal deformation to the downstream, vertical deformation is uplift; on the other hand, the temperature dropped dam's horizontal deformation to upstream, the vertical deformation for settlement. Deformation delay temperature change 2 ~ 3 months.

(3) Low dam height contact seam is closed or compression, high elevation in addition to the individual contact joints was slightly open state outside, and the others are in the state of compression; open joint measurement meter open values were small and open process occurred in the instruments installation in the early period of adjustment. In the dam body, the induced fracture of the joints is open, and the effect of the induced fracture is better.

(4) The uplift pressure strength coefficient of the main drainage hole before the dam is 1, and the other is between 0.042 ~ 0.134 and the other is between PB1 – 1, and the intensity coefficient of the residual uplift pressure is between 0.031 ~ 0.214. From the perspective of the uplift pressure measured head and base curve and regression results, and no obvious uplift pressure of the dam and the head relation;, downstream of anti – seepage curtain and drainage holes effect is good. The dam seepage pressure gauge in addition to the 599.0m elevation osmometer. The remaining parts of the measured value with the longitudinal depth (from the dam surface distance) rapid reduction, at 7.5 m is zero, indicating good cemented dam roller compacted layer, impermeable layer thickness and material selection more reasonable. 599.0m elevation seepage pressure gauge seepage pressure coefficient is larger, the embedding level connectivity of better, the latter should strengthen observation on the site.

(5) The relationship between the dam seepage flow and the reservoir water level is not obvious, the total water flow is 21.10 L/s, which is much less than the dam. On both sides of the curtain osmometer except PMR1 – 2, the remaining points of seepage pressure coefficient between 0.063 ~ 0.442, curtain reduction effect. Located in the end of the curtain PMR1 – 2 and the seepage pressure coefficient was 0.745. In addition, near the left bank of the F1 fault seepage pressure coefficient is also relatively large, the measured value consistent with the reservoir water level variation, on the premises should strengthen observation.

(6) The layout of dam heel unit strain gauge measured stress is compressive stress, the

stress range is −0.33 MPa and −2.13 MPa; from strain gauge monitoring results show that, perpendicular to the direction of the stress is compressive stress, lower stress, small in the upper part; dam body temperature drops due to upstream and downstream direction influence, concrete shrinkage tensile stress; stress measurement within the scope of the design. The change law of concrete pressure in each part of the dam is normal, and no abnormal signs are found.

Innovative Technologies for Construction of RCC Arch Dams

Chongjiang Du, Bernhard Stabel

(Lahmeyer International GmbH, Germany)

Abstract: While roller – compacted concrete is used worldwide for construction of gravity dams to an ever greater extent, its application in arch dams is still a more complicated and challenging task for dam builders. Through fundamental research, testing and practice in the last three decades, the dam builders successfully created new ways to apply the RCC to arch dams. A number of innovative technologies have been developed and successfully applied in engineering and construction of RCC arch dams. This paper reviews the current state of knowledge and technology in this subject. After a brief retrospect of the development of RCC arch dams, their special characteristics are presented in terms of the arch formation and performance distinction in contrast to that of the conventional concrete arch dams. The essential technologies and procedures for the construction of RCC arch dams are summarised. In particular, several key aspects of the technologies are highlighted and discussed, including forming and grouting of transverse contraction joints, post – cooling of RCC dams and options for delivery of RCC mix on steep abutments, which are termed the indispensable steps for the construction of RCC arch dams. The described technologies are demonstrated at example of construction of the 133 m high Gomal Zam RCC arch – gravity dam in Pakistan.

Key words: RCC dam, RCC arch dam, forming of transverse contraction joint, grouting of transverse joint, post – cooling of RCC dam

1 Introduction

Application of roller – compacted concrete (RCC) to build dams is growing at a rapid pace worldwide. However, the RCC is predominantly applied in the construction of gravity dams, while application of RCC in arch dams (including arch – gravity) is still a challenge faced by dam builders. Of 637 RCC dams completed worldwide up to the end of 2013, merely around 35 arch dams have been constructed of RCC. The key issues responsible for this may include the intrinsic properties of the RCC material, its construction methodology and ambient conditions. Accordingly, special technology and construction methods for RCC arch dams are necessary, besides the elementary requirements of conventional arch dams.

Since three decades, researchers, designers and contractors, all have contributed to the brainstorming of possible solutions. The fortitude, innovation and collaboration in this regard allow the problems to be solved and a number of novel technologies and construction procedures have been developed. New technologies are arriving one following another at a fast pace. The authors of this paper are confident that with the new advancements in technology, RCC arch dams are providing a new prospect in dam engineering and opening in a new era of construction.

2 Development of RCC arch dams

The development and application of RCC technology to construction of arch dams initiated in 1980s, which have been attracting and evolving a high level of interesting and innovation, when the development and construction of RCC gravity dams were also in the infancy. The first RCC arch dams with the type of arch – gravity were completed in the late of 1980s by the South African dam builders. The 70 m high Wolwedans and 50 m high Knellpoort RCC arch – gravity dam were the pioneer works in this field[7,10].

Thereafter, China leapfrogged over South Africa and has systematically developed the technology and built the world's most of the RCC arch dams, in which the 132 m highShapai single curvature arch dam[2,3] is to be highlighted. The China's technology is also profited for projects out of China including the Gomal Zam of 133 m height in Pakistan[6] and Nam Ngum 5 dam of 99 m height in Laos, both are RCC arch – gravity dams. Besides South Africa and China, a few of countries and organisations pursued to the technology development. The Portugues RCC arch dam completed at the end of 2013 is the important milestone in extension of the RCC technology into arch dams in USA.

The research and practice resulted in deepening the understanding of properties of the RCC material and behaviours of the RCC arch dams, so that innovative technologies and construction methods have been developed. The benefits have been realised by dam owners. Thanks to the noteworthy pioneering efforts and persistent hard work of the RCC ach dam builders, nowadays the height of RCC arch dams completed is increasing from about 50 m in 1980s to 134.5 m at Dahuashui in 2006. The on – going Wanjiakouzi double curvature RCC arch dam has even touched a record height of 167.5 m. The RCC has been applied in all types of arch dams to date from the arch – gravity to the double curvature thin arch dams[13].

3 Special characteristics of RCC arch dams

An arch dam is curved in the shape of an arch and primarily uses the arch action and the strength of the dam's material to resist loads acting on it. The arch dam is most suitable for and usually constructed in narrow and steep – sided gorges or canyons of stable rock to support the structure and stresses. The arch action is essential for arch dams to transmit loads to foundation and abutments, which requires the arch dam to be a monolithic structure.

Conventional concrete arch dams are normally constructed in blocks. The contraction joints between blocks are grouted prior to initial reservoir impounding, for which the blocks need to be cooled to the final stable temperature using post – cooling. After the joint grouting, the respective dam blocks are integrated into an essentially monolithic structure, gaining the necessary arch action. Another effect of the cooling and grouting is to force the sides of the dam into close contact with the canyon walls.

An RCC arch dam performs principally similar to the conventional concrete arch dams. The primary differences lie in the construction. The RCC arch dam is constructed as if it was a single monolithic structure, i. e. the dam height is raised more or less uniformly – notblock – wise. Corresponding to the RCC construction technology, special technologies for design and construction of the contraction joints and joint grouting as well as the associated post – cooling of the dam should be developed and applied. Moreover, owing to the differences in construction, the stress distribution in an RCC arch dam is different from that in a conventional concrete arch dam.

For an RCC arch dam only with induced joints, the joints will not be grouted before they are open, and thus the post – cooling system may notabsolutely be necessary. Because the RCC arch dam in this case is uniformly raised without clear contraction joints, the arch is formed up to the finished elevation of the dam during construction, so that the arch action is achieved and its potential will be increased as the strength and elastic modulus of the concrete is developed over time with age.

For an RCC arch dam with full conventional transverse contraction joints, the joint conditions can be described as follows:

- In the part below the joint starting elevation, the arch is formed during construction, the same as that of the induced joints before joint opening, as described in the above paragraph.

- In the part above the joint starting elevation, the arch is formed at first as RCC placement, gaining of the arch action. However, as the dam body is cooled to the final stable temperature by using post – cooling prior to the reservoir impounding, the joints are opened and the initially formed arch action is lost. Only after joint grouting, the arch action of the whole dam is mobilised again, and then the working principle of the joints of RCC arch dams is in common with that of the conventional concrete arch dams. Nevertheless, joint – forming and groutingsystem as well as the post – cooling systems for the RCC arch dams are considerably different from those of the conventional concrete arch dams, because of the fast construction and thin – layer placement of RCC using bulldozers and vibratory rollers, and these must be carried out concurrently with RCC placing.

Owing to the construction procedure and arch forming mechanism of RCC arch dams, the stresses induced by temperature variation andautogenous volume change of concrete could remain and be accumulated in the dam structure[3], and the stress induced by self – weight of concrete is not linearly distributed. The remaining stress in the arch dam is somewhat similar to the residual stress remaining in a steel structure induced by the welding process. For this reason, a detailed finite element analysis with simulation of the layer – by – layer construction procedure has to be performed to properly evaluate the stresses in the RCC arch dams.

Because arch dams rely on the arch action and strength of the dam's material to resist loads, stress levels in arch dams are usually higher than that in gravity dams and hence RCC of higher strength is normally required, which in turn results in higher content of thecementitious materials.

In addition, delivery of the RCC mix on the steep abutments is a difficult task for the construction. To overcome this difficulty, various construction methods have been developed in the course of construction of RCC arch dams.

In fact, the RCC arch dams share the same advantages with the conventional concrete arch dams in behaviours and operation, while they are frequently competitive to both gravity dams and conventional concrete arch dams, if the topography and geology allowbuilding an arch dam. In the first place, concrete volume of an arch dam is substantially less than that of the gravity dam, and the RCC construction is usually faster than the placement of conventional concrete. These may result in the construction of the RCC arch dam within one or two low temperature seasons. The fast construction process and short construction period can also result in simplified river diversion works and earlier commissioning of the project. All these factors may offer a significant reduction in project costs.

4 Joint – forming technology

4.1 Type and spacing of transverse contraction joints

Different from the joint forming for RCC gravity dams, the joint forming for an RCC arch dam should be considered and constructed together with the construction procedure and grouting system used. There are various techniques on the joint design. Accordingly, various joint types and joint forming technologies have been studied and successfully applied in the construction of RCC arch dams. These techniques were presented in a previous paper of the firstauthor[4] of this paper, and can be summarised as follows.

- induced transverse contraction joints,
- conventional transverse contraction joints,
- short structural joints, and
- hinge joints.

Among these systems, the induced and conventional transverse contraction joints are mostly used in the RCC arch dams either separately or combined. In the following paragraphs, the two types of joints are discussed in details.

As well – known, arch dams are vulnerable to uncontrolled cracking. Transverse cracks without grouting may destroy the monolithic nature of the arch structure and thus impair the arch action, affecting the dam stability. Controlling cracking is therefore a primary task in design and construction of an RCC arch dam. The transverse contraction joints may be regarded as artificial and controlled cracks arranging in the dam to avoid uncontrolled cracking.

Research and experience indicated that the stresses induced by temperature variation andautogenous volume change of concrete are critical factors to be considered in arrangement of the joints and selection of the joint types[14,15]. For an RCC arch dam up to 70 m height without the conventional contraction joint, such stresses could be held without affecting its stability, whereas for an RCC arch dam above this height, the conventional contraction joints or combination with the induced joints should be constructed to avoid the excessive stresses and thus ensure its monolithic nature.

In principle, number of the transverse contraction joints foran RCC arch dam should be limited to a minimum, and their starting elevation to the foundation level should be arranged as high as possible, in order to reduce the construction costs, because construction of the joints will interfere with the RCC construction process. In other words, the spacing of the transverse contraction joints should be arranged as larger as possible, and the lower portion of the dam with no transverse contraction joint should be placed as higher as possible. Thus, RCC placement should be commenced and arranged in the low temperature seasons so as to increase the height from foundation with no contraction joint.

The exact spacing and starting elevation of the joints should be individually determined after detailed thermal studies for the dam. Reviewing most of the completed RCC arch dams it can be found that the spacing of the induced contraction joints ranges from 10 to 40 m and conventional-contraction joints from 30 to 70 m, respectively. It is worthwhile to note that the joint spacing is usually a compromise between the temperature control measures, construction procedure and prescribed construction schedule. Moreover, no longitudinal joint is necessary to the RCC arch dams.

4.2 Induced transverse contraction joints

The induced joint can be defined as a weakenedplane that is created by provision of bond breakers at the pre-arranged location to reduce the effective cross sectional area of the dam, so that the tensile strength along the weakened plane is significantly lower than that in the other locations. As a result, cracking will occur along the weakened plane, when the tensile forces induced by the tensile stress exceed the tensile capacity of concrete along the weakened plane. Before the weakened plane is pulled off, the structure including the weakened plane can normally transmit forces. Once the weakened plane is open, the tension in the other parts of the dam concrete is released, preventing from cracking.

Experience indicated that the induced joints shall be arranged in the areas potentially with higher tensile stresses where crack may occur without stress release through the joint[8,16]. As a general rule, the cross section at the joints should be reduced by roughly 1/6 to 1/3 times its full sectional area. In the course of RCC arch dam development, a plenty of systems for the joint-inducers have been established.

In the RCC arch-gravity dam at Wolwedans, the joint-inducers and the grout feed/return pipes were made of HDPE sheets and pipes[11,12], as shown in Fig. 1. A grout compartment consisted of 8 RCC layers of 25 cm thickness, in which 3 layers were installed with the joint-inducers in a vertical spacing of 50 cm (2 layers). The grouting system was arranged in the lower two layers of the joint-inducers. In the upstream and downstream facing concrete, 150 mm wide 2 mm thick HDPE sheets were placed on both sides of the PVC water stop to ensure that the joints could be formed passing through the centre of the water stops. The reduction of the cross sectional area was about 35%.

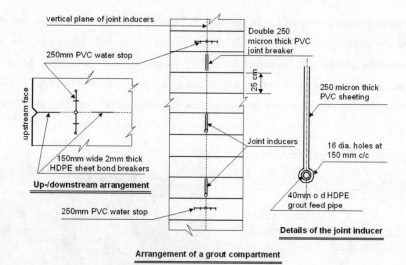

Fig. 1 Joint inducer and grouting system used at Wolwedans RCC arch – gravity dam

In China, the system using pre – cast concrete blocks are more preferred to create the induced joints by the dambuilders[19]. The pre – cast concrete blocks are the same as that used for forming the conventional contraction joints, as will be presented in Section 4.3. Difference is merely in the number of the concrete blocks to be installed. Usually, the pre – cast concrete blocks will be placed in every 2 or 3 layers and in a net spacing of 0.5 to 1.0 m. This system is used in majority of the RCC arch dams constructed by Chinese. The so – called re – injectable grouting system (Section 5) is integrated in the induced joints. Apparently, the work for forming the induced contraction joints is significantly less than that for forming the conventional contraction joints.

Once an induced joint is open, contact grouting for the joint shall be performed. During operation of the dam, the grouted joint may be open again and thus the re – injectable grouting shall follow. The practice demonstrated that most of the induced joints in an RCC arch dam are not open. However, there are exceptions that cracks occur in other locations where no joint is arranged, while the induced joints remain closed. This case happened where the spacing of the induced joints isprobably too large.

4.3 Conventional transverse contraction joints

The name "conventional transverse contraction joint" was coined to define a fully disconnected joint in an RCC arch dam. Its function is the same as that in conventional concrete arch dams. For an RCC arch dam in large size, say over 70 to 100 m in height, the conventional contraction joints are usually considered necessary. Up to now, application of the conventional contraction joints can be found almost merely in the Chinese built RCC arch dams, because alllarge RCC arch dams are built by them.

Zhu[19] proposed a joint – forming system with the pre – cast concrete blocks. The joints are formed with two types (A and B) of the pre – cast concrete blocks of 1 m long and 0.3 m high (equal to the layer thickness) with a bottom width of 0.3 m. The sloped rare side is formed with "teeth" to promote bonding with RCC. In the type A blocks, the holes for installation of the grout feed pipes and the vent pipes are additionally built. The two types of blocks will be alternately placed. The concrete blocks will be tied and fixed by means of steel bars in the blocks in the alignment of the contraction joints. In every five or six layers, one layer of the type A blocks is installed, while in other layers the type B blocks are used in alignment with the transverse joints, as shown in Fig. 2.

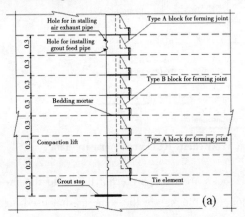

Fig. 2　Transverse contraction joint (a) and pre-cast concrete blocks (b)

Bedding mortar is applied beneath the concrete blocks to improve their bonding effect andimpermeability. The grouting pipes are installed on site during construction. Thereafter, the RCC placement follows.

The conventional contraction joints for RCC arch dams shall be grouted prior to the initial reservoir impounding as that for conventional concrete arch dams. To this end, the post-cooling system is usually required.

5　Joint grouting technology

The joint grouting has the primary objectives to uniformly fill the transverse contraction joints to integrate the arch dam blocks as a monolithic structure so as to restore the arch action in the dam. In the course of development of the RCC arch dams, a lot of strategies and methodologies have been developed. In essential, the following three systems are predominantly applied[4]:

- Once grouting system with post-cooling;
- Double grouting system; and
- Re-injectable grouting system.

In principle, the conventional contraction joints may require only once grouting, if the post-cooling system is installed in an RCC arch dam. Provisions of the grouting system and procedure are very similar to that for conventional concrete arch dams.

For the induced joints as well ascautiously for the conventional contraction joints, the double or re-injectable grouting system shall be installed. As the name suggests, the double grouting system means to install two independent grouting systems in a contraction joint. The first system is used for the first time grouting while the second for the late grouting if required. As an alternative, a novel re-injectable grouting system has been developed in China and applied specially for grouting of the transverse contraction joints of RCC arch dams[1]. The re-injectable grouting system can be used more times for the joint grouting. The principle of this grouting system is similar to that of the re-injectable grouting systems used in Europe, e.g. FUKO System.

As shown in Fig. 3, the key component of the re-injectable grouting system is the grout outlets. This consists of a rubber sleeve, a perforated steel pipe and pipe couplings on both ends of the steel pipe to connect the perforated steel pipe along the joint line with a series of the feed/return pipes in tandem, forming the grouting system. The highly elastic rubber sleeve tightly wraps around the steel pipe and functions as an irreversible valve, preventing the ingress of water or other substance outside the grouting pipe system. Only when the internal pressure in the grouting pipe exceeds about 60 to 150 kN/m^2 (0.6 to 1.5 bar), will the rubber sleeve be disconnected with the perforated pipe forming an open channel, so that the grout in the feed pipe can penetrate through the outlets into the contraction joint. After grouting operation, the grouting pipe system is washed with low pressure water for its next use. The joints with an opening of 0.2 mm or wider are groutable. The grouting operation for the transverse contraction joints shall be completed one month before starting the reservoir impounding.

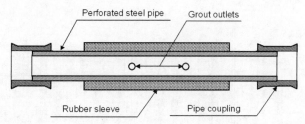

Fig. 3 Sketch showing the grout outlets of re-injectable grouting pipe

In Gomal Zam RCC arch-gravity dam, four conventional transverse contraction joints were constructed, two of which in the central portion were grouted while the other two on abutments at higher level were kept open to reduce some arch action and thus increase the gravity action to balance the excessive vertical tensile stresses at dam heel. Each of the grouting compartments was 6.0 m high containing 20 layers of 0.3 m thickness. Grout stops formed the boundaries of the compartments. The joint-forming system is schematically illustrated in Fig. 4 and Fig. 5. The grouting of the two contraction joints were completed one month before the reservoir raising.

6 Post-Cooling for RCC dams

Post-cooling of RCC dams by circulating cooled water through embedded pipes while placing has proved more difficult than that for the conventional concrete dams, because

• installation of the cooling pipes during RCC construction shall not affect the fast RCC placement operations; and

• the embedded thin-walled pipes shall not be damaged by rollers and/or other heavy machines during placement and compaction of the RCC mix.

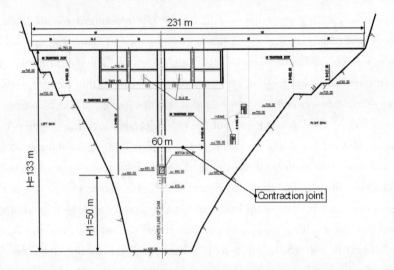

Fig. 4　Arrangement of contraction joints in Gomal Zam RCC arch – gravity dam

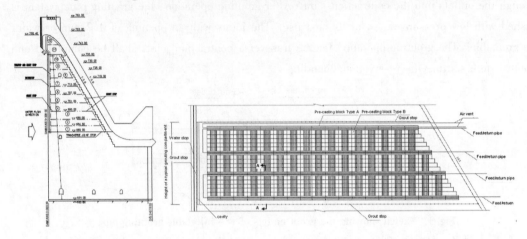

Fig. 5　Conventional transverse contraction joints for Gomal Zam RCC arch – gravity dam

Thanks to the recent innovative developments from research and practice in RCC construction technology, the above pre – requisites can be met by proper design and construction management as well as selection of the cooling pipes of suitable material. The details of this technology were presented in a previous paper of the first author[5] of this paper. In the following, the essential aspects of the post – cooling which was successfully applied in the construction of Gomal Zam RCC arch – gravity dam are itemised.

6.1　Selection of the pipe material

High – density polyethylene (HDPE) pipes should be selected as the cooling pipes, whereas the steel pipes are not suitable because of their time – consuming installation of all the steel parts (pipe pieces, fittings, elbows and couples etc.). The important characteristics of the HDPE pipes are:

● Light in weight: The specific gravity of HDPE pipes is only 860 to 1 000 kg/m^3. Thus, a 200 m long HDPE pipe weighs barely 35 to 40 kg. This significantly facilitates the transport to and fast installation at site.

• Flexible and able to be coiled: The HDPE pipes have a minimum bending radius of 20 to 25 cm and therefore a section can fully satisfy the flexibility requirements of cooling pipes.

• Long single length: A single reel of HDPE piping is approximately 200 to 250 m long. Thus the connection between cooling tubes is rarely if ever necessary.

• High strength: HDPE cooling pipes have a fairly high tensile strength and break at forces exceeding 20 to 25 MPa as well as a high Mullen Burst Strength of not less than 3 to 10 MPa.

• High elongation capacity: The minimum elongation at break is excellent with a value of 200 per cent.

• Costs effective: The HDPE pipe is much cheaper than the steel pipe.

6.2 Partition of the construction area into two or more units

The whole RCC placing area, in plan, should be divided into at least two units, as shown in Fig. 6. This will permit installation of cooling pipes at one unit while RCC placing at the other and as such, installation of the cooling pipes will not interfere with RCC construction. Cooling pipes in the second unit should be installed on the RCC surface two or three layers higher than those on the first unit.

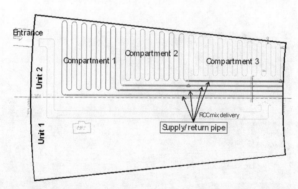

Fig. 6 Partition of construction units and cooling compartments

6.3 Arrangement of compartments

The construction units should be divided into several cooling compartments, as shown in Fig. 6, such that the total length of the cooling pipe in each of the cooling compartments does not exceed the length of one reel (240 m at Gomal Zam). In this way, no coupling is required and thus installation time is reduced. In addition, the cooling effect is not affected as using excessively long cooling pipes.

6.4 Connection of the distribution pipes to supply/return pipes

At locations where the distribution pipes shall be connected to the supply/return pipes, a piece of three-way (Tee) steel pipe should be inserted into the three pipe ends. The HDPE pipe ends shall be firstly heated for example using a blow torch. Then the soften pipe ends are fastened on the three-way pipe piece with wire. In addition, the connection area should be wrapped with several layers of Polytetrafluoroethylene (PTFE) thread seal tapes.

6.5 Erection of the cooling pipes

Before installation and after being covered with a layer of RCC mix (30 cm thick), the HDPE pipes should be checked for leaks by the passing of water or air with a pressure of 0.1 MPa (1 bar) as part of the checking procedure, and any leaks shall be repaired. After an early experience of minor concrete damage resulting from leakage of cooling water from a cooling pipe at a connection joint at Gomal Zam dam, the connection procedure was amended as described in paragraph 6.4.

The cooling pipes was carried and placed by labourers on the surface of the freshly compacted but not hardened RCC layer. The U-shaped reinforcing bars of 4 to 6 mm diameter may be used to tie the pipes onto the fresh RCC surface at a spacing of 2 to 4 m for straight portions and with three pieces for each elbow portion.

6.6 Covering the installed pipes

After installation, the pipe grids shall be covered with a layer of overlaying RCC mix not less than 25 to 30 cm thick. Dumping and spreading the RCC mix should be started from one side of the cooling pipe network. Only then should heavy machines, such as bulldozers, trucks and rollers, work on them so that heavy machinery never operates directly on the bare cooling pipes, as shown in Fig. 7.

Fig. 7 Heavy machines working on the covered cooling pipes

This is a critical step to successfully apply the pipe systems into RCC dams. Unless the pipes are covered with the RCC mix, heavy machines may cause serious plastic deformation of the pipes, leading to leakage or hindrance of cooling water passing through them.

At Gomal Zam dam, the horizontal and vertical spacing of the cooling pipes was 1.5 m × 1.5 m. The cooling pipes were installed both on the flat layers and on the sloping layers without difficulty as the sloped-layer method was used for RCC placement. Two typical types of the pipes were used for the post-cooling of dam concrete, namely, the pipes with an outside-diameter of 32 mm and a wall thickness of 2.0 mm were used as the distribution pipes, while the pipes with an outside-diameter of 40 mm and a wall thickness of 3.0 mm were used for supply and return.

6.7 Operation of the cooling pipes

At Gomal Zam dam, circulation of cooling water of 14 ℃ in the grid was started 6 hours after compaction of the RCC layers for duration of 14 days. The last stage of post – cooling was performed at least one month prior to the grouting operation for the transverse contraction joints, so that the RCC temperature dropped to the specified closure temperature. In some areas, the cooling system was also used in an intermediate stage in autumns to accelerate cooling the RCC, so as to decrease the temperature differential between the interior and facing concrete and reduce the thermal stresses in the dam while the modulus of elasticity of RCC was relatively low. In addition, the cooling water was left in the cooling pipes after a cooling stage and stood for several days to fully utilise the residual cooling effect of the water.

The flow rate through a single pipe system was approximately 0.8 to 1.2 m^3/h. The direction of flow was reversed every 12 or 24 hours to reduce the temperature gradients within each lift as it is cooled. The acceptable rate of temperature drop shall not exceed 1 ℃ per 24 hours. The temperature differential between the RCC and the cooling water through the pipes did not exceed 25 ℃ to decrease the so – called thermal shock on the concrete in contact with the cooing pipes when circulation of cold water was turn on. Nonetheless, there was no evidence of damage inflicted by such effect.

7 Delivery of RCC mix on steep abutments

7.1 Special methods for delivery of RCC mix

Because arch dams are usually located in narrow and steep – sided gorges or canyons, transport of RCC mix on the steep abutments is a difficult task the dam builders face. As for RCC gravity dams, the end – dump trucks are the predominant mean to deliver RCC mix from batching plant to placing locations for the construction of RCC arch dams by far, commonly in combination with other transport means, such as various belt conveyor systems, slow drop elephant trunks, vacuum chutes, full – tube ducts, M – Y boxes and M – Y box – pipe systems. In principle, the slow drop elephant trunk, vacuum chute, full – tube duct, M – Y box and M – Y box – pipe system are the devices to convey concrete mix relying on gravity, so that lower energy is consumed, resulting in low costs. All the methods depend on the local conditions and have advantages and disadvantages. The important criteria for selection of the methods and equipment are that segregation of the RCC mix shall be kept to a minimum and the RCC mix can be quickly, reliably and effectively transported with least costs.

Topography at the dam site is a critical factor influencing the selection of the delivery methods. As a rule of thumb, trucks can generally be used to deliver RCC mix to the lower portion of a dam. When the abutments are not very steep, say with a slope up to approximately 40°, trucks and/or conveyor belts can be used by cutting berms into the slopes downstream and/or upstream of the dam seat to access each elevation.

The slow drop elephant trunk is a flexible rubber hosepipe and can be installed on the outlet end of a conveyor belt to vertically deliver the RCC mix. However, the transport height of concrete mix using the slow drop elephant trunk is limited to a maximum of 15 to 20 m. When the abutment slopes are between 40° to 70° (with the ideal slope of 45° to 55°), the vacuum chute or full – tube duct may be the best candidate, whereas the M – Y box or M – Y box – pipe system should be applied to the very steep abutments with a slope of 60° up to vertical (90°). For construction of an RCC arch dam, the delivery methods are usually combined.

7.2 M – Y box and M – Y box – pipe system

An M – Y box, also called M – Y mixer, is a vertical fall – damping and mixingbox that was developed and firstly applied in Japan as a concrete mixer and transport device for continuously mixing and conveying concrete mix. It features with a series of box – shaped units comprising two aligned twisted boxes. Each unit of the M – Y box has two parallel vertical inlets and two parallel horizontal outlets separated by steel plates. From the inlet to the outlet, the vertical dimension of cross section gradually decreases while the horizontal dimension increases with the same proportion, resulting in the same cross – sectional area all the way. When the material is discharged into the box units, the inner structure of the M – Y box makes the material to be kneaded by gravity while passing through each of the box units, allowing continuous mixing the mix while transporting on very steep or even vertical slopes[9, 18]. Meanwhile, the falling speed of the concrete mix is damped.

The M – Y box – pipe system is a modification and further development of the M – Y box, specially used for vertical transport of concrete mix. In this system a series of the M – Y boxes and steel pipes of 6 to 15 m length (depending on the slope steepness) are alternately connected in tandem. The concrete mix passing through the M – Y box is buffed and re – mixed, reducing the fall speed and preventing segregation[18]. Then it passes through the steel pipe section. This process is repeated during the transport of the concrete mix, until the mix reaches down to the outlet. Now the M – Y box and M – Y box – pipe system are used in vertical transport of conventional concrete mix, RCC mix and hardfill or cemented sand and gravel in an increasing extent. The construction practice demonstrated its applicability and capacity. The equipment is cost – effective and can be repeatedly used. It is believed that application of the M – Y boxes or M – Y box – pipe systems to the construction of RCC arch dams is the best solution to transport of the RCC/concrete mix on the very steep abutments in a narrow valley. For more details of this device refer to the relevant literature [9, 18].

7.3 Vacuum chute and full – tube duct

A vacuum chute is a closed semi – circular conduit system with necessary accessories, primarily consisting of:

• An inlet hopper with a radial valve: the hopper usually has a volume of 6 ~ 10 m^3 to deposit RCC mix and adjust the mix delivery intensity to the placing locations;

• A transit section: this section is not restricted by the flexible cover to accelerate the handling of the RCC mix;

• The chute body with a flexible cover: This section is the main part of the vacuum chute, in which vacuum is formed while transporting the RCC mix; and

- An outlet elbow: The function of the outlet elbow is to alter the direction and reduce the speed of the RCC mix, so that the RCC mix can be discharged into a truck below it.

The whole vacuum chute system is supported on rigid frame columns of steel, as shown in Fig. 8. The RCC mix is firstly dumped into the hopper. By opening the valve, the RCC mix slides down into the transit section by gravity where the mix movement is accelerated. As the mix enters into the chute with the flexible cover, its speed is further increased by gravity. Meanwhile, the pressure in the closed chute is decreased, forming vacuum. The pressure differential between the chute interior and exterior in turn counteracts the mix movement to mitigate the speed. As the RCC mix slides down along the chute, this process is repeated resulting in a wave-like outlook, so that the falling speed of the RCC mix is controlled in a reasonable range (usually 10~15 m/s). The magnitude of the vacuum and the speed of the mix movement can be adjusted by means of adjusting the valve opening. Using a vacuum chute to transport RCC mix, segregation of the mix is prevented. Vacuum chute has a high efficiency to vertically deliver the RCC mix with a transport capacity of 200~550 m^3/h. Moreover, its manufacturing costs are low and it is convenient in operation and maintenance.

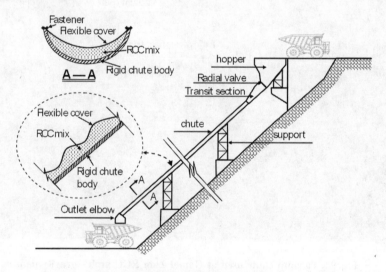

Fig. 8　Sketch showing the vacuum chute

The full-tube duct (also called full-bin system)[17] is a further development of the vacuum chute. Similar to the vacuum chute, the full-tube duct contains an inlet hopper, the full-tube duct body, a radial valve and an outlet elbow. Same as the vacuum chute, the full-tube duct is supported on the rigid frame columns of steel on the slope. The full-tube duct body has a quadrate or circular cross section in size of 40×40 to 80×80 or ϕ40 to ϕ80 cm. The radial valve is installed near the outlet at lower end of the chute to control the mix movement, in contrast to the vacuum chute. During the delivery, the concrete mix is fully filled in the tube. Through regulation of the opening of the radial valve, the speed of the concrete mix is adjusted, so that the mix falling speed is damped, preventing the segregation. The application condition of a full-tube duct is the same as a vacuum chute. Both devices have high efficiency to vertically transport concrete mix and widespread used in the construction of RCC dams with steep abutments.

7.4 Example

The Gomal Zam RCC arch – gravity dam is 133 m high from its low foundation level at El. 630.0 m with a crest length of 231 m at El. 763.0 m. The maximum width at base amounts to 78 m. A four – bay spillway of 17.5 m length each is integrated in the central part of the dam, abutting non – overflow sections on both sides. A bottom outlet of 3.0 m diameter with an invert at El. 680.0 m is constructed in the centre of the dam to flush sediment. The dam is located in the 800 m long Khajuri gorge. At dam site the gorge is V – shaped with a bottom width of 25 to 40 m. The narrow canyon is slightly asymmetrical and has very steep flanks with an average slope of 75° for the left and 65° for right flank, respectively, while its upper flanks tend to flatten with an average slope of 40° to 45°. The total RCC quantity of the dam is 408 760 m^3 with additional conventional concrete of 84 660 m^3.

Fig. 9　Vacuum chute used at Gomal Zam RCC arch – gravity dam

For the construction of the arch – gravity dam, various conveying methods were successfully used:

● From dam – foundation interface at El. 630.0 to El. 696.6 m, the concrete mix was transported by end – dump trucks;

● From El. 696.6 to 736.0 m, the trucks were used for the horizontal transports from the batching plant to a vacuum chute as well as on the dam placing area, while the vacuum chute of 54 m length with a slow drop elephant trunk of 10 m length at its outlet elbow was used for the vertical transport, as shown in Fig. 9. The vacuum chute inclined to 70°. The total volume of the RCC mix transported through the vacuum chute was 125 000 m^3;

● From El. 736.0 m to the dam crest, a belt conveyor was used in combination with trucks to overcome the hindrance of the spillway section; and

- The conventional concrete was conveyed by means of a tower crane and concrete pumps.

8 Conclusions

The successful construction practice of RCC arch dams demonstrates that RCC is a suitable material and technology for the construction of arch dams owing to the economic advantages and the time – saving benefits. Essential issues for this include forming transverse contraction joints and their grouting, the temperature control for the RCC arch dams using post – cooling, delivery of RCC mix to the dam on steep abutment slopes and construction management in limited space considering the site – specific ambient conditions. The state – of – the – art knowledge and state – of – the – practice technology summarised in this paper pave the way to extend the application of RCC technology into arch dams. It is reasonable to expect that more innovations in technology will be developed, bringing a boom in application of RCC technology into the construction of arch dams.

References

[1] G. X. Chen, G. J. Ji, G. X. Huang. Repeated grouting of RCC arch dams[J]. Roller Compacted Concrete Dams ed. by L. Berga, J. M. Buil, C. Jofre & C. G. Shen, Madrid, Spain, 2003:421 – 426.

[2] Q. H. Chen. New design method of RCC high arch dam[J]. Roller Compacted Concrete Dams ed. by L. Berga, J. M. Buil, C. Jofre & C. G. Shen, Madrid, Spain, 2003. 427 – 430.

[3] Q. H. Chen, Y. T. Ding. Study of structural joint design of Shapai RCC arch dam[C]. Proceedings of International Symposium on Roller Compacted Concrete Dams, Chengdu, P. R. China, April 21 – 25, 1999: 560 – 571.

[4] C. J. Du. Transverse Contraction Joints and Grouting Systems for RCC Arch Dams[J]. International Journal on Hydropower and Dams, Issue 1, 2006:82 – 88.

[5] C. J. Du. Post – cooling of RCC Dams with Embedded Cooling Pipe Systems[J]. International Journal on Hydropower and Dams, Issue 1, 2010:93 – 99.

[6] C. J. Du. Structural Features of Gomal Zam RCC Arch – Gravity Dam[C] // Proceedings of the 6th International Symposium on Roller Compacted Concrete (RCC) Dams, Zaragoza, Spain, Paper No. C008, 23 – 25 October 2012.

[7] J. J. Geringer. The design and construction of the groutable crack joints of Wolwendons dam[C] // Proceedings of the 1st International Symposium on Roller Compacted Concrete Dams, Santander, Spain, Oct. 2 – 4, 1995:1015 – 1036.

[8] A. J. Gu, et al.. Analysis of effectiveness of crack directors of RCC arch dams[J]. Journal of Yangzhou University, Vol. 6, No. 2, 2003:66 – 70. (in Chinese)

[9] T. R Gyawali, K. Yamada, M. K. Maeda. High productivity continuous concrete mixing system[C] // Proceedings of the 17th ISARC, Taipei, Taiwan, 2000: MA3. doc – 1 to 6.

[10] L. C. Hattingh, W. F. Heinz, C. Oosthuizen. Joint grouting of an RCC arch/gravity dam: practical aspects, Proceedings of International Symposium on Roller Compacted Concrete Dams, Santander, Spain, Oct. 2 – 4, 1995:1037 – 1052.

[11] F. Hollingworth, D. J. Hooper, J. J. Geringer. Roller compacted concrete arched dams[J]. International Water Power and Dam Construction, Nov. 1989:29 – 34.

[12] R. E. Holderbaum, D. P. Roarbaugh. Trends and innovation in RCC dam design and construction[J]. International Journal on Hydropower and Dams, Issue 3, 2001:63-69.

[13] ICOLD, Roller-compacted concrete dams - State of the art and case histories. Paris, France: International Commission on Large Dams, Bulletin 126, 2003.

[14] G. T. Liu, P. H. Li, S. N. Xie. RCC arch dams: Chinese research and practice[J]. International Journal on Hydropower and Dams, Issue 3, 2002:95-98.

[15] G. T. Liu, P. H. Li, S. N. Xie. Research and practice of roller-compacted concrete arch dams[C] // Proceedings of International Conference on RCC Dam Construction in Middle East, Jordan, April 7-10, 2002:68-77.

[16] G. S. Sarkaria, F. R. Andriolo, M. A. C. Juliani, et al. Contraction joints and monolithic behaviour of RCC arch dams[C] // Proceedings of 22nd Annual USSD Conference, California, USA, June 24-28, 2002:47-57.

[17] X. R. Wu, Z. R. Chen. New technology on vertical conveyance of concrete by full-bin system in Guangzhao dam project[C] // Proceedings of the 6th International Symposium on Roller Compacted Concrete Dams, Zarokoza, Spain, Oct. 2-4, 2012: Paper C0005.

[18] W. H. Zhong, G. Yu. New RCC transporting device and its construction technique[J]. Water Conservancy & Electric Power Machinery, Vol27, No.3, June 2005:23-27. (in Chinese)

[19] B. F. Zhu. RCC arch dams: temperature control and design of joints[J]. International Water Power and Dam Construction, Aug. 2003.26-30.

The Soil Treated Dam (STD) a New Concept of Cemented Dam

Michel Lino[1], François Delorme[2], Daniel Puiatti[3]

(1. ISL Ingénierie, France; 2. ELECTRICITE DE FRANCE (EDF), Hydro Engineering Center, France; 3. DPST Consulting, France)

Abstract: Soils, e.g. clay containing materials, treated with lime and / or cement are commonly used in the construction of transport infrastructure (roads, railways, airports, etc.). Some applications also exist in hydraulic structures, particularly in the USA, Australia, South Africa and Europe. This presentation deals with a new concept of dam using that type of material, the Soil Treated Dam (STD). It starts with a description of the techniquefollowed by a presentation of the difference with hardfill/CSG, in terms of characteristics ofmaterials as well as of the binders used and the level of mechanical performance. Then, a newconcept of cemented dam is developed. The typical STD has a height lower than 50 m, its profile is symmetrical and it is sealed by an upstream facing. STD may be an economical solution on many embankment dam sites. The paper finally how treated soils can fill a gap by giving the possibility to enhance the mechanical performanceof clayey materials and to reduce the duration of earthworks and construction cost.

Key words: soil, treatment, lime, cement, soil treated dams, STD.

1 Introduction

Soil treatment with lime and / or cement is a profitable technique, widely and successfully used in transport infrastructures[2]. Applications are also existing in hydraulic works (USA, Australia, South Africa, Europe). Soils treated (ST) are different from cemented sand and gravel used in hardfill. The characteristics of the materials before treatment (soil versus sand and gravel for hardfill) as well as the mechanical performance of the mixturesjustify the definition of a new concept and therefore a new design for hydraulic structures and dams.

2 Compatibility between binders and soils

2.1 Binders

In this paper, the following mineral binders are considered:
- Hydraulic binders (like cement) : by definition, they harden when mixed with water,

---

1. For the need of this paper, hydroulic binders will be Called" cement".

both in the air and under water, without addition of another reactant[1]. Hydraulic binders contain all the necessary components (see Fig. 1) to behave as a glue able to bind the particles of a granular material.

• Pozzolanic binders : they need lime to set and harden (natural pozzolanas, siliceous fly ashes, etc.) (see Fig. 1). Once mixed with lime, they behave like hydraulic binders.

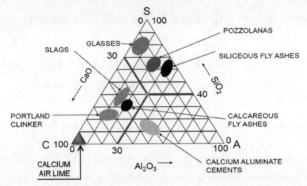

Fig. 1 Positioning of the binders according to their composition

• Calcium air lime[2] : product obtained by calcination of limestone (see Fig. 1). It can be either in the form of quicklime (CaO) or hydrated lime (Ca(OH)$_2$). Lime reacts differently than a cement, particularly in presence of a clay containing soil :

– Short term actions : reduction of the moisture content, flocculation of the clay minerals and modification of the geotechnical characteristics (modification of the Atterberg limits and of the Proctor curve, immediate increase of the bearing capacity).

– Long term actions : slow combination with the clay minerals of the soil ("pozzolanic" reaction) and increase of the mechanical performance (Rc and cohesion).

2.2 Lime or cement

The answer depends on the clay content of the material to be treated as well as on the objective of performance. Cement is effective in presence of "clean" materials (very low content of clay, like sand and gravel). Thanks to its combination with clay, lime is effective in presence of clay containing materials. The limit between the fields of application of cement and lime depends on the proportion and the activity of the clay. As it cannot be predetermined, performance tests are necessary to chose the right binder and the right content.

In general, the mechanical performance of a cement treated granular material is higher than the performance of a lime treated clayey material. However, it is possible to enhance the performance of a clayey material thanks to a double treatment : lime treatment first, to flocculate clay and reduce its activity, cement treatment after, to obtain a higher level of performance rapidly.

3 Difference between STD and hardfill

Soil Treated Dams (STD) are one of the 3 types of the Cemented Material Dams (CMD), as

---

2. For the need of this paper, calcium air lime will be called "lime".

indicated in Table 1, for which ICOLD Technical Committee P is preparing new bulletins.

Table 1 Types of Cemented Material Dam

Natural material		type	Grading	Material combined with cementitious binder					
Earth & Rockfill	Earth Fill	Soils	Clay	ST					
			Silt						
			Sand						
	Gravel Fill	Soils or from Rock	Fine Gravel	SC	HF or CSG	CSGR	Mortar SCC	RCC	CVC
			Coarse Gravel or Stone				Masonry		
	Rock Fill	Roc	Block				RFC		CYC
Embankment Dam				ST Dam	Hardfill Dam		Masonry Dam / RFC Dam	RCC Dam	Concrete Dam
				Cemented Material Dam (CMD)					
				3	4	5		6	7

Legend of Table 1:

ST: Soils Treated with lime and/or cement   RCC: Roller Compacted Concrete

HF: HardFill   CYC: CYclopean Concrete

SC: Soils Cement   RFC: RockFill Concrete

CSG: Cemented Sand Gravel   CVC: Conventional Vibrated Concrete

CSGR: Cemented Sand Gravel Rock   SCC: Self Compacted Concrete

3. This type of dam is with symmetrical faces. Minimum material processing before mixing with lime or/and cement.

4. This type of dam is most of time with symmetrical faces. Minimum material processing before mixing with cementitious binder. Hardfill Dams are proposed to be classified considering total sum of upstream and downstream slopes larger than $1:1(H:V)$.

5. This type of dam can be mostly of gravity, arch – gravity, arch type. Minimum processing of stones or rock blocks before placement of mortar or self compacted concrete.

6. This type of dam can be of gravity, arch – gravity, arch type and also with trapezoidal shape orwith symmetrical faces. Various conditions of material processing before mixing with cementitious binder from minimum to maximum. RCC Dams are proposed to be classified considering total sum of upstream and downstream slope smaller or equal to $1:1(H:V)$.

7. This type of dam can be of all type of concrete dams (gravity, arch – gravity, arch, buttress). Maximum material processing (several classes of aggregate and sand, washing, strong aggregate strength).

STD use natural soils materials, which can be composed mainly of clay and silt (see Fig. 2) with almost no processing except eventually screening at the Maximum Size of Aggregatebefore to be mixed with adequate content of lime and/or cement, and, when necessary, with water. STD also use other materials composed mainly of sand mixed with cement, previously known as Soils Cement (SC cf. ICOLD Bulletin n° 54).

Generally Hardfill (HF) or Cemented Sand Gravel (CSG) materials accept only limited quantity of clay elements. When some blocks are used, in some cases, it can constitute what Chinese Engineers are calling Cemented Sand Gravel Rock (CSGR). For the new bulletin to be published, the ICOLD CMD Technical Committee (P) has recently decided to call Hardfill dams all the CMD dams of this type, as originally named by P. Londe and M. Lino who first proposed such a type of dam[1].

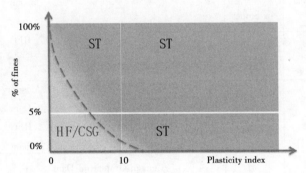

Fig. 2 General characteristics of ST materials compared to HF/CSG

4 Expected functionalities for the "soils treated" (ST)

In order tooptimize the use of the soils treated (ST) as well as the design of STD, the different functionalities of the ST component have been listed, based on the requirements for either the construction or the behaviour of the dam. For each functionality, the requested performance are given. Table 2 synthesizes the results.

Table 2 should be read according to the following : a given project may have one requirement or several amongst the list in the table. The functionalities are described in the lignes below as well as the performance to be checked. For instance, if the soil is too wet, one may only look for the workability and will choose the right binder and dosage – generally lime in this case – to reach the necessary bearing capacity and compacity. If workability, stability and surface protection are requested, the performance to reach will be bearing capacity, compacity, cohesion, compressive strength, homogeneity and resistance to surface erosion. The content and procedures of the studies shall be established in order to quantify these parameters.

Table 2 Requirements for the project and functionalities for the component "treated soil"

Requirements	Workability	Stability	Impermeability	Surface protection	Evacuation
Function(s) of the treated soil	• Facilitated works	• Improved mechanical stability	• Impermeabilility • Resistance to internal erosion	• Resistance to surface erosion when steady overtopping	• Resistance to surface erosion in the spillway
Performance	• Bearing capacity • Compacity	• Shear strength (cohering) • Compressive strength	• Homogeneity • Permeability • Resistance to Holr Erosion Test (HET)	• Homogeneity • Resistance to surfce erosion when steady overtopping	• Homogeneity • Resistance to surfce erosion in the spillway
Application	• Execution procehure	• Fill	• Watertightness • Filters and drains	• U/S and D/S faces protection	• Spillway
Studies procedure	Description of the organization and the content of the studies (on going)				
Execution procedure	Description of the technology and the procedures for production and placement (on going)				

5 Levels of performance

From the kinetics point of view, the hardening of the lime treated soils is slower than the kinetics of the cement (or lime + cement) treated soils. Fig. 3 gives examples of increase of the unconfined compressive strengthRcwith time for a silty soil (PI = 7; 24% of clay). A threshold between 4 to 5 MPa at 90 days is shown. It corresponds to the minimum performance commonly accepted for hardfill / CSG.

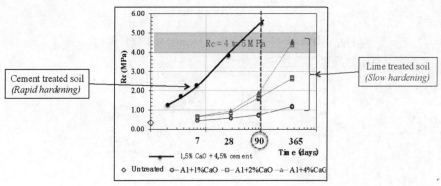

Fig. 3 Example of hardening kinetics of a silty soil treated with lime and lime + cement

In parallell with the increase of Rc, the cohesion is also increasing with time, as shown in Fig. 4 with a clayey silt (PI = 16) treated with 3% quicklime.

Several tests have been performed in the lab and in the field to compare the permeability of non treated soils with the same soilstreated with lime. The results (see an example in Fig. 5) show that we can get the same order of magnitude for a lime treated soil as the natural soil provided the compaction is made on the wet side of the Proctor curve ($W = 1.15$ OMC) with a sheep

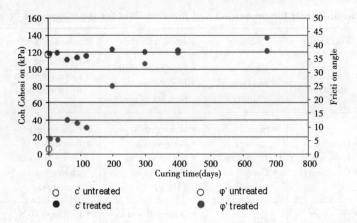

Fig. 4 Modification of c' and φ' with time of a clayey silt ($PI = 16$) treated with 3% quicklime foot roller.

The resistance to internal erosion has also been measured according to the Hole Erosion Test (HET)[3] on treated mixtures[5]. In spite of the slow kinetics of reaction between lime and clay, the critical stress increases rapidly with time, even in the case of a silty soil ($PI = 9$) treated with 2% quicklime only (Fig. 6).

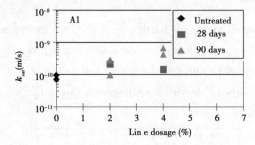

Fig. 5 Permeability of a lime treated soil ($PI = 7$)

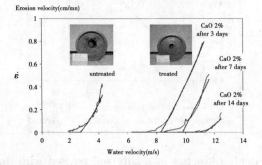

Fig. 6 Hole erosion test on a lime treated soil ($PI = 9$)

6 Soil Treated Dam concept

The question is: How to design and build a dam using soil treated material?

6.1 Height of the dam

The first issue is about the height of a STD. For instance, the lime treatment of a soil increases the cohesion of the material. Fig. 4 here above showsthe results on clayeysilttreated with 3% of quicklime: the cohesion grows from 10 kPa before treatment to 20 kPa just after treatment and 100 kPa after one year. It should be noted that the friction angle is not modified by the lime treatment and is in the range [28° ~ 35°]. As a consequence, only small to medium height dams shall be in the scope of STD with lime treated material, because, as we know, the stability of small dams relies upon the cohesion. But the stability of high dams relies mainly on friction angle and cohesion becomes marginal. In this paper we will discuss dams with height up to 50 m as a reference.

6.2 STD mechanical behaviour

Fig. 4 shows that the cohesion develops slowly, from 20 kPa after treatment to 100 kPa after one year, and continues to increase after.

This is a main difference with granular materials treated with cement like hardfill or CSG where the cohesion is increased in hours or days following the placement. A consequence is that the slopes of the ST dam will be basically determined by the stability during construction and therefore the profile will be symmetrical.

6.3 Pore pressure building

The stability of an embankment built with fine materials depends greatly on the pore pressure building during construction. Pore pressure building is also a concern for STD. Pore pressure building depends on the permeability of the material and its deformability. To reduce the risk of pore pressure building, it is convenient to place the ST close to the optimum Proctor, let's say [OMC − 1, OMC + 1]. In this range, the quantity of free water in the soil is reduced and also the permeability of the treated material is higher than for the naturel soil, which is favorable for low pore pressure building.

Oedometric tests have been performed on ST [4]. The general trend is as follows:

- The expansion index Cs of the natural clay soil is divided by a factor of 5 to 10 after treatment,
- The yield strengthps of the natural clay soil is multiplied by a factor of approximately 5 to 10 after lime addition (from 50 kPa to 400 ~ 500 kPa with 2% lime addition, in the example cited)
- Compressibility index C_c of the natural clay soil is not explicitly affected by thetreatment.

So, the ST deformability is low for a chargeof fillup to 20 ~ 25 m, compared to 2 m,5 m for untreated soil. This low deformability tends to limit the pore pressure building for dam height lower than 50 m. This preliminary analysis has to be confirmed by further laboratory and in situ tests.

6.4 Proposed typical section for STD

The profile is symmetrical with upstream and downstream slopes in the range 1:1($H:V$) to 1.5:1($H:V$) A watertight facing, with drainage underneath, is provided on the upstream face of the dam (Fig. 7). In this manner, the ST dam body is mainly out of water and has no function of

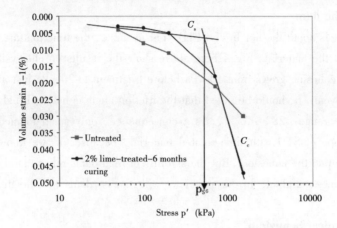

Fig. 7 Compressibility of ST (oedometric tests from [4])

watertightness. Cracking of the dam body during construction or first filling is not a problem, provided that the cracking of the dam can be accomodatedby the upstream facing.

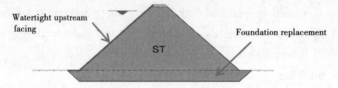

Fig. 8 Soil treated dam section

The upstream facing can be a concrete slab as in hardfill dam but the sliding stability of the slab is questionable if the foundation is deformable. Geomembrane anchored in the dam body can also be considered, as it is designed for the Filiatrinos (Greece) and in the Quatabian (Irak) hardfill dams.

6.5 Foundation improvement

Generally, the sites where a STD will be appropriate may have soil or weak rock foundation. The slopes of the STD must be compatible with the mechanical properties of the foundation. A partial replacement of the soft part of the foundation by the same ST material has to be considered.

The objective of this replacement is to improve the stability of the dam and also to limit the settlement of the foundation. It is considered as a basic component for STD to accommodate poor foundation conditions.

6.6 Stability analysis

The stability of the STD of 30 m high built with the ST material described above (see Fig. 4) has been checked by circular slope analysis as an embankment dam as well as by limit equilibrium analysis along horizontal planes as a concrete or hardfill dam. It proves that the slope stability analysis is the most relevant.

The stability of a 30 m high STD dam, 300 m crest length, being placed in one year has been checked. With the shear strengthof the material described in Fig. 4 and with the hypothesis of a limited pore pressure building (ru = 0.2), the stability of the profile is satisfied during construction and operation.

7 Application procedures and technology

Two procedures are possible to achieve soil treatment : in place (or "in situ") and in central plant.

The most common way is in place, layer by layer, either in the cut, followed by earthmoving, or in the fill, after earthmoving. The thickness of each layer depends on the capacity and performance of the mixer and the roller. It is currently limited to 35 cm.

The execution procedure consists in (see Fig. 8) :
- Moisturing when necessary;
- Spreading of the binder;
- Mixing (sometimes several mixing operations may be necessary);
- Compacting.

The technology has improved dramatically during the last fifty yearsand allows for good quality mixtures with a good accuracy in the binder dosage. The output depends on the type and number of equipments used (mainly the number of mixers). One mixer is able to mix 200 to 300 m^3/h.

Since 10 to 15 years, it is also possible to treat with lime or cement humid and/or clayey soils in central plant. In that case, moisturing, spreading and mixing are achieved by the plant (Fig. 9).

This procedure allows for very homogeneous mixtures with a high accuracy in the binder dosage as well as the water content. The output depends on the size of the plant. It can be from 50 to more than 500 m^3/h.

(a)Moisturing

(c)Binder spreading

(b)Mixing

(b)Rolling

Fig. 9 **Treatment in place**

Fig. 10　**Treatment in central plant**

8　Cost comparison

The cost comparison is based on a 30 m high, 300 m long dam with a crest width of 6m.

8.1　Homogeneous embankment dam (us/ds slope: 3 to 1 (H/V))

- earth for foundation (depth: 0.5 m → 9 900 m^3): excavation + backfill = 18.60/m^3
- earth for embankment (864 000 m^3): 7.20/m^3
- filter & drain (thickness: 3 m → 56 700 m^3): 25.00/m^3
- rip-rap (thickness: 1.25 m → 35 600 m^3): 35.00/m^3

Total cost: 9.1 M – Duration of earthworks: 123 days

8.2　Symmetrical STD (us/ds slope: 1 to 1 (H/V))

- ST (3% lime) for foundation (depth: 3.0 m → 59 400 m^3): excavation + backfill = 24.60/m^3
- ST (3% lime) for embankment (324 000 m^3): 13.00/m^3
- concrete facing (thickness: 0.25 m → 182 m^3): 500/m^3

Total cost: 7.3 M – Duration of earthworks: 55 days

8.3　Comparison

Cost reduction with STD: 1.8M (20%) – Duration reduction: 68 days (55%)

9　Conclusion

There is a large international experience of use of soil treated with lime and cement in the field of transport infrastructure (roads, railways, airports, etc.). Some applications also exist in hydraulic structures for canals and dykes. Soil treatment with lime and/or cement dramatically improves the shear strength and the resistance to erosion of the soil and reduces its deformability. This paper is a tentative proposal of a new dam design in order to use ST for dam construction. The typical STD has a height lower than 50 m, its profile is symmetrical and it is sealed by an upstream facing. Foundation improvement by substitution of the soft surface layer by ST is considered. The main question mark may be the development of pore pressure during construction. There are some indications that ST will not develop high pore pressure but it still has to be confirmed by additional laboratory tests and in situ measurements.

Acknowledgements

The authors are members of the French mirror group of the ICOLD Committee P. They gratefully acknowledge the other members of the group for their support and contributions: P. Agresti (Artelia), S. Bonelli (Irstea), J. -J. Fry (EDF), V. Mouy (Coyne & Bellier), N. Nerincx (ISL).

References

[1] P. Londe, M. Lino. The faced symmetrical hardfill dam: a new concept for RCC – Water Power & Dams Construction – February 1992.

[2] LCPC – SETRA. Soil treatment with lime and/or hydraulic binders: Application to the Construction of fills and capping layers – Technical Guide – LCPC Editions, Paris (France), 2000.

[3] S. Bonelli, N. Benahmed. Piping flow erosion in water retaining structures [J]. International Journal of Hydropower and Dams, 2011, 18(3): P4 – 98.

[4] G. Herrier, D. Puiatti, S. Bonelli, et al.. Froumentin, Lime treated soils properties for application in railways infrastructure with hydraulic constraints – Georail International Symposium, Paris, 2014.

[5] G. Herrier, D. Puiatti, S. Bonelli, et al.. Nerincx, M. Froumentin, Le traitement des sols à la chaux : une technique innovante pour la construction des ouvrages hydrauliques en terre – 25th ICOLD Congress (Q96, R39), Stavanger, June 2015.

A Study of Lift Joint Shearing Strength of RCC Dams

Zhang Jianguo, Zhou Yuefei, Xiao Feng

(Power China ZhongNan Engineering Corporation Limited, Changsha, 410014, China)

Abstract: Roller compacted concrete (RCC) is different from conventional concrete in terms of strength property due to its layering system. Usually the lift joint shearing strength is lower than the strength of the body for RCC, so the lift joints are the controlling sections for dam's stability against sliding in most cases. Consequently, it's worth discussing the current practice of applying the processing method for conventional concrete gravity dams to sectional design of RCC gravity dams. The paper sorts out the reliability – theory – based partial factor limit state design process according to the physical properties of RCC, focuses on the study of probability distribution pattern of RCC lift joint shearing parameters, characteristic value selection principle, material partial factor calculation and structural coefficient conversion through project investigation and research, expert assessment and statistical analyses of test data of Longtan and other existing projects, works out the equations suitable for limit state design of lift joint stability against sliding for RCC gravity dams, and carries out reliability calibration for various types of typical gravity dam sections designed by the proposed equations based on analyses of the statistical characteristics of basic variables affecting gravity dam's reliability, determines the target reliability indicator of lift joint stability against sliding of RCC gravity dam on the basis of the results of analyses and calculations above, and thus judges the rationality of the proposed equations for ultimate limit state design of stability against sliding.

Key words: RCC, lift joint shearing strength, partial factor, reliability calibration

1 Background

Since 1970s, roller compacted concrete (RCC) dam building technology has experienced rapid development across the world due toits advantages of quick construction, short construction time and low cost. Through years of development, the RCC dam has become into one of the most competitive dam types, and its building technology has been widely applied in the construction of hydropower projects. Since the completion of Fujian Kengkou Dam (the first RCC gravity dam in China) in 1986, there have been almost one hundred RCC dams completed or under construction in China. Marked by Longtan and Guangzhao dams, the RCC dam building technology in China has made new breakthrough through innovation based on technology introduction, digestion and absorption, and RCC dam building technology with Chinese characteristics has been developed affirmed by the world dam engineering circle.

At present, in most cases the design of RCC gravity dams in hydropower industry is following the design philosophy for conventional concrete gravity dams, stressing on the common characters of these two types of dams and ignoring the characters of RCC to some extent. To meet the demand of building high RCC dams and reflect the technological advance in RCC dam building, it's necessary to carry out a special study on sectional design of RCC gravity dams according to their characteristics.

2 Content and Methodology of Study

Due to the layering system of RCC gravity dam, usually the lift joints are critical to the control of dam's stability against sliding, so many experimental studies have been carried out in China and other countries. Following the stipulations of the Unified Design Standard for Reliability of Hydraulic Engineering Structures[1] and the design philosophy of probability limit state, the paper presents the sectional design and calculations for RCC dams based on partial factor limit state design equations. The study focuses on the lift joint shearing design for RCC gravity dams less than 200 m high and proposes the probability limit state design equations and the corresponding partial factors. The study mainly covers:

(1) Determination of bearing capacity limit state design equations;
(2) Valuing of lift joint anti-shear material partial factor;
(3) Derivation of structural coefficient of lift joint;
(4) Reliability check for lift joints of RCC gravity dams designed on a trial basis.

The method adopted in the study is to determine the distribution model and characteristicfractile of concrete lift joint shearing parameters through investigation, research and expert assessment, ascertain the anti-shear material partial factor as per the Unified Design Standard for Reliability of Hydraulic Engineering Structures, carry out sectional design for RCC gravity dams of various height orders separately based on reliability theory, conduct structural coefficient conversion based on limit state design theory, make calibration through comparison with the safety factors against sliding stated in the pertinent industry specifications, then work out the correlation coefficients in the limit state design equations, and finally do trial design for sections of RCC gravity dams of various height orders by the proposed equations and make reliability calibration.

3 Bearing Capacity Limit State Design

In the bearing capacity limit state designprocess for lift joint stability against sliding of RCC gravity dams, following the calculation principles stated in the Unified Design Standard for Reliability of Hydraulic Engineering Structures and DL 5108 – 1999 the Design Specification for Concrete Gravity Dams, the bearing capacity limit state design equation[2] may be described by the following equation:

$$\gamma_0 \psi S_d(\,\cdot\,) \leq \frac{1}{\gamma_d} R_d(\,\cdot\,) \tag{1}$$

where, γ_0 is coefficient for importance of structure; ψ is design condition coefficient; $S_d(\,\cdot\,)$ is effect design value function of action combination; $R_d(\cdot)$ is design value function of structure resistance; γ_d is structural coefficient.

The effect design value of basic combination of bearing capacity limit state shall be calculated by the equation below:

$$S_d(\cdot) = S(\gamma_G G_k, \gamma_P P, \gamma_Q Q_k, a_k) \tag{2}$$

$$R_d(\cdot) = R(\frac{f_k}{\gamma_m}, a_k) \tag{3}$$

where, G_k is standard value of permanent action; P is representative value of prestressed action; Q_k is standard value of variable action; f_k is standard value of material strength; γ_G is partial factor of permanent action; γ_P is partial factor of prestressed action; γ_Q is partial factor of variable action; γ_m is partial factor of material property; a_k is standard value of geometrical parameter (it may be treated as a constant value).

The design equations above show that the bearing capacity limit state design equations are described by six partial factors, i.e., coefficient for importance of structure γ_0, design condition coefficient ψ, partial factors of permanent and variable actions γ_G and γ_Q, material performance partial factor γ_m and structural coefficient γ_d, as well as the standard values of design variables.

In those factors, coefficient for importance of structure γ_0, design condition coefficient ψ, partial factors of permanent and variable actions γ_G and γ_Q are unrelated to which dam building technology is applied, RCC or conventional. Therefore, the study of this paper emphasizes on:

(1) Material partial factor γ_m, which reflects the adverse variation of actual material strength on the adopted standard value of material strength;

(2) Structural coefficient γ_d, which reflects the uncertainty of action effect calculation mode, uncertainty of material resistance calculation mode, and other uncertainties that cannot be reflected by the action partial factors and material performance partial factor above-mentioned.

4 Study of Material Partial Factor

In the limit state design, the conversion between standard value and design value is made through material partial factor, so the determination of material partial factor is directly related to the material strength probability model, variability and fractile selection of standard value and design value.

4.1 Determination of characteristic value

According to the stipulations of theUnified Design Standard for Reliability of Hydraulic Engineering Structures, discrete degree of shearing parameters "f" and "c", and selection of dam foundation shearing parameter probability distribution model, normal distribution and lognormal distribution were selected as the probability distribution of concrete lift joint shearing parameters "f" and "c", respectively.

The standard values can be selected by various mathematical statistic methods. The methods stated in the relevant specifications include mean value, small mean value, 0.2 fractile, lower limit of predominant slope method, and so on, while in practice the standard values obtained from various mathematical statistic methods differ from each other to some extent. This difference has brought about considerable inconvenience to the designers in parameter selection. In accordance with the experiences from preparing the Design Specification for Concrete Gravity Dams and expert assessment comments and referring to the principle on selection of shearing parameter stated in SDJ21—78 the Design Specification for Concrete Gravity Dams[3], the small mean value of peak strength was selected as the standard value in the study.

Normally the design value should beselected close to the design check point, and when it is difficult to do so, a fractile of its probability distribution may be adopted referring to the experiences. 0.977 3 was taken as the fractile in the study through overall consideration (so - called "2σ" criterion).

4.2 Statistics of coefficient of variation

The data for statistical processing of variation coefficients of RCC dam body lift joint shearing parameters "f" and "c" were mainly sourced from the RCC shearing tests of Longtan Hydropower Project. In the statistics of the test data, RCC specimens were categorized into 19 types, i.e., approach channel, left bank and right bank with various concrete grades and curing conditions. The mean values of variation coefficients of shearing parameters "f" and "c" were 0.117 and 0.212, respectively.

In consideration of the particularconditions of Longtan HPP and through comparison with the variation coefficients of lift joint shearing parameters of Yantan[4] and Dengke[5] RCC gravity dams, 0.2 fractile of statistical values of variation coefficients of 19 types of Longtan RCC specimens was taken as variation coefficients of RCC lift joint shearing parameters. The statistical values of variation coefficients of lift joint shearing parameters "f" and "c" were 0.17 and 0.32, respectively.

4.3 Material performance partial factor

Material performance partial factor is utilized to describe the unfavorable variation of material performance against the standard value. Proceeding from the statistical test data of material specimens and taking the variability of specimen material performance into consideration, it is a coefficient reflecting the variation of material performance. The material partial factor was calculated by the equation below:

$$\gamma_m = \frac{f_k}{f_d} \tag{4}$$

where, f_k is standard value of material performance; f_d is design value of material performance.

Since standard value and design value are not single constant, it's difficult to work out the material partial factor simply by the above equation. It's necessary to determine the fractile of characteristic value based on probability model, and the key in calculation is the standard value based on the small mean value.

4.3.1 Fractile corresponding to standard value of shear friction coefficient "f"

Shear friction coefficient "f" was considered as in normal distribution, and its density function was:

$$f(x) = \frac{1}{\sqrt{2\pi}\sigma} e^{-\frac{(x-\mu)^2}{2\sigma^2}} \tag{5}$$

Since the friction coefficient "f" was considered as in normal distribution and the accumulative probability below mean value was 50%, theoretically the small mean value may be expressed as:

$$E(x') = \int_{-\infty}^{\mu} \frac{x'f(x')}{0.5} dx \tag{6}$$

Fractile of small mean value of the friction coefficient "f" was derived from:

$$\alpha_f = \varphi\left(\frac{2}{\sqrt{2\pi}}\right) \tag{7}$$

The result shows that the assurance rate of small mean value of friction coefficient "f" is a fixed value unaffected by other factors, so the fractile corresponding to small mean value of friction coefficient "f" $\alpha_f = 0.212$.

4.3.2 Fractile corresponding to standard value of shear cohesion "c"

Shear cohesion "c" is considered as in lognormal distribution[6], and its density function was:

$$f(x) = \int \frac{1}{\sqrt{2\pi}\sigma_1} e^{-\frac{(\ln x - \mu_1)^2}{2\sigma_1^2}} \tag{8}$$

The fractile corresponding to small mean value of shear cohesion "c" was derived from:

$$\alpha_c = \varphi\left(\frac{\ln(\sqrt{1+\delta^2} \cdot y(\delta))}{\sqrt{\ln(1+\delta^2)}}\right) \tag{9}$$

Where,

$$y(\delta) = \varphi\left(\frac{\ln\left(\frac{1}{\sqrt{1+\delta^2}}\right)}{\sqrt{\ln(1+\delta^2)}}\right) \bigg/ \varphi\left(\frac{\ln\sqrt{1+\delta^2}}{\sqrt{\ln(1+\delta^2)}}\right) \tag{10}$$

The result suggests that the fractile corresponding to small mean of cohesion "c" only relates to the variation coefficient of the specimens and is unrelated to the mean value. According to the statistical value of variation coefficient of RCC cohesion, the fractile corresponding to small mean value of cohesion "c", $\alpha_c = 0.260$.

Based on the determined fractile corresponding to small mean value, the material partial factor for lift joint shear resistance can refer to the stipulations of the Unified Design Standard for Reliability of Hydraulic Engineering Structures, and the material partial factors for RCC lift joint shearing parameters "f" and "c" were assumed at 1.3 and 1.5, respectively.

5 Calculation of Structural Coefficient

The structural coefficient is a coefficient of colligation describing the uncertainty of calculation mode in limit state design. With target reliability as the design criterion in the calculation of structural coefficient for dam lift joint stability against sliding, the structural shape design was carried out, and the safety factor was checked and compared with that stated in SDJ 21—78 the Design Specification for Concrete Gravity Dams.

Various dam heights, structural security levels, and rich/poor cementitious materials for lift jointswere considered in the structural shape design, and 14 groups of dam sections up to target reliability indicator were proposed for structural coefficient conversion. The statistical results are presented in Table1.

Table 1 Determination of structural coefficient in anti – sliding stability limit state design for RCC gravity dam foundation

No.	Dam height	Structural security level	Coefficient for importance of structure	Type of cementitious material	Dam body structural section		Design comparison		Structural coefficient
					U/S slope	D/S slope	Target reliability	Safety factor in SDJ 21-78	
1	200	I	1.1	Rich	0.15	0.67	4.2	2.329	1.53
2	200	I	1.1	Poor	0.15	0.84	4.2	2.379	1.56
3	170	I	1.1	Rich	0.15	0.65	4.2	2.445	1.56
4	170	I	1.1	Poor	0.15	0.82	4.2	2.491	1.6
5	150	I	1.1	Rich	0.12	0.63	4.2	2.474	1.62
6	150	I	1.1	Poor	0.12	0.79	4.2	2.502	1.6
7	120	I	1.1	Rich	0.10	0.55	4.2	2.497	1.57
8	120	I	1.1	Poor	0.10	0.69	4.2	2.494	1.58
9	100	I	1.1	Rich	0.05	0.5	4.2	2.501	1.57
10	100	I	1.1	Poor	0.05	0.64	4.2	2.538	1.6
11	70	II	1	Rich	0	0.34	3.7	2.164	1.48
12	70	II	1	Poor	0	0.43	3.7	2.164	1.49
13	50	II	1	Rich	0	0.27	3.7	2.134	1.44
14	50	II	1	Poor	0	0.34	3.7	2.134	1.45

Notes: (1) The shape design presented in the table is limited to the anti – sliding stability limit state design for lift joints.

(2) Lower limit strengths of poor/rich cementitious materials stated in DL 5108 – 1999 the Design Specification for Concrete Gravity Dams were taken as the shearing strength of lift joints.

The design results of 14 groups ofvarious types of sections up to target reliability indicator presented in Table 1 suggest that:

(1) According to the comparison with the stipulations of SDJ 21—78 the Design Specification for Concrete Gravity Dams, the margin is greater when lift joints for RCC gravity dam are designed based on the safety factor method stated in the industrial specifications, especially for the dams lower than 100 m high;

(2) The structural coefficients converted from design sections of various dam heights, structural security levels and cementitious materials are concentrated in the range from 1.44 to 1.62. That shows it's basically feasible to take a fixed structural coefficient to replace the uncertainty of various calculation modes.

6 Reliability Calibration

According to the study results above and the RCC gravity dam characteristics, the proposedlift joint shear bearing capacity limit state calculation equation and the relevant coefficients are as follows:

$$\gamma_0 \psi S_d(\gamma_G G_k, \gamma_Q Q_k, \alpha_k) \leq \frac{1}{\gamma_d} R_d\left(\frac{f_k}{\gamma_m}, \alpha_k\right) \quad (11)$$

where, γ_0 is coefficient for importance of structure, taken at 1.1, 1.0, and 0.9 for structures or structural members with structural security levels of I, II and III, respectively; ψ is design condition coefficient, taken at 1.0 for permanent situation; γ_m is material performance partial factor, 1.3 for partial factor of lift joint shear friction "f" and 1.5 for partial factor of cohesion; γ_d is structural coefficient, 1.6 for comprehensive statistics.

In order to analyze the reliability of the gravity dams designed based on the partial factors proposed herein, firstly, a structural safety factor of 1 was taken to work out the sectional dimensions of gravity dam under various conditions, and then the reliability indicator of the corresponding sections of gravity dam was calculated based on the reliability theory.

In the process of reliability calibration, comprehensive statistical analysis was conducted for various dam heights, lift joint anti-shear materials and structural security levels. The statistical results are listed in Table 2.

Table 2 Statistics of reliability calibration results

No.	Dam height	Structural security level	Coefficient for importance of structure	U/S slope	D/S slope up to proposed equation	Reliability	D/S slope up to SDJ21-78	Ratio of material utilization amount
1	190	I	1.1	0.20	0.76	4.17	0.92	83.41%
2	170	I	1.1	0.18	0.71	4.18	0.87	82.75%
3	150	I	1.1	0.15	0.67	4.2	0.82	83.53%
4	130	I	1.1	0.12	0.63	4.24	0.76	85.16%
5	110	I	1.1	0.10	0.58	4.27	0.70	85.99%
6	90	II	1	0.05	0.50	3.93	0.66	80.82%
7	70	II	1	0	0.45	4.05	0.58	80.84%
8	50	II	1	0	0.348	4.10	0.44	82.74%

Note: The ratio of material utilization amount refers to the sectional area designed by the equations proposed herein and that designed as per SDJ 21—78.

The results of sections designed by the proposed lift joint shear bearing capacity limit state design equations show that the reliability of all the trial design sections with structural security level I is close to 4.2, and the reliability of all the trial design sections with structural security level II is over 3.7. In general, the sections designed by the proposed lift joint shear bearing capacity limit state design equations could meet the reliability demand. Through comparison with the material utilization amount stated in SDJ 21-78 the Design Code for Gravity Dams, the design based on the proposed bearing capacity limit state is obviously more efficient economically.

7 Conclusions

According to the characteristics of RCC gravity dams and based on the test data of Longtan and other existing projects, a set of new parameters suitable for lift joint shear limit state design for RCC gravity damswere sorted out, and lift joint anti-sliding reliability of RCC gravity dams was calibrated through trial design based on the proposed correlation coefficient proposed. It's verified that the proposed bearing capacity limit state design equations could meet the structural reliability demand stated in the Unified Design Standard for Reliability of Hydraulic Engineering Structures.

Theresults of comparison with the shearing design as per SDJ 21-78 the Design Specification for Concrete Gravity Dams suggest that lift joint stability check based on dam foundation shearing is conservative somewhat and not economically efficient. However, the lift joint shearing test data collected in the study is limit, further collection of material test data for lift joints from existing projects is therefore necessary to work out limit state design equations appropriate for lift joint design for RCC gravity dams.

References

[1] Unified Design Standard for Reliability of Hydraulic Engineering Structures (GB 51099—2013) [S]. Beijing: China Planning Press, 2013. (in Chiese)
[2] Design Specification for Concrete Gravity Dams (DL 5108—1999) [S]. Beijing: Water Resources and Electric Power Press, 2000. (in Chinese)
[3] Design Specification for Concrete Gravity Dams (SDJ 21—78) [S]. Beijing: Water Resources and Electric Power Press, 1979.
[4] Miao Qinsheng. Statistical Analysis and Determination of Shearing Parameters of Lift joints of Dam Foundation and Dam Body and Contacts between Rock Mass and Concrete for Concrete Gravity Dams [J]. Standardization of Survey and Design for Water Conservancy and Hydropower Projects, 2000.
[5] Zhu Junsong, Zi Jinjia, et al. Reliability Analysis of Lift Joint Anti-Sliding Stability for Dengke RCC Gravity Dam [J]. China Rural Water and Hydropower, 2011.
[6] Hehai University, et al. Hydraulic Reinforced Concrete Structures (3rd Edition) [M]. Beijing: China Water & Power Press, 1996.

Kahir Dam: First FSHD Experience in Iran

Ebrahim Ghorbani[1], M. Lackpour[2], A. Mohammadian[3]

(1. Independent concrete specialist, Iran; 2. Pajouhab Consulting Engineers, Iran;
3. Abfan Consulting Engineers, Iran)

Abstract: The Kahir dam will be the first faced symmetrical hard Engineering fill dam to be constructed in Southeast of Iran with maximum height of 54.5 m and 370 m crest length with $1(V):0.8(H)$ upstream and downstream slope. According to the kahir dam designer, due to weak foundation and high dynamic loading, the required compressive strength at the age of 180 day, is 7 MPa. Also, with regard to the results of petrography, long and short term tests, borrow area have alkali – silica reactions. Both of them, increase the cement content in mix design, while, a low cement content is a basic characteristic of a hard fill dam, So, in this paper, after description about general aspect of the project and main design features of the Kahir dam, the procedure of making mix designs with low cement content and economic is presented.

First steps, to simplify the construction procedure and reducing the aggregate production process and the cost, using one proportion of aggregate (0 ~ 50 mm) were studied. On one hand, since in this method of material processing, the sand content of each batch may be changed, the variance of sand content is derived. On the other hand, the effect of increase in sand content over the mechanical properties of the concrete is tested and studied. The results show that it may not be possible to use just one proportion because of high fluctuations in strength and needs to add more cement, So, mix designs were made with two proportion of aggregates.

The second step, the results of supplementary tests show that 15% pozzolan is sufficient for ASR control. In step three; several mix designs with different proportions of cement and admixtures (retarder & water reducer) were examined to obtain optimum mix designs. The cement content in mix designs were assumed 90 kg/m³, 100 kg/m³, 110 kg/m³ and 120 kg/m³. For ensuring that all voids are filled, V_p/V_m ratio considered greater than 0.42. The results show that using two proportion of aggregates (0 ~ 12 mm & 12 ~ 50 mm) and 15% pozzolan, mix designs with cement content about 100 kg/m³ are acceptable in terms of strength and economic.

Key words: FSHD, design, one portion aggregate grading, ASR, mix design

1 General aspect of the project

Kahir dam, the first faced symmetrical hard Engineering fill dam to be constructed in Southeast of Iran. The dam body, which has a height of 54.5 m from the foundation level and symmetric slopes of $0.8(H):1.0(V)$ on both upstream and downstream faces (Fig. 1 and Fig. 2) is under construction.

The execution of the civil engineering works for the construction of Kahir dam, includes the following items: Temporary diversion works, Cofferdam with height of 20 m from the riverbed, The dam and all associated hydraulic structures, Access roads, Auxiliary installations.

The structure of the Kahir dam shall be a straight gravity structure 370 meters long, including a 160 m long free surface spillway.

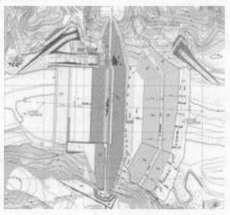

Fig. 1   Plan view of kahir dam

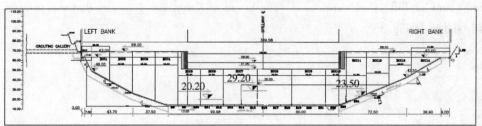

Fig. 2   Elevation from the upstream face

The main quantites involved in the Project are summarized in Table 1.

Table 1   Main quantities of the Project

Excavation	12 000 m³	Dam height from foundation	54.5 m
Reinforcement	6 140 ton	Reservoir Vol	314 million m³
Conventional concrete	179 000 m³	Crest length	370 m
Formwork	76 600 m²	Crest width	5 m
RCC (hard fill)	485 000 m³	Spillway width	160 m

2   Main design features of the Kahir dam

2.1   Introduction

FSHD is a fairly new type of dam, first proposed in 1992[1,2], which is called CSG (cemented sand and gravel) in Japan. This type of dam is built using a low-cost cemented sand and gravel material known as hardfill. A FSHD has several advantages, including a high degree of safety, strong earthquake resistance, low demands for the foundation (proper for poor foundation), simple and quick construction and minimal negative effects on the environment.

Too few hardfill dams are built in the world to establish general criteria for their foundation, but some guidelines can be proposed, on the basis of the yet scarce experience, and theoretical considerations.

In kahir dam, investigations show that foundation is poor with modulus of elasticity between 0.5 and 0.8 GPa (about 5 000 to 8 000 kgf/cm^2). Therefore, FSHD option is a good choice for kahir dam.

2.2 Stress analysis

One of the major challenges has been the need for exhaustive studies to define its geometry in order to guarantee its stability and resistance against various stress conditions, among which are those of its own weight, water pressure, seismic loads and thermal actions. In this respect several analyses where carried out using two dimensional finite element models (by ANSYS Software), considering static, pseudo – static, dynamic and thermal conditions. For each analysis, the dam geometry, edge conditions and stress loads for spillway and non – overflow section with the highest elevations and also three different concrete (CVC1(downstream face), CVC2 (upstream face and leveling) and hardfill) were considered. The layout of Kahir dam geometry that was used in analyzes is shown in Fig. 3.

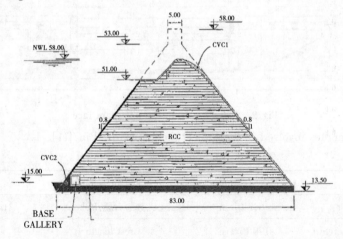

Fig. 3 **Typical cross section**

Also, two levels of earthquakes DBE (including records of earthquakes Kahak, Sibsooran and Silkahor) and MCE (including records of earthquakes Manjil, Bam and Montenegro) to investigate the seismic behavior were considered. The structural design considered a peak ground acceleration of 0.221g for design basis earthquake (DBE), which corresponds to an earthquake expected to occur during the useful life of the dam, and a peak ground acceleration value of 0.753g for the maximum probable earthquake (MCE).

Behavior of concrete dam body and foundation is assumed to be plane strain and four – node isoparametric plane element is used for modeling dam body, foundation and reservoir. The total numbers of plane element of this finite element meshes in thespillway and non-overflow models are equal to 1 153 and 1 180 elements, respectively.

The computational mesh, maximum principal stresses in hardfill, CVC1 anad CVC2 inspillway and non-overflow section are presented in Fig. 4.

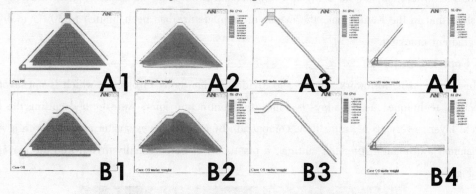

Fig. 4 A: non-overflow section , B: spillway section
(1: meshing, 2,3,4: pricipal stress in hardfill, CVC1 and CVC2)

The concrete strength requirements were established by structural analysis, taking into account different combinations of load and safety factors, determined a maximum value to Compressive strength of 7 MPa at 180 days of age for hardfill, 25 MPa and 20 MPa at 28 days for CVC1 and CVC2, respectively. However as this value is only produced in a very specific sector of the dam, a corresponding design maximum tensile strength for hardfill of 0.62 MPa was defined.

The results show that the absolute maximum horizontal displacement of dam crest for DBE and MCE are 3.09 cm and 7.16 cm, respectively. These values are acceptable for concrete dams with 54.5 m height.

2.3 Upstream face of the hard fill dam

The impervious upstream facing system is essential in the hardfill dam as CFRD type rockfill dam. In the beginning stage of designing Kahir dam, reinforced concrete upstream facing was considered. The facing would be after hardfill dam body complete. But after assessment of shrinkage potential cracks in concrete face andin order to reduce the the execution time, three different facing alternative were tried.

After evaluating the advantage and disadvantage of this various option, PVC membrane faced concrete panels have been adopted.

2.4 Downstream face of the hard fill dam

The important consideration in downstream face is durability and appearance. Because of that durable, high quality cast in place concrete used for downstream facing in the dam.

2.5 Maximum placing temperature

The mean annual air temperature in the dam-site is 27.7 ℃ with minimum monthly average temperature of 18 ℃ in the January; the maximum monthly average temperature is 37 ℃ in June. In this project the RCC will be placed in 0.3 m lifts that is conventional for RCC dams.

There are two major type of cracking in the RCC dams including surface gradient cracking and mass gradient cracking. There is no risk of surface gradient due to the tropical region and covers the upstream, so, surface gradient cracking has not been investigated. In this respect several analyses where carried out using two dimensional finite element models (ANSYS 5.4). Results show that in the Kahir dam, the maximum temperature has been limited to 30 ℃ to control mass gradient cracking.

2.6 Contraction joints

Instead of using formwork, that is, to divide the dam body in blocks, which is not favorable to the RCC technique, as nowadays is done; the contraction joints were formed cutting the fresh concrete layer, every 36 m, after RCC compaction by means of a excavator mounted with a vibrating hammer and blade. After the cutting, a plastic sheet was manually driven into the performed slot (Fig. 5).

Fig. 5 Procedures for casting the contraction joint (insert of a plastic element)

2.7 Bedding mixes

In FSH dams, no bedding mixes need to be provided after cold joints or in the upstream dam area[3]. However, all layers was performed a bedding – mix within the limits specified in the upstream and downstream area (Fig. 6). Its mix design include 350 kg/m^3 kash pozzolan cement, 242 L/m^3 water, 1 640 kg/m^3 sand (0 ~ 5 mm) and 2.1 L/m^3 additive (Tard CHR).

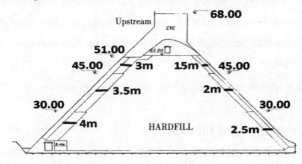

Fig. 6 Bedding – mix within the limits specified in the upstream and downstream area

3 Investigation of alternative proposals

3.1 Feasibility study of using one portion of aggregate

At the request of the contractor, for simplify the construction procedure and reducing the aggregate production process and the cost, using one proportion of aggregate (0 ~ 50 mm) was studied. The effect of using one proportion of aggregate and increase in sand content over the mechanical properties of the concrete test was studied in cofferdam. Required compressive Strength of RCC face in cofferdam at the age of 28 days is 10 MPa. Its mix design include 160 kg/m^3 kash pozzolan cement, 112 L/m^3 water, 2 197 kg/m^3 one portion of aggregate and 1 L/m^3 additive.

Since in this method of material processing, the sand content of each batch may be changed, if the sand proportion of the aggregate is less than specified (43%), the lack of sand is added in the batching plant.

Maximum size aggregate (MSA) was used 50 mm. Grading curve of one portion of aggregate (0 ~ 50 mm) is given in Fig. 7. The fluctuations range in sand (0 ~ 5 mm), fine gravel (5 ~ 25 mm), coarse gravel (25 ~ 50 mm) and passing #200 for each gradation in every month are given in Table 2. As can be seen in the Table 2, the minimum and maximum sand content (0 ~ 5 mm) are 25% and 48%, respectively. Also, the difference between the maximum and minimum content of fine gravel, coarse gravel and passing #200 can be seen significant.

The effect of the high fluctuations of aggregate grading (0 ~ 50 mm) on compressive strength of concrete executed in the cofferdam, have been shown in Fig. 8. Maximum, average, minimum and standard deviation of concrete strength at the age of 7 days, 28 days, 90 days and 180 days are given in Table 3.

According to Table 2 and Table 3, it seems that fluctuations of aggregate grading (0 ~ 50 mm) and concrete strength are very high.

Therefore, the use of one portion of aggregate (0 ~ 50 mm) in the construction of Kahir dam was not proper assessment and instead two portion of aggregates including 0 ~ 12 mm and 12 ~ 50 mm were used.

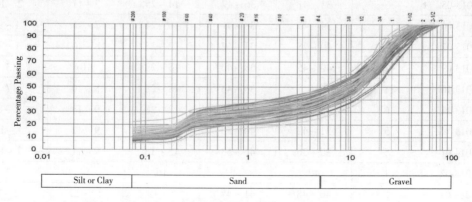

Fig. 7　Grading curve of one portion of aggregate (0 ~ 50 mm)

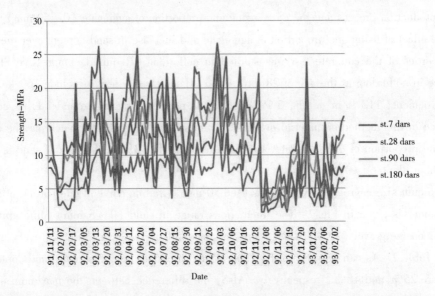

Fig. 8 Fluctuations of compressive strength of concrete executed in the coffer dam

Table 2 Fluctuations range in sand, fine gravel, coarse gravel and passing #200

Date	Number of sample	Fluctuations range in culves (0 ~ 50 mm)				Percentage increase sand to 43%
		Sand(%) (0 ~ 5 m)	Fine Gravel(%) (5 ~ 25 m)	Coarse Gravel(%) (25 ~ 50 m)	Passing #200	
2013-01	3	28 ~ 38	36 ~ 41	20 ~ 36	8 ~ 9	5 ~ 1
2013-02	9	25 ~ 48	34 ~ 41	14 ~ 34	5 ~ 11	3 ~ 18
2013-03	2	31 ~ 37	41 ~ 47	22 ~ 23	6 ~ 7	6 ~ 12
2013-05	6	33 ~ 41	36 ~ 47	14 ~ 25	7 ~ 11	2 ~ 10
2013-06	9	34 ~ 45	38 ~ 41	16 ~ 24	7 ~ 11	1 ~ 9
2013-07	9	40 ~ 48	33 ~ 42	12 ~ 27	8 ~ 10	1 ~ 3
2013-09	2	40 ~ 45	38 ~ 40	15 ~ 23	10	3
2013-10	3	36	36 ~ 46	18 ~ 28	8 ~ 10	7
2013-11	6	35 ~ 46	36 ~ 44	12 ~ 28	9 ~ 13	1 ~ 8
2013-12	5	31 ~ 42	40 ~ 50	16 ~ 22	9 ~ 12	1 ~ 12
2014-01	5	30 ~ 36	46 ~ 53	15 ~ 18	9 ~ 11	7 ~ 13
2014-02	1	40	44	16	14	3
2014-03	5	40 ~ 48	41 ~ 47	7 ~ 19	12 ~ 22	2 ~ 3
2014-04	4	35 ~ 41	47 ~ 55	10 ~ 15	12 ~ 18	3 ~ 8
2014-05	3	34 ~ 42	46 ~ 50	10 ~ 16	11 ~ 16	1 ~ 9
Min – Max	70	25 ~ 48	33 ~ 55	7 ~ 36	5 ~ 22	1 ~ 18
Ave	70	39	42	20	10	6

Table 3 Maximum, average, minimum and standard deviation of concrete strength

Strength(MPa)	7 days	28 days	90 days	180 days
Max. St	15.16	23.61	24.79	26.64
Ave. St	6.8	9.8	12.8	15.2
Min. St	1.45	1.66	1.94	3.55
SD	2.9	4.3	5.1	5.7

3.2 Feasibility study reducing the percentage of pozzolan

In the second phase of studies, the results of petrography, long and short term tests (according to ASTM C1260 and ASTM C1293) indicate that river aggregates of borrow area have alkali-silica reactions (ASR). Also, resultes of short term test show that adding 30% natural pozzolan is effective on control of ASR expansion. The results of long term tests at 7 days, 14 days, 28 days, 56 days, 90 days and 180 days and short term tests with and without adding 30% natural pozzolan are presented in Fig. 9 and Fig. 10. As can be seen in Fig. 10, the expansion at the age of 6 months is much more than 0.04% (expansion criteria). In the construction phase, client laboratory sampling method (accordance to ASTM C1293) was reviewed and we found that there are some problems in sampling method. So, it was decided that long and short term tests repeated again after correction sampling method and results of them are given in Fig. 11 and Fig. 12. The results show that the percentages of expansion of new samples are significantly less than the old samples. Also, adding 15% natural pozzolan is sufficient and effective on control of ASR expansion.

Additionally, in the construction phase, Preparation of cement with 30% pozzolan was not easy. In Sistan and Baluchistan state, there are only two cement factories including Zabol cement with more than 900 km distance from Kahir dam and Kash cement with more than 450 km. Both of them produce cement with 15% kash pozzolan and don'ttend to produce cement with 30% pozzolan. So, it must add in the batching site that has some executive problem. However, a reduction in the amount of pozzolan from 30% to 15%, Because of no need to supply of 30% pozzolan cement, help To the excution of the project.

4 Hard fill mix design

After confirming the use of cement with 15% pozzolan and lack of using one portion of aggregate (0~50 mm), the client asked the contractor to develop a trial mix program. The objective of this program was to find an optimized hard fill mix that would meet the design criteria. We include here below some details of the present situation of the hardfill mix design process.

4.1 Design criteria

The average compressive strength of the hard fill at an age of 180 days should be at least 7 MPa. The mean target compressive strength in laboratory is assumed 10 MPa at the age of 180 days. The minimum in-situ density of the hard fill assumed in the stability analysis is 2 400 kg/m^3.

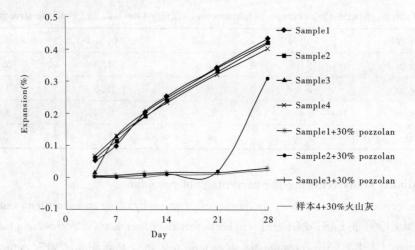

Fig. 9 Short term results (ASTM C1260)

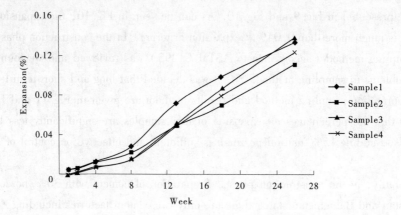

Fig. 10 Long term results (ASTM C1293)

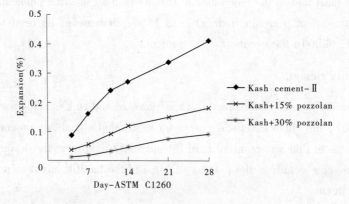

Fig. 11 Short term results (New tests)

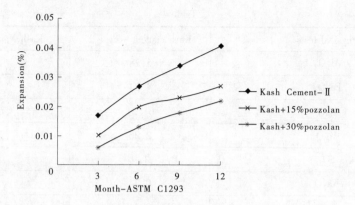

Fig. 12 Long term results (New tests)

4.2 Investigation of the materials

A first step in the mix design program was the analysis of the properties of the material available. Specific gravity and absorption percent for fine aggregates (0 ~ 12 mm) are 2.57% and 3.38%, also, for coarse aggregates (12 ~ 50 mm) are 2.67% and 0.5%.

Khash natural pozzolans obtained from pozzolanic deposits located in South – East of Iran complying with Class N of ASTM C 618-03 were used. Physical and chemical properties of the natural pozzolans are shown in Table 4. Pozzolanic activity of the pozzolan was also measured according ASTM C 311 and is presented in Table 4.

4.3 Gradation curve

The specification of the aggregate has been re – adapted to thenatural grading of aggregate from borrow area in Kahir project. New gradation limits have been established for each size and a new combined gradation for the hard fill has been specified (Fig. 13). The combination of aggregates sizes to meet the gradation curve within the optimum range of density is 52% of 0 ~ 12 mm and 48% of 12 ~ 50 mm.

Table 4 Properties of cement & pozzolan

Physical tests	Kash ceuleut – Type 2	Kash Natutal Pozzolan
Fineness		
Blaine, (m^2/kg)	3 030	
Passing 45 mm(%)		87.5
Compressive strength of 51 mm^3		
3 d(MPa)	14.8	
7 d(MPa)	23.7	
28 d(MPa)	34	
Chemical analysis(%)		
SiO_2	21.82	60.5

Table 4

Physical tests	Kash ceuleut – Type 2	Kash Natutal Pozzolan
Al_2O_3	5.28	18
Fe_2O_3	4.24	5
CaO	61.21	6.75
MgO	2.26	2.75
SO_3	1.9	0.1
Na_2O	0.24	1.6
K_2O	0.46	1.4
Loss on ignition	2.3	2
C_3S	36.3	
C_2S	35.2	
C_3A	6.8	
C_4AF	12.9	
Pozzolanic activity with Pottland cement		92

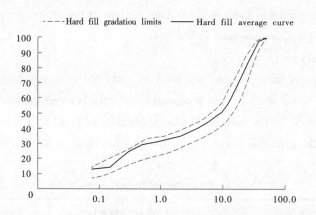

Fig. 13 Combined gradation of the aggregates

4.4 Design of the paste

The quantity of paste in the hardfill mix has been designed for a minimum past/mortar ratio 0.42. The composition of this paste has been calculated after testing of both consistency and strength.

A free – water content of 97 L/m^3 with an acceptable range between 92 ~ 102 L/m^3 has been selected. Preliminary results of cylinder compressive strength are shown in Fig. 9. It is expected that RCC mix with 93 kg/m^3 of cement and 17 kg/m^3 of kash natural pozzolan will meet all the design requirements at an age of 180 days.

4.5 Laboratory trials

Low cement content is a basic characteristic of a hard fill. More than 120 mixes were evaluate since middle of 2013 in four stages, looking for hard fill mixes with VeBe time in between 10 and 20 seconds, and good amount of paste and mortar.

The cement with 15% Kash natural pozzolan content in mix designs were assumed 90 kg/m^3, 100 kg/m^3, 110 kg/m^3 and 120 kg/m^3. In mix designs containing 90 kg/m^3 and 100 kg/m^3 pozzolan cement was used 55% (0 ~ 12 mm) and 45% (12 ~ 50 mm) and also, it consider 52% and 48% in 110 kg/m^3 and 120 kg/m^3 pozzolan cement.

In the following, some samples without additive and some samples with water reducer(Plast 209) and retarder (Tard CHR) additives were made and their results are given in Fig. 14.

The type and dosage of retarder that has been finally selected (CHR at 0.6%) creates an initial set between 3 ~ 5 hours and a final set at more than 20 hours. The admixture will be used for all the selected mixes. In the warm months, to prevent a sharp drop in the workability of RCC, dosage of retarder increase up to 1% and in the cold months decrease to 0.5%. All mix designs with results are given in Table 5.

At the start of concreting in the Kahir dam, Mix design CK3 was selected, and after four month, the amount of pozzolan cement decreased to 100 kg/m^3 (CK2).

Table 5 Mix designs and results with retarder additive Tard CHR(0.6%)

Mix No.	Cement kg/m^3	Pozzolan kg/m^3	W/C	(0 ~ 12 mm)/(12 ~ 50 mm) (%)/(%)	V_p/V_m	VeBe (Sec)	Compressive Strength(MPa)			
							7 days	28 days	90 days	180 days
CK1	76	14	0.93	55/45	0.42	13	47	71	80	94
CK2	85	15	0.88	55/45	0.43	12	51	77	94	111
CK3	93	17	0.8	52/48	0.443	14	65	95	114	116
CK4	102	18	0.8	52/48	0.457	14	89	107	117	145

4.6 Full scale trial

After the laboratory testing stage, the optimum RCC mixture proportion will be selected for the construction of a 12 layers full scale trial (FST). This will be a good opportunity to train the equipment and personnel on site and find out the in - situ properties achieved when the RCC is manufactured, Transported, placed and compacted with the plants that are planned for the main dam. The results show all the mix design with single static pass, two light vibratory passes and single heavy vibratory pass, achieve the specified target density.

5 Conclusions

The first FSHD dam in Iran is under construction at the present moment. The principal goals achieved in the dam design and mix design process were:

(1) FSHD dam construction in weak foundation and high dynamic loading conditions, is

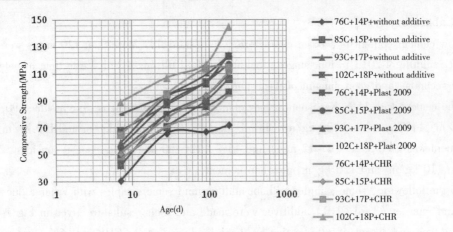

Fig. 14 Preliminary results of compressive strength testing for different mixture proportions 4.5

caused to increase the volume of dam and required of compressive strength. So, in this conditions, there are the possibilities of increasing the amount of cement content in comparison with stronger foundation.

(2) The results show that it may not be possible to use just one proportion because of high fluctuations in strength and needs to add more cement, So, mix designs were made with two proportion of aggregates

(3) The results show that using two proportion of aggregates (0 ~ 12 mm, 12 ~ 50 mm) and mix designs with pozzolan cement content about 100 kg/m^3 are acceptable in terms of strength and economic.

Acknowledgements

The authors wish to thank to project managers Eng. Alireza Shariat and Eng. Saeed Samadi for technical support through the project and also to technical advisor of project Eng. Francisco Rodrigues Andriolo.

References

[1] T. Hirose, T. Fujisawa, et al. . Design Criteria for Trapezoid – Shaped CSG Dams[C] // Icold – 69th Annual Meeting, 2001.
[2] F. Wei, J. Jinsheng, M. Fangling. Study design of mixing proportion for cemented sand and gravel[C] // The 6th International Symposium on Roller Compacted Concrete Dams, Spain 2012.
[3] Icold Bulletin 126, Roller Compacted Concrete Dams, 2003.
[4] Andriolo, F. R. The use of roller compacted concrete[M]. 1998.

The Paradox of RCC Reseach in Developing Countries (Libyan Example)

S. Y. Barony

(Libyan Academy of PG. Studies, Libya)

Abstract: The paper presents a short note on the problem of research activities in developing countries, particularly when research requires more than the traditional laboratory facilities as in the case of roller compacted concrete (RCC). The paper presents the Libyan case as an example of paradox facing research work in developing countries, particularly in the field of RCC. The academic research in the area was triggered by the failure of the Gattara earth dam near Benghazi on 1979, by currying an initial experimental research about the adaptability of local materials to internal specifications of RCC. Research continued on the academic level for more than 25 year without recognition from dam and water authorities, as yet to design and construct at least one dam using RCC technology. The introduction of such paradox example is intended to highlight the necessity of finding ways and means to support research and make use of its finding in a harmonious way between the academic institutions, government and private industry in different fields of science and technology.

1 Introduction

Research activities in developingcountries usually lacks connection with government strategies if available. Due to the fact that research requirementsare usually constrained with time and finance, where governments think of solving bottle nicks in an eargent way autside a common strategic objectives and goals.

Particularly, in construction industry they resort to turn key and BOT (Build, Operate and Transfer) projects which they think that such contracts are safer and less time consuming where they rely on experienced international companies. Forgetting that such process leaves the local scientific and technological groups and personal outside the activity and contribution.

The author was lucky tochair many different national committees dealing with structural problem and large projects such as Tripoli harbor and Brega break water problems as well as Tripoli metro project and different university faculty projects.

As early as 1974 Tripoli break water problems of over toping started even before end of construction, then concrete deterioration and collapse of different parts of parapet on 1981 took the attention of national committee, and then extended to Bregha harbor [1-3]. After wards the general conditions of concrete technology in the country started to attract the attention of research grouping in such area which started with an MSc. thesis on produced concrete strength in the country [4], as well as leading to sponsor the 2nd and 3rd Int. Conference of concrete technology in developing countries were held in Tripoli(1986 and 1993) and setting up the international committee to supervis the future conference series [4]. As well as series of local symposiums on building materials and structural engineering.

All these activities were under the sponsor ship of Libyan scientific society, where the society played a major factor in dissemination of science and Technology in the country. It should be noted that by forwarding such research activities in the last thirty years or so is not intended to inform the outside world the agony and paradox facing research people and their research activities, particularly in few developing countries, by lacking the full recognition and support as well as utilization of research finding and trying to build national strategies and objectives after assessing such activities.

2 Research progress in Libya

Any wise authority or government must take into consideration research spending and encouragement of research and development (R&D) in its polices, strategies and development programs due to the fact that research activities are very important to the nation's economy and well fare of the society. Therefore Libya as a developing country, as early as the end of seventies worked to expand universities from two to six universities by the end of eighties and twelve by the end of thenineties, as well as a number of specialist research institutes with the aim of making the research activities to work in harmony in different areas and institutes. The government set up the national research organization at the end of the seventies, after wards being discontinued and replaced by the ministry of science research, which in turn coupled with it higher education. But at the nineties the national research council was reestablished, where research priorities in the field were set up as well as trying to set up the policy of scientific research in a form of general frame work of science and technology in Libya [5,6]. But the objectives did not materials due to changes in administration, lack of funding, and focusing the attention of the authority in other orientation. In the early days UNESCO tried to help by setting up program to explore the science and technology potential (STP) in the country, but followed the same paths of neglect, delay, and cancelation.

3 RCC research in Libya

The failure of theGattara earth dam near Benghazi on 1979 due to heavy rain storm has triggered the academic enthusiasm and concern to pay attention to research in the area and particularly roller compacted concrete (RCC). At the same time through scanning and reviewing ICOLD literature and activities was attracted by technology development in the area of RCC, which then was taught to be a good alternative to earth dams, particularly the favorable geographic and environmental conditions of the country. But an initial research has to be carried out to prove to the authorities the possibility of adapting such technology. The first MSc thesis research was carried by an engineer working in the national water and dam authority to study the adaptability of local materials to RCC technology [7,8] then the work was extended to field study with specific attention to horizontal layer connectivity and bonding [9-11].

Afterwards the research activities were forwarded to a PhD program dealing with a more detailed use of local materials in RCC technology, and extended to two areas of international interest by studying the used procedures and introduced alternatives, mainly the area of biding types and boding of horizontal layers as well as an alternative direct tension and shear tests [12,13]. Most recently an MSc thesis was concluded presenting a new alternative test for direct tension of concrete cylinders [14].

It is essential to note that no one should progress in reviewing or discussing RCC research in any country or region without giving at least a short indication of RCC technology development around the world, as well as giving credit pioneers and outstanding figures and their contribution to such developments.

It is clear that ICOLD congresses and international conferences and symposiums havepaved the ground for such technology development in the area of design, testing and construction. But due to the fact that one cannot give credit to each and every individual who made a renowned contribution in developing RCC technology, it is thought that the following references summarize RCC development during the last thirty years. [15-19]

4 Discussion

In reflectionof the presented modest research contribution in the area of RCC, when appraised from the point of view of small developing country, which is small in number of inhabitants but large in area being located in the arid region of north Africa, one may think it should have an effect on decision makers and national related authorities. But the paradox and the agony surfaces to the top, that during, a time span of more than thirty years, all these research trials and activities in specific area of RCC, that the local authorities up to now did not include RCC technology as an alternative to earth dam design and construction, even with few suggestions by international companies as in the case of Gattara dam rehabilitation and after wards projects.

This condition of RCC research which presents what may be termed as a neutral research area of technology, clearly it reflects the paradox and agony of other areas of research fields which may be related to health, food, education, and industry, or general areas of sustainable development when coupled with controversialissues of sociology and environment. To conclude, one cannot despair and must continue search for ways and means to activate cooperation between academic and research organization from one side international, national and private authorities from another side. This road must be followed due to the importance of research to developing countries in general and particularly sustainable development, which they really lack insight and vision for the welfare of present and future generation locally and worldwide. Such policies and strategies when set and followed are in the benefit of mankind.

References

[1] S. Y. Barony, D. V. Mallick, M. Gusbi, and F. Sehery. Tripoli Harbor nw Break Water and Its Problems

Procd. Int. Conf. On Coastal and Port Engineering in Developing Countries, Colombo, 1983.

[2] S. Y. Barony, R. A. Mohammed. Concrete Deterioration of coastal Structures: BregaPrblem, Procd. 2nd Inte Concrete cof. on Technology for Developing countries, Tripoli – Libya, 1986.

[3] S Y. Barony, D. V. Mallick, and F. Sehery. Tripoli Harbor nw Break Water New Design Criteria and long-term Solution, Procd. Int. Conf. On Coastal and Pert Engineering in Developing Countries, Bejing, China, 1987.

[4] S. Y. Barony, M. A. Ghabar, and Z. A. Hatoush. An Assessment of Produced Concrete Strength in Jamahiriya, Procd 2nd Int. conf. On Concrete Technology for Developing countries Tripoli – Libya, 1986.

[5] S. Y. Barony. Requested Review Paper . Procd. of Expert meeting on Research Development in Arab Countries, Tripoli, Libya, 1998.

[6] National Science Foundation. General Framework of Science and Technology in Libya, Procd. 2nd Symposium on Technology Transfer, Tripoli, Libya, 1996.

[7] T. S. Sifaw. Roller CompaetedConcrete Technology And Application. MSc thesis in civil Eng. Dept. , Tripoli Uniu, Tripoli, Libya, 1999.

[8] S. Y. Barony, T. S. Sifaw. Laboratory Study to Find Physical Properties of RCC Procd. Int. Conf on RCC Dam Constructed in Middle east, Jordan, 2002.

[9] S. Y . Barony, I. M. Rouis , A. M. Mansor. The Potential of Using Local Materials in RCC: An Experimental Study, prod. 1st Int. Conf. for Technology and Durability of Concrete, Algiers, 2004.

[10] A. M. Ali, M. I. Rouis, S. Y. Barony. Roller Compacted concrete: An Appropriate Technology for Developing Countries, Procd 8th Int. Conf. On Concrete Technology for Developing countries, Tunesia, 2007.

[11] A. M. Mansour, M. Bneni. Field Study For RCC Using Local Materials, Procd. 5th National Conference on Building Materials and Structural Engineering School of Eng Applied Science Libyan Academy. Tripoli, 2010.

[12] A. M. Mansour. Experimental Study for Roller Compacted Concrete and Material Technology Using Local Material PhD These, sfax University, Tunisia, 2009.

[13] S. Y. Barony, J. M. Rouis, A. M. Mansour. New Alternative method for bonding RCC Layers, Proc. 5 th Int. Symposium on RCC Dams, Guiyang, China, 2007.

[14] M. S, Khalifa. Development of New Meth for Direct Tension Concrete Test, MSC Thesis, Civil Eng. Dept. Libyan Academy Tripoli, Libya, 2015.

[15] J. Hall, D. Hongton. Roller Compacted Concrete studies at Lost Creek Dams, U. S. Army Engineering District, oregn, USH, 1974.

[16] A. T. Richardson, Upper Still Water. Roller Compacted Concrete Design and Construction Concepts, Procd. 15th ICOLD Congreid, Lausanne, 1985.

[17] B. A. Forbes. The Develop mint and Testing of Roller Compacted Concrete For Dame, Procd. 16th ICOLD Congress, Sanfransisco, USA, 1988.

[18] C. Shen, D. Malcolm. The Development of RCC dams through The World, Procd. Int. Symposium on Roller Compacted concrete, Beijing, China, 1991.

[19] E. Schrader. Experience and lessons Learned in 30 years of Design, Testing, Constructing of RCC Dams, prod. Int. Conf. on RCC Dam Construction in Meadle East, Irbid, Jordan, 2002.

The Quality Control of Lai Chau RCC Dam

Nguyen Hong Ha[1], Marco Conrad[2], David Morris[3], Pham Thanh Hoai[1]

(1. Son La Management Board (EVN), Vietnam; 2. AF – Consult Switzerland Ltd, Switzerland; 3. AF – Consult Switzerland Ltd, Vietnam)

Abstract: The 131 m high Lai Chau RCC dam is the second large RCC dam in Vietnam constructed on the river Da. Its construction started in 2010 with excavation of the diversion channel and RCC placement started in March 2013. Completion of the dam containing 1.884 Mm3 of RCC was scheduled for June 2015. An extensive Quality Control (QC) plan was developed and operated to assure construction quality and to provide the necessary evidence for acceptance by Government Agencies. Overall QC planning included laboratory trials, trial embankments, site testing of manufactured specimens and coring and testing of the RCC. This paper describes the testing programme and some of the results obtained. The Lai Chau RCC dam, like the Son La RCC dam downstream, was completed ahead of programme.

Keywords: RCC gravity dam, quality assurance, quality control, testing

1 Introduction

It is often thought that the construction of RCC dams is easy but appearances are deceptive, and it is precise and difficult even though it looks easy, if proper attention is not paid to quality control.

The construction of a RCC dam can be likened to a continuous industrial process with an almost constant flow of materials from quarry and crushers; cement producer; pozzolan producer and admixture producer to the RCC batch plant established for that project, and further flow of the batched RCC to the dam where it is spread and compacted. Because it is, or should be, a continuous process that any quality control system adopted must be reliable, robust, auditable and able to provide records of what was placed where and when in the event that testing of cores from the dam show low strength results.

RCC dams must be designed in such a way that the dam can be constructed in a continuous and speedy manner to make use of the advantages of the RCC technology, i.e. saving time and providing a better economy compared to other types of dams, and to complete a dam body of high quality. A dam or RCC block raise of 10 m per month, or placement of a multiple of 30 cm thick RCC layers per day is typical in RCC dam construction. In combination with the fact that the control age (or design age), at which the RCC reaches its design strength, is typically defined in excess of 91 days (often 182 days and even 365 days) this creates a significant challenge to the quality assurance and quallity control (QA/QC) during dam construction. In the case of the Lai Chau RCC dam more than 1.0 Mm^3 of RCC were placed during the first 12 months of dam construction and in some sections the dam was raised by more than 100 m above the lowest foundation level, where the RCC just reached the control age of 365 days. It is clear that methods have to be provided with the RCC QA/QC plan that allow to understand extremely early, ideally before the fresh RCC lift under construction is covered by the subsequent lift, if the present fresh RCC layer will reach its design strengths or if any corrective measures have to be implemented in case of any non – compliance.

For the Lai Chau RCC dam a comprehensive quality system with detailed records was used to provide evidence of the proper construction of a high quality RCC dam that the State Acceptance Committee of the Vietnamese Government would accept and allow the reservoir to be impounded.

2 The project

The Lai Chau Hydropower Project is located on the river Da in Northwest Vietnam. It is the uppermost project in a cascade of three projects all owned by EVN. Together the three projects have a combined installed capacity of 5 520 MW. The Lai Chau RCC dam is a 131 m high gravity structure with a volume of some 1.884 Mm^3 of RCC and the installed capacity is 1 200 MW.

The layout of the RCC dam was complicated by three things but had been approved earlier by the Government, thus, making changes difficult and time consuming. The three factors were:

(1) The need to construct the central blocks in the river channel as quickly as possible to allow a start to construction of the conventional concrete works of the gated spillway and as early a start as possible to the low level outlet which resulted in the need for a separate section of RCC as it could not be combined with the spillway blocks;

(2) the intake which split the left section into two separate zones;

(3) and the diversion channel and culvert on the right side.

The dam was thus split into four distinct and separate sections requiring three separate conveyor lines[1]. This affected the overall speed of construction because of the need to move the conveyor line and swinger to a new delivery point for each section.

3 RCC mix design

The quality control for the construction of the Lai Chau RCC dam already started in the design stage, when the RCC mix design was developed to satisfy the strength and durability requirements for the dam. The RCC mix design started in the Technical Design Stage in 2010 and comprised a series of four sets of trial mix programmes in the laboratory. The laboratory trial mix programmes were followed by two full scale trial embankments in 2011 designed to test joint preparation methods, provide training for the Contractor's staff, pre – qualify equipment and provide initial in – situ RCC properties to support the Higher Authorities' process for approval of the Technical Design. The initial in – situ RCC lift joint direct tensile strength results allowed lift joint treatmentsto be defined in terms of exposure time and Modified Maturity Factor (MMF) and the findings were used to supplement the Specification. A third full scale trial embankment was constructed in 11/2012, just prior to the start of RCC placement and using the actual Contractor's RCC plant and equipment to be used on the dam to confirm RCC parameters, e. g. retarder dosage, lift joint treatments and exposure times, and to provide more training to the Contractor's and Supervisor's staff. In regard to the quality control plan, respectively verification of the RCC quality in the dam, an essential outcome from the full scale trial embankments was a baseline of data correlating the fresh and hardened RCC properties to judge RCC quality during the initial RCC placement in the dam. Fig. 1 shows the timeline of RCC laboratory programme and full scale trial embankment stages as part of the quality assurance at Lai Chau.

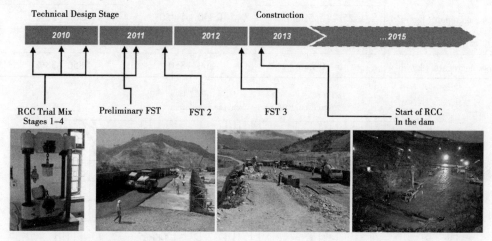

Fig. 1 RCC mix design and full scale trial embankment stages for the Lai Chau RCC dam

The RCC mix design focused on the design of a contractor friendly mix with minimized segregation and a very good workability at a modified loaded Vebe time of 8 – 10 s. The mix also contained a set time retarder to delay the initial set of the RCC to between 18 and 24 hr to maximise the number of hot joints without any bedding mix, i. e. fresh on fresh placement of the RCC while minimizing the lift joint preparation efforts to only cleaning and removal of water puddles. The Specification allowed the retarder dosageto be adjusted as a function of the RCC layer volume to minimise the load on formwork. The mix design specified and finally selected is given in Table 1. This table also shows the proportions of individual aggregate groups specified to meet the specified combined gradation envelope (Fig. 2), which is one of the most important itemsto be quality controlled during construction. The aggregate proportions were established in the Technical Design stage as one of the first steps in the RCC trial mix programme.

Table 1 Lai Chau RCC mix design (SSD) in specification and as prescribed for placement

Item		Specification	Actual placement
Nominal Maximum Size of aggregate (mm)		50	50
Portland cement (kg/m^3)		60	60
Pha Lai treated flyash (kg/m^3)		160	160
Free added water (kg/m^3)		110 ~ 120	110
Conplast R retarder (% total cementitious)		For initial set 18 ~ 24 hr	0.75
Coarse aggregate (kg/m^3)	50 ~ 25 mm	1 350 ~ 1 450	529 (23%)
	25 ~ 12.5 mm		502 (22%)
	12.5 ~ 5 mm		385 (17%)
Coarse aggregate (kg/m^3)	5 ~ 0 mm	800 ~ 850	856 (38%)

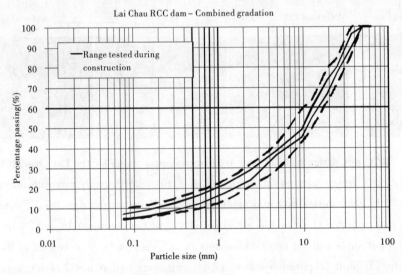

Fig. 2 Aggregate combined grading limits and range of test results during dam construction

Since focus by the Supervision and State Authorities was very much on strength for the quality monitoring and acceptance of the dam construction, the findings from the laboratory trial mix tests and the full scale trial embankments were used to develop lower limit guide strengths for earlier RCC ages than the control age of 365 days (see Table 2). These guide strengths were implemented into the Specification, however, the core strengths at the control age of 365 days being the final strength acceptance criteria. For further guidance that the design strengths at the control age would be achieved, accelerated curing of manufactured specimen was carried out as part of the laboratory trial testing to obtain baseline data for the correlation between the accelerated and normal curing strengths, and during construction to provide a guidance for long-term strengths developing. It was seen from this testing that accelerated and normal curing direct tensile strengths resulted in a different correlation than compressive strengths (see Table 4) and that the latter appeared more reliable.

The characteristic strengths specified were based on an assumed allowable failure rate of 20% and a Coefficient of Variation of 25% giving a Standard Deviation of 3.6 MPa and a margin for the design strength of 3.1 MPa and thus a design strength rounded to 17.5 MPa at the control age of 365 days.

Table 2 Specified strength requirements.

Test/Sample Type	Age[d]	Compressive Strength (MPa)		Direct Tensile Strength (MPa)	
		Characteristic	Sample Design	Characteristic	Sample Design
Manufactured cylinders	7	2.9	3.3	not specified	not specified
	28	6.4	7.8	0.5	
	56	8.8	10.6	0.7	
	91	10.3	12.5	0.8	
	182	12.6	15.2	1.0	
	365	14.5	17.5	1.3	
Cores taken from the dam	91	10.0	not specified	not specified	not specified
	365	13.3		0.8 (lift joints)	
	365	13.3		1.1 (parent)	

4 Quality control plan

A comprehensive quality control and reporting plan was developed based on the one used for the construction of the Son La RCC dam downstream of Lai Chau. This was divided into four basic sections of which essentially three were related to sampling and testing and one related to inspection on site:

(1) Material testing comprised: aggregate grading from each crusher and in the batch plant bins, flakiness and elongation, specific gravity and absorption, moisture content; cement and fly ash certificates of chemical and physical properties; retarder admixture uniformity; RCC Vebe time and Vebe density at batch plant and at dam; RCC fresh temperature at batch plant and at point of placement; initial and final setting times of the RCC mix; in-situ RCC fresh compacted density and moisture (nuclear densitometer); compressive and direct tensile strength of manufactured control cylinders from each layer including accelerated curing;

(2) Data concerning the placement including the time and finish, not only of each layer but also of each lane or strip making up that layer; lift joint type and treatment; volume of RCC and GEVR in the layer; hourly ambient temperature and rainfall; placement journal (e.g. plant stops, observations);

(3) Visual inspection of RCC placement processes including compaction, curing, lift joint surface preparation, etc.;

(4) Compressive and tensile strength (joints and parent) from cores taken from the dam body.

Sampling and testing schedules were based on the volume of RCC and the control ages of the various specimens. Since the RCC essentially has to be approved before it is covered by the subsequent lift of RCC, emphasis is on the constituent materials, on the RCC fresh properties and on the RCC placement and compaction processes to comply with the Specification and the limits defined therein. Table 3 details the RCC sampling and testing plan initially applied at Lai Chau.

Table 3 RCC sampling and testing plan at Lai Chau

Type of sample	Frequency	Location	Tests carried out	Test age(d)
Cylinder	2 000 m^3	Point of placing	Compressive strength	7, 28, 56, 91, 182, 365, 1000 (56 d and 1000 d sampling alternate)
	6 000 m^3		Compressive & direct tensile strength	Accelerated testing 6 d normal curing + 7 d curing @90℃ +1 d normal curing
	10 000 m^3		Direct tensile strength	91, 182, 365, 1000
	100 000 m^3		Elastic modulus in compression & Poisson's ratio	7, 28, 56, 91, 182, 365, 1000 (56 d and 1000 d sampling alternate)
Bulk	250 m^3	Batching Plant	Vebe time & density	Immediate
	250 m^3	Point of placing		
	250 m^3	Batching Plant	Fresh temperature	
	250 m^3	Point of placing		
Nuclear Densitometer	Two locations for every 500 m^2	Point of placing at 150 and 300 mm depths	Compacted fresh density & moisture	
Cores	Three cores from each RCC section	From galleries in dam, dam crest and / downstream face	Compressive strength	91, 365, 1000 (actual test ages to date are between 110 and 470 d)
			Direct tensile strength of parent and jointed specimen	
			Elastic modulus in compression & Poisson's ratio	

As can be imagined this amounts to a significant quantity of data which allows a detailed evaluation of the strength of the RCC within the dam body down to the level of the characteristics of the material (parent or joint) in any particular lane in any layer given the location of the core. It also enables the Modified Maturity Factor (MMF) to be calculated for any lift joint and, together with the joint tensile strength and age allows 3D models[2] of strength, MMF and age to be created to gain a better understanding of lift joint treatments obtained.

Data collection is carried out by the staff of the Owner's supervision team as well as by the Contractor's staff. The Contractor's input covers the material characteristics and site laboratoty testing of manufactured samples. The Owner's staff are responsible for collecting the data at the placement and for collating all of the results on site. Cores are logged, cut and tested in an independent laboratory in Hanoi.

5 Results from the quality control

The hardened properties of the Lai Chau RCC are discussed in reference[3]. A total of 778 cylinder specimens have been tested to date for compressive and direct tensile strength. The results are summarised below in Fig. 3 and Table 4. It is seen that the cylinder compressive strength exceeded the guide strengths (see Table 2) at each intermediate testing age and exceeded the specified cylinder compressive strength well before the control age at 365 days. Also the cylinder direct tensile strengths exceeded the guide strengths at the intermediate testing ages. The ratio between cylinder direct tensile strength and the cylinder compressive strength is 6.5% at 365 days.

Table 4 Summary of manufactured cylinder strengths achieved

	Cylinder compressive strength (MPa)							Directt tensile strength (MPa)			
Age(d)	7	28	56	91	182	365	A14(1)	91	182	365	A14(1)
Max	4.9	9.1	12.9	17	20.3	22.8	26.4	0.99	1.27	1.49	1.08
Min	3	7.2	10.7	13.4	16.3	20.9	22.3	0.89	1.06	1.37	0.84
Average	4.4	8.1	11.8	15.5	17.9	21.3	25.1	0.9	1.1	1.4	1.0
CoV	13.6%	6.1%	4.9%	8.8%	7.0%	2.8%	6.5%	2.8%	5.6%	2.8%	8.6%

Note: A14(1) refers to specimens tested after accelerated curing for 14 days equivalent to testing at about 365 days.

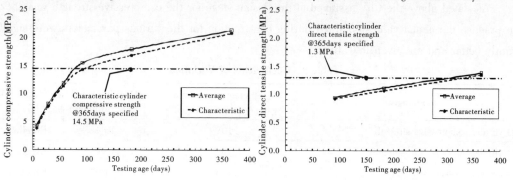

Fig. 3 Cylinder strengths achieved

To date, a total of some 1 400 sections of core from the RCC dam have been tested for strength or modulus. Given that some 480 m of core have been extracted to date from the dam, equivalent to 1 600 300 – mm long specimens, this is a very high percentage (almost 90%) particularly as the core barrel was only 1.5 m long and therefore some of the cores would have had to be broken in order to extract the cores. The RCC core specimens have been tested at various ages ranging from 110 to 468 days allowing the development of compressive and direct tensilestrength with age to be determined with a degree of accuracy. The results are summarised in Table 5 and Fig. 4 (reference is also made to [3]).

Table 5. Summary of in – situ hardened properties tested on cores

Parameter	Unit	No.	Age(d)		Value			
			Max	Min	Max	Min	Average	CoV
Density	kg/m³	1 343			2 603	2 543	2 577	0.9%
Compressive strength	MPa	307	468	110	24.1	14.4	17.02	12.6%
Direct tensile strength (Parent)	MPa	340	468	115	1.55	0.97	1.30	15.4%
Direct tensile strength (Joint)	MPa	677	468	116	1.34	0.93	1.15	18.1%
Elastic modulus	GPa	164	466	114	26.8	17.9	22.2	12.3%
Poisson's Ratio	—				0.18	0.13	0.15	12.6%

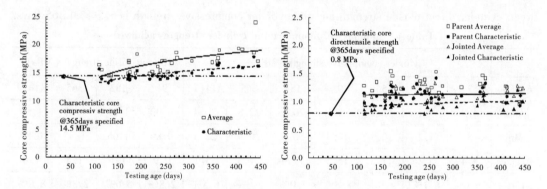

Fig. 4 Core strengths achieved

As stated above the CoV assumed at the design stage for the compressive strength was 25%. In practice the actual values obtained during construction for the various parameters were significantly better and are given in Table 6.

Table 6 Coefficients of Variation

Property	Coefficient of Variation upper range achieved	Quality control acc. to ICOLD Bulletin No. 126	
Cylinder compressive strength	10%	10%	Excellent
Cylinder direct tensile strength	15%	15%	Excellent
Core compressive strength	12.5%	<15%	Excellent
Direct tensile strength of unjointed cores	15%	<25%	Excellent
Direct tensile strength of jointed cores	20%	<30%	Excellent

The required tensile strength was derived from the design requirements and a compressive strength to ensure that the tensile strength was met using an empirical correlation factor. The actual relationship between the unjointed core direct tensile strength and the core compressive strength is 7.6%, between the jointed cored direct tensile strength and the core compressive strength 6.8% and between the jointed and unjointed core direct tensile strength 89%, all better than assumed in the dam design.

The excellent level of quality control with regard to materials, fresh RCC mix and RCC placement and compaction processes is also reflected in the various densities generally monitored in RCC dam construction (see Table 7).

Table 7 Densities achieved

Theoretical air free density (kg/m^3)	Vebe density (kg/m^3)	Compacted fresh density (kg/m^3)	Density of manufactured cylinders (kg/m^3)	Core density (kg/m^3)
2 603	2 547	2 568	2 578	2 577
100% TAFD	97.8% TAFD	98.7% TAFD	99.0% TAFD	99.0% TAFD

6 Conclusions

The quality control plan adopted for the Lai Chau RCC dam was comprehensive and ensured that the RCC met the specified and design requirements in all respects. Although it required considerable effort the end result showed that the dam was constructed to a very high standard and met all of the requirements. It also enabled the relationships between age at test, Modified Maturity Factor and direct tensile strength of joints to be examined in detail as is explained in [2].

The quality assurance applied at Lai Chau mainly contributed to the fact that 98% of all lift joints were hot joints, i.e. where the RCC has been placed fresh on fresh with a minimum of lift joint surface preparation. Scuh placement condition not only contributes to high RCC placement rates, but also to a very good quality of the RCC mass and to effectively monolithic RCC blocks.

Acknowledgements

The authors would like to thank Viet Nam Electricity (EVN), the Son La Management Board and PECC1 for permission to publish this paper and the fruitful cooperation during the design and construction stages of the Lai Chau HPP. This publication is dedicated to Mr. Nguyen Hong Ha, Vice Director of EVN, who passed away during the preparation of this paper.

References

[1] N.H. Ha, M. Conrad, D. Morris, et al. Speed of construction of EVN's Son La and Lai Chau RCC dams, Proceedings of 7th International Symposium on Roller-Compacted Concrete (RCC) Dams, Chengdu, China, September 2015.

[2] M. R. H. Dunstan, M. Conrad. The relationships between the in – situ tensile strength across joints of an RCC dam and the maturity factor and age of test, Proceedings of 7th International Symposium on Roller – Compacted Concrete (RCC) Dams, Chengdu, China, September 2015.

[3] N. H. Ha, N. P. Hung, D. Morris, et al. The in – situ properties of the RCC at Lai Chau, Proceedings of 7th International Symposium on Roller – Compacted Concrete (RCC) Dams, Chengdu, China, September 2015.

Laboratory Study of Shear and Compressive Strength of Construction Joints in two Ways Categorization Maturity Factor and Time of Setting in RCC Dams

Sayed Bashir Mokhtarpouryani, Shahram Mahzarnia

(Iran Water & Power Resources Development Company, Iran)

Abstract: In RCC dams, different layers are executed with the thickness of 30 to 35 cm according to bulk of sources, available facilities, methods of execution, providing and assuring density of the RCC. The execution of RCC especially in hydraulic structures should be done in a way that the executed structure have consonant blocks for the intended goals.

It is clear that the characteristics and the conditions of boundaries between layers is the most effective parameter in determining the boundaries between layers. Although this kind of execution causes higher speed in execution in comparison of usual concrete dams, and it also reduces problems of voluminous concrete placing in sites significantly but the stability of dams against side forces because of relative weakness in shear strength between layers in the place of joints needs especial attention and it is one of the important factors that is important to dam designers.

According to technical codes in RCC dams, joints are categorized into four groups, hot, warm, cold, and very cold. Necessary works are done according to the kinds of joint. The standard to define these joints is the amounts of maturity factor which is the product of temperature of the surface of the concrete and time based on Fahrenheit that will be defined after the experimental study and defining time.

It seems that this definition is too broad and the most important problem of it is, not mentioning the time of setting of concrete in calculation which is the important factor in determining time span for any kind of joints. In this artical the researcher tries to propose ways to proof the subject according to codes and to find new ways to categorize these joints to promote shear and compressive strength in the place of joints in real conditions.

Key words: roller Compacted Concrete, maturity factor, time of Setting, shear Strength, compressive Strength

1 Introduction

RCC dams can be considered as the most important development in dam construction technology over the 40 years, which its recognition can turn back to 1970s decade during two meetings of 1970 and 1972 in Asilomar state in California America under the title of "Economic Construction of Concrete Dams" [1]. Experts in these two meetings were seeking new type of materials for dam construction, which can meet safety of concrete dam and can also include speed of implementing embankment dams. This could result in creation of roller – compacted concrete dams as a new type of dams in 1974.

Roller – compacted concrete, before being a new type of material, has been a modern method for implementation and has been similar to classic concrete dams in terms of safety. The method could cause implementation of concrete dams with lower costs and could also play vital role in regard with rapid implementation of the project [2]. Over the years, roller – compacted concrete has been widely applied for constructing pavement of roads, highways, airports and implementation levels of industries such as petrochemistry, fineries, ports and so on. Now, the technology has found special position in different constructional domains.

Permeability of implemented dams with RCC has been constantly one of the main challenges existed for implementation of these dams. The problem has been specifically important in regard with joints and this has been because of increase in number of horizontal joints. Hence, permeability of these joints can have considerable effect on amount of permeability of these dams.

According to existing technical codes for RCC dams, joints can be classified in 4 groups of hot, warm, cold and vary cold and required conditions for implementation of the next layer would be defined adjusted with type of the joint. General criterion for defining these joints, except for some dams constructed in China, can be regarded as maturity factor, which can be defined in form of multiplication of surface temperature of concrete in time based on °F. After examining the joint in laboratory conditions, temporal limitation would be determined and then it would be named.

It seems that definition of maturity factor is general and the most important weakness of it can be lack of considering setting time of concrete in calculations, which can be regarded as one of the most important factors in determining time periods for attributing to each type of joints.

Mokhtar Pouryani, Siose marde and Mikail Zadeh (2013) [3] have investigated classification of implemented joints in RCC dams based on setting time through constructing samples in three different environmental modes including processing in curing room, inside the room and under open air based on maturity factor. Obtained results from their studies indicated that the concrete can result in different setting times under different environmental conditions and this issue can indicate that concrete surface humidity is more important than concrete surface temperature. According to physic of this issue, this mode may be resulted from manner of transferring temperature of concrete under different environmental conditions. In addition, their results indicated that with a constant mixing pattern, in the first setting, different values would be obtained for maturity factor. In other words, only temperature and time can't affect setting time and maturity of joints, but also humidity has also important and effective role. The study indicated that method of classification of joints based on setting time can present more acceptable results for different environmental conditions.

At the present study, as a result of these results, new series of examinations have been provided in framework of existed regulations to investigate and compare both attitudes carefully and to consider setting time in classification of joint, so that they can evaluate amount of shear and compressive strength in space of joints in both classification systems with two different levels of cement materials. In addition, in order to make the research applicable, applied materials in one of the roller – compacted concrete dams has been applied, which is under construction in West of Iran.

2 Mixed RCC model

2.1 Cement materials

Achieving low exothermal action in massive concrete can be met mainly through using pozzolanic materials and overloads in cement. At the present study, pozzolanic cement type II from Kurdistan's Cement Factory (Bijar) with 25% pozzolan has been applied. In addition, for purpose of enhancing validity of comparisons, two levels of 125 kg/m^3 and 150 kg/m^3 of cement materials have been applied.

2.2 Aggregates

Rock materials have been considered in 4 levels of aggregation including a group for fine aggregates, 2 groups for coarse aggregates and one group for combining fine and coarse materials based on Table 1. Maximum size of aggregates has been also considered to 50 mm [4].

Table 1 Mixed pattern of RCC applied in this study

| Type of cement | Cement (kg/m^3) | W/C | D_{max} (mm) | Aggregate Percen(%) | | | | Vebe time (sec) | Unit weight (kg/cm^3) | V_p/V_m (Ave.) |
				0~3 (mm)	0~5 (mm)	5~25 (mm)	25~50 (mm)			
Special Pozzolanic Bijar	125	0.9	50	13.5	31.5	27.5	27.5	15	2 455	0.45
	150	0.9	50	13.5	31.5	27.5	27.5	15	2 455	0.45

According to research subject, adherence of joints is so important. Roller – compacted layers with weak adherence include lower caulk quality and lower shear strength. In this regard, bed making mortar was used for 2 – layer joint for the first time by 1980 in Shimajigawa Dam Japan[1]. In first series of examinations of the study, 4 types of mortar mixing patterns for inter – layer spaces have been applied according to Table 2 for purpose of investigating impact of increase in cutie of cement materials in inter – layer mortar on quality of joints. However, in the second group of examinations, according to the main objective of the present study for purpose of classifying joints and necessity of applying a uniform pattern for achieving desired target, cutie of cement materials was assumed fixed.

Table 2 Mixing pattern of inter – layer mortar applied in first and second series of examinations

	Cement (kg/m³)	W/C	Aggregate Percent(%)		Air (%)	Fresh Concrete		Compressine Strength (kg/m²)	
			0~3 (mm)	0~5 (mm)		Slump (cm)	Unit weight (kg/m³)	7	28
First series	400	0.7	30	70	2.5	17	2 230	118	180
	450	0.63	30	70	2.5	21	2 240	130	183
	500	0.59	30	70	2.5	16	2 200	178	241
	550	0.56	30	70	2.5	16	2 200	181	245
Second series	400	0.66	0	100	2.5	20	2 240	130	236

3 Making samples

RCC samples can be made for different purposes and in different forms. At the present study, in all conducted tests, cubic frames with size of 15 cm × 15 cm have been applied for purpose of making 34 samples (2 samples as control samples) in order to test depth of penetration based on European Standard EN – 12390 – 8. Also, cylindrical frames with size of 30 cm × 15 cm have been also applied for purpose of making sample to determine adherent strength based on ASTM C1245.

Density of samples was measured based on type of joint with two rectangular cubic overloads with weight of 9.1 kg and 11.6 kg (choosing weight of overloads has been based on the stress created in the test of determining web time).

Because of nature of the test, in order to investigate the joint between two layers, the test was implemented in hot joints in cubic frame each sample and ion two layers with thickness of 7.5 cm and in cylindrical frame in two layers with heights of 15 cm. In other types of joints, it was implemented in cubic frame of each sample and in two layers with thickness of 7 cm and in cylindrical frame in two layers with heights of 14.5 cm with a layer of inter – layer mortar with thickness of 1 cm.

Fig. 1 First layer of concrete in the cubic and cylindrical frames

In order to optimize surface of joints in warm, cold and supercool joints before implementation of the second layer, some actions have been taken as follows:

(1) Ridged surface: according to presented explanations in regard with removing mortar from surface of joint, the surface has been ridged (Fig. 2).

(2) Implementation of bed mortar: after ridging and fastening the frame again, firstly bed mortar with thickness of 1 cm was implemented based on mixture pattern and then the second layer was implemented.

Fig. 2 **Ridging layer surface**

At the present study, in order to determine time periods for attributing to each type of joint in method of maturity factor, defined ranges in technical specifications of one dam, which is under construction in west of Iran, has been applied based on second row of Table 3.

In addition, in order to achieve classification system based on setting time, proposed values of Mokhtar Pouryani et al (2013) have been applied based on third row of Table 3.

Before starting sample making and using ASTMC403 test, primary and ultimate setting times have been determined for each mixing pattern. Based on obtained results from these tests for primary and secondary setting times, with a suitable approximation for both values of mixing cutie, primary setting time has been considered to 5 hours and secondary setting time has been considered to 12 hours and tests have been also conducted through considering these values.

Table 3 **Defining types of joint based on both classification systems**

Type of joint	Defining joint based on		Conditions of implementing next layers
	Maturity factor (°F – hr)	Setting time	
Hot Joint	<1 022	Before primary setting	Can be implemented with no arrangement
Warm joint	1 022 ~ 2 012	Before ultimate setting	Weak zones would be removed from surface of implemented layer using water or wind pressure and there is no need to ridge aggregates
Cold joint	2 012 ~ 3 994	4 times higher than primary setting	Weak zones would be removed from surface of concrete using pressure of water or wind in a manner that aggregates can be ridged and sand – cement mortar with thickness about 2cm can be implemented
Supercool joint	>3 994	8 times higher than primary setting	Similar to cold joint

4 Results

4.1 Obtained results from structured tests based on maturity factor index

Fig. 3 illustrates results of compressive strength of tests in vitro. Obtained results indicate properly maturity law. N. J. Carino and H. S. Lew (2001) have quoted from Saul (1951) that "concrete with uniform mixing pattern and maturity would have uniform strength and there would be no matter that temperature and time are in what level while maturation"[6]. As it was expected and Fig. 3 also indicates, only with rise of cutie of cement materials, compressive strength would be enhanced to some extent.

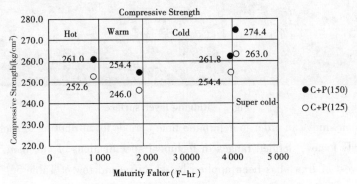

Fig. 3 Diagram of compressive strength of samples in place of joint

In regard with shear strength, results based on diagram of Fig. 4 have been as follows:

(1) When inter-layer mortar is applied, the mortar can determine shear strength of joint. Mean shear strength of samples without ridging has been about 3% ~7% and in samples with ridging it has been about 7% ~12% compressive strength has been observed.

(2) Using inter-layer mortar can pave the way for increase in layering strength, compared to the mode without mortar.

(3) Comparing depth of penetration of samples and shear strength indicates that in some permeable samples, shear strength is even lower than expected level.

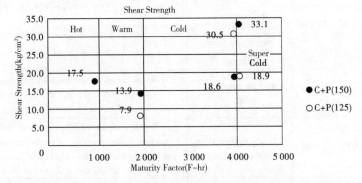

Fig. 4 Diagram of shear strength of samples in place of joint

4.2 Results of structured tests based on setting time index

In all experiments for both values of mixture cutie, primary setting time has been equal to 3 hrs and secondary setting time has been considered to 12 hrs and tests have been conducted with these assumptions.

Fig. 5 indicates test of compressive strength of structured samples based on setting time. Clearly, increase in cement materials to 25 kg in RCC can have no tangible effect on compressive strength of concrete.

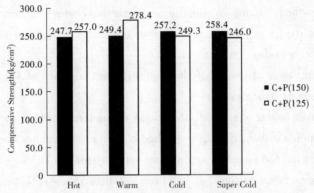

Fig. 5 Diagram of compressive strength of samples in place of joint

In addition, Fig. 6 indicates items as follows:

(1) Mean shear strength in samples without ridging has been obtained relatively to 3% ~ 12% and in ridged samples; compressive strength has been obtained to 12% ~21%.

(2) Increase in cutie of cement materials in concrete mixture pattern can cause increase in shear strength of joint, which can be because of increase in compressive strength of concrete.

(3) Increase in cement materials to 25 kg/m^3 in concrete mixture, mean value of shear strength of concrete mass in samples without ridging has been obtained about 40% and it has been enhanced to 60% in ridged samples.

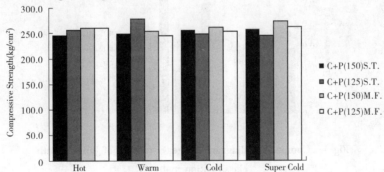

Fig. 6 Comparison diagram of compressive strength of joint in both methods of joint classification

5 Comparing two joint classification methods

5.1 Comparing two methods of joint classification in terms of compressive strength

The tests for determining compressive strength have been conducted on 15 cm × 15 cm cubic samples after 180 days humid farming. Results of structured tests based on two methods of classification have been presented in Fig. 6.

As ratios of mixture have been fixed in this study, the aim by conducting compressive strength test has been controlling unity of structured mixtures and investigating damages resulted from core – making during the tests. Changes in strength can be resulted from some factors such as changes in W/C, changes in required water, changes in features of concrete's components, inadequate sampling method, change in construction methods, changes in farming and weakness in conducting tests. As it was expected, according to the maturity law, no specific mutation has been observed in two methods.

Comparing mean values of results of shear strength test in place of joint based on Fig. 7 in two classification methods indicates that:

When inter – layer mortar is applied, the mortar can determine shear strength of joint. Mean shear strength in samples without ridging in method of maturity factor has been about 3% ~ 7%. Also, in classification method based on setting time, it has been about 3% ~ 12% and in samples with ridging in maturity factor method, it has been equal to 7% ~ 12% and in method of joint classification based on setting time, it has been obtained relatively to 12% ~ 21% of compressive strength. This can indicate better performance of classification system bed on setting time.

In the classification method based on setting time, with the increase in 25 kg/m^3 of cement materials in the concrete mixture, mean value of shear strength of concrete mass in samples without ridging has been relatively equal to 40% and in ridged samples it has been enhanced to 60%. However, in maturity factor method, samples without ridging have been increased to 60% and ridged samples have not been changed significantly. The factor indicates better performance of maturity factor method in regard with shear strength.

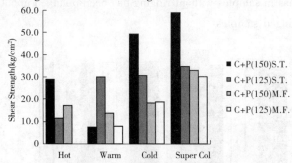

Fig. 7 Comparison diagram of shear strength of joint in both classification methods

6 Conclusion

Obtained results and findings from the study have been presented as follows:

(1) Based on maturity law, no specific mutation has been observed in diagram of compressive strength of both methods.

(2) Through replacing setting time based on tables and diagrams for defining joint, one can achieve higher values of shear strength compared to maturity factor.

(3) There is significant difference among values of shear strength of warm joint in two methods, which can be because of being exposed to layer surface. More delay in working time can result in dryness of layer surface and as a result, more decrease in water than optimized humidity ratio (according to fixed values of water to cement ratio of the mortar in each mode) and increase in porosity. The value has been equal to 29 hrs based on presented diagrams for maturity factor method and has been equal to 12 hrs for setting time method.

(4) Shear strength in place of hot and warm joints in setting time method has been higher than maturity factor method, which can be because of better penetration of existed mortar in inter-layer joint.

(5) Obtained results indicate that definition of warm joint and required arrangements for implementation of concrete under conditions of warm joint can't be sufficient. Hence, it would be necessary to revise definition of warm joint and its arrangements. Dunsten et al (2012) [7] has also proposed scarifying using broom on layer syrface in warm joint.

7 Acknowledgments

The present study has been conducted using equipment of technical laboratory of soil mechanics. Hence, I would appreciate Mr. Vakili Ayyob, respected custodian of the laboratory, and his personnel, who have helped authors of this paper during all steps of the study.

References

[1] Warren hard Henson. Roller – compacted Concrete dams[J]. Power Ministry, national Committee of Great Dams of Iran,1997(14):283.

[2] Chasemi Hooman. Investigating effective factors in RCC permeability[D]. Tehran University, Collage of Civil Engineering, Technical University,2001.

[3] Mokhtar Pouryani Sayed Bashir, Siose Marde, maroof, et al. Classification of implementing joints in RCC dams based on setting time, 7th national Congress of Civil Engineering, Engineering Collage of Shahid Nikbakht, Zahedan University,2014.

[4] Mokhtar Pouryani Sayed bashir, Siose Marde, maroof, et al. Pattern of RCC dams mixture with attitude to dams with average cement materials, 7th national Congress of Civil Engineering, Engineering Collage of Shahid Nikbakht, Zahedan University,2013.

[5] Moukhtar Pouryani Sayed Bashir. Investigating permeability off implemented joints in RCC dams, MA thesis of water – civil engineering, Islamic Azad University Mahabad Branch, Engineering and Technical University,2014.

[6] N. J. Carino and H. S. Lew. The Maturity Method: From Theory to Application, Structures Congress & Exposition, Washington, D. C., American Society of Civil Engineers, Reston, Virginia, Peter C. Chang, Editor,(2001):19.

[7] Dunsten, et al. . "The Tensile Strength of RCC Dams", 6th International Symposium on Roller Compacted Concrete (RCC) Dams, Zaragoza, Spain,2012.

Part II : The Key Technology and Management of High Dam Construction

Part II: The Key Technology and Management of High Dam Construction

Rapid Construction Method of High Gravel – soil Core Wall Rockfill Dam of Changheba Power Station

Wu Gaojian

(Sinohydro Bureau 5 Co. , ltd, Chengdu, 610066, China)

Abstract: During Changheba Power Station construction, the following technologies have been deployed, such as anti – seepage of spraying polyuria materials; accurately mixing of gravel – soil and filter materials; intelligent water – adding technology and transportation across core wall; digital dam control and digital construction; quick quality test; hydraulic sliding rail system of steep side slope concrete; steep side slope supporting and cantilever trusses; copper water stop molding machine; copper water stop sweat soldering technology, etc. By now, the work progress, quality, and safety are under control, and the construction is moving forward generating of First Unit in May, 2017 (15 months in advance of programme).

Key words: Changheba Power Station, Spraying polyuria materials, Accurately mixing, Intelligent water – adding technology and transportation across core wall, Digital, Quick construction

1 General

Changheba Power Station is located in Kangding County, Ganzi Tibetan Autonomous Prefecture, Sichuan Province, China. It is the 10th project in the hydropower cascade development of Dadu River basin. It's a super – huge hydraulic & hydroelectric project with additional performance such as flood control, etc. The barrage is located 2.5 km downstream from Jintang River. Main works include gravel – soil core rockfill dam, underground diversion system, and flood discharge system. The project is a large (1) engineering work. The maximum height of gravel – soil core rockfill dam is 240 m. The total reservoir capacity is 1.075 billon m³. The catchment area is 55 880 km². The total installed capacity is 4 ×650(2 600) mw and the annual generation capacity is 1.083 billion kWh.

The slopes at both banks of dam are very steep, forming to a V – type valley. The slope next to the river is around 700 m high. Angles of left bank slope below EL. 1 590 m and right bank below EL. 1 660 m are both between 60°and 65°. The rock at dam site is Jinning – Chenjiang Period granite and (quartz) diorite. The geologic structure is characterized with minor fault and long joint. From bottom to the top, the overburden of river bed is respectively over – size gravel layer, over – size sand gravel layer with silt (with silt and medium to fine sand in the middle course), and over – size gravel layer. It has multiple layers with partial porosity. The overburden has a great permeability with a thickness of 79.3 m. The dam is located at the north of intersection among Xianshui River fracture, Qianning – Kangding fracture, Zheduotang fracture, Shimian fracture, and Longmenshan fracture. The regional basic seismic intensity is 8 degree, while the anti – seismic intensity of the dam is designed to be 9 degree.

The gravel soil fill quantity of Changheba Power Station is around 4.28 million m^3, and the rock fill quantity is 33 million m^3. The project is commenced on December 1st, 2010, and is proposed to be ready for generation of the first Unit in May, 2017. The completion date is proposed to be April 30th, 2018. The project embodies four difficulties, i.e. super – high core rockfill dam, deep river bed overburden, high seismic intensity, and steep & narrow valley. There is no completed project of the same size regarding to anti – seepage treatment of dam body and foundation, sedimentation deformation, and anti – seismic safety, which brings challenges to design, construction and operation.

2 Project characteristics and technology challenges

2.1 Dam body structure

The gravel – soil core rockfill dam of Changheba Power Station has a crest elevation of 1 697.00 m. The max. dam height is 240.0 m, and the dam crest length is 502.85 m, with its width of 16.0m. Both the grades of upstream and downstream dam slopes are 1:2. The elevation of core wall bottom is 1 457.00 m, and its bottom width is 125.70 m. The elevation of core wall crest is 1 696.40 m, and the crest width is 6.0m. Both the grades of upstream and downstream wall slopes are 1:0.25. Filter materials is placed at upstream and downstream of core wall. The filter layer at upstream is 8.0 m thick. There are totally two filter layers at downstream, with each thickness of 6.0 m (total 12.0 m). Transition layer is arranged between U/s & D/s filter materials and rockfill materials. The thickness of U/s & D/s transition layers is respectively 20.0 m. Please see dam structure in Fig.1, and the typical profile in Fig.2.

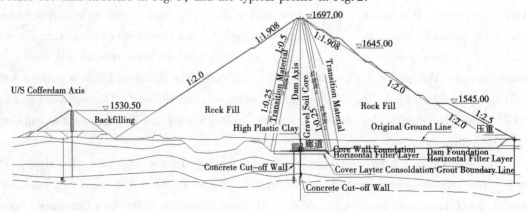

Fig.1 Dam structure (unit: m)

2.2 Foundation treatment

Two concrete cut – off walls with 14.0 m interval are set in the overburden of river bed at the bottom of core wall. The U/s cut – off wall is 1.4 m thick, while the D/s cut – off wall is 1.2 m

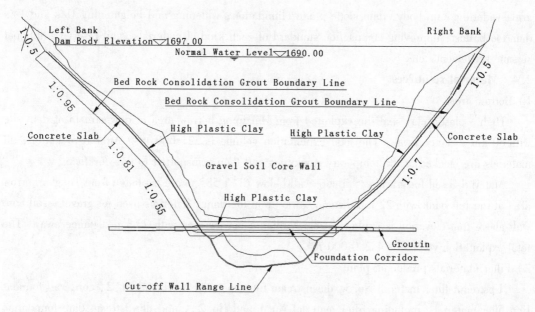

Fig. 2 Typical Dam Profile (unit: m)

thick. The max. depth of cut – off wall is around 53.8 m. The U/s cut – off wall is connected to core wall with an integrated grouting gallery. The lowest elevation of curtain grouting of cut – off wall is 1 290.00 m, and the max. depth is around 117.00 m. The D/s cut – off wall is connected to core wall with concrete key wall inserting into the core wall for a depth of 15.00 m. The lowest elevation of curtain grouting of cut – off wall is 1 397.00 m, and the max. depth is around 10.00 m. Consolidation grouting shall be carried out for overburden layer with thickness less than 5m. High – plastic clay with thickness not less than 3 m shall be placed between the foundation grouting gallery and concrete key wall. 3 m – thick high – plastic clay shall be placed at the contact surface between core wall and base rocks above EL. 1 597 m at left bank and above EL. 1,610 m at right bank. For the contact surface below those elevations, the placing thickness of high – plastic clay shall be 4 m. The sand prism of U/s dam foundation shall be jet grouted with high pressure for around 3 000 m^2 in area.

Five curtain grouting adits shall be set respectively at both banks of abutment at elevation 1 697.00 m, 1 640.00 m, 1 580.00 m, 1 520.00 m, and 1 460.00 m. The adit axis at each elevation shall be in the same plane. The grouting adit of left bank abutment shall be integrated with the grouting adit of underground power house. The axis of grouting adit of right bank abutment shall be the same as dam axis.

2.3 Aseismic strengthening

The aseismic design of dam is Grade A. For a 100 years design reference period, the seismic peak acceleration of bedrock beyond 2% probability is 0.359 g. For a 100 years calibration reference period, the seismic peak acceleration of bedrock beyond 1% probability is 0.430 g. Aseismic strengthening measures include removing sand layers of dam foundation; high – pressure jet to the dam foundation prism; reinforcing the dam slope within 6 ~ 50 m below dam crest with geogrid; widening dam body, dam slope grade elimination; widening and heightening U/s and D/s dam toe ballast; improving compaction standard of each kind of materials; and reserve additional seismic settlement, etc.

2.4 Material resources

1) Borrow areas

High – plastic clay shall be exploited from Haiziping borrow area at downstream of dam site with 60 km distance away. The total exploitation volume is 221 000 m^3. After exploitation, all materials are stocked at the temporary stockpiles and then transported to the workface.

Materials used for gravel – soil core wall below EL. 1 585 m are exploited from Tangba borrow area at the left bank with 22 km distance away from the dam. Materials used for gravel – soil core wall above that elevation is exploited from Xinliantu borrow area with 23 km distance away. The total exploitation volume is 4 280 000 m^3.

2) Filter materials producing plant

Upstream filter material No. 3, downstream filter material No. 1 and No. 2, core wall foundation filter material (including filter material No. 1 and No. 2), and downstream dam foundation filter material No. 4 are all mixed and produced through the aggregate processing system which is located at Mozigou of Dadu River, around 6 km away from downstream of dam site. The aggregate processing system is mainly used for producing dam filter materials and concrete aggregate, with a processing capacity of 1 000 t/hour. The raw materials are come from Jiangzui quarrt and Mozigou borrow area. The total quantity of filter material is 1.66 millon.

3) Quarry

The rockfill materials, transition materials, and ballast are all exploited from Xiangshuigou quarry at upstream of dam site with 3.5 km distance away, and Jiangzui quarry at downstream of dam with 6km distance away. The Jintang Hekou quarry is used for spare. The total exploitation quantity is 33 millon m^3, in which 2.88 millon is used as transition materials.

2.5 Fill materials standard

1) High – plastic clay and gravel soil

High – plastic clay is come from Yeba stockpiles exploited from Haiziping borrow area. The gravel materials exploited from Tangba borrow area is screened without any over – size gravel, and then mixed and adjusted with moisture content. Then it is transported to the workface. Please see the control standard of high – plastic clay and gravel in Table 1.

2) Filter materials

Filter layer is designed at upstream and downstream of gravel soil core wall. The upstream filter layer consists of filter material No. 3 of 8 m thick. Downstream filter layer includes filter materials No. 1 & 2 of 6 m thick respectively. The filter layer of downstream dam foundation consists of filter material No. 4. All these materials are mixed and produced through the aggregate processing system. Please see the control standard of filter materials in Table 2.

Table 1　Control standard of high – plastic clay and gravel

Soil control index		High – plastic clay	Gravel soil	Note
Water – soluble salt content		≤1.5%	<3	
Organic matter content		≤1.0%	<2	
Max. particle (mm)		<5	≤150 and 2/3 h	h = placing thickness
Over – size particle content (%)		≤5	—	
Particle content (%)	>5 mm	—	<50	continuously graded
	<0.075 mm	—	>8	
Clay content (%)	<0.005 mm	>25	—	
Plasticity index		>15	10~20	
Permeability coefficient (10^{-6} cm/s)		1.0	10	
Seepage gradient		>12	>5	
Max. dry density (g/cm³)		around 1.69	—	
Compaction dry density (g/cm³)	<5 mm fine material	—	>1.82	
	P5 content 30%	—	>2.07	
	P5 content 40%	—	>2.10	
	P5 content 50%	—	>2.14	
Degree of compaction (%)	<5 mm fully – graded material	92%~95%	—	592 kJ/m³
	<5 mm fine material	—	100%	592 kJ/m³
	≤150 mm fully – graded material	—	>97%	2 688 kJ/m³
Fill moisture content (%)		$\omega_{0p}+1\% \sim \omega_{0p}+4\%$	$\omega_{0p}-1\% \sim \omega_{0p}+2\%$	

Table 2　Control standard of filter materials

Control Index		Filter No.1	Filter No.2	Filter No.3	Filter No.4	Note
Saturated compressive strength (MPa)		>45	>45	>45	>45	
Max. grain size (mm)		≤20	≤80	≤40	≤100	
Characteristic grain size (mm)	D15	0.15~0.50	1.40~5.00	0.25~0.75	2.00~5.00	
	D85	2.80~7.80	15.00~46.00	8.00~19.00	34.00~52.00	
Particle content (%)	<5 mm	—	—	—	<8	
	<0.075 mm	<5	<2	<5	—	
Permeability coefficient (10^{-3} cm/s)		≥1.0	≥10.0	≥2.0	≥10.0	
Compaction dry density (g/cm³)		≥2.08	≥2.14	≥2.20	≥2.25	
Relative compaction density		≥0.85				

3) Transition materials and rockfill materials

Transition materials shall be exploited from quarries. Weak, sheet, needle-shaped rocks shall be avoided. Rocks shall be weather resistant and hardly dissolve in water. The rockfill materials shall be continuously graded. The maximum and minimum side ratio shall be not more than 4. Please see the control standard in Table 3.

Table 3 Control standard of transition and rockfill materials

Control Index	Transition materials	Rockfill materials	Note
Saturated compressive strength (MPa)	>45	>45	
Softening coefficient		>8	
Frost thawing loss ratio (%)		<1	
Max. grain size (mm)	≤400	≤900	
Max. and min. side ratio		≤4	
Characteristic particle size (mm)	200	29	
0.075 mm particle content (%)	≤3	≤3	
5mm particle content (%)	4~17, ≥10	≤20%	
Permeability coefficient after compaction (10^{-2} cm/s)	5	10	
Relative density	≥0.9		
Porosity (%)	≤20	≤21	
Compaction dry density (g/cm³)	≥2.33	≥2.22	

2.6 Fill parameters

Based upon rolling test on all fill materials, the decided fill parameters are as Table 4.

Table 4 Decided fill parameters

Material	Placing thickness (cm)	Placing method	Rolling machine (self-propelled)	Rolling speed (km/h)	Rolling passes
Rockfill materials	100	Advancing Method	26t smooth wheel roller	2.7 ± 0.2	Static 2 + Vibration 8
Gravel soil	30	Advancing Method	26t padfoot roller	2.5 ± 0.2	Static 2 + Vibration 12
Filter materials	30	Backward Method	26t smooth wheel roller	2.7 ± 0.2	Static 2 + Vibration 8
Transition material	50	Backward Method	26t smooth wheel roller	2.7 ± 0.2	Static 2 + Vibration 8
High-plastic clay	100	Advancing Method	26t smooth wheel roller	2.5 ± 0.2	Static 2 + Vibration 8

3 Application of rapid construction method

3.1 Construction of anti – seepage coating and deformation transition zone for foundation gallery and key – wall concrete

Two concrete anti – seepage walls with 14.0 m intervals shall be deployed at the deep overburden layer of lower riverbed of Changheba core wall. The thickness of the U/S wall is 1.4 m. Concrete foundation gallery shall be used to connect the U/S wall and core wall. The thickness of the D/S wall is 1.2 m. Concrete key – wall plug – in connection structure shall be used to connect the D/S wall and core wall. The thickness of key – wall which plugs into the core wall shall be 15 m. The high plastic clay with the thickness not less than 3 m shall be placed around the concrete foundation grouting gallery and concrete key – wall. Due to the gallery on the overburden layer foundation and key – wall is at the bottom of dam foundation. The force situations of structure are very complex. Three – dimensional nonlinear finite element calculation indicates that because the transverse joint is not deployed along the dam axis of the riverbed section, and the normal stress values of the gallery is vertical to the river which appears at a quarter of full cross location on the left and right bank is big, the tensile stress of the gallery floor along the river is big which will easily cause the longitudinal crack of the floor (during recent years, the crack in different degree exists in building high core wall foundation gallery concrete during the dam filling process and water storage operation period). In order to avoid the cracks may appear in gallery, key – wall under the overburden layer foundation differential settlement, temperature changes during construction period, load changes during construction period, the stress changes of operation condition and the seismic stress condition, these cracks may result in water course which will damage the core wall. Therefore, the cracks need to be sealed by chemical grouting treatment method. Integral structure without open seam shall be adopted for dam foundation gallery. The construction shall be carried out by continuous placement method. The concrete strength and grade is high. Concrete by pumping method shall consume a large amount of cement. The hydration heat is big. Shallow crack is found in gallery during construction, interpenetrate cracks appeared in newly – cast concrete during Lushan earthquake on 20th April. After detailed survey analysis, surface seal and chemical grouting treatment shall be adopted for interpenetrate cracks. For the choice of sealing material, the requirement is that the cracks of the gallery under adverse conditions can suffer 230 m water head positive pressure, and the slurry can be stopped during chemical grouting processing. Method such as outsourcing modified asphalt waterproofing materials, the Perth waterproof base material is normally used for sealing in other similar gallery treatment works. Although the waterproof roll material can meet the high water head seepage control requirements, the adhesion stress on concrete face is not strong which could not ensure the quality of the connection part and the chemical grouting confining pressure requirement. The Perth waterproof base material is a brittle material which can't meet the dam deformation requirements of the deep overburden layer. Finally, the Spray Polyurea Elastomer is selected to be sprayed on the surface of the gallery, and the special asphalt impervious membrane is selected to be deployed on the surface of the key – wall to eliminate possible potential safety hazard in the future.

Spray Polyurea Elastomer is a newly coating structure waterproof technology with high strength, high permeability, ageing resistance, corrosion resistance, good thermal stability, flexible, impact resistance, strong adhesion with concrete, jointless, insoluble, pollution – free, green coating merits etc. Spray Polyurea Elastomer can effectively simplify the structure of waterproof layers and reduce the thickness, with simple process, convenient construction, a high construction efficiency merit, which shows incomparable superiority of the traditional waterproof protection technology. Polyurea CW730 anti – seepage coating with 4 mm thickness shall be used for high water head water tightness test and contact permeability test, when the crack width is 5 mm, non – leakage appears under 300 m water head. The coefficient of permeability of high plasticity clay contact area shall be not more than 3×10^{-7} cm/s, and the contact surface seepage failure grade shall be not less than 13. The main physical properties is: solid content 100%; hardness A30 ~ D50(Hardness JIS); breaking elongation 400%. The 28 d tensile strength, tear strength, and bond strength is 10 MPa, 30 MPa, 3.5 MPa respectively. Gelation time is 5s ~ 30s. Concrete base surface polishing treatment, epoxy mortar defect repair, prime paint spraying and spray polyurea coatings process shall be used for construction. The spraying thickness is 4 mm. The spraying area is 3 150 m^2. The seepage control method is a great innovation in the domestic hydropower station construction.

Special asphalt impervious membrane (CF – 16 hydraulic modified asphalt impervious coils) is with anti – aging, flow resistance, convenient construction, and fast speed advantages, which is usually used for concrete face crack repairing. Specifications performance is: thickness 5mm, width 1.05 m, length 15 m. The density is greater than 1.3 g/cm^3. The elongation percentage is greater than 20%. The permeability coefficient is less than 1×10^{-9}. For tensile strength, the longitudinal strength is greater than 500 N/mm, and transverse strength is greater than 400 N/mm. The environmental temperature is 90℃ and non – flow appears under grade 1:0. No blisters, flowing, crack and wrinkling phenomena appears under 200 times freezing and thawing cycles. the material is flexible, curved without brittle below －5℃. The method as first asphalt cement, then impermeable membrane shall be improved during paving period. The paving of asphalt cement shall be carried out together with the paving of impermeable membrane, that is to say, the seepage control coiled material shall be cut by sizing. The melting asphalt cement shall be paved by fix quantify. Concrete surface shall be cleaned and grinded, and cold primer oil shall be brushed. Impermeable membrane shall be paved at fixed location to ensure the quality of the paved asphalt membrane. Mobile trolley and movable scaffolding can be used when the wall is high and long.

In order to improve the work condition of the cut – off wall and reduce the major principal stress, high plastic clay filling transition zone shall be used around the concrete gallery and key – wall. The paving thickness shall be 14 m. In order to prevent contact scouring between high plastic clay and dam foundation, the geomembrane shall be paved within 30 m at the U/S of the key – wall. Gap shall be reserved for feedstock at the key – wall concrete. Advance method shall be used for high plastic clay material paving. The paving thickness shall be 30 cm, and 18 t smooth drum vibratory roller shall be used to vibrate 2 times in static rolling method and 6 times in vibration grinding method. Series of test and research shall be carried out on filling compaction degree and moisture content of the high plastic clay, critical compaction degree concept is promoted, and the suggested compaction degree shall be controlled in 98% or more. Principle of wet without dry shall be used for core wall contact clay filling in order to increase the plasticity and viscosity of contact clay of the core wall, so as to reduce the permeability and dam settlement after reservoir impoundment, and improve the contact erosion resistance.

3.2 Gravel soil mix adjustment technology

The material for Changheba Core wall is mainly from Tangba borrow area. The Tangba borrow area is formed by glacial, diluvial and clinosol terrains. The slope range is between 20° and 30° and the elevation range is between 2 050 ~ 2 260 m. The total area is 575 000 m^2.

After the re-investigation, the borrow area can be divided into glacial area and diluvial area-a. Refer to the following table 5 for the physical-mechanical index of the materials in different areas.

Table 5 Physical - mechanical index of materials in different areas

Physical – mechanical index		Glacial area	Diluvial area	Remark
Natural density (g/cm^3)		2.06	2.12	
Dry density (g/cm^3)		1.86	1.98	
Natural moisture content (%)		10.7	7.2	
Void ratio		0.45	0.36	
Plasticity index		14.3	11.0	
Clay content (%)	Range	4 ~ 18	2 ~ 12	
	Average	9.86	6.31	
<5 mm Particle content (%)	Range	35 ~ 74	25 ~ 55	
	Average	53.17	37.15	
0.075 mm Particle content (%)	Range	19 ~ 45	6 ~ 33	
	Average	28.6	22.2	
Non – uniform coefficient		1 800	2 473	
Curvature coefficient		0.2	16	
Maximum compaction dry density (g/cm^3)		2.194		2 000 kJ/m^3 Compaction
Optimum moisture content (%)		7.6		
Permeability coefficient		0.86 ~ 1.05		
Thickness of available layer (m)	Range	7 ~ 16	2 ~ 7	
	Average	10.7	5.5	
Area (m^2)		45.7	11.8	
Reserves (m^2)		450	64.4	

Diluvial area is located at the downstream side of borrow area. For the big change of physical and mechanical characteristics, large content of P_5, coarse grain size and less available materials in this area, it goes against concentrated material excavation and it is not worth exploitation.

The glacial area is located at upstream side of borrow area, where the soil material has good anti – seepage and anti – permeability performance and high physical strength, and the quality meets the standard requirements. However, for the different formation causes of the borrow area, inhomogeneous distribution of the soil material, mix of available and unusable materials, physical – mechanical characteristic differences, large variation range of material grading, inhomogeneous content of oversize rocks, inhomogeneous distribution of P_5 (26% ~ 65%) and 70% moisture content (higher than the optimum moisture content), it presents the features of randomness and dispersion of the materials distributed at upper and lower layers as well as at the horizontal and vertical direction. The material source for gravel – earth core wall is uneven distributed, which is the key factor to affect the high gravel – earth core wall construction quality and differential settlement.

As per the designed P_5(30% ~ 50%) index, the borrow areas can be divided into the following types, coarse material area ($P_5 > 50\%$), qualified material area (P_5 30% ~ 50%), fine material area ($P_5 < 30\%$) and spoil material area. According to the contour of P_5 content, the exploratory points on the borrow area plan and profile plan shall be connected into the smooth lines to form the contour map of different P_5 indexes. The grade of material in the area divided by the same contour or by the contour and boundary line shall be the same. By the detailed division of the borrow area, Tangba borrow areas can be divided into 4 areas (Area I, II, III, IV), which can be also divided into sub – regions as per the depth (ground, 0 ~ 6 m underground, below 6m underground), including 4 coarse material concentrated areas and 3 fine material concentrated areas. The qualified material can be directly transported to the dam and the coarse and fine materials shall be mixed before transportation, so as to meet the design standard of high rock – fill dam. Refer to the Fig. 3 for Tangba borrow area geological plan. Refer to the Fig. 4 for Tangba borrow area different P_5 content regions.

The stick – type vibration sieving machine shall be used to screen out the oversize rock and the gravel in a certain grade. Five 2 500 t/h screening systems shall be installed to screen out the oversize rocks with diameter more than 150 mm. The coarse and fine materials shall be mixed and based on the detected P_5 content in coarse and fine materials, content of particles with diameter less than 5 mm and the content of 0.075 mm diameter particle content, the mix proportion shall be dynamically calculated and converted into volume ratio. The paving layer thickness control method shall be adopted to carry out the material paving in the order of coarse material first and then fine ones. When it is paved 3 ~ 4 layers, the materials shall be turned over for 4 times with excavator or bulldozer and loaded after the material is mixed evenly.

For the problem of natural moisture content of most gravel – earth is higher than the optimum moisture content, the tests, such as conventional drying, tedding with four – share mounted plow, and tedding with scarifier, are carried out and finally it is decided to adopt tedding with improved bulldozer scarifier (five – tooth – hook inclined moldboard) to adjust the moisture content.

The distribution of gravel – earth in Tangba borrow area is much complicated and the materials of different particle grades are unevenly distributed with only 1/3 qualified materials. By effective material preparation technology, the different materials are fully used and the quality is also guaranteed.

3.3 Accurately mixing and fine construction of filter materials

Filter material layer is the most direct and efficient way to prevent soil stratum seepage failure. Whether the seepage failure of the seepage proof material will happen under the high water head or not to a large extent depends on the protection of the filter material behind the seepage proof material. Except meeting the requirements of filtering soil, drainage and self – healing of the protected soil after cracking, it shall also be able to adapt the large shear deformation to play an important role of deformation transition.

The aggregate processing system is utilized to produce the filter material of Changheba Power Station. The methods of controlling the opening size of vertical crusher and reducing the stone powder content shall be adopted to adjust the particle size of filter material's grading, ensure the continuous grading and reduce the content of stone powder, etc. The automatic mixing computer system will be used for making the semi – finished materials to qualified fill materials. The traditional mixing process is changed, which makes the production and mixing of filter materials even uniform, accurate and high efficiency, and which reduces the occupied area. During the stockpiling of filter material, the measures including separately stockpiling, slowing down, mixing & loading, etc., are used for ensuring the quality of filter materials.

During the fill construction at filter materials area, the automatic vehicle identification system for different kinds of filter materials and accurately spread process of boundary line between two kinds of filter materials are used for avoiding the mixed loading & unloading of filter materials, approaching the boundary line and cross contamination. And the advantages are accurately spread size of each filter material, saving the levelling equipment, reducing the materials separation, less construction disturbance and high fill efficiency.

3.4 Blasting & mechanical crushing process of transition materials

For the transition materials of gavel – earth core rockfill dam at Changheba, the quantity of fine materials with less than 5 mm is generally larger. So the research works about directly exploiting the transition materials through the blasting method have been carried out. By means of adjusting the blasting parameters, such as hole diameter, hole depth, hole spacing, explosive quantity, detonation method, etc., the unit explosive consumption is increased from 0.75 kg/m^3 to 2.5 kg/m^3, the drilling diameter is reduced from 120 mm to 90 mm, the bench height is reduced from 15 m to 10 m, and the hole mesh area is reduced from 11.7 m^2 to 1.3 m^2. The comparison tests of explosive type have been made to the 2# rock, emulsion explosive and ANFO. The tests result shows that for directly exploiting the transition materials through the blasting method at granite quarry, the quantity of fine materials is generally lower; one of the coarse materials is generally larger, the indicators and grading curve cannot meet the design requirements. Therefore, the combined filter materials production process of blasting and mixing has been adopted. The unit explosive consumption of 1.85 kg/m^3 is used for the blasting of coarse materials and some fine materials from aggregate production are mixed into the coarse materials. After the fine materials and coarse materials being staggered spread, the mixed materials are vertically excavated and transported to the dam.

After the dam being filled to El. 1 536 m, the grading design of core wall's transition materials will be adjusted. According to the results of many blasting tests, when the unit explosive consumption is 2.2 kg/m^3, the transition materials can meet the grading requirements after the removal of over – size rock and graded discontinuity materials. Besides the utilization ratio is 73.5% and the economic performance is poor. At present, the research works about the deficiency of fine materials from the production of overall mobile crushing system are being carried out.

3.5 Intelligent water – adding system and transportation across core wall

Rolling with water is the key measure to improve the compaction effect of rockfill materials. By adding water, masonry will be soaked; fine material will be softened; compressive strength will be decreased; frictional force and apparent cohesion of particle will be decreased; meanwhile, angular and weak part of particle will be softened and broke to improve compaction density and efficiency and reduce the afterward settlement which happened after completion of works. The watering of rockfill materials shall be carried out on the workface and outside of it. Intelligent adding water system shall be adopted to add water before rockfill material entering the filling surface, the weight of dam materials shall be recognized by detecting on – board RFID of transportation vehicles, the adding water quantity shall be calculated as preset proportion, and liquid flow sensor and solenoid – controlled valve shall be adopted to control add water flow and time to realize intelligent adding water.

In order to guarantee the quality of high gavel – earth core rockfill dam core wall, heavy trucks shall not be allowed to pass through the core wall as requirements of design. During construction, the unbalanced strength of up and down stream rockfill quarry, the economical of different load distance, obstructed ventilation exhaust of truck in long tunnel and transportation technology for heavy truck passing through core wall shall be resolved immediately. The stress and strain effect of three measures of passing through core wall such as paving gavel – earth padding singly, padding with paved steel plate, and paving land box pier shall be researched and compared. The paving land box pier principal shall be adopted to design and manufacture articulated trestle to resolve the transportation technical problems of heavy trucks passing through core wall. It guarantees that balance of material source; rockfill material transporting to dam; general environment benefit, economy benefit improvement; and construction progress acceleration. After adjusting the plan, Xiangshuigou quarry which is 6.3 km distance away will provide 64% of the total (45% in original plan) rockfill material, and Jiangju quarry will provide 36% (55% in original plan) of the total rockfill material.

3.6 Digital control of dam and digital construction

In order to guarantee filling quality of high gavel – earth core rockfill dam, 3S technology

(GPS, GIS, and RS), high-volume database management technology, network technology, multi-media and virtual reality (VR) technology etc. shall be adopted to establish integrated digital information platform and 3D virtual model to dynamically collect and digitally process the information that is involved in construction quality and progress during design, construction and operation of Changheba Power Station. Sharing all kinds of project information and data, and dynamic update and maintenance of integrated information will be realized in the whole product life cycle of project, which provides information application and support platform for project decision making and management, dam safety operation and health diagnosis. Online real time monitoring and feedback control for construction quality (material excavation and transportation, core wall material mixing, material placing, adding water for rockfill material, dam surface construction, foundation grouting operation) and progress of core rockfill dam will be realized. Integration management for construction parameter, quality test and progress information etc. will be realized, which provides information application and support platform to construction quality and progress control, and dam safety operation and health diagnosis. It also has the function that employer and consultant could participate in construction quality, progress and precision management of project. Through system automatic monitoring, construction quality and progress shall be controlled effectively, and the fast response on dam construction quality and progress will be realized. The management level of Changheba Power Station will be improved and the innovative management of project construction will be realized to provide the strong technical guarantee for building an excellent project. It also provides data information platform for the completion acceptance, safety appraisal and future operation management.

Integrated dam digital information system consists of real time monitoring & analyzing system for quarry exploitation, core wall material mixing and transporting to dam; automatic monitoring and feedback control system of rockfill dam compaction quality; and rockfill dam construction information collection and analysis system based on PDA etc. The pavement equipment, compaction equipment and dam material transportation vehicles shall be positioned precisely (precision reaches cm, m, respectively) by using GPS, wireless network transmission technology and computer real-time analysis technology, and establishing digital control station, network relay station, random GPS terminal network. The compaction parameters such as compaction thickness, compaction track, times, driving speed, excitation force condition and the information of offloading material that is transported to dam by vehicles shall be online real time monitored and analyzed. When the compaction condition is over the preset alert limits, or transportation vehicles occurring offloading mistakes, the monitoring system will report to site Engineer and construction personnel to handle with the situation timely.

Digital construction will be realized by engineering machine combined with earth moving machinery control products and dam surface mechanical control products which is integrated with laser aiming, sonar control, angle transducer, GNSS, total station etc. Traditional machinery construction will be changed through digital construction to reach the effect that machine guides the operators to construct. The functions of machinery control products consist of slope trimming, trench excavation, and complex modeling which is carried out by excavator during excavation; layer thickness, elevation control, slope control works which is carried out by bulldozer during pavement; layer thickness, elevation control, slope control works which is carried out by grader during leveling; track monitoring, times monitoring, compaction thickness quality inspection which is carried out by vibration compaction equipment during construction.

Unmanned driving technology of vibration compaction equipment and remote control technology of pavement and leveling equipment is another innovation on digital construction, which will make great progress for construction technology.

3.7 Rapid quality test technology

Since requirements on deformation and anti – seepage of super – high core wall is very strict, and the constitution and distribution of gravel soil particles is quite uneven, special experimental study shall be done for gavel – soil core wall material inspection method and evaluation criterion. It is proposed that water content of gravel should be replaced by saturated surface dry water content of gravel which will be used for calculating the weight of gravel in dry density of earth material. This accelerates test speed. It is also proposed that double control method (fine aggregate dominates first and fully – graded aggregate supplements) will be adopted for core wall compaction site inspection and control. Three points compacted method is adopted in site rapid test, and software of three points compacted method has been developed to facilitate the site rapid test. The 800 mm diameter ultra – large electric compaction apparatus which is the largest in China has been developed to review the compaction degree of fully – graded aggregate. Large microwave drying equipment which could dry 50 kg material once has been developed. Movable laboratory equipped with large microwave drying equipment, vehicle – mounted controller, high precision metering tank used for water – filling method, and work platform equipped with sufficient experiment and office instruments is invented to reduce test time and facilitate the rapid test. Based on lots of indoor experiment and site compaction test for gravel soil from Tangba quarry of Changheba Power Station, the site quality testing methods and evaluation standards of gave soil materials has been put forward.

3.8 Other construction technologies

Lots of practical and innovative technical studies such as quick construction of deep cut – off wall high earth – rock cofferdam, concrete tracked hydraulic sliding mode system of high steep slope, support suspension shelving system of high steep slope without berm, copper water stop molding machine, copper water hot melt welding technology, mud mechanical spraying construction technology etc. have been researched and applied in the engineering construction. At present, deformation compatibility study of 300 m level gavel – earth core wall dam, production scheduling assistant decision system, damming machinery optimization allocation system, LNG environment friendly vehicle transportation system study etc. are under researching and developing, which brings higher technological content to the 300 m level earth – rock dam. Modernization of earth rock dam will be realized through industrialization and informatization.

4 Comment and conclusion

Changheba gavel – earth core wall dam has been constructed to the elevation of 1 557 m, which has risen 100 m (exceeds 1/3 of total dam height) accumulatively. 2 060 thousand m^3 gavel – earth and 12.47 million rockfill have been placed accumulatively. The schedule, quality and safety of dam are under control. And project department strides forward towards the goal of first unit generating on May, 2017.

References

[1] National Energy Administration. DL/T 5129—2013, Specifications for Rolled Earth – rockfill Dam construction[S].

Reviewing and Enlightenments of more than one Million Resettlers System Engineering Management in Three Gorges Project

Liang Fuqing[1], Sun Yongping[1], Zhou Hengyong[2]

(1. Resettlement Management Consultation Center of the State Council Three Gorges Project Construction Committee Executive Office[1], Yichang, 443003, China;
2. China Three Gorges university, Yichang, 443003, China)

Abstract: First, the brilliant results of more than one million resettlers project construction and management in Three Gorges Project are reviewed in this thesis. Then, nine main practices of more than one million resettlers system engineering management are summarized. The first was system innovation, which accorded with the resettlement management of Three Gorges and improved the benefits of the resettlement system management continually. The second was theory innovation, which enriched and improved the policy and theory of developmental resettlement. The third was policy innovation, which ensured the resettlers' benefits. The fourth was method innovation, which was to resettle the resettlers in accordance with laws and regulations. The fifth was mode innovation, which was to organize the whole country to support the resettlers in Three Gorges areas as one province or municipality to one county or district. The sixth was supervision innovation, which was to strengthen resettlement supervision in the whole range and in all directions. The seventh was concept innovation, which promoted to construct a harmonious society in the resettlement. The eighth was scientific and technological innovation which promoted the sustainable development of economy and society in the reservoir area. In the end, three pieces of enlightenment is produced to do a good job for more than one million resettlers system engineering management. One is that the fundamental guarantee is to uphold the leadership of Party and government and strengthen the top-level design. Another is that the power and source is to carry forward the Three Gorges resettlers' spirits and initiative. The last one is that the important key is overall improvement and systems innovation in the management.

Key words: Three Gorges, resettlement system engineering, management, reviewing and enlightenment

1 Introduction

The Yangtze River Three Gorges Project is a symbol project for realizing Chinese Dream of controlling water to be making Chinese Nation prosper. The key of Three Gorges Project's successful construction is more than one million resettlers' resettlement. More than one million resettlers' resettlement in Three Gorges project is a complex and mega system engineering, which involves 20 districts or counties of Hubei Province and Chongqing Municipality. There are 1.296 4 million people to be resettled. There are 12 cities or counties, 114 towns, 1 632 factories and many professional facilities had to be moved and rehabilitated. The resettlement involves over 20 aspects, such as resettlement, industrial development, environmental protection, funds management, project management, laws and regulations, protection of resettlers' rights and interests, stability of reservoir, and so on. The compensatory funds for resettlers are up to 52.901 billion Yuan. (This is the static price of May, 1993.) The resettlement time is up to 17 years (from 1993 to 2009). The resettlement management is unusually arduous. So we can say, overall improvement and systems innovation is very important for improving the quality and benefits of the resettlement management project, for ensuring that the task of more than one million resettlers' resettlement is finished on time and the Three Gorges Project is constructed smoothly, for exerting the huge benefits of Three Gorges Project in time, for realizing Chinese Dream, and so on.

2 The magnificent Three Gorges resettlement in a pioneering and innovative spirit to have been realizing Chinese Dream

From 1993, our country and governments at all levels in the reservoir area have led the resettlement work in according to Deng Xiaoping Theory, Three Represents Theory and Scientific Development View. They led broad resettlers' cadres and masses to forge ahead continuously. At the same time, they improved totally and innovated systematically the resettlement management project. As a result, the construction of the resettlement project has been made brilliant achievement. They have written a magnificent Three Gorges resettlement chapter to have been realizing Chinese Dream. By the end of 2009, the resettlement investment and project construction in the Three Gorges area had been completed. 1.296 4 million urban and rural resettlers in the reservoir area had been resettled. Their production, living and housing conditions had been improved obviously. The resettlement was stable as a whole. 2 cities, 10 counties and 114 towns had been moved and rebuilt. Various professional facilities projects had been rebuilt. The appearance of cities and towns and infrastructure had been improved obviously. 1 632 industrial and mining enterprises had been moved and rebuilt. 310 thousand staff had all been resettled properly. The protection and excavation task of 1 087 projects for cultural relics had been completed. The work of controlling geological hazard, preventing and treating water pollution and protecting environment had been strengthened. The geology and drinking water was safe in the resettlement areas. The main stream water quality of Yangtze River had remained in II – III class standard. Plan Outline of Economic and Social Development for the Three Gorges Reservoir Area had been implemented fully. The industries had been adjusted and optimized in the reservoir area. A number of characteristic enterprises with certain advantages had been formed. The economic and social development was in a good situation in the reservoir area. Some work about urban and rural resettlers had been developed fully, such as training and employment, vocational education and later support, social security. The conditions of resettlers' employment, income, and basic living security had been improved continuously. The social undertaking, such as education, healthy, culture, etc, had been developed considerably. From 1993 to the end of 2009, the scale among primary, second and tertiary industry had reduced from 39:35:26 to 12.5:54.5:33 in the 20 resettlement districts or counties in the reservoir area. Regional GDP had increased from 20.44 billion Yuan to

275.555 billion Yuan, up to 13.5 times. Per capita GDP had increased to 19 734 Yuan, up to 10 times. Local financial revenue had increased from 1.281 billion Yuan to 18.971 billion Yuan, up to 14.8 times. The disposable income per urban resident had reached 10 700 Yuan, up to more than 6 times. The rural per capita net income in the reservoir area had increased from 616 Yuan to 4 473 Yuan, up to 7.26 times. The resettlers' legitimate rights and interests had been guaranteed. Society in the reservoir area was stable as a whole.

At the same time, more than one million resettlers system engineering management had made remarkable achievement. In addition to have completed the resettlement successfully, it had achieved a lot of innovative results attracting worldwide attention. For example, new resettlement system had been established and improved, that is united leadership by central government, in charge by province or municipality, based on counties. Democratic and scientific decision – making mechanism such as first consultation, then decision had been established and improved. The policy of developmental resettlement had been used and improved creatively. Resettlement had been implemented in accordance with laws and regulations. The resettlement policy had been improved and adjusted. Resettlement planning principle had been innovated, and the new was fixed amount of funds according to area and quota planning. The management method had been innovated, and it was double fixed. One was that the resettlement funds were fixed, the other was that the resettlement task was fixed. Furthermore, the management mode of resettlement investment had been innovated, and that was static control and dynamic management. The whole country supporting the resettlers in the Three Gorges reservoir area as one province or municipality to one county or district had been creatively organized and implemented. The modern engineering consultation and comprehensive supervision system had been introduced and innovated. The network system of supervising resettlement work in the whole process and in all directions had been established. Some national standards had been established and implemented. They covered reservoir bottom cleaning, price difference control of resettlement funds, elimination and stage acceptance of resettlement project. The management style that was comprehensive supervision and first planning, then implementation had been first carried out in protecting the Three Gorges cultural relics. The view of scientific development had been adhered to. The environmental protection, economic and social sustainable development had been promoted. The resettlers' legitimate rights and interests had been maintained entirely. Three Gorges resettlers' spirits had been carried forward. These above but not limited these were all in the management levels that are advanced at abroad and in the lead at home. They were often used for reference in the work of domestic and foreign large reservoir resettlement.

Chinese Academy of Engineering organized and published publicly Stage Assessment Report of the Three Gorges Engineering, Comprehensive Volume in sept. 2010. The resettlement brilliant achievement and more than one million resettlers management caused by the Three Gorges project were affirmed fully, too[7].

3 Main practices and basic experience of more than one million resettlers system engineering management

3.1 System innovation, establish a new resettlement management system which accorded with the Three Gorges actual situation

The central committee of the communist of China and State Council paid high attention to the resettlement project management in Three Gorges project. According to the actual situation of Three Gorges resettlement work, they established a new system of resettlement management that was called united leadership by central government, in charge by province or municipality, based on counties. They established and improved the administrative mechanism which was that the party committee led in a unified way, the government took all responsibilities, the resettlement departments provide comprehensive administrative, and the relevant departments took its own responsibilities. They clarified the administrative responsibilities between central government and local governments in Three Gorges resettlement work. These mobilized the enthusiasm and initiative of local governments at all levels in the reservoir area to do the resettlement work well. All these ensured the task of more than one million resettlers in Three Gorges resettlement to be finished on time.

3.2 Management innovation, improve the management benefits of resettlement system continuously

The first was to establish democratic and scientific decision – making management mechanism of first consultation and then decision – making. Major decisions and projects must not enter into decision – making process before experts' consultation or review. The second was to establish principles of programming resettlement planning that was fixed investment and quota planning, and to handle the relationship between compensation and development properly. The third was to innovate and implement the double fixed management administrative measure. One was fixed resettlement funds; the other was fixed resettlement task. The resettlement investment was controlled strictly. The fourth was to implement the administrative mode of resettlement investment that was static controlling, dynamic management and spread calculation in order to ensure the benefits of resettlement funds and the resttlers' rights. The fifth was to establish operational mechanism to manage resettlement. The mechanism was in charge of the government, involved in immigrants, supervised by society. Its purpose was to guarantee the resettlers' rights and benefits effectively. The sixth was to establish and improve the three systems of resettlement funds management in townships. The three systems were that the accounts of villages were managed by townships, townships appointed accountants and accounts were open. The systems were conscious to accept the resettlers supervision, so that the resettlement funds were ensured to use safely. The seventh was to introduce and innovate modern project consultation system and comprehensive supervision system in order to promote the construction of resettlement projects standardization. The eighth was to establish and improve a series of administrative systems, including inspection of resettlement projects, audit of resettlement funds, inspection of annual plan execution, inspection of engineering construction quality etc. These systems ensured the quality of resettlement project construction and the safety of using funds. The ninth was to first innovate and implement an administrative way of first planning and then implementing and comprehensive supervision in the protection of Three Gorges cultural relics in the whole country. It improved the quality and effectiveness of protecting cultural relics.

3.3 Theory innovation, enrich and improve the policy and theory of developmental resettlement

The first was to introduce the analysis of environmental capacity into the work of resettlement planning, coordinate and solve the problem of balance between resettlement and bearing capacity of resources and environment, so the theory of developmental resettlement was enriched and improved. The second was to integrate the theory of sustainable development into resettlement practice, apply and develop creatively developmental resettlement policy, properly handle the relationship among resettlement, the economic development and the protection of ecological environment in the reservoir area. Thus these created condition for resettlement stability and sustainable development. The third was to enrich the connotation of developmental resettlement policy and theory by organizing and carrying out our whole country to support the immigration work of Three Gorges reservoir area one to one. The fourth was to transform all of the involuntary resettlers of Three Gorges to voluntary immigrants. This successful practice embodied the overall unity of investment and benefit on resettlement, helped to improve the theory system of water conservancy project resettlement with Chinese features.

3.4 Policy innovation, protect the benefits of resettlement

The first was to implement the policy of compensating in the early stage and combining compensation with production supporting in the later stage in resettlers' compensating. It could combine compensation with development organically and promote resettlement to be stable. The second was to adopt pluralistic and flexible policy in resettlement. Rural resettlement modes had local resettlement and remote resettlement, centralized resettlement and decentralized resettlement, resettlement by the government and resettlement by oneself. The purpose was to improve the quality of rural resettlement hard. The relocation of enterprises was closely integrated with the adjustment of economic structure and industrial structure. Multiple ways were taken to achieve the objective of relocation, such as relocation, upgrading, shutting down, etc. Urban and town relocation took a way of new urbanization. The city relocation was integrated with promoting the process of urbanization and improving the quality of urbanization, so that the good situation that the level and quality of urbanization increased at the same time could come into being. The third was to comprehensively use various supporting policies creatively, such as late support for resettlers, developmental fund for the Three Gorges industry, preferential conditions for relocating submerged industrial and mining enterprises, returning farmland occupancy tax, deducting tax and fee, etc. These measures promoted resettlers to resettle smoothly and economy to develop sustainably in the reservoir area. The fourth was to put forward two – adjustment policy according to the situation of reservoir area. That was to encourage rural resettlers to resettle remote places largely and adjust the structures of industrial and mining enterprises in the process of relocation.

3.5 Method innovation, resettlement in accordance with laws and regulations

Our state formulated and issued Immigration Ordinance for Project Construction of the Yangtze River Three Gorges. The ordinance stated the guide line, compensatory policy, resettlement mode, preferential policy, etc. for Three Gorges resettlement. It provided support for the resettlement by law. Our state still introduced millions of policies, laws, regulations, and management systems one after another in almost twenty aspects which covered the compensation for resettlers, planning and design, plan management, fund management, spread management, project management, resettlement inspection, resettlement audit, project acceptance, enterprise relocation, cultural relics protection, geological disaster prevention and control, environmental protection, one – to – one support, reservoir management, resettlers' petition, and so on. They ensured the resettlement work to go on in the way of laws and systems orderly.

3.6 Mode innovation, organize and implement one – to – one support for the resettlement work of Three Gorges Reservoir area in our whole country

Our Party and state organized and implemented creatively one-to-one support for the resettlement work of Three Gorges Reservoir area in the whole country. It promoted effectively the successful resettlement of more than one million resettlers and sustainable development in the reservoir area. It demonstrated entirely the superiority of socialist cooperative system. By the end of 2009, all kinds of funds of 69.463 billion Yuan had been introduced to the Three Gorges Reservoir area in recent 17 years, in which the free aid social charity project accounted for 3.664 billion Yuan. In addition, the funds were used to resettle resettlers 29 142 person-times, arrange resettlement labours 93 774 person-times, train various of personnel 42 055 person-times in one – to – one support projects. They promoted effectively the reform and opening, advance of human resource, development of characteristic industries, construction of infrastructure and resettlement in the reservoir area. Meanwhile, 29 provinces or cities in the whole country stressed politics, paid attention to the whole situation, and overcame difficulties. They helped to resettle 196 thousand rural resettlers in remote places properly. This promoted to finish the task of the resettlement. [3]

3.7 Supervision innovation, strengthen the resettlement work in the whole process and in all directions

The supervision network for resettlement work was established and improved generally around the reservoir area. It was led by the department of supervision and made up of the departments and units of resettlement, audit, inspection, finance, bank, etc. It took resettlement funds and quality management for resettlement project as core. It carried on the whole process supervision for resettlement work in all directions. Meanwhile, our state and the local governments in reservoir area also established and improved supervision network systems for resettlement that consisted of supervision network for the constructive quality and safety of resettlement projects, supervision network for resettlement funds, supervision network for the quality of resettlement, and so on. Four supervisions in the resettlement work were formed. They were administrative supervision, financial and disciplinary supervision, resettlers' supervision, news and public opinion supervision.

They had unique characteristics of the Three Gorges and ensured the resettlement benefits, construction quality of resettlement projects, and safety of using resettlement funds.

3.8 Concept innovation, promote to construct resettlers' harmonious society

Our state and the governments of reservoir area adhered to the people – oriented concept. They adopted comprehensive administrative measures to guarantee the resettlers' rights and interests and promote to construct a harmonious society in the reservoir area. The first was to adhere to regard resettlers as fundamental, take into account the interests of the state, collective and individual when formulating resettlement policy and plan. Don't damage the resettlers' legitimate rights and interests absolutely. The second was to fully respect the resettlers' rights of know, participation and supervision in the work of resettlement, listen to carefully and adopt the resettlers' reasonable views. The third was to properly tilt the policy to special population of them and ensure their legitimate rights and interests, such as the elderly, women, children, persons with disability, etc. The fourth was to make the policy clear and implement measures to ensure that the resettlers had all the same rights with residents in the resettlement area and did not be discriminated. The rights covered election, joining the army, going to school, marriage, welfare, relief, culture, etc. The fifth was to treat seriously and deal with in time the resettlers' complaints and petitions. More than 200 thousand resettlers' letters or visits had gotten reply and implementation by and large around the reservoir area in 17 years. This ensured the resettlers' legitimate rights and interests and the social safety.[4]

3.9 Science and technology innovation, promote the sustainable development of economy and society in the reservoir area

The first was to establish a set of technical systems and technical specifications. They consisted of the survey outline for indicators of submerged matters around the Three Gorges Reservoir, compiling outline for resettlement plan, technical requirement for cleaning the reservoir bottom, calculation method for the spread of resettlement funds, elimination method for resettlement projects, acceptance outline for the resettlement projects, and so on. This provided forceful technical support for strengthening resettlement management, doing resettlement well and the sustainable development of reservoir area. The second was to innovate science and technology, protect environment in the reservoir area with technical, engineering and ecological methods, carry out scientific and technological research and treatment work to change traditional productive way, save energy and reduce emission, prevent and treat water pollution, treat soil erosion, control rural non – point source pollution, protect biodiversity, protect and surveil population health, and so on. Thus the harmony between people and nature was promoted. The third was to establish implementary network for protecting the environment of Three Gorges. A large monitoring system had been established to monitor the ecology and environment of Three Gorges in the whole process. According to the annual monitoring bulletin released at home and abroad for 18 years, the society and economy in Three Gorges area developed rapidly. The resettlement work, adjusting the structure of relocated enterprises and protecting environment progressed smoothly. The ecological environment that included the quality of water, water and soil conservation, and so on in the reservoir area was good on the whole.[5]

4 The enlightenment of doing well more than one million resettlers system engineering management

4.1 Upholding the leadership of our Party and government and strengthening the top-level design were the fundamental guarantee

Our Party and government had been upholding to lead the resettlement work all the time. They strengthened the work of top-level design. They guided and innovated the management of Three Gorges resettlement continuously. For example, they issued Resettlement Ordinance for Project Construction of the Yangtze River Three Gorges. They established new system and mechanism for resettlement management. They organized and mobilized the whole nation to support the work of Three Gorges resettlement one to one. They presented two-adjustment policy that was adjusting the resettlement policy for rural resettlers and the relocation policy for enterprises in the reservoir area. They strengthened the work of two preventions and treatments that were preventing and treating geological disaster and water pollution in the reservoir area. They introduced some policies, such as developing the advantage industries of reservoir area, implementing the late support for resettlers, promoting resettlers to be resettled stably, and so on. These measures improved the benefits of resettlement management, ensured that the task of resettling more than one million resettlers was finished successfully and got brilliant achievement. The government at all levels in the reservoir area strengthened resettlement management, signed the responsibility agreement of resettlement task at every level, formulated countdown to finish the resettlement task. These mobilized the responsibility and enthusiasm of broad cadres fully. Chongqing municipality regarded the Three Gorges resettlement as fundamental of establishing city, did the resettlement work well with all strength of the city, so it promoted effectively the resettlement tasks of various stages to be finished on time.

4.2 Carrying forward the spirits of Three Gorges resettlement and exerting the resettlers' initiative were the power and source

The spirits of Three Gorges resettlement were produced and carried forward in the great practice of more than one million Three Gorges resettlers. The spirits were patriotic spirit of paying attention to the whole interests, dedicatory spirit of making personal sacrifice for the public good, cooperative spirit of uniting million people as one man and fighting spirit of starting an undertaking painstakingly. The spirits became the power and source of turning involuntary resettlers to voluntary immigrants. 1.296 4 million urban and rural resettlers carried forward the spirits of Three Gorges resettlement, overcame difficulties and moved on time. Therefore, it ensured that the Three Gorges project was constructed smoothly. It highlighted the era feature of patriotism. After resettlement, 560 thousand rural resettlers, including 196 thousand people moved to remote places gave full play to their subjective initiative. They wrote one after another song of starting an undertaking painstakingly through their own efforts. They settled down gradually. A number of typical persons of getting rich stably and getting rich by working even appeared. This displayed fully their great independent ability. They wrote a magnificent Three Gorges resettlement chapter to have been realizing Chinese Dream.

4.3 Improving on the whole and innovating systematically management work was the important key

The practice of more than one million resettlers in Three Gorges resettlement proved that improving on the whole and innovating systematically management work was the key to improve the management quality and benefit of resettlement system engineering. For complex resettlement system engineering, we should pay attention to optimize the structure of management system as a whole, coordinate and manage the relationship among all elements systematically. We should advance the improvement and innovation of management work in the round, such as concept, decision, system, mechanism, policy, law and regulation, method, mode, technique, and so on. We should implement overall coordination, integrated linkage, systematical promotion, coordinated development in the management of resettlement system engineering. Only in this way can we improve the management quality and benefit of resettlement system engineering fully and on the whole, realize the management optimization and established target and task of resettlement system engineering successfully.

References

[1] Liang F Q. Solving Preliminarily the World Problem of Three Gorges Project Million Reseetlers' Relocation [J]. Journal of China Three Gorges University (Humanities & Social Sciences), 2009.

[2] Chinese Academy of Engineering. Phased Assessment Report of the Three Gorges Project (Comprehensive Volume)[M]. Beijing: China Water & Power Press, 2010.

[3] The State Council Three Gorges Project Construction Committee Office. Overview of One – to – one Support [EB/OL]. http://www.3g.gov.cn,2011.

[4] Liang F Q. Research on Resettlers' Petitions in the Three Gorges Project[J]. Journal of Chongqing Three Gorges University, 2010.

[5] Ministry of Environmental Protection of the People's Republic of China. Bulletin of Ecological and Environmental Monitoring for the Three Gorges of Yangtze River[R],Beijing.

Real-time Intelligent Tracking Method of Fresh Concrete Vibration Status

Tian Zhenghong[1], Bian Ce[1], Xiang Jian[2]

(1. College of Water Conservancy and Hydropower Engineering, Hohai University, Nanjing, 210098, China;
2. Sinohydro Bureau 7 Co., Ltd., Chengdu, 610081, China)

Abstract: Based on GNSS (Global Navigation Satellite System) technology, an integrated visual monitoring system to determine trajectory and duration on each vibrating motion of vibrator during concrete consolidation on construction site was realized which could graphically display vibrating status of fresh concrete in real-time. Defects on concrete placing process such as missed vibration, insufficient vibration, or excess vibration were quantifiably assessed, and therefore, precise and suitable remedy could be offered timely. Tests results from outdoor and application in-situ proved that the system could continuously monitor the vibration process in a reliable and quantifiable way. An innovative visualization program was developed to displaying vibration location and duration in real-time, besides, ongoing modifications of the system were also introduced. In all, this system as a new concrete consolidation tool would allow contractors to proactively address concrete consolidation issues in time, a problem common to many concrete construction projects.

Key words: Visual monitoring system, Concrete consolidation, Vibrating trajectory, GNSS, Real-time tracking

1 Introduction

During casting concrete structures in situ, vibration has to be applied on fresh mixture to ensure compaction. Insufficient and missed vibration may result in defects, such as honeycombs, voids, or vacancies, in reinforced concrete members. On the contrary, excess vibration can cause mixture segregation and aggregate heterogeneous distribution[1,2].

Among various vibration methods such as internal, external and both[3,4], internal ones are the most common ones used to introduce vibrations to freshly placed concrete. Related research and construction manuals[5,6], typically specify how vibration should be done in order to produce a dense concrete without segregation. Nevertheless, the suitable means to quantified control the vibration of fresh concrete in construction sites are still few at present. In other words, there is virtually no record of when and where a concrete vibrator has been inserted and for what duration after the placement of concrete. This results in the random vibrating manners and once it could not meet the normal technological requirements, the defects mentioned above would inevitably be involved, and fail to be found and repaired in time. Therefore, how to assess and control the in-situ vibration effect quantifiably in real-time is significant and meaningful.

Based on GNSS (Global Navigation Satellite System), this paper presents an innovative visual monitoring system of concreting which transfers collected internal vibrator – operated trajectory and vibrating duration simultaneously via wireless way and then display them continuously. In addition, ongoing modifications are also introduced. The system allows real – time 3D visualization of the spatial and temporal distribution of vibration efforts, therefore it will significantly improve concrete consolidation quality control practices.

2 Research method

2.1 Principle

A novel real – time visual monitor system on concrete casting process to track trajectory of vibrator head through GPS (Global Position System) & GLONASS (Global Navigation Satellite System) satellite receiver and RTK (Real Time Kinematics) measure mode was designed. Meanwhile, based on the potential variation of electrode assembly attached to vibrator from embedding vibrator into concrete to pulling out it, an approach to determine vibrating duration simultaneously was also developed. After that, MCC (Mono – Chip Computer) embedded with PLC (Programmable Logic Controller) integrated the vibrating trajectory and duration and sent them to the remote terminal computer. Finally, the developed calculating and assessment program in computer continuously displayed the vibrating effect in graphical visualization way and it would help operators and managers understand the cast situation vividly. The working principle of system is shown in Fig. 1.

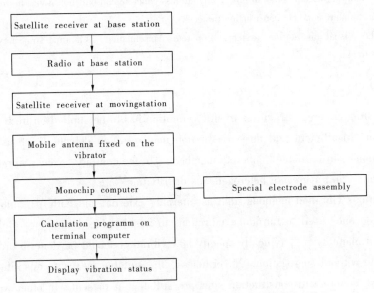

Fig.1 Sketch of system working principle

2.2 System configuration

The visual monitor system configuration is presented on sketch shown in Fig 2, and the whole components scene pictures are illustrated in Fig. 3, Fig. 5, Fig. 6, Fig. 7, and Fig 9. There are three comparative independent components, i.e., reference satellite station devices, mobile vibration signal sending – receiving device set, and terminal received analysis and display program, which communicated each other by wireless transceiver.

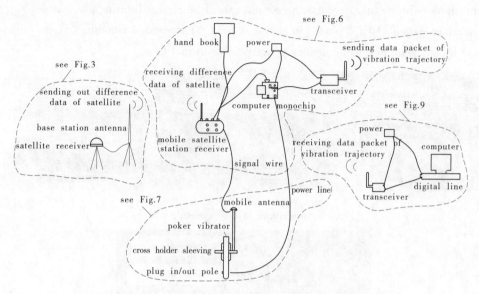

Fig. 2　Sketch of visual monitor system configuration

2.2.1　Vibrating trajectory collected subsystem

The vibrating trajectory collected subsystem consisted of reference station and mobile station. The reference station (Fig. 3) sends its standard position to mobile station with the goal to get a more accurate location of mobile one by difference algorithm.

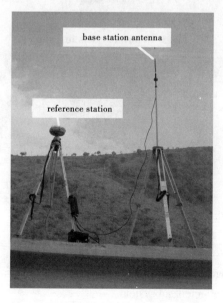

Fig. 3　Appearance of reference station

To get stable solutions accurately on canyon or narrow construction area where have no clear access to satellite signals, GPS & GLONASS double satellite receivers are selected (Fig. 4) and

RTK mode is served. The antenna of mobile station is screwed on light carbon – fiber bar which is fixed tightly to vibrator by clamp. The bar must be paralleled with vibrator, moreover, the height from antenna to vibrator tip is predetermined and cannot be changed during placement (Fig. 5). Once vibration started, MCC (placed in knapsack, Fig. 6) integrated vibrating trajectory into data packet by and send to remote monitor by transceiver in real time.

Fig. 4 **Appearance of dual system receiver**

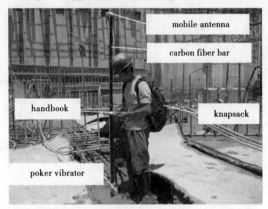

Fig. 5 **Vibratory sets in – situ**

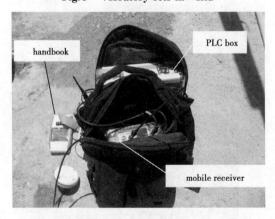

Fig. 6 **Appearance of mobile receiver knapsack**

2.2.2 Vibrating duration collected subsystem

Vibration duration is collected by special electrode assembly fixed on head of poker (Fig 7). The assembly could measure potential divergence value due to variation of residue paste amount while vibrator is introduced into or extracted of concrete, then the duration of each vibration from start to end is figured out (Fig. 8).

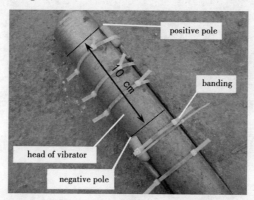

Fig. 7 Appearance of assembly for vibrated duration collection

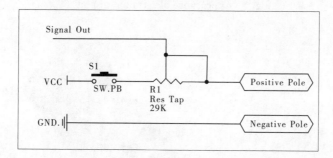

Fig. 8 Schematic circuit diagram for collection of vibrated duration

2.2.3 Data process subsystem

Knapsack (Fig. 6) is the main functional parts of data process system, which integrate many parts such as PLC, transceiver, dual system receiver, power into a portable unit, which may adapt to complicated construction site. It is carried by operator and accompanying with the controller of handbook type to meet easy operation.

2.2.4 Display subsystem

Through transceiver, the developed software in computer receives the data packet from PLC, calculates each real valid vibrating position and scope compared with threshold duration, effect boundary, and displays vibration status online marked on different color (Fig. 9).

3 Verification

Vibration process is carried out on a rectangular cast layer in steel formwork of fresh concrete (5.0 m × 2.5 m × 0.3 m). The effective vibrating radius and appropriate duration are set as 50 cm and 20 ~ 40 s respectively. The visualized vibrating effect (Fig. 10) show the vibrating modes such as normal vibration and insufficient one are very clear and the average difference between real vibrating time recorded by stopwatch and calculated one by software at each position is controlled within 1.5 s so that electrode assembly proved to be rather sensitive and reliable.

In addition, the system is successfully applied in casting process at different layer of fresh

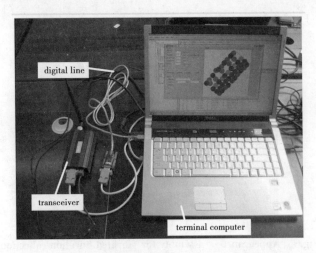

Fig. 9 Appearance of terminal display

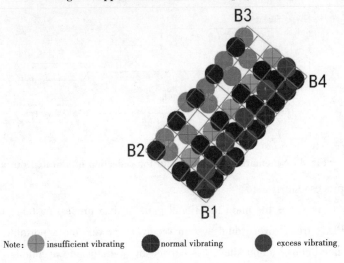

Fig. 10 Resulting picture of vibration effect

concrete from the powerhouse superstructure of Guanyinyan hydropower station, Sichuan province, China. Visualized effect from three cutting depths at same plane is shown in Fig. 11. It could be seen that the quality of vibrated concrete is very easy to be recognized and placing efficiency could be assessed quantifiably.

4 Ongoing modifications

Until now, some modifications have been going on all the time according to test results and operation effects.

(1) Considering work efficiency of operator, the devices of mobile station have become more simplified and integrated to reduce weight and extensive connecting cables (Fig. 12).

(2) The double satellite – positioning system i. e., GPS & GLONASS is replaced by the three one (GPS & GLONASS & Beidou) to greatly improve capabilities such as receiving satellite signals especially in congested areas and acceleration of calculating speed. Besides, much smaller

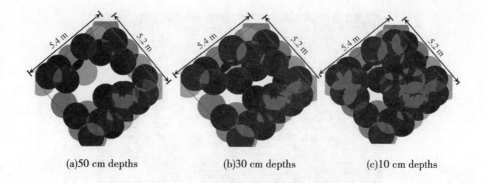

(a) 50 cm depths   (b) 30 cm depths   (c) 10 cm depths

Fig. 11   Visual results of vibration monitor on construction situ

Fig. 12   Appearance of new mobile station

and lighter satellite antenna is used and fixed to helmet of operator for convenient construction (Fig. 13). The real-time accurate determination of distance and direction from the helmet to tip of vibrator is under research.

Fig. 13   Appearance of new satellite antenna

(3) Vibrating duration is realized by new developed equipment non-contacting with vibrator based on principle of working current variation in vibrator instead of electrode assembly to enhance durability and convenience (Fig. 14). Preliminary test proved that the equipment could calculate vibrating time stably.

(4) Data communication is changed into internet instead of radio station considering the pos-

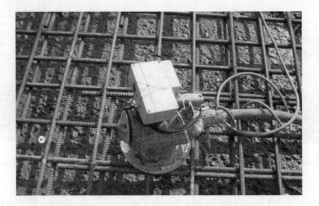

Fig. 14 Appearance of equipment computing vibrating time

sible demand of long – distance transmission in future. Moreover, its storage, calculation and visualized display will be achieved based on cloud technology which not only help to speed up big data processing of multi – engineering simultaneously, but also realizes construction information sharing among related organizations involved. Other performances such as real – time feedback and software interface which is more friendly, simply and conveniently are also under research. Result of preliminary in – situ test shows vibration quality could be visually displayed although some works still need to be done (Fig. 15).

Fig. 15 Vibration effect of modified system

5 Conclusion

The system can real – time accurately measure and display with color image spatial trajectory and duration of each vibrating motion during the fresh concrete placement. Therefore, the continuous data of vibrating status can be obtained expediently and visual monitor system invented for concrete placing status is valid. In order to further improve work efficiency and construction convenience, corresponding modifications are under way and some of them have accomplished and proved to be effective in preliminary test.

Acknowledgement

This study is financially supported by National Natural Science Fund of China (No. 51279054).

References

[1] Domone P L. Self-compacting concrete: an analysis of 11 years of case studies[J]. Cement and Concrete Composites, 2006, 28 (2): 197-208.

[2] Supernant B. Concrete Vibration[M]. Concrete Construction Publications, 1988.

[3] Soutsos M, Bungey J, Brizell M. Vibration of fresh concrete: experimental set-up and preliminary results [C]. Proceedings of the International Seminar on Radical Design and Concrete Practices, University of Dundee, Scotland, 1999:91-101.

[4] Davies R. D. Some experiments on the compaction of concrete by vibration[J]. Magazine of Concrete Research, 1951, 8(3):71-78.

[5] Liu Yang. Research on vibration forming test of slope concrete[D]. Shandong University of Science and Technology, 2010. (in Chinese)

[6] ACI committee 309. Report on behavior of fresh concrete during vibration[R]. Farmington Hills :American Concrete Institute, 2008.

Study on a Distressed High CFRD with Face Slab Rupture

Xu Yao, Jia Jinsheng, Hao Jutao, Li Rong

(China Institute of Water Resources and Hydropower Research (IWHR), Beijing, 100044, China)

Abstract: In recent years, face slab rupture has occurred on several high CFRDs, resulting in large leakage and even potential safety problems. In this paper, a case study on a distressed high CFRD (H =135.8m) in China is conducted to illustrate mechanism analysis of the face slab rupture by using numerical analysis. It is found that large deformation and differential deformation of rockfill, cavity zones under face slab, and structural defects of joints are the most important causes of the face slab rupture. Based on the monitoring data, the deformation and leakage of the dam are analyzed for evaluating the dam performance before the rehabilitation and predicting that the leakage of the dam after the rehabilitation will decrease by at least 80%. Rehabilitation measures are proposed and adopted for repairing the ruptured slabs and joints as well as the damaged waterstops. The leakage of the dam has decreased to less than 20% compared with the value for the same water lever before rehabilitation, which is consistent with the previous prediction. Finally, conclusions are drawn, including the suggestions for dealing with the face slab rupture issue of high CFRDs.

Key words: face slab rupture, CFRD, safety evaluation, leakage

1 Introduction

The concrete face rockfill dam (CFRD) has become a popular and competitive dam type due to its good performance in safety, economy, and environment. By the end of 2012, the total number of the world's CFRDs higher than 30 m constructed and under construction is over 500 (Atlas 2013). The maximum height of CFRD has reached 233 m at Shuibuya Dam of China. Meanwhile, face slab rupture has occurred on several high CFRDs (i.e., Barra Grande, Campos Novos, Mohale, Tianshengqiao – 1, Sanbanxi), resulting in large leakage and even potential safety problems. The causes of the face slab rupture may be multiple, complex, and interrelated. A few studies have been devoted to this issue (e.g., Pinto 2007; Cao et al. 2008). To better understand the mechanism of such structural damage to the face slab, a case study on a distressed high CFRD of China is conducted in this paper (Jia et al. 2013). The dam performance is also evaluated by analyzing the deformation and leakage data. Finally, rehabilitation measures are proposed and adopted for repairing the ruptured slabs and joints as well as the damaged water stops of the dam.

2 Observed distresses at the adm

The dam has a height of 135.8 m, a crest length of 271 m, and a storage capacity of 2.36 ×

10^8 m^3. The upstream slope is 1.4 H:1V while the downstream slopes are 1.45H:1 V and 1.5H:1V with the separating boundary at 3 263 m. The dam crest is at. 3 305.8 m; the normal water level and the dead water level are at 3 300 m and 3 240 m, respectively. The thickness of the slabs uniformly tapers from 30 cm at the crest to 77.9 cm at the lowest portion of the toe. The dam was constructed between March 2008 and February 2011. The slabs were constructed in the two stages with the construction joint at 3 259 m.

2.1 Leakage of the dam

Since the first impoundment started in September 2010, the dam had performed well until June 2012. In July 2012, the water level increased quickly in the flood season. For instance, the average daily rise was 0.69 m during July 1~17; and the maximum daily rise reached as large as 1.25 m. At the end of July 2012, the leakage increased sharply, from 0.1 m^3/s to more than 1 m^3/s. Then, the leakage continuously increased, reaching the maximum value of 1.982 m^3/s at the water level of 3 291 m on January 16, 2013. After that, the leakage decreased as the water level dropped. When the water level was below 3 259 m, the leakage decreased to 0. In other words, no leakage occurred if the water level was under the construction joint of the face slabs. Fig. 1 shows the relationship between the water level and the leakage.

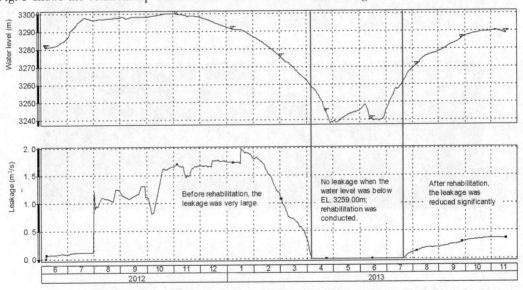

Fig. 1 **Relationship between the water level and the leakage of the dam**

2.2 Displacement of the dam

Similar to the seepage performance, the displacements of the dam were small until June 2012. The dam was in good condition with no cracks and ruptures on the slabs. After entering into July 2012, the settlement of the middle dam part became larger than that of the two side parts; and both side parts showed movements to the centre. The maximum settlement of the dam was 87.5 cm, 0.64% of the dam height. The ruptures occurred at the vertical joints of the middle slabs; and the cracks occurred at most of the slabs. Fig. 2 shows the ruptures along the vertical joint of Slabs 15# and 16#. It is indicated in Fig. 2 that the surface concrete was damaged due to the extrusion and uplift under compression while the reinforcing bars remained well with no deformation. As the water level decreased, more cracks and ruptures revealed. On April 9, 2013 when the water level decreased to below the construction joint, serious ruptures were found along the construction joint, ranging from Slabs 7# to 27# with a horizontal length of about 200 m and a width of about 1m along the slab longitudinal direction, as shown in Fig. 3. The longitudinal reinforced bars were installed in the double rows, of which the upper row was seriously bended and the lower row was also bended. In contrast, no bending occurred with the horizontal reinforced bars. The upper concrete was intensely damaged while the lower concrete was almost intact without any damage. It was also found that the face slab above the construction joint moved outward along the joint interface.

Fig. 2   Ruptures occurred along the vertical joint of Slabs 15# and 16#

Fig. 3   Ruptures occurred along the construction joint

3   Numerical analysis

3.1   Analysis of dam performance with using designed material properties

The numerical analysis uses the Duncan – Chang E – B nonlinear model to describe stress – strain relationship. The contours of the settlement and horizontal displacement of the dam under the normal water level are calculated with using designed material properties, as shown in Fig. 5 and Fig. 6, respectively. The maximum settlement is 83.3 cm; and the maximum horizontal displacement towards downstream is 36.1 cm while that towards upstream is 17.7 cm. Compared with the calculated values, it is found that the measured settlement is larger while the measured horizontal displacement towards downstream is smaller. In other words, the actual volumetric compression deformation of the embankment was larger than the expected value.

The contour of the deflection of the face slab is shown in Fig. 7. The maximum deflection of

Fig 4 Bending of longitudinal reinforcing bars at the construction joint

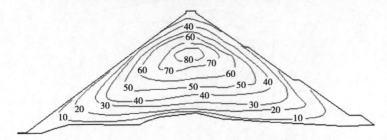

Fig. 5 Contour of settlement

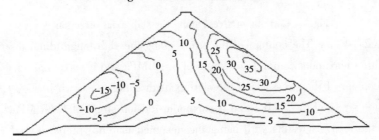

Fig. 6 Contour of horizontal displacement

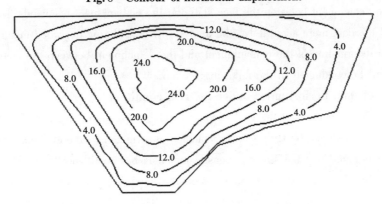

Fig. 7 Contour of deflection of the face slab

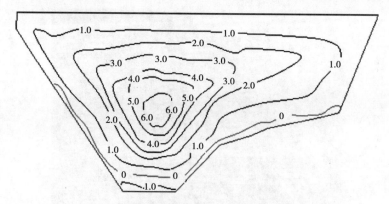

Fig. 8 Contour of stress along the slab longitudinal direction

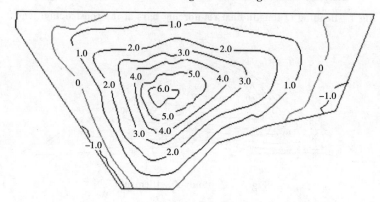

Fig. 9 Contour of stress along the slab axial direction

the face slab is 27.4 cm. The contour of the stress along the slab longitudinal direction is shown in Figure 8, with a maximum compression stress of 6.12 MPa and a maximum tension stress of 1.73 MPa. The contour of the stress along the slab axial direction is shown in Fig. 9, with a maximum compression stress of 6.77 MPa and a maximum tension stress of 1.68 MPa.

Based on the analysis results with using the designed material properties, it can be judged that neither detachment nor rupture is associated with the face slab. In fact, both the face slab rupture and the detachment between the face slab and the embankment occurred. To evaluate the performance of the dam more accurately, the analysis by considering degraded material properties and rheological deformation of the embankment and the detachment between the face slab and the embankment will be conducted in the following section.

3.2 Analysis of dam performance under the real condition

Fig. 10 shows the contour of the stress along the slab axial direction by considering degraded material properties (reduced by 15%) and rheological deformation of the embankment. Compared with Fig. 10, it is found that the stress increases obviously but is still not large enough to damage the face slab. Therefore, the detachment between the face slab and the embankment is further considered at the three levels of the detachment ratio (10%, 20%, 30%), which is defined as the percentage of the detached area to the whole face slab area.

Fig. 11 shows the contour of the stress along the slab axial direction by considering degraded material properties (reduced by 15%) and rheological deformation of the embankment and the detachment between the face slab and the embankment (the detachment ratio of 30%). It is indicated that the maximum compression stress in the central area is over 30 MPa, leading to the face slab rupture. Therefore, the detachment has great influence on the performance of the face slab. Therefore, the rupture at the vertical joints is caused by large deformation of the embankment and partial detachment between the face slab and the embankment.

It is calculated that the longitudinal strain at the construction joint is 580×10^{-6} while the longitudinal compression stress is 12.7 MPa. Although the stress and strain are large, they are still under the limit values of the C30 concrete. As mentioned before, serious face slab rupture did occur along the construction joint. Another important reason for the rupture is that the construction joint interface is horizontal and not perpendicular to the face slab, making the joint interface as structural weakness.

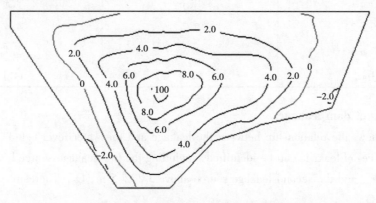

Fig. 10 Contour of stress along the slab axial direction by considering degraded material properties and rheological deformation

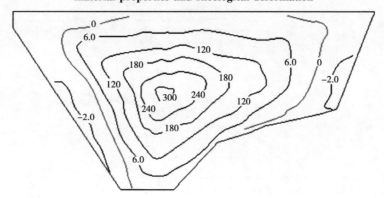

Fig. 11 Contour of stress along the slab axial direction by considering degraded material properties and rheological deformation and the detachment

In general, the rupture along the construction joint can be explained from the two aspects: ①the structural weakness with the joint interface reduces the resistance;②the large displacement of the embankment leads to a large eccentric load on the face slab; besides, the differential deformation of the upper part and the lower part of the upstream embankment results in a horizontal shearing load on the construction joint.

4 Aanlysis of dam performance before rehabilitation

4.1 Analysis of dam settlement

The increased dam settlements in the following three years are predicted by using the numerical analysis, as shown in Table 1. The cumulative increased dam settlement in the following three years is 23.6 cm. The additional longitudinal strain at the construction joint in the following three years is also calculated as 260×10^{-6}.

Table 1 Prediction of increased dam settlement in the following three years

Time	Prediction before rehabilitation (cm)	Prediction after rehabilitation (cm)
Year 1	13.6	8.3
Year 2	7.7	4.4
Year 3	2.3	1.4
Total	23.6	14.1

4.2 Analysis of dam leakage

Fig. 12 shows the relationship between the leakage and the water level before rehabilitation. Two main sources of leakage can be identified, including the first leakage source between 3 259 ~ 3 272.5 m, Q_1, and the second leakage source above 3 272.5 m, Q_2. Therefore, the total leakage $Q_t = Q_1 + Q_2$.

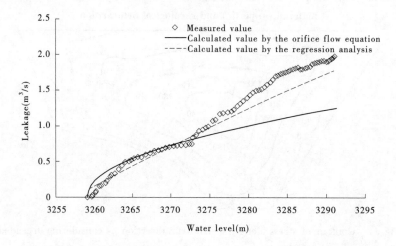

Fig. 12 Relationship between leakage and water level before rehabilitation

As mentioned earlier, the construction joint at 3 259 m is seriously damaged; hence, it is the primary leakage source of Q_1. In this study, the leakage through the damaged construction joint is calculated by using the orifice flow equation, as shown the solid line in Fig. 12. For the maximum leakage of 1.982 m³/s at the water level of 3 291 m, the contribution through the construction joint is calculated as 1.242 m^3/s, taking 63% of the total amount. Based on the monitoring data, the regression analysis is also conducted to fit the measured curve of Q_1, as shown the dash line in Fig. 12. For the maximum leakage of 1.982 m^3/s at the water level of 3 291 m, the contribution through the construction joint is calculated as 1.793 m^3/s, taking 90% of the total amount. Therefore, based on the results by using the orifice flow equation and the regression analysis, it is predicted that the leakage will decrease by 63% ~ 90% after the rehabilitation of the construction joint.

Referring to the second leakage source above 3 272.5 m, it is probably related to the peripheral joint at the right bank, where the maximum settlement of the face slab is as large as 30 cm and the water stops are partially damaged. In addition, it is found that the curtain grouting under the plinth at the right bank is defective based on the water pressure test results. Therefore, Q_2 is treated as the contribution from leakage channels along the peripheral joint at the right bank above 3 272.5 m. It is estimated that Q_2 takes about 20% of the total leakage.

In total, it is predicted that the leakage will decrease by at least 80% after the rehabilitation of both the construction joint and the peripheral joint at the right bank.

4.3 Rehabilitation measures

Based on the analysis results and engineering experience, rehabilitation measures were proposed by IWHR for repairing the ruptured slabs and joints as well as the damaged water stops of the dam. Fig. 13 shows the rehabilitation for the face slab rupture at the vertical joints. The damaged concrete is removed and then backfilled by using new concrete with the original designed grade. If the copper water stop and reinforced bars are damaged, they are restored in terms of the original design requirements. The 2 cm wide joints are filled with two 10mm thick rubber sheets. The GB plastic filler and GB EPDM composite cover board is used for repairing the surface water stop. Figure 14 presents the rehabilitation for the face slab rupture at the horizontal construction joint. The damaged concrete reinforced bars and cushion layer are removed and then and then backfilled by using new cushion layer and concrete with the original designed grade as well as restoration of the reinforced bars. The water stop treatment is conducted at joint parts between the new and old concrete. Figure 15 gives the rehabilitation for the damaged surface water stop at the peripheral joints. The corrugated rubber water stop is firstly installed above the supporting rubber rod, which is followed by using the GB plastic filler and GB EPDM composite cover board. In June and July 2013, the above rehabilitation measures were adopted for repairing the dam.

5 Analysis of dam performance after rehabiliation

Based on new monitoring data after the rehabilitation, the increased dam settlements in the following three years are again predicted by using the numerical analysis, as shown in Table 1. The cumulative increased dam settlement in the following three years is 14.1 cm. The additional longitudinal strain at the construction joint in the following three years is also calculated as 175 × 10^{-6}. Both the settlement and strain values are distinctly smaller than those before the rehabilitation.

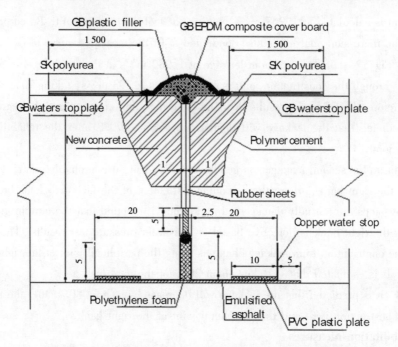

Fig. 13 Rehabilitation for face slab rupture at vertical joint

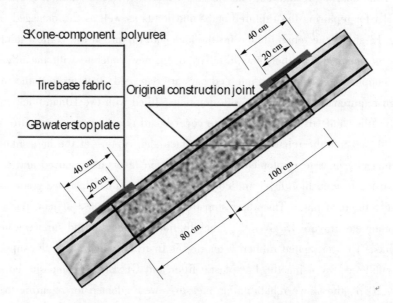

Fig. 14 Rehabilitation for face slab rupture at horizontal construction joint

After the rehabilitation, the water level reached the maximum of 3 290 m when the leakage was 0.36 m³/s. Compared with the value of 1.92 m³/s for the same water lever before the rehabilitation, the leakage has decreased by 81%. This is consistent with the prediction (decreasing by at least 80%) before the rehabilitation in Section 4.2. Also it is proved that the rehabilitation measures are very effective for the dam.

In general, the settlement has become gradually stable; and the leakage has decreased a lot. The dam performance has improved significantly after the rehabilitation. The current dam safety

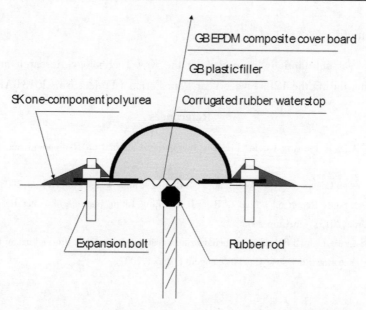

Fig. 15 Rehabilitation for damaged surface waterstop at peripheral joint

generally meets the requirements.

6 Conclusions

The causes and mechanisms of the face slab rupture of high CFRDs is complex, involving both the influence of embankment deformation and the interface between the face slab and embankment. For the dam studied in this paper, the face slab ruptures at both vertical and horizontal joints are fundamentally caused by large and non – coordination deformation of the embankment as well as cavity zones under the face slab; and the second reason is the structural defects.

After the adoption of effective rehabilitation measures, the leakage of the dam has decreased to less than 20% compared with the value for the same water lever before rehabilitation, which is consistent with the earlier prediction based on physical equation analysis and regression analysis. The dam performance has improved a lot with the basically satisfied safety status.

Finally, several suggestions for dealing with the face slab rupture issue of high CFRDs are proposed as follows:

(1) The key issue of the face slab rupture is large deformation and non – coordination deformation of the dam. Therefore, it is necessary to conduct deformation control and deformation coordination in the design phase by considering the embankment and the face slab as a whole, in order to avoid the adverse stress and strain conditions.

(2) Appropriate measures should be used in advance to improve the structural resistance; and structural defects such as the interface of the horizontal construction joint being not perpendicular to the face slab should be avoided.

(3) At the lower part of the upstream face slab, fly ash or fine sand can be used for replacing of clay blanket, in order to increase the abilities on silting and filling and self – healing of the overall anti – seepage system.

Acknowledgement

This study was substantially supported by the Key Technology Research and Development Program of China during the 12th Five – Year Plan Period (Project No. 2013BAB06B00).

References

[1] Atlas. World Atlas & Industry Guide[J]. The International Journal on Hydropower and Dams, 2013, London, UK.
[2] Cao, K. M. et al. Concrete face rockfill dam[M]. Beijing: China Water Power Press, 2008. (in Chinese)
[3] Jia, J. S, et al. Study Report on a high CFRD [R]. Beijing: China Institute of Water Resources and Hydropower Research, 2013. (in Chinese)
[4] Pinto N. L. S. Very high CFRD dams – behavior and design features[A]. Proceedings of the 3rd Symposium on Concrete Face Rockfill Dams[C]. Florianopolis, Brazil, 2007.

Structural Characteristic Analysis on Hardfill Dam Based on Model Test Method

Yang Baoquan, Zhang Lin, Chen Yuan, Dong Jianhua, Chen Jianye

(State Key Lab. of Hydraulics and Mountain River Eng., College of Water Resources and Hydropower, Sichuan Univ., Chengdu, 610065, China)

Abstract: Hardfill dam is a new type of dam, which has a symmetrical trapezoid – shaped cross section and a concrete impervious facing or other impervious facilities in the upstream. The dam is filled up with cemented sand – gravel material called Hardfill, which is inexpensive and low – strength. In order to investigate the structural behavior under normal working conditions and the failure mode and mechanism in the process of overloading of Hardfill dam, and to analysis the advantages and characteristic of the dam structure, and also to promote the development and application of the new dam construction technique, a stress model test and a geo – mechanical model failure test of the typical Hardfill dam were carried out respectively. Results show that the Hardfill dam dam body mainly endured the compressive stress and without obvious stress concentration under the condition of normal water load. Compared with the conventional gravity dam, Hardfill dam has lower stress level and more uniform stress distribution. The failure mode was concluded as dam heel cracked firstly and then dam toe damaged, finally the dam integrally slid along the interface of the dam and foundation under the condition of water load overloading. Meanwhile, the overloading safety factors derived from the geo – mechanical model failure test are as follows: the crack initiation factor K_1 ranges from 6.0 to 7.0, the nonlinear deformation factor K_2 ranges from 7.5 to 8.5, and the ultimate load factor K_3 ranges from 9.0 to 10.0. Analysis showed that the Hardfill dam has good work performance, strong ultimate bearing capacity and big overload coefficient. It is a kind of high – quality dam type.

Key words: Hardfill dam, structural characteristic, model test, failure mode, overloading safety coefficient

1 Introduction

In water resources and hydropower engineering, the earth rockfill dam and the gravity dam are the most widely used type of dam. Two kinds of dam type have advantages and disadvantages respectively and has been developing along a different path. The emergence of RCC (roller compacted concrete) materials is the first combine of the two type of dam. The RCC dam uses the roller compacted construction method of the earth rockfill dam to construct the rigid concrete dam, which plays their respective advantages and develops rapidly. But because of the problems of complex seepage control, level stability and temperature control measures, etc, the dam engineering and technical personnels and researchers have to explore a better type of dam. And the excellent dam type should meet the high safety, low cost, rapid construction and small disturbance and destruction to the environment, etc. Hardfill dam[1,2] is proposed by constant exploration and practice of the predecessors. The new type of dam has a symmetrical trapezoid – shaped cross section and a concrete impervious facing or other impervious facilities in the upstream. The dam is filled up with cemented sand – gravel material called Hardfill, which is inexpensive and low – strength. The material is obtained by simple mixing of the riverbed sandy gravel near the dam site and easy to get, or excavation abandoned slag, adding water and a small amount of cement. Since the 1990 s, the design concepts and damming technology of this new dam type had put into practice in foreign countries, and many dams had been built in Japan, Greece, France and Turkey, etc. At present, the highest Hardfill dam is named Cindere in Turkish with 107 m dam height[3]. The practice of the Hardfill dam in China began in 2004, successively used in Daotang Reservoir's upstream cofferdam in Guizhou province[4], Jiemian hydropower station's downstream cofferdam and Hongkou engineering upstream cofferdam in Fujian province[5], and used in the Shatuo hydropower station's downstream cofferdam in Guizhou province[6].

Although a batch of Hardfill dams have been built abroad, the design criteria and safety standards of it frequently refer to the related design theory of the gravity dam and the earth – rock dam partly or totally, and it's be considered that its structure failure mode is similar to the gravity dam and have no further discuss[7]. In China, the research of Hardfill dam is started late, it's mainly used in some temporary cofferdams engineering. Currently, the design researchers have mainly done research on the mechanical properties of the materials, the static and dynamic characteristics of structure and the safety of the Hardfill dam, etc[8-10]. But research of the failure modes, the design rules and safety standards, and the structure design method of the Hardfill dam are still in groping[11]. In order to investigate the structural behavior under normal working conditions and the failure mode and mechanism in the process of overloading of Hardfill dam, and to analysis the advantages and characteristic of the dam structure, and also to promote the development and application of the new dam construction technique, a stress model test and a geo – mechanical model failure test of the typical Hardfill dam were carried out respectively. The stress distribution status and performance, deformation characteristics, and the failure pattern and the process can be obtained by the test. And the overloading safety coefficients are put forwarded. Evaluation of the stability and security of the dam type has been done. It can provide a reference for the formation of a set of suitable design method of the Hardfill dam.

2 Overview and mechanics parameters of the experimental prototype

In order to make the test results more representative, a typical section of the Hardfill dam was designed to carry out experiment research. The dam height is 70 m, and the dam slope is taken as 1:0.7 according to the Oyuk dam and Cindere dam of Turkish. The main loads are the dam body gravity, the hydrostatic pressure and the uplift pressure. The upstream water level is flush with the dam crest, and no downstream water. It is assumed that the uplift pressure in dam heel drain screen reduction for half of the head, and linear distribution of uplift pressure in dam base.

Considering the large discreteness of the actual Hardfill materials' properties because of the influence of aggregate, water – binder ratio and construction technology, and considering the certain limitations of the loading capacity and model making process in the practical experiment, so 5 set of parameters have drewed up in the actual possible range for test selection. And according to the actual test loading capacity and model making conditions, a set of parameters has chosed eventually to carry out test. In addition, the corresponding material parameters of the dam foundation are selected. The main physical and mechanical parameters of the typical Hardfill dam and the foundation are shown in Table 1. The shear strength parameters of the joint surface of dam body and dam foundation are as follows: the c ranges from 0.5 MPa to 0.6 MPa, the f ranges from 0.65 to 0.8.

Table 1 Main physical and mechanical parameters of a typical Hardfill dam (prototype)

Material	Density (kg/m³)	Elasticity modulus (GPa)	Poisson's ratio	Compressive strength (MPa)	Strength of extension (MPa)	Shearing strength	
						Cohesion (MPa)	Frictional angle(°)
Dam foundation	2 400	15.0	0.2	8.0	1.0	1.0	50
Hardfill – E	2 350	16.0	0.2	9.0	0.8	1.0	50

3 Stress model test of Hardfill dam

3.1 Model similarity relation and simulation scope

The main research object of the stress model is the elastic stage of the prototype structure, so the prototype and the model should meet the basic equations of elastic mechanics and boundary conditions. By the balance equation, the geometric equations, the physical equation of the elastic mechanics and boundary condition equations, the main similarity index of the stress model experiment[12] can be deduced and it can be expressed as: $C_\varepsilon = 1, C_\mu = 1, C_\rho = C_P C_L^{-3}, C_\sigma = C_E = C_L C_\rho$, among them, $C_\varepsilon、C_\mu、C_\rho、C_P、C_\sigma、C_E、C_L$ is strain ratio, poisson's ratio, density ratio, concentration force ratio, stress ratio, elastic ratio and geometric ratio of the prototype and model respectively.

According to experimental conditions and requirements, the geometric scale of the stress model was taken as 280. And according to the experience of the experiment, the range of the experiment simulation can be expressed as: The upstream side length of the dam foundation surface takes more than one times dam height, the downstream side length of the dam foundation surface takes 1.5 times dam height, the depth below dam foundation surface takes more one times dam height. Using gypsum material to produce the stress model, water paste ratio is decided by the model parameters which is calculated by the similarity relation, the photo of completed model is shown in Fig. 1.

Fig. 1 Photograph of the Hardfill dam structure stress model

3.2 Model loading system and measurement system

The main simulation load for the experiment include the weight of the dam and the surface water pressure of the dam upstream and the uplift pressure of the dam bottom. Because the gypsum material is lighter, it can't meet the similarity of the density, so we should apply the external force to achieve the similarity gravity load. The uplift pressure on dam is by decreasing the body weight to equivalent simulate. The combination of vertical load is added by jack on the dam crest. The surface water load of upstream dam is translated to the model load according to the similarity principle, and perpendicularly applying on the surface by jack. The jack load applying is shown in Fig. 1. The pressure of the jacks is from the 8 channel self – control hydraulic pressure and stabilivolt device of WY – 300 / Ⅷ.

The calculation of the structure stress is through measuring the strain of the model. There are three rows of strain measuring point in the experiment, respectively in the dam foundation surface, 1/3 high dam and 2/3 high dam. And on the dam foundation where near the foundation surface also decorate a row of measuring point. Each measuring point has three piece strain gauge to measure the level strain, vertical strain and 45 ° direction strain respectively. There are total 16 strain measuring points and 48 strain gauge. The strain is tested by the UCAM – 8 BL universal digital test device. At the same time, we set the compensator to eliminate the temperature effect. We layouted displacement points to test surface displacement of the dam body, they are respectively on near 1/3 high dam and dam downstream, and with SP – 10A digital display monitor to test the displacement. The layout of the displacement points and strain gauge are shown in Fig. 2.

3.3 The test results and analysis of the stress model

Before the test, the model is prepressing to eliminate the additional deformation, then through the incremental method to exert pressure step by step. First exert the vertical load, that exert the horizontal load until the design load is reached. Maintain 8 to 10 minutes after loading at each level, then record the strain and displacement values of each point in the loading process to get the stress and deformation of the dam under the condition of self – weight and normal water level.

The test using the weighted average method to eliminate deviations and using the arithmetic average of multiple readings as the average strain value of each measurement point, then apply the

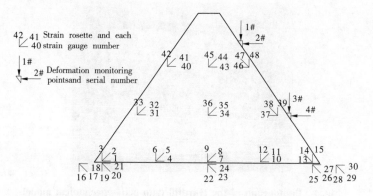

Fig. 2 Arrangement diagram for the strain and deformation monitoring points of the Hardfill dam

Hooke's law and the similar relationship make the strain was converted into the stress of the prototype dam. Under normal working condition, the principal stress distribution of Hardfill dam as presented in Figure 3. The figure shows that the dam body is in a state of compression. With the increase of dam height, the main pressure decreases and the pressure stress in the downstream face of dam is larger especially in dam toe. Due to the influence of the loading of the jack, the stress in the upstream face and the height in the 2/3 dam height is larger. At the same time, compared with the results of the gravity dam structure model[13] in the same height, we know that under normal water load condition, the stress level of Hardfill dam is lower than conventional gravity dam. The stress distribution uniform and is largely compressive stress also has no obvious stress concentration, so the Hardfill dam has a good working performance.

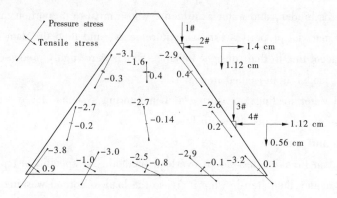

Fig. 3 The principal stress distribution diagram of the Hardfill dam under normal working condition

4 Geo – mechanical model failure test of Hardfill dam

4.1 Design of the model test

Geo – mechanical model test belongs to failure test, which makes the structure beyond the elastic state through continuous overloading to analyze the failure process, mode and mechanism of the dam and find weakness zone. The similarity relations between the prototype and the model should be satisfied, such as the relations of geometric, mechanics, boundary and initial condi-

Fig. 4 Photograph of the Hardfill dam geo – mechanical model

tions: $C_\gamma = 1, C_\varepsilon = 1, C_f = 1, C_\mu = 1, C_\sigma = C_\varepsilon C_E, C_\sigma = C_E = C_L, C_F = C_\sigma C_L^2 = C_\gamma C_L^3$. Here, $C_E, C_\gamma, C_L, C_\sigma$ and C_F are the deformation modulus ratio, the bulk density, the geometric ratio, the stress ratio, and the concentration force ratio, respectively. In addition, C_μ, C_ε, and Cf are Poisson's ratio, the strain and the friction factor ratio, respectively. Combined with the test conditions, we selected a geometric ratio of $C_L = 150$. And the model simulation dimensions are: 75 m of the upstream (about 1.1 times of dam height), 140 m of the downstream (about 2 times of dam height), 58 m depth under the riverbed (about 0.83 times of dam height). The load combinations of geo – mechanical model test are coordinated with that of stress model test, except weight which is same as that of the prototype in order to meet the similarity requirement ($C_\gamma = 1$). The physical – mechanical parameters of geo – mechanical model test material are the same as the stress model, but geo – mechanical experimental material of the dam body and foundation selects barite powder as aggregate, gypsum as binder, and water as diluent, whose mixture proportion ratios can be determined by different material property's various requirements, and then the dam body and foundation were created according to their designed shape, which were highly processed and glued after air – dry, the final model as presented in Fig. 4.

The upstream water loading, measuring and monitoring systems of geo – mechanical model are similar to stress model test as shown in Fig. 4.

4.2 Test results and analysis

The Hardfill dam breaking test adopt overload method. Its specific test process is: first of all, preloading the model, then step by step increase the load to normal working condation, on this basis to overload the water load, every load level add to 0.3 ~ 0.4 P_0 (P_0 is the load of normal operating conditions), until the dam was broken, dam and foundation appeared overall buckling or instability trend. Recording strain and deformation data in all levels of load in the test, observing of dam and foundation deformation characteristics, failure process and failure pattern. The main results and analysis are as follows:

(1) Deformation distribution characteristics: From the test result, the Hardfill dam mainly generate the deformation down the river, and the vertical deformation is smaller and given priority to with subsidence. When the load is small, the dam deformation is small, then with the increase of overload ratio, the deformation down the river starts to increase gradually, deformation in the

dam crest is greater than the part, it conform to the conventional. When the overload ratio $K_p \geqslant 4.0$, the whole dam deformation began to increase; When the overload ratio $K_p \geqslant 7.0$, the dam deformation increase further; When $K_p = 9.6$, the vertical displacement began to decrease; When $K_p > 11.0$, the dam vertical displacement reverse, from sinking to uplift, the deformation down the river increase rapidly, which indicate the dam was instability. Relation curves regarding the deformation along the river δ_L and vertical deformation δ_V downstream of the dam body surface and the overloading coefficient are shown in Fig. 5.

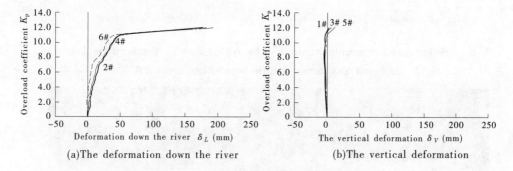

Fig. 5　Relation curves regarding the deformation downstream of the dam body surface and the overloading coefficient

(2) Dam strain distribution characteristics: Due to the limitation of the material nonlinearity properties of the dam, the measured strain can't be use to converse for the dam stress, but the relationship curve of strain could be a basis for determine the dam stability safety. Through the analysis of the overload characteristics such as fluctuation, inflection point, swerve of strain curve to get a different phase of overload failure process and safety factor. Relation curve of typical strain and overload factor $\mu_\varepsilon - K_P$ as shown in Fig. 6.

Many strain curves indicated that when $K_p = 1.0$, the dam deformation is small, and during the overload phase, the dam strain increases with the increasing of the overload coefficient K_p; When $K_p = 1.3 \sim 4.0$, the whole strain is small and linear growth; When $K_p \geqslant 4.0$, the strain of a lot of points appear nonlinear increase rapidly or reverse. There is an obvious stress and strain distribution characteristics as the dam heel is in tensile and the dam toe is in compression; When $K_p = 6.0 \sim 7.0$, strain near the dam foundation surface increase faster, most of the curve appear larger fluctuations, forming a larger inflection point, the strain of a lot of strain measuring point showed a sharp increase. Combining with the cracking of the model, it can estimate that the dam has been appear large deformation instability. When $K_p = 9.0 \sim 10.0$, the strain was keep the same or reduced, and the strain releasing. The dam heel pull shear fracture and toe pressure shear cracks, the dam cracks are in a local extension, the dam lose bearing capacity gradually; When $K_p = 12.0$, the cracks has been cut – through completely, the dam slip to downstream along the whole foundation, the test stopped.

(3) Model failure modes and mechanisms: by destruction test, the final failure destroyed shape of the model as shown in Fig. 7.

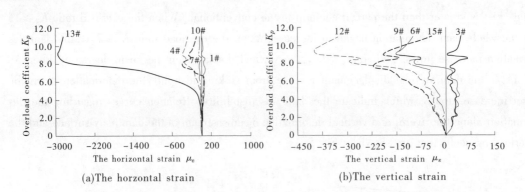

Fig. 6 Relation curves regarding the strain of the typical measure points on the dam near the foundation surface and the overloading coefficient

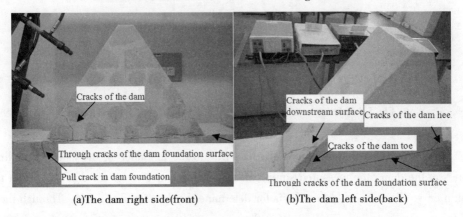

Fig. 7 The final destroyed shape of the typical Hardfill dam model

The failure regions of model are mainly in dam heel, dam toe and dam foundation base. First of all, with the increase of overload, the foundation base on the dam heel occurred shear failure. When $K_p = 6.0 \sim 7.0$, the crack appeared on dam foundation base in the dam heel. When $K_p = 7.6$, the right (front) side of dam foundation base cracked about 8 cm (model values) from the dam heel to downstream, the left (back) side of dam foundation base cracked about 6cm (model values) from the dam heel to downstream. With the increasing of overload ratio, the dam foundation base crack in the dam heel continue to expand to the downstream, at the same time, the dam heel and toe of the two deltas are the weak position of this dam type. When $K_p = 7.5 \sim 8.5$, the crack occurred at the triangle location of dam section, which about 13 cm (model values) downstream of the dam heel on the right side and about 9 cm (model values) on the left. The dam toe around the middle position downstream surface of the dam also cracked. When $K_p = 9.0 \sim 10.0$, the cracks have been extended. When $K_p = 12.0$, the dam foundation base completely shear failure as a whole. Finally, the dam was downstream slip as a whole, the upstream dam heel triangle crack has been extended to the upstream dam surface, the dam and foundation demonstrated plastic instability as a whole.

(4) Overload safety evaluation: the safety evaluation of the three dimensional semi-global geo-mechanical model overload method test for the typical dam section of Hardfill dam is mainly

according to the surface deformation of dam, the strain of the dam and foundation, and the failure pattern, etc to comprehensive assessment. Especially according to the fluctuation of each curve, a turning point, growth, steering, overload characteristics, to comprehensive analysis the overload coefficient of each overload stage. According to the analysis results, the overloading safety factors of the Hardfill dam derived from the geo-mechanical model failure test are as follows: the crack initiation factor K1 ranges from 6.0 to 7.0, the nonlinear deformation factor K2 ranges from 7.5 to 8.5, and the ultimate load factor K3 ranges from 9.0 to 10.0. Results show that the Hardfill dam has strong ultimate bearing capacity and big overload coefficient.

5 Conclusions

In this paper, a stress model test and a geo-mechanical model failure test for a typical Hardfill dam were carried out respectively to research the structural behavior under normal conditions and the failure mode and mechanism in the process of overloading of Hardfill dams. Based on experimental investigations, the following conclusions were drawn.

(1) The stress model testresults show that the Hardfill dam has good work performance. The dam body mainly endured the compressive stress and without obvious stress concentration under the condition of normal water load. Compared with the conventional gravity dam, Hardfill dam has lower stress level and more uniform stress distribution.

(2) The geo-mechanical model failure testresults show that the failure mode of a typical Hardfill damwas concluded as dam heel cracked firstly and then dam toe damaged, finally the dam integrally slid along the interface of the dam and foundation under the condition of water load overloading. Meanwhile, the overloading safety factors derived from the geo-mechanical model failure test are as follows: the crack initiation factor K_1 ranges from 6.0 to 7.0, the nonlinear deformation factor K_2 ranges from 7.5 to 8.5, and the ultimate load factor K_3 ranges from 9.0 to 10.0. That means the global stability of the Hardfill dam is high.

(3) The experimental research shows thatthe Hardfill dam has good work performance, strong ultimate bearing capacity and big overload coefficient. It is a kind of high-quality dam type. It should be noted that: the foundation conditions of the two experiments in this study were all simplified, that is only considered homogeneous foundation; in addition, the Hardfill material possesses obvious heterogeneity, but it does not take into account in the model test. These are still issues that need further study in the future.

Acknowledgements

This work was financially supported by the National Natural Science Foundation of China (Nos. 51109152, 51379139 and 51409179).

References

[1] Londe P. LinoM. The faced symmetrical hardfill dam: a new concept for RCC[J]. International Water Power & Dam Construction. 1992,44(2):19-24.

[2] Jia Jinsheng, Ma Fengling, Li Xinyu, et al. Study on material characteristics of cement – sand – gravel dam and engineering application[J]. ShuiliXuebao,2006,37(5):578-582 . (in Chinese)

[3] S. Batmaz. Cindere dam – 107 m high roller compacted Hardfill dam (RCHD) in Turkey[A]// Proceedings 4th International Symposium on Roller Compacted Concrete Dams[C], Madrid, 2003: 121-126.

[4] Yang Zhaohui, Zhao Qixing, Fu Xiangping, et al. Study on CSG dam construction technique and its application to Daotang Reservoir Project[J]. Water Resourcesand Hydropower Engineering,2007, 38(8): 46-49. (in Chinese)

[5] Yang Shoulong. Charateristics and load carrying capacity of CSG dam construction materials [J]. China Civil Engineering Journal,2007, 40(2): 97-103. (in Chinese)

[6] Wel Jianzhong, Wu Zuting, Wu Youwang, et al. Researchand application on a new type lean cemented material and hard filler damming technology [A]//Proceedings of Roller Compacted Concrete Dams in China [C], Guiyang, 2010:164-171. (in Chinese)

[7] Hirose T,Fujisawa T,Kawasaki H,et al . Design concept of trapezoid – shaped CSG dam[C]//Proceedings 4th International Symposium on Roller Compacted Concrete Dams [C]. Madrid,2003:457-464.

[8] Li Yongxin, He Yunlong, Yue Zhiji. Analysis of the Stress and Finite Element Stability for the Cemented Sand &Gravel Dam[J]. China Rural Water and Hydropower,2005(7):35-38. (in Chinese)

[9] He Yunong, Peng Yunfeng, XiongKun. Structural characteristic analysis on hardfill dam[J] . Journal of Hydroelectric Engineering, 2008, 27(6):68-72. (in Chinese)

[10] Xiong Kun, He Yunlong, Liu Junlin. Integral stability of Hardfill dam[J]. Journal of Hohai University(Natural Sciences), 2011,39(5):550-555. (in Chinese)

[11] Xiong Kun, He Yunlong, Wu Di. Study on structure failure test of Hardfill dam[J]. Jomal of Hydraulic Engineeting,2012,43(10):1214-1222 . (in Chinese)

[12] Zhang Lin, Chen Jianye. The engineering application of model text about hydraulic dams and foundation [M]. Chengdu:Sichuan University Press,2009. (in Chinese)

[13] Deng Ziqian, ZhangLin, Chen Yuan, et al. Contrastive Analysis on Structural Characteristic of Hardfill Dam and Gravity Dam [J]. Journal of Sichuan University(Engineering Science Edition), 2014,46(Sup1):63-68. (in Chinese)

Technical Features of Lower Reservoir for Panlong Pumped Storage Power Station and its Design

Liu Chun, Shi Hanxin, Xia Yueyi, Xie Liang

(Power China Zhongnan Engineering Corporation Limited, Changsha, 410014, China)

Abstract: at the dam site of the lower reservoir of Panlong Pumped Storage Power Station, the geological condition is unfavorable and the sources of dam filling materials vary greatly. The river course downstream of the spillway outlet is narrow and the condition for discharged flood to return to chute is poor, which may lead to direct scouring to the mountains on the opposite bank. This paper mainly discusses the key technique issues like different dam body filling zones and energy dissipation and erosion control of downstream water releasing structures.

Key words: Panlong Pumped Storage Power Station, Lower Reservoir, Dam body filling by zone, Energy dissipation and erosion control

1 Project overview

Chongqing Panlong Pumped Storage Power Station is located in Zhongfeng Town, Qijiang District of Chongqing Municipality. It is about 80 km away from Yuzhong District of Chongqing in straight-line distance and about 50 km away from the urban area of Qijiang. The power station has a total installed capacity of 1 200 MW, being a Class I Great (1) project.

The main structures of this station are an upper reservoir, a lower reservoir, a water conveyance system and a powerhouse. The permanent main structures are designed as per Class I structure and the secondary structures are designed as per Class III structure.

The upper reservoir has a normal pool level of 995.50 m and is mainly composed of the main dam, an anti-seepage body on the right bank of the main dam, Dahuangou saddle dam and Xiaohuangou anti-seepage saddle dam. The lower reservoir has a normal pool level of 549.00 m and is mainly composed of a concrete faced rockfill dam, a flood releasing tunnel on the left bank and a spillway on the right bank.

2 Structure layout of lower reservoir and main technical features

2.1 Geological condition

The lower reservoir is at the upper reaches of Shijia Gully and is in a Y-shaped in the plane. A dam is constructed in a V-shaped valley about 400 m downstream of the Lianghekou Hydropower Station. The mountains around the reservoir are high and strong and no low saddle exists. The area is of good topographical condition. The overburden of the slopes on both banks is comparatively thin and the fully weathered rock – soil body or above is less than 10 m thick. The bedrock of the riverbed in the dam site area is of PenglaizhenFm top (J_{3p}^{2-3}) consisting of purple – grey and grey – green sandstone, siltstone and mudstone. The slopes on both banks belong to Jiaguan Formation Section 1 (K_{2j}^{1}) consisting of fine to medium sandstone, conglomerate and siltstone, with the latter two accounting for 70%.

At the intake section of the flood releasing tunnel, the bedrock is exposed and the lithology belongs to Jiaguan Formation ($K_{2J}^{1-1} \sim K_{2J}^{1-2}$), consisting of purple – red fine to medium sandstone, conglomerate, gravel – bearing coarse sandstone, siltstone and mudstone etc. The geological structures along the tunnel are simple and no fault is detected to have across the tunnel.

The bedrock of the spillway is mainly composed of sandstone, intermingled with argillaceous siltstone, silty mudstone and mudstone. It is weakly weathered.

2.2 Layout of hydraulic structures and main technical features

The lower reservoir is mainly comprised of a dam, a flood releasing tunnel on the left bank and a bank – run spillway on the right bank. The lower reservoir is of high excavation quantity. To achieve a balance between cut and fill while taking account of the feature of large water head variation of a pumped storage power station, concrete faced rockfill dam is selected as the dam type, with a dam crest elevation of 552.30 m and a max. dam height of 79.30 m.

For the lower reservoir, the catchment area at the dam site is about 100 km². Regarding to the requirement on flood releasing and sand flushing, a spillway and a flood releasing tunnel are used in combination for flood discharging. The tunnel is on the left bank of the dam and is reconstructed from a diversion tunnel, with orifice dimensions being 4.5 m × 4.5 m. For the lined non – pressure tunnel section, the interior clear dimensions are 6 m × 8 m (width × height). The spillway, with a net width of 10 m, is arranged next to the dam on the right bank, with its left side wall at the control section used concurrently as the toe wall of the dam.

For the dam site area at the lower reservoir, the geological condition is poor and the sources of the dam filling materials vary significantly. Besides, the river channel downstream of the spillway's outlet is narrow and it is difficult for the chute to retain the discharged flood, it is likely that the mountain body of the opposite bank will be subject to direct scouring. Therefore, the difficulties to be tackled with at the lower reservoir mainly include how to divide up the dam in terms of different types of filling materials and how to achieve energy dissipation and erosion control downstream of the water releasing structures.

3 Dam design and its design features

3.1 Structure layout

The dam of the lower reservoir is a concrete faced rockfill dam, having a dam crest elevation of 552.30 m. The dam has a foundation elevation of 473.00 m at its plinth in the riverbed, a max. dam height of 79.30 m, a dam crest width of 10 m, an upstream slope of 1:1.4 and a downstream integrated slope of 1:2.256, The dam crest is 162.10 m wide and the max. width at the dam bottom is 280 m. At the top of the dam, an L – shaped wave wall is provided, having a

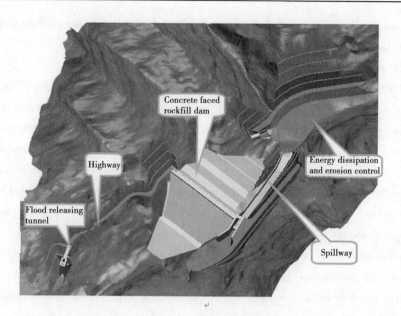

Fig. 1   Layout plan of lower reservoir structures

wall crest elevation of 553.60 m, A 0.7 m wide maintenance platform is set upstream of the wave wall.

3.2   Technical features

3.2.1   Complicated in dam material lithology, inhomogeneous in strength and poor in water permeability

For the lower reservoir, the dam filling quantity is about 850 000, with the dam filling materials mainly coming from the excavated materials of the main structures and the mountain body opposite the outlet of the spillway. The dam body is designed following the principle of "deciding the filling quantity based on the excavated amount". The lithology of the lower reservoir is sandstone interbedded with mudstone. According to the results of rock physical − mechanics test, the medium − coarse sandstone has a saturated compressive strength of 42.2 ~ 60.6 MPa, an average softening coefficient of 0.58 and a dry density of 2.45 ~ 2.61 g/cm^3; the fine − medium sandstone has a saturated compressive strength of 21.8 ~ 45.3 MPa, an average softening coefficient of 0.58 and a dry density of 2.43 ~ 2.52 g/cm^3. As soft rock (less than 30 MPa) exists in part of the medium − coarse sandstone and the mudstone cannot be removed thoroughly, an aquiclude layer with poor permeability will develop at the compacted layer after compaction. Therefore, how to divide the dam body into reasonable zones based on the situation of the dam materials is one of the key issues in dam design.

3.2.2   Water draining of spoil dumping area

As there is no appropriate spoil yard at the lower reservoir and to cut down land acquisition, spoils need to be dumped at the slope downstream of the dam. Water draining shall be considered in dam body zoning due to the low water permeability of the spoils.

3.3   Control in design

The following measures for quality control are taken in the design based on the mechanical properties of material sources and dam filling materials and in regard to the existing technical challenges in the dam design for the lower reservoir.

3.3.1 Strict requirement on the quality of dam filling materials

To deal with the inconsistency in the lithology and the strength of the sources of dam filling materials, strict requirements are proposed on the quality of these sources. According to the geological conditions of the dam site area and material yards and the results of laboratory rock physical – mechanics test, the average saturated compressive strength of medium – coarse sandstone is 31.5 MPa, the average saturated compressive strength of fine – medium sandstone is 48.7 MPa. The fine – medium sandstone could meet the requirement on the strength of rockfill materials for the dam. Therefore, the moderately and weakly weathered fine – medium sandstone, conglomerate and pebbly sandstone are excavated as the sources of rockfill material and rubble material. The materials are not allowed to be filled to the dam before the mudstone is stripped and removed.

3.3.2 Reasonable division of dam structure

Due to the diversity in material source lithology and the poor permeability in filling materials, the dam structure is divided up according to conventional practice with one more zone for water draining considered. To ensure the unobstructed draining inside dam body and coordinated deformation and following the principle of hydraulic transition, the dam of the lower reservoir is divided up into the following zones from upstream to downstream: riprap zone (IB), clay blanket zone (IA), cushion layer zone (IIA), special cushion zone (IIB), transitional zone (IIIA), drainage zone (IIIBA), upstream main rockfill zone (IIIBB) and downstream rockfill zone (IIIC). The design for these zones is as follows.

3.3.2.1 Cushion layer zone (IIA)

A cushion layer serves as the foundation of the anti – seepage face plate and also plays the role of transmitting hydraulic pressure and resisting seepage. The layer is 3 m in horizontal width; it has a design dry density of 2.20 t/m^3, corresponding to a porosity of 16% to 18%. The materials of this layer have a max. particle size of 80 mm and the fine grains less than 5 mm account for 30% to 45%. The permeability coefficient stands between 1×10^{-3} cm/s and 1×10^{-4} cm/s.

3.3.2.2 Special fine cushion layer zone (IIB)

A special filling zone to be filled up with fine materials shall be set next to the peripheral joints of the weak part where concentrative seepage may occur. For this zone, the design dry density is 2.20 t/m^3, the max. particle size is 20 mm and the fine grains less than 5 mm account for 35% to 60%. Besides, 42.5 MPa Portland cement (around 5%) will be added with each lift being 20 cm thick. The cement will then be compacted by a frog hammermanually. The permeability coefficient stands at $(1 \sim 5) \times 10^{-3}$ cm/s. This layer, together with the upstream clay blanket, forms a self – repairing filter system.

3.3.2.3 Transitional zone (IIIA)

The transitional zone is mainly used to prevent the loss of the fine particles from the cushion layer and the grading of this zone is considered as per filtration. The zone is 4 m in horizontal width and is filled with fresh and slightly weathered rocks. The design dry weight is 2.15 t/m^3, corresponding to a porosity of 18% to 20%. The max. particle size is 300 mm and the uniformity coefficient Cu > 15. The filling layer is 40 cm thick and the water spraying volume is 25%. The layer will be compacted for 8 to 10 times by a 26 t vibrating roller.

3.3.2.4 Drainage zone (ⅢBA)

Test results show that the rock here has high porosity and low saturated compressive strength and softening coefficient. Due to the low strength of the sandstone of the material yard, in site compaction, part of the rockfill material is subject to second crushing and an aquiclude layer of poor permeability will get formed at the contact face between the layers. Therefore, upstream of the rockfill zone close to the transitional zone is set with a vertical drainage zone (3.0 m in top horizontal thickness, 1:1.4 in upstream slope gradient and 1:1.2 in downstream slope gradient). For the part of drainage zone below 494.50 m in the riverbed (1.2 m higher than the downstream check flood level), slightly weathered to fresh fine sandstone of comparatively high strength from Shengjigang material yard will be used or slightly weathered to fresh fine sandstone excavated elsewhere will be used for filling (with a saturated compressive strength ≥40 MPa). Requirements on both its grading and design indices shall be no lower than those of the upstream rockfill area.

3.3.2.5 Main rockfill zone (ⅢBB)

The upstream main rockfill zone plays the main role in carrying hydraulic pressure of the dam. It is also a sensitive part subject to deformation after it is compressed. It is required that the rocks be well graded and of low compressibility. The weakly weathered to fresh, fine to medium sandstone (with a saturated compressive strength ≥30 MPa) from the material yards and the materials excavated from the main structures will be used for filling. The design dry weight.

3.3.2.6 Downstream rockfill zone (ⅢC)

The hydraulic pressure acting on the downstream rockfill zone is relatively small and will have little impact on face plate deformation. Therefore, the requirement on filling is not strict. For this zone, the top elevation is 533.80 m. With dam axis as the datum line, the upstream slope is 1:1.03 and the lower slope is 1:1.2. The downstream rockfill zone is filled with the slightly weathered and part of the scattered weakly weathered materials (with a saturated compressive strength≥30 MPa) excavated from the material yards and the main structures. It is required the materials should be well graded with low compressibility. The design dry density is 2.05 t/m^3, corresponding to a porosity of 23% ~ 25%. The max. particle size is 800 mm. The nonuniform coefficientCu > 10. The filling layer is 120 cm thick. Water spraying volume is 25%. This layer will be compacted for 6 to 8 times by a 26 t vibration roller.

3.3.2.7 Upstream clay blanket zone (1A)

Upstream of the face plate below an elevation of 509.00 m, a blanket (mainly consisting of fine silt, sandy soil and clay) with a top width of 4.0 m and an upstream slope of 1:1.6 is laid as supplementary measures for seepage resistance.

3.3.2.8 Rock ballast coverage zone (1B)

The clay blanket is covered with rock ballast to keep the upstream blanket zone stable and protect it. This zone has the same elevation with the clay blanket, with a top width of 6 m and an upstream slope of 1:2.5.

3.3.3 Filtering layer and draining prism provided at downstream spoil dumping area

This project is featured by high quantity of rock excavation, therefore, a downstream spoil dumping area is provided on the basic profile of the rockfill dam to reduce the amount of construction spoil. The filling materials of this project mainly come from the strongly and weakly weathered soft rocks excavated from tunnels and spillway. The rocks with a saturated compressive strength of 5 MPa or above are allowed to be compacted on the dam before it is filled; the max. thickness of a filling layer is 1.2 m, the max. particle size is 800 mm; the design porosity is 25% ~28%, the design dry density is 1.9 t/m^3; the filling quantity of spoils is about 200 000 m^3.

To make sure that the compaction of the soft rocks in the downstream rockfill area will not affect the permeability performance of the horizontal drainage layer at the bottom of dam foundation, a filter layer is set at the contact face between the bottom of downstream rockfill area and the horizontal drainage layer. The filter layer is 3 m thick and is graded in a way to meet the requirement on filtering. Meanwhile, drainage prisms are provided downstream of the drainage layer with a top elevation of 495.300 m, higher than the downstream check flood level. For the typical profile of concrete faced rockfill dam at lower reservoir, see Fig. 2.

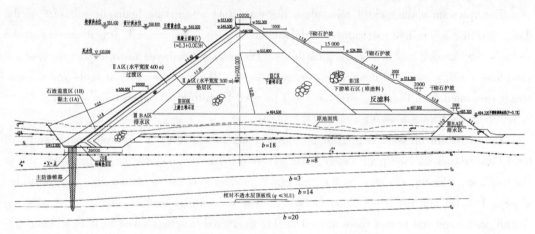

Fig.2 Typical profile of concrete faced rockfill dam

4 Technical features and design of water releasing structures

4.1 Layout of water releasing structures at lower reservoir

The lower reservoir has a catchment area of about 100 km^2 at the dam site, with a mean annual suspended sediment transport modulus being 500 t/km^2. Therefore, the catchment area is large and sediment content is high. With the requirements on flood releasing, sand flushing and operation scheduling taken into consideration as a whole, water releasing structures at the lower reservoir are consisted of the open spillway on the right bank and the flood releasing tunnel on the left bank.

The spillway is a gate-controlled one and consists four parts, which are an intake, a gate chamber, a releasing chute and a downstream energy dissipation facility. The intake has an invert elevation of 533.00 m and weir is a WES type, which is 10 m wide and 539.00 m in weir crest elevation. The chute is 10 m in net width and the bottom slope is 1:1.4. Flip bucket is adopted for energy dissipation.

The flood releasing tunnel is located on the left bank and is reconstructed from the diversion tunnel. It is used for flood discharging, emptying and sand flushing. The tunnel is composed of an intake section, a gate chamber section, the body of the tunnel and an outlet channel. The intake is of tower type and has an invert elevation of 518.00 m; it is a pressure short tube inlet with orifice dimensions of 4.5 m × 4.5 m. The intake tower is connected to a non - pressure reverse - U - shape tunnel. Flip bucket is adopted for energy dissipation at the outlet.

4.2 Main technical features

Restricted by topographical condition, the river channel at the spillway outlet is narrow and the anti - scouring velocity of the rock in it is low (3 m/s ~ 5 m/s). The angle between the axis and the downstream river channel is small and the condition for water to flow back to the chute is poor. The strong water flow produced in flood releasing will result in downstream riverbed scouring and large areas of backflow, undermining the slopes on both banks. Besides, as the flood releasing tunnel and the spillway outlet are arranged on the left and right banks respectively and the two structures are located nearby, the water discharge from them may get contacted and crossed in the mid - air, resulting in energy dissipation. Therefore, in designing water releasing structures at the lower reservoir, one of the key emphasis should be placed on energy dissipation and erosion control at spillway outlet.

4.3 Control measures in design

4.3.1 Flared bucket used by spillway

As proved by a number of hydraulic model tests, a flared bucket is used by the spillway. The top elevation of the bucket is 497.69 m, the bucket's radius is 25 m, the angle of the jet leaving the bucket on the left sidewall is 16.2° and that on the right sidewall is 34°. Besides, dentate sill is provided at the end bottom of the bucket. In this scheme, the nappe of discharges becomes longer and energy is more scatteredly dissipated. The deepest point of the scour pit is in the middle of the river, making the drop - point of the nappe far from the slopes on both banks.

4.3.2 Expansion and protection of downstream river channel

Based on the comparison of the spillway arrangement alternatives and the studies on energy dissipation and erosion control measures, it is decided that the downstream river channel should be expanded and excavated to form a plunge pool to deal with the scouring of riverbed and bank slopes.

4.3.2.1 Left bank slope protection at the outlet

The left bank excavation slope falls between 1:0.25 and 1:0.75. Overburden at the surface is cleared up. For the slope below the downstream check flood level, it is supported by the concrete placed on the "slope + patterned" anchor bolt; for the slope above the downstream check flood level, patterned anchorage shotcrete (anchor pile) "support + patterned" drainage holes are adopted; the middle and upper slope uses patterned anchor cable for support.

4.3.2.2　Riverbed bottom protection

For the riverbed 220 m to 320 m downstream of the dam axis, it is excavated to an elevation of 476.00 m with an average excavation width of 50 m. For the riverbed within a scope of about 80 m in the area 320 m downstream of the dam axis, it is excavated to an elevation of 482.00 m as per a slope of 1:2. A base plate is made of reinforced concrete and provided with anti-floating anchor bolts and drainage holes to lower down the uplift pressure acting on the plate.

4.3.2.3　Right bank slope protection

For the right bank slope, excavation is made from the spillway outlet along the riverbed with an excavation slope of 1:1. The max. slope height is 25 m. To avoid scouring, the slope below the downstream check flood level is supported by concrete placed on the "slope + patterned" anchor bolt. For the area above the downstream check flood level to the excavated slope, shotcrete is used for support.

4.3.2.4　Model test results

According to the analysis on the test results under various working conditions, the main scouring pit of the movable bed basically stays on the center line of the spillway in the middle of the riverbed. The slope of the scouring pit is no higher than 1:6.

5　Conclusion

In this project, the lower reservoir is featured by the lithology of fine to medium sandstone, pebbly coarse sandstone, siltstone, silty mudstone, argillaceous siltstone and mudstone. The rock is of low softening coefficient yet high porosity. The dam building materials mainly are the excavated materials from other structures and slopes and therefore are poor in quality. To minimize the impact of the project on the surrounding environment and the unfavorable effect brought about by land acquisition and resettlement, reasonable design on the dam structure was carried out based on geological condition and following the principle of "deciding the filling quantity based on the excavated amount". Besides, requirements were also proposed on the technical parameters in dam filling material compaction, which basically realize the goal of using the excavated material as the dam building material as much as possible, thereby reducing the quantities of spoils to be transported away from the project site.

In addition, for the water releasing structures, the flared bucket of inclined type is adopted. The outlet of the spillway is model tested for type selection. Besides, the riverbed downstream of the spillway and the outlet of the flood releasing tunnel is properly expanded and excavated and other measures for support and excavation are taken in a certain area at the outlet on the left bank. In this way, energy dissipation and erosion control of these water releasing structures in the narrow valley is basically addressed, serving as helpful reference for the design of similar projects.

References

[1] Chen Shuguang. Study on Test to Optimize Energy Dissipating Devices for Spillway at Lower Reservoir of

Panlong Pumped Storage Power Station[J]. South China Hydropower, 2005.

[2] Ning Yongsheng. Technical Challenges and Countermeasures of Upper Reservoir of Liyang Pumped Storage Power Station[J]. Design of Pumped Storage Power Station, 2013.

Durability Test on Concrete Using Fiber – reinforced Air – entraining Fly Ash

Zhang Jinshui, Chen Meng, Yang Jingliang

(Xiaolangdi multi – purpose dam project management center of the ministry of water resources, Zhengzhou, 450000, China)

Abstract: A test was conducted to compare the compressive strength, resistivity and chloride diffusion coefficient between ordinary air – entraining fly ash concrete and concrete using polypropylene fiber and cellulosic fiber air – entraining fly ash. The result of test shows that, the use of cellulosic fiber can enhance resistivity of concrete, but the influence of adding polypropylene fiber to concrete resistivity was not significant. The resistivity of the different types of concrete almost rises along with the increase of cubic compressive strength, and shows a significant linear correlation as well. The chloride diffusion coefficients of ordinary air – entraining fly ash concrete and concrete using cellulosic fiber air – entraining fly ash are decreased along with the increasing of cubic compressive strength, and shows a significant linear correlation.

Key words: concrete using fiber – reinforced air – entraining fly ash, cubic compressive strength, resistivity, chloride diffusion coefficient, correlation.

Durability of concrete is a capability of concrete to resist various physical and chemical actions while exposed to outside environment. To enhance the durability of concrete, in recent years, the scholars have devoted in the study of using different fiber – reinforced material in concrete. Due to their favorable split – resistant and strengthening effects, fibers can reduce remarkably the structural cracks of concrete and also effectively prevent the cracks from being extended. So the breaking tenacity and shock resistance are well improved, the brittle failure is similarly as the ductile facture.

At present, the scholars both domestic and abroad have conducted many tests research and theoretical analysis, most of which focused on physical and mechanical properties of fiber concrete and study of early – stage split resistance. The long term properties and durability of fiber concrete were studied as well, but the study is not sufficient and needs to go deeper.

In this article, a test was conducted to compare the compressive strength, resistivity and chloride diffusion coefficient between ordinary air – entraining fly ash concrete and concrete using polypropylene fiber and cellulosic fiber air – entraining fly ash. The objective of test is to study the change laws of durability properties, including resistivity and chloride diffusion coefficient, along with the age of concrete, and to analyze the correlation between mechanics and macroscopic durability properties.

1 Test plan

1.1 Raw materials of test

Cement: P · O 42.5 ordinary Portland cement, specific surface area is 354 m²/kg;

Fine aggregate: river sand, fineness modulus is 2.8, apparent density is 2 630 kg/m³;

Coarse aggregate: gravel, particle size 5~20 mm with continuous grading, apparent density is 2 708 kg/m³;

Water: drinking water;

Fly ash: Grade II, Type F, fineness is 19%;

Water reducing agent: liquid HT – HPC polycarboxylate super plasticizer, water reducing rate is 25%;

Air – entraining agent: powder air – entraining agent is YF – HQ.

The physicochemical properties of polypropylene fiber are listed in Table 1, and the same of cellulosic fiber are listed in Table 2.

Table 1 Physicochemical properties of polypropylene fiber

Type of fiber	Sarciniform monofil	Strength of Extension(MPa)	≥500
Acid – base resistance property	≥96%	Elongation at break (%)	≥15
Equivalent diameter(μm)	15~45	Elasticity modulus(MPa)	≥3 850
Proportion (g/cm³)	0.91 – 0.93	Melting point(℃)	160~180
Length(mm)	20	hydroscopicity(g/cm³)	≤0.000 1

Table 2 Physicochemical properties of cellulosic fiber

Proportion(g/cm³)	1.0~1.2	Diameter(μm)	15~20
Length(mm)	2~3	Strength of extention(MPa)	500~1 000
Elasticity modulus(GPa)	8~10	Specific surface area(cm²/g)	20 000~30 000
Spacing of fiber(μm)	500~700 (0.9 kg/m³ for amount of admixture)	Numbers of fiber (100 million)	12~15 (0.9 kg/m³ for amount of admixture)
Acid – base resistance property	≥95%	Bond stress	Strong

1.2 Mix proportion of concrete

According to *Design Code for Mix Proportion of Ordinary Concrete*, the mix proportion of concrete using air – entraining fly ash with and that without mixture of fiber are given in Table 3. In the table N1 indicates concrete with use of ordinary air – entraining fly ash, N2 indicates concrete with use of polypropylene fiber air – entraining fly ash, N3 indicates concrete with use of cellulosic fiber air – entraining fly ash.

Table 3 Mix proportion of concrete using air – entraining fly ash with and without fiber

(unit:kg/m³)

No. of Sample	Cement	Fine aggregate	Coarse aggregate	Water	Fly ash	Water reducing agent	Air – entraining agent	Polypropylene fiber	Cellulosic fiber
N1	250	765	1 148	140	63	3.432	0.1	—	—
N2	250	765	1 148	140	63	3.432	0.1	0.9	—
N3	250	765	1 148	140	63	3.432	0.1	—	0.9

1.3 Test content

The cubic compressive strength of concrete is tested in accordance with Standard of Test Method for Mechanical Properties of Ordinary Concrete.

The test of resistivity of concrete is carried out by Wenner method, the spacing of probe is 50 mm.

The chloride diffusion coefficient of concrete is tested in accordance with Technical Specification for Anticorrosion of Concrete Structures in Road Projects (JTG/T B07 – 01 – 2006), by the RCM method.

2 Analysis and discussion

2.1 Cubic compressive strength

The test results of cubic compressive strength of N1, N2 and N3 are shown in Table 4,

Table 4 Cube crushing strength of concrete (unit:MPa)

No. of Samples	7 d	14 d	28 d	56 d	84 d
N1	20.5	25.5	27.4	28.2	33.1
N2	19.6	24.7	26.7	27.0	31.3
N3	22.5	29.3	30	31.3	37.9

2.2 Resistivity

The test results of resistivity of N1, N2 and N3 are shown in Table 5.

Table 5 Resistivity of concrete (unit:kΩ · cm)

No. of Samples	7 d	14 d	28 d	56 d	84d
N1	7.97	9.63	17.57	31.13	56.90
N2	6.00	9.47	13.33	35.13	53.30
N3	8.40	12.33	16.87	34.37	63.43

The changing curve of resistivity of N1, N2 and N3 along with the age is shown in Fig 1.

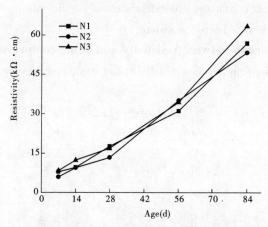

Fig. 1　Time – varying curve of resistivity

Table 5 and Fig. 1 show that the resistivity of N1, N2 and N3 rises along with the age, and the increasing trend keeps similar. The resistivity of N3 with use of cellulosic fiber is larger than that of N1, which indicates that the use of cellulosic fiber can increase the resistivity. However, the influence of adding polypropylene fiber to concrete resistivity is not great significant which even declines at some ages.

2.3　Chloride diffusion coefficient

The rest result of chloride diffusion coefficients of N1, N2 and N3 is shown in Table 6.

Table 6　Chloride diffusion coefficient of concrete　(unit: 10^{-11} m²/s)

No. of Samples	28 d	56 d	84 d
N1	1.79	0.90	0.55
N2	1.97	1.12	0.66
N3	1.78	1.11	0.71

The time – varying curve of chloride diffusion coefficient of N1, N2 and N3 is given in Fig. 2.

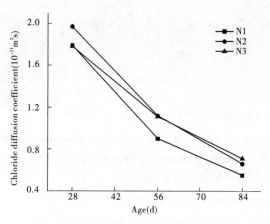

Fig. 2　Time – varying curve of chloride diffusion coefficient

Table 6 and Fig. 2 show that the chloride diffusion coefficients of N1, N2 and N3 decline with time, and the decline rate keeps consistent.

2.4 Regression relationship between resistivity and cubic compressive strength

The regression relationship between resistivity and cubic compressive strength of N1, N2 and N3 is shown in Fig. 3.

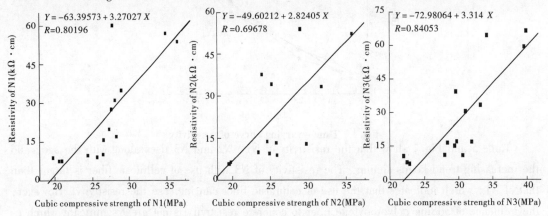

Fig. 3　Regression relationship between resistivity and cubic compressive strength for N1、N2、N3

Fig. 3 shows that the resistivity of N1, N2 and N3 all rises along with the increase of cubic compressive strength, which shows an apparent linear correlation.

2.5 Regression relationship between chloride diffusion coefficient and cubic compressive strength

The regression relationship between chloride diffusion coefficient and cubic compressive strength of N1 and N3 is shown in Fig. 4.

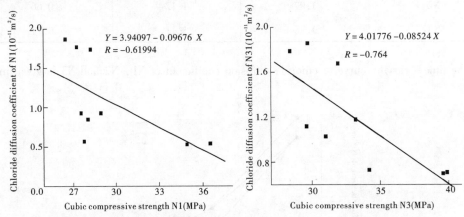

Fig. 4　Regression relationship between chloride diffusion coefficient and cubic compressive strength for N1、N2

Fig. 4 shows that the chloride diffusion coefficients of N1 and N3 decline along with the increase of cubic compressive strength, which indicates an apparent liner negative correlation. The larger the cubic compressive strength, more dense and lower porosity the concrete, so is the chloride diffusion coefficient smaller.

3 Conclusion

(1) Resistivity of the different types of concrete rises along with the age, and the growing trends keep consistent. The mixture of cellulosic fiber may enhance resistivity of concrete, while the mixture of polypropylene fiber does not have big impact on resistivity of concrete.

(2) Resistivity of the different types of concrete rises along with the increase of cubic compressive strength, which all shows an apparent linear correlation.

(3) Chloride diffusion coefficient of the different types of concrete declines along with the age, and the decline rate keeps consistent. The chloride diffusion coefficients of concrete with use of ordinary air – entraining fly ash and with use of cellulosic fiber air – entraining fly ash both decline along with the increase of cubic compressive strength, which shows an apparent linear correlation.

References

[1] Chen Runfeng, et al. Research and Application of Synthetic fiber in China [J]. Journal of Building Materials, 2001, 4(2):167 – 173.

[2] Wei Jinfeng, et al. "Research on Mechanical Property Test for Synthetic fabric concrete [J]. Concrete, 2010(3):67 – 70.

[3] Zhu Yunhua. Effect of Polypropylene fiber on Mechanical Properties of Concrete [J]. Marine Traffic Engineering, 2011(5):63 – 66.

[4] Deng Zongcai, et al. Tensile Mechanical Properties of High – tenacity Cement – based Composite Material Enhanced by Cellulosic Fiber [J]. Journal of Beijing University of Technology, 2009,35(8):1069 – 1073.

[5] Qian Hongping, et al. Study on Anti – cracking Property of Fiber Concrete and Application in Engineering [J]. Concrete, 2011(6):128 – 130.

[6] Li Dong, et al. Test and Study of Early – stage Anti – cracking Property of Polypropylene Fiber Concrete [J]. Concrete and Cement Product, 2009(6):39 – 42.

[7] Zhang Peng, et al. Effect of Fly Ash and Polypropylene Fiber on Anti – cracking Property of Concrete [J]. Building Science, 2007,23(6):80 – 83.

[8] Dong Yun, et al. Study of Anti – cracking Property of Fabric Concret [J]. The People's Yangtze River, 2006,37(8):89 – 90.

[9] Wang Kai, et al. Effect of S – P Hybrid Fiber on Long – term Property and Durability of Concrete [J]. Journal of Harbin Institute of Technology, 2009,41(10):206 – 209.

[10] Yang Chengjiao, et al. Mechanical Property and Seepage Resistance of Hybrid Fiber Concrete [J]. Journal of Building Materials, 2008,11(1):89 – 93.

[11] Wang Chenfei, et al. Test and Study of Durability of Polypropylene Fiber Concrete[J]. Concrete, 2011(10):82 – 84.

[12] JGJ 55—2011,Design Code for Mix Proportion of Ordinary Concrete[S].

[13] GB/T 50081—2002,Standard of Test Methods for Mechanical Properties of Ordinary Concrete[S].

[14] Zhao Zhuo, et al. Inspection and Diagnosis on Durability of Eroded Concrete Structures [M]. Zhengzhou: Yellow River Conservancy Press, 2006.

[15] JTG/T B07 – 01 – 2006,Technical Specification for Deterioration Prevention of Highway Concrete Structures [S].

Risk Analysis of Dams during Construction Based on Catastrophe Evaluation Method

Ge Wei, Li Zongkun, Li Wei, Guan Hong yan

(School of Water Conservancy & Environment, Zhengzhou University, Zhengzhou, 450001, China)

Abstract: To address the accident-prone problem of dams during construction, the difficulty of supervise and control construction goals such as safety, quality, process, costs, etc., and the adverse effect that traditional risk assessment would cause when determine the weight of risk factors subjectively, this article, based on the establishment of comprehensive evaluation index system of dams during construction, the calculation of each goal and the comprehensive goal of value at risk using the catastrophe e-. valuation method, is to achieve the objective of comprehensive evaluation. Through applying the method to risk evaluation of the construction program of Yanshan Reservoir, the author verifies rationality of the method, which preferably offers instructions to engineering construction.

Key words: dam, risk, catastrophe evaluation method, construction, goal

1 Introduction

Compared to other large-scale civil projects, dam projects require larger amount of investment, longer construction period, more construction work, and meet more complex construction conditions, so it's difficult to the control objectives such as safety, quality, process and cost comprehensively. Researchers have carried out a lot of researches on risk analysis for each of the construction goals, and have achieved remarkable results. But there are two main issues: the first is that mostly just one or two goals are analyzed, which is not conducive to the practical application of the research achievements; the second is that some assignment of risk factors that influence construction goals are determined by subjective opinions, so the objectivity of the results is questionable.

Catastrophe theory is based on the inherent contradictions and mechanisms of each goal in normalization formula to quantify its relative importance[1-3], which can effectively reduce the subjective factors in evaluation. The catastrophe evaluation method, based on this theory, can be effectively applied to multi-objective evaluation and decision. Therefore, based on the construction of comprehensive evaluation index system of dams during construction, using the catastrophe evaluation method for multi-objective risk analysis has certain theoretical and practical significance.

2 Basic theory of the catastrophe evaluation method

2.1 Catastrophe theory

The French mathematician Rene Thom systematically described the catastrophe theory in 1972 in the 《The stability of structure and morphogenesis》, which marked the birth of the catastrophe theory[2]. Catastrophe theory focuses on the leap forward from one steady state to another stable state of certain system or process when the control variables change. By studying changes of minimum value of function describing the system state, potential function F(x), it can determine the characteristics of the noncontiguous changing state nearby the critical point. The basis of catastrophe theory involves very esoteric mathematic knowledge such as calculus, the singularity theory, topology and structure stability and so on. However, its application model is relatively simple and has been applied widely in many fields such as environment, disasters, rock and soil[4, 5].

1.2 Catastrophe evaluation method

1.2.1 Catastrophe models

Catastrophe evaluation method determines the number of evaluation index according to the intrinsic system mechanism, and uses normalization formula to quantify state variables (X, Y) to get catastrophe fuzzy membership value which are similar to the fuzzy membership function, and then recursively calculates using potential function to get the comprehensive evaluate value for the system[6]. Its greatest advantage is that it only requires determining the relative importance of the evaluation index without the need for accurate weight assignment, which can effectively avoid the negative influence brought to the evaluation of the objectivity of subjective factors. Rene Thom proved that when the control variable is less than four, there are at most seven kinds of potential function, the seven types are collectively referred to as the elementary catastrophe.

Three catastrophe models commonly used are shown in Table 1.

Table 1 Three catastrophe models in common use

Category	Potential function	Normalization formula	Style
Cusp	$x^4/4 + ax^2/2 + bx$	$x_a = a^{1/2}, x_b = b^{1/3}$	
Swallowtail	$x^5/5 + ax^3/3 + bx^2/2 + cx$	$x_a = a^{1/2}, x_b = b^{1/3}, x_c = c^{1/4}$	
Butterfly	$x^6/6 + ax^4/4 + bx^3/3 + cx^2/2 + dx$	$x_a = a^{1/2}, x_b = b^{1/3}, x_c = c^{1/4}, x_d = d^{1/5}$	

2.2.2 Standardized handling of indexes

The dimension of quantitative indexes differs, and qualitative indexes also need to use mathematical methods to analyze and evaluate, so the indexes need to be standardized. For easier calculation of risk values, the "the-more-the-better" principle can be adopted to process indexes.

For quantitative indexes,

"the-bigger-the-better" principle can be used:

$$R_i = \frac{r_i - r_{min}}{r_{max} - r_{min}} \tag{1}$$

"the – smaller – the – better" principle can be used:

$$R_i = \frac{r_{max} - r_i}{r_{max} - r_{min}} \tag{2}$$

For qualitative indexes, the rating system can be used:

Table 2 Value range of qualitative indexes

Rating	Terrible	Bad	General	Good	Excellent
Index	[0,0.2]	(0.2,0.4]	(0.4,0.6]	(0.6,0.8]	(0.8,1.0]

2.2.3 Numerical calculation

Numerical calculation of catastrophe evaluation method should employ recursive formula, calculation process is:

(1) The evaluation index system should be firstly established according to the project construction situation;

(2) The indexes should be standardized based on 2.2.2, as the bottom index membership value;

(3) The bottom index membership value should be normalization processed according to corresponding potential function in table 1;

(4) catastrophe evaluation value can be calculated based on recursive formula: If there is obvious mutual effect among control variables of the same object, the use of "complementarity" principle should be adopted and take the average; if there is not, use "minimax" principle.

3 Risk assessment system of dams during construction

Dam project goals can be affected by many factors, including two aspects: (a) the design plan, the settled project amount, construction schedule and other conditions lead to certain goals; (b) hydrology, flood, weather, construction management and other uncertain factors affecting the construction goals uncertainly – the risk that is referred in this article.

1) Safety risk

Accidents that may occur during construction periods of dams include overtopping failure and structural damage. Overtopping failure depends mainly on the flood level and the elevation of the retaining building top[7, 8]; structural damage depends mainly on the quality of basis processing, material properties and mechanical properties and the downstream water level difference[9-11].

2) Quality risk

The construction quality of dams depends largely on the quality of filler, on – site management, construction technology and machinery[12, 13].

3) Schedule risk

The schedule risk mostly depends on the process of the upper stream flood process, the weather conditions, and the material factory, funding and other factors mainly affects [14, 15].

4) Cost risk

The construction cost risk of dams can be divided into emergency cost and repair cost [16, 17].

Emergency cost mainly depends on the temporary reinforcement measures when flood process exceeds the design criteria; repair cost mainly depends on the loss caused by accidents, which needed to be evaluated based on the design of specific project.

Based on the above-mentioned analysis, according to the relative importance of project construction goals, multi-objective comprehensive evaluation index system of dams during construction can be established, as shown in Fig. 1.

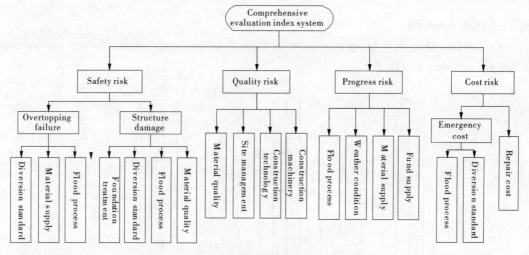

Fig. 1 Comprehensive evaluation index system

4 Example analysis

Yanshan Reservoir Dam project, located in Yexian County, Henan province of China, is one of the nineteen key projects renovating the Huaihe River determined by the China's State Council [1]. It was planned that the river would be dammed in the November of 2006, and dam started to retain water in the June of 2008. In the process, according to the overall goal of the Huaihe River renovation, the project should be finished one year ahead of schedule from the end of 2008 to the end of 2007. According to the construction goal, design units provided new construction schemes:

Scheme 1: Cofferdam is used to retain water with the standard of 20 years' return period during flood season.

Scheme 2: Temporary dam elevated 91 meters is used to retain water, and water is allowed to flow over dam surface temporarily.

Scheme 3: Temporary dam is used to retain water with the standard of 100 years' return period during flood season.

4.1 Risk value calculation

Adopting catastrophe evaluation method to calculate the risk values of three construction schemes respectively, the calculation results are shown in Table 3 (the calculation process is omitted).

Table 3 Risk value of three construction schemes

Scheme	Safety risk	Quality risk	Progress risk	Cost risk	Comprehensive risk
Scheme 1	0.956 8	0.316 2	0.951 5	0.563 1	0.884 7
Scheme 2	0.974 4	0.316 2	0.962 5	0.647 6	0.893 9
Scheme 3	0.874 6	0.316 2	0.907 5	0.580 7	0.872 4

4.2 Result Analysis

Comparison analysis is being conducted for these risk calculation results of three construction schemes, as shown in Fig. 2.

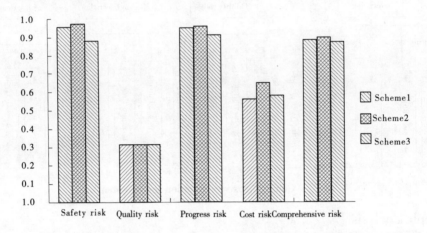

Fig. 2 Risk comparison of three construction schemes

According to Fig. 2:

(1) Catastrophe evaluation method can calculate the single construction goal risk value (such as safety, quality, progress and cost risk) and comprehensive risk value effectively, in the case of the relative importance of the risk factors, without accurate weight assignment. The result is clear, and easy to compare and analyze.

(2) The quality risk values of three schemes are the same, it states that construction quality of different construction schemes can be assured effectively, and the main distinctions are in terms of safety, progress and cost risk. Scheme 2 which results in the biggest risk value and comprehensive risk value in terms of safety, progress and cost risk, is the worst scheme; while Scheme 3 has larger cost risk value compares to scheme 1, but with smaller safety, progress risk value and lower comprehensive risk value. Considering all these factors, the scheme consequence selected from the best to the worst is: Scheme 3 > Scheme 1 > Scheme 2. Yanshan Reservoir construction authority adopted Scheme 3 in actual construction, achieved periodical project construction task on time, meanwhile assured project security effectively and controlled project investment reasonably, acquired good economic and social efficiency.

5 Conclusions

It's difficult to control safety, quality, process and cost of dams during construction comprehensively, and the weight of all the risk factors is hard to secure. This research adopted catastrophe evaluation method to calculate each and comprehensive goal risk on the basis of analyzing the risk factors which influence construction goals, and establishing comprehensive evaluation index system. This method was applied on the comprehensive selection of Yanshan Reservoir Dam construction schemes, and is consistent with expertise demonstration result and actual project construction situation. It shows that this method possesses good accuracy and practicability, provides a brand new thought for dam risk analysis during construction period.

Acknowledgements

This work was supported by the National Natural Science Foundation of China (Grant No. 51379192).

References

[1] Li Z K, Ge W, Wang J, et al. Improved catastrophe theory evaluation method and its application to earth – rock dam risk evaluation during construction [J]. Journal of Hydraulic Engineering, 2014, 45(10): 1255 – 1260. (in Chinese)

[2] Nabivach V E. Catastrophe Theory and Risk Control: Conceptual Framework [J]. Journal of Automation and Information Sciences, 2013, 45(5): 13 – 24.

[3] Murnane R J. Catastrophe risk models for wildfires in the wildland – urban interface: What insurers need [J]. Natural Hazards Review, 2006, 7(4): 150 – 156.

[4] Zhao Z, Ling W, Zillante G. An evaluation of Chinese Wind Turbine Manufacturers using the enterprise niche theory [J]. Renewable and Sustainable Energy Reviews, 2012, 16(1): 725 – 734.

[5] Su S L, Zhang Z H, Xiao R, et al. Geospatial assessment of agroecosystem health: development of an integrated index based on catastrophe theory [J]. Stochastic Environmental Research and Risk Assessment, 2012, 26(3): 321 – 334.

[6] Michel – Kerjan E, Hochrainer – Stigler S, Kunreuther H, et al. Catastrophe risk models for evaluating disaster risk reduction investments in developing countries [J]. Risk Analysis, 2013, 33(6): 984 – 999.

[7] Goodarzi E, Shui L T, Ziaei M. Risk and uncertainty analysis for dam overtopping – Case study: The Doroudzan Dam, Iran [J]. Journal of Hydro – environment Research, 2014, 8(1):50 – 61.

[8] Marengo H, Arreguin F, Aldama A, et al. Case study: Risk analysis by overtopping of diversion works during dam construction: The La Yesca hydroelectric project, Mexico [J]. Structural Safety, 2013, 42:26 – 34.

[9] Peyras L, Carvajal C, Felix H, et al. Probability – based assessment of dam safety using combined risk analysis and reliability methods – application to hazards studies[J]. European Journal of Environmental and Civil Engineering, 2012, 16(7): 795 – 817.

[10] Zhong D H, Sun Y F, Li M C. Dam break threshold value and risk probability assessment for an earth dam [J]. Natural Hazards, 2011, 59(1): 129 – 147.

[11] Lienhart D A. Long – Term Geological Challenges of Dam Construction in a Carbonate Terrane[J]. Envi-

ronmental & Engineering Geoscience, 2013, 19(1): 1 - 25.
[12] He H Q, Huang S X, Wu G. Height fitting by radial neural network for the construction quality control of face rockfill dam[J]. Geomatics and Information Science of Wuhan University, 2012, 37(5): 594 - 597. (in Chinese)
[13] Liu D H, Sun J, Zhong D H, et al. Compaction quality control of earth - rock dam construction using real - time field operation data [J]. Journal of Construction Engineering and Management, 2011, 138(9): 1085 - 1094.
[14] Zhong D H, Chang H T, Liu N, et al. Simulation and optimization of high rock - filled dam construction operations [J]. Journal of Hydraulic Engineering, 2013, 44(7): 863 - 872. (in Chinese)
[15] Xu Y, Wang L, Xia G. Modeling and Visualization of Dam Construction Process Based on Virtual Reality [J]. Advances in Information Sciences and Service Sciences, 2011, 3(4).
[16] Wang Z F, Liu J Y, Ding J Y. Cost - risk analysis of earth dam construction with considerations of schedule uncertainty [J]. Journal of Hydroelectric Engineering, 2011, 30(5): 229 - 247. (in Chinese)
[17] Hattori A, Fujikura R. Estimating the indirect costs of resettlement due to dam construction: A Japanese case study [J]. Water Resources Development, 2009, 25(3): 441 - 457.

The Application of Underwater Construction Techniques in the Project of Adding the New Maintenance Gate Slots in Yaotian Hydropower Station

Shan Yuzhu, Chen Ye

(Qingdao Pacific Ocean Engineering Co., Ltd, Qingdao, 266100, China)

Abstract: There are 27 overflow weir holes in the spillway section of the dam in Yaotian hydropower station. Now the work is to add a maintenance gate at the front of each radial gate, the maintenance gate access for 10 m × 9 m plane door. In August 2014, Qingdao Pacific Ocean Engineering Co., Ltd. contracted and implemented the project. This is the first time in the domestic construction through the underwater construction techniques to build a new gate slot.

This project is completed by divers underwater, no need to set the cofferdam to create dry land construction conditions, not be limited to reservoir operating conditions, without river diversion, without emptying reservoir, it can save water resources and power resources. Not only save the construction period, but also greatly reduce the project cost. Divers underwater equipped with some necessary tools which include surface supplied dive equipment, underwater light, underwater camera, dive telephone and underwater pneumatic equipment, hydraulic equipment and so on. According to the instruction of surface surveillance officers, divers can complete the underwater inspection, underwater drilling, underwater casting and other work. At the same time, underwater camera is able to continuously transport each process scene to the surface monitor so that the surface technical personnel can watch it. Engineering and technical personnel can also supervise, inspect and guide divers for their underwater work by the monitor and dive telephone, in order to make the surface and underwater work unity and synchronization.

The key points of this construction project include underwater overall chiseled of the pier, underwater concrete pouring and the precise installation of gate slot and embedded parts. This paper describes the major advanced construction methods and key technology, such as underwater positioning, overall slotted pier, installation the sill underwater, introduced the application of underwater diving construction techniques in the project of adding the new maintenance gate slots in Yaotian hydropower station.

Key words: Yaotian hydropower station, the pier integral slot, underwater positioning, underwater concrete pouring

1 Project profile

1.1 Hydropower general situation

YaoTian hydropower station is located in the downstream of Hunan province Lei River, which is 26 km away from Leiyang City. It is the twelfth cascade of Lei River casade hydropower planning. This power station is comprehensive benefit project, mainly for electricity generation, shipping and water supply is assistant, total drainage area is 11 905 km². The drainage area above the dam toe is 10 470 km², average annual precipitation is 1 580.3 mm, average annual water flow is 269 m³/s. Yao Tian hydropower station has been preliminary designed in 1977 and been formal constructed in October in the same year. The first unit was put into operation on December 31, 1988, all four units have been putted into operation until Nov18,1993[1].

The main hydraulic structure include river dam, power house, transformer station, navigation lock, approach channel, tailrace. This project belongs to the third – class engineering. In this hydro – complex, permanent water retaining structure belongs to third class, secondary building belongs to fourth class, temporary building belongs to fifth class[1].

1.2 Project content

There are 27 overflow weir holes in the spillway section of the dam in Yaotian hydropower station, which are surface outlet discharge, including 24 high weir holes, fitted 10 m × 6 m curved steel gate, crest elevation is 67.0 m; there are 3 low weir holes, equipped with 10 m × 9 m curved steel gate, crest elevation is 64.0 m, using underflow energy dissipation with stilling pool. Now the work is to add a maintenance gate slot at the front of each radial gate, the maintenance gate access for 10 m × 9 m plane door[1]. See Fig. 1.

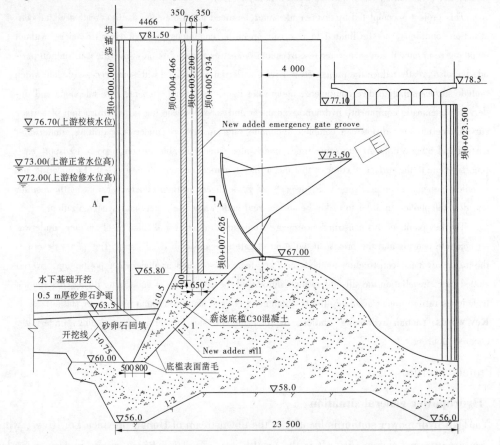

Fig. 1 Additional emergency gate slot in weir hole in Yaotian Hydropower Station   (unit:mm)

Depending on the different of the construction site, the main project is divided into three sub-projects: foundation excavation and backfill engineering, sill project and gate slot project.

1.2.1 Foundation excavation and backfilling

Firstly, excavating the stone pitching at piers and the clay blanket under water accordance the design drawings, the excavation slope is 1:1. The excavation depth is to the design elevation. We use a crane in conjunction with a grab to clean(Fig. 2).

Grab type excavation equipment(Fig. 3), which is made of 12 tons Crane, dredging grab, rope, etc., is attached to the excavation ship. Excavation ship is the subject of excavation equipment, is a water-bearing structure. Excavation the ship is fixed by four-point mooring positioning.

Fig. 2 Grab excavating

Fig. 3 Mechanical Grab equipment

In the process of underwater excavating, the properly slope margin should be left in the actual slope construction. Then use manually trim to meet the slope and flatness requirements of the construction drawings.

After the gate slot project is finished, backfilling the base slab in front of the pier according to the requirement of the design drawings.

1.2.2 Bottom sill project

Bottom sill project includes pouring of underwater concrete and underwater installation of embedded parts(Fig. 4).

First, pouring bottom sill concrete after the foundation excavation. The new bottom sill concrete is used underwater non-dispersible concrete, the strength grade is C30. Then install embedded parts in the bottom sill concrete according to design requirements. In order to effectively limit the dispersion and segregation of the underwater concrete in pouring process, we need to mix the concrete with anti-dispersants. To ensure the combination of old and new concrete, especially the combination of new concrete and the pier concrete, we need to add micro expansion agent in concrete. Making construction joints in new bottom sill concrete at the original construction joints, the slot seam width is 10 mm. Using asphalt cedar board to fill structural joints.

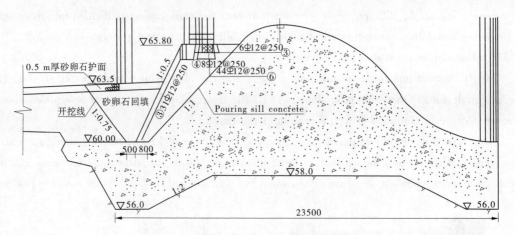

Fig. 4　Bottom sill schematic diagram

1.2.3　Gate slot project

Gate slot project mainly includes pier overall slotted, the installation of gate slot steel bar and embedded parts, gate slot concrete pouring. First, cutting and grooving the pier concrete overall, after installing the steel bar and embedded parts in accordance with the design drawings, pouring secondary concrete(Fig. 5).

Underwater concrete using non-dispersible fine stone concrete, the strength class of the concrete is C30. Above water section using ordinary concrete, the strength class of the concrete is C25.

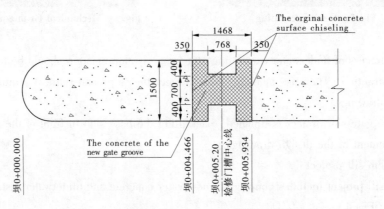

Fig. 5　Gate slot concrete schematic diagram

2　Project features and the key technology of construction

2.1　Project features

Two schemes are compared in this project: cofferdam construction and underwater construction.

Cofferdam construction is to layout steel sheet pile cofferdam in front of sluice piers. Each period includes three holes. After each hole of sluice gate repair construction is completed, the steel sheet pile cofferdam will be pulled out and used in the next stage of the project. Each steel sheet pile cofferdam weight 145.7 t. Firstly, using a vibratory hammer to punch the steel sheet pile cofferdams by - piece to the design elevation and fixing them. The bottom front the pier should be cleaned before the steel sheet pile cofferdam punched. After the cofferdam installation, drain the water in the cofferdam, suck away the sediment, install internal support and back cover foundation pit, in order to create dry conditions for the construction of maintenance gate slot.

This project is completed underwater by divers. Underwater construction uses surface supplied diving, divers carrying surface supplied dive equipment, underwater light, underwater camera, dive telephone and underwater pneumatic equipment and hydraulic equipment, follow the instruction of surface surveillance officers to complete the underwater operations. There are a light and a camera on the diver's helmet, which can shoot down the whole process when divers working in the water and everything that can be seen by divers. Underwater camera is able to continuously transport each process scene to the surface monitor so that the surface technical personnel can watch it. Engineering and technical personnel can also supervise, inspect and guide divers for their underwater work by the monitor and dive telephone, in order to make the surface and underwater work unity and synchronization.

Since the cofferdam construction needs to comprehensively consider the performance design, such as the arrangement, structure, anti - scour, anti - seepage and so on, it requires a lot of preparatory work on the construction operation. The project cost is very high, and construction is too difficult. Compared with the cofferdam construction, underwater construction has the following several advantages:

(1) Not limited by the operating conditions of the reservoir, no diversion, no abandoned water construction, does not affect the normal operation of the reservoirs and power plants, saving water and power resources.

(2) Normally flood discharge in the flood season, does not affect the safety of the normal release flood water.

(3) Compared with the cofferdam construction, the preparatory work of underwater construction is less than cofferdam construction, it can maximize the construction schedule.

(4) Working underwater directly, without setting a large cofferdam to create dry land construction conditions, greatly reducing the project cost.

Eventually, through a comprehensive comparison of various aspects, underwater construction program, with its obvious advantages, won the final approval of the owner.

2.2 The key technology of construction

In the project of adding the new maintenance gate slots in Yaotian hydropower station, the accurate installation of the gate slot embedded parts is the most important in this project. So, the gate slot and the sill construction are the most critical parts of this project.

The critical process of gate slot construction and sill construction will be described in detail in this paper, including underwater overall slotted of the pier, the pouring of the underwater concrete and the precise installation of gate slot embedded parts.

3 Pier overall slotted process

Pier slotted is not only the first step but also the key process in gate slot construction. There are 28 piers need to be slotted, the engineering quantity is very large, and the construction period is very tight. So how to cut and slot the pier rapidly and accurately is a major focus of this project.

In the construction site, two methods were tested at the same time, in order to choose the best option for Yaotian Station reconstruction project.

(1) Method One: The combination of static crushing and wire saw.

The first method used the way of drilling the broken holes and filling static fracturing agent in the holes, to crush and remove the pier by section. First, we used the wire saw to cut out slot edge lines, and then used $\phi 42$ hydraulic drill to drill the static crush holes according to the position shown in Fig. 6 and filling static fracturing agent in the holes after cutting (Fig. 7). The depth of the hole was 2 m. After the concrete pier was crushed (Fig. 8) by the static fracturing agent, we chiseled the pieces of concrete by artificial.

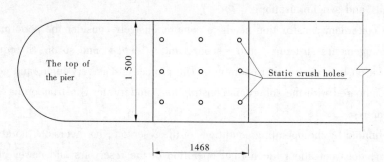

Fig. 6 Diagram of static crush holes (unit: mm)

Fig. 7 Field static crush test

Fig. 8 The effect of Static crushing

(2) Method two: Using wire saw to cut integral pier.

The second method, we used wire saw (Fig. 9) to cut overall pier and cut the pier into several pieces according to the site's lifting capacity (Fig. 10). Then used the crane to lift out the concrete blocks overall.

The wire saw consist of drive, flywheel, guide wheels, diamond wire saw chain[2]. Wire Saw can cut reinforced concrete structures, thick walls, and underwater cutting jobs. The cutting depth that the wire saw achieved is deeper than the wall saw. The cutting operating depth of wire saw is not limited, its operating environment is more adaptable, and operating efficiency is higher.

Hydraulic wire saw has the following salient features: it can reduce labor intensity, has safe and reliable operation, with overload protection function, strong power, improve cutting ability and labor productivity. It is an advanced equipment for the demolition construction project. Its linear cut may make construction cross-section neater. It is capable of doubling the pace of work to shorten the construction period, and reduce the labor costs. As the hydraulic system itself is safe, reliable and stability, it can greatly reduce the loss cost of construction equipment. In addition, it can save the stability and security of the existing structure to the maximum extent. As an advanced cutting construction equipment, wire saw is widely used in construction reinforcement, replacing traditional way of construction such as the assault chisel and rig row of holes to break[3].

(3) Slotted plan finalized.

According to the objective comparison of the two schemes in all respects, we believed that using wire saw to cut overall pier was faster than static crushing, and the cost of manpower resources was far less than static crushing. But wire saw cutting required a higher lifting capacity. After comprehensive comparison, in order to ensure the Yaotian project completed on schedule, we decided to use wire saw to cut overall pier.

Fig. 9 **Installation of hydraulic wire saw**

Fig. 10 **Hydraulic wire saw cutting**

4 Underwater concrete pouring process

The concrete of bottom sill is used the way of underwater construction to pour, the underwater concrete pouring quality is the key point of the construction in this project.

4.1 Formwork installation underwater

The pouring height of low weir hole bottom sill is about 3.6 m, the pouring height of high weir hole bottom sill is about 5.8 m. Therefore, we decided to install formwork and pour concrete by layer. The height of each pouring layer is 2.5 m. Formwork is mainly assembled by wooden template and steel ribs, and fixed by anchorages. The weight of wooden template is light. The installation and demolition of the wooden formwork is very convenient and flexible, and it is very easy to underwater construction.

4.2 The material properties of underwater non-dispersible concrete

The underwater part of this project used underwater non-dispersible concrete, which is prepared by UWB-II type flocculent. UWB-II type flocculent is a powder material. The concrete which is formulated by UWB-II type flocculent, have powerful anti-dispersible, appropriate liquidity and satisfactory performance of construction. It fundamentally solves the contradiction among the dispersion performance, the construction performance and mechanical performance of underwater concrete, truly achieve self-leveling and self-compacting of the underwater concrete [4]. It can be used for the repair of wharf, dam and reservoir; caisson sealing, cofferdam, caisson, riprap grouting, underwater continuous wall casting, underwater foundation leveling and filling, RC panels and other large-scale no construction joints underwater engineering; large diameter bored piles, drain water impact reinforcement plate, water bearing platform, seawall revetment, slope protection, sealing pile plugging and the underwater engineering which is difficult for ordinary concrete construction.

The performance characteristics of UWB type underwater non-dispersible concrete:

(1) Anti-dispersible: Even under the erosion of water the concrete still has a strong resistance to dispersion, which can effectively inhibit the PH and turbidity increased during underwater concrete construction.

(2) Excellent construction: UWB type underwater non-dispersible concrete, mortar are very viscous and plastic, with excellent leveling property and filling. It can be filled by gravity, without vibration in the gap between the reinforcing bar, skeleton and template.

(3) Good water retention: UWB type underwater non-dispersible concrete can improve the water retention of concrete; there has no bleeding or floating pulp.

(4) Safety and environmental protection: UWB type underwater non-dispersible concrete flocculent has been detected by the health and quarantine departments. It is non-toxic harmless to human body, can be used for drinking water project.

(5) The UWB underwater non-dispersible concrete, which is used in this project, has erosion and abrasion resistance, is suitable for reinforcement project, such as underwater scour hole repair and other damaged engineering.

4.3 Concrete pouring way

Underwater non-dispersible concrete is centralized mixed by large mixing station. The UWB concrete semi-finished products are mixed by the advanced equipment, in strict accordance with the norms and procedures in a predetermined proportion. It is transported by concrete trucks to the pouring site to ensure maximum stability of concrete quality.

The main sites of underwater non-dispersible concrete pouring are bottom sill and gate slot underwater concrete pouring. The pumping method (Fig. 11 and Fig. 12) is mainly used in the pouring of underwater non-dispersion concrete pouring. This method has the characteristics of fast construction, good pouring quality, saving labor, convenient and so on. It not only improves productivity, but also can greatly improve the construction schedule, which can help to shorten the construction period.

Fig. 11 Pumping concrete

Fig. 12 Pumping concrete

5 Critical process of embedded parts installation – underwater positioning

Bottom sill embedded parts should be installed after the installation of gate slot embedded parts is carried out, therefore, before installation, we must first the measure and check the accuracy of the gate slot embedded parts installation.

5.1 Measurement and setting-out

Before the installation of bottom sill embedded parts, it is necessary to measure and set the center line of the gate slot, the center line of the hole, the mounting reference point of the elevation. Set the installation points (center point of the gate slot, mounting axis of bottom sill embedded parts, etc.) on the bottom sill embedded parts, and mark them clearly with yellow paint.

The reference point lofting (Fig. 13): First set control points on the top of the pier based on the construction drawings. According to the reference points, baselines and standard points provided by the owners, use total station to set up slot center line on the top of the pier. Use heavy hammer line to lead the control point to underwater. It is necessary to take measures to fix the heavy hammer line after the hammer is steady underwater.

The control point of installation lofting: The control point of bottom sill embedded parts mounting axis can be staked out in detail underwater, with the center lines of the gate slot and the hole as the baselines, using a special tool for local measure. To ensure accuracy, the control points of installation need to be measured in the same horizontal plane. Before measurement, set up a steel plate on the heavy line whose size is similar with the gate slot, after ensuring that the steel plate is level, fix it to the installation elevation of the bottom sill. The installation elevation of the steel plate can be measured by the heavy line and combined with special steel ruler. Each control point can be measured by the special measuring tools at the steel plate, which can be used as a measurement plane, according to the points of bottom sill mounting axis on the construction drawings.

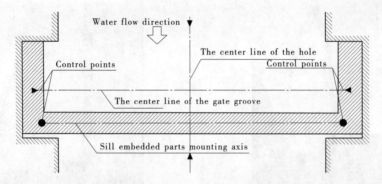

Fig. 13 Floorplan of the control points

5.2 The difficulties of underwater positioning and measures

Since bottom sill need to be installed underwater, while installing, we use hammer throw-point method to lead the control points from land to underwater, set out the control points of the slot center line and the hole center line. Hammer throw point method is commonly used to measure the vertically of building on the land, the measurement can be done using a principle that the direction of gravity is always vertical downward. However, when the hammer method is used underwater, it is susceptible to be disturbed by flow, diver's breathing bubble and other factors, so that the heavy hammer can not easy to be stable, thereby increasing the difficulty of the operation.

Therefore, how to make the heavy hammer line stable and fixed in the water, is the focus of the hammer line installation. In order to make the heavy hammer slowly stabilize in the hydrostatic and to prevent the flow of water and divers breathe air bubbles cause disturbance to hammer line, we install underwater cameras on the bottom sill, observe the stability of hammer in real-time. At the same time, we fix a special positioning plate on the bottom sill under the heavy hammer, make the tip of heavy hammer as far as possible close to the positioning plate. After supervisors on the land observe and determine the hammer is stable through the camera, slowly lower the hammer, so that the tip of the hammer can touch the positioning plate and keep a mark. Then, the divers fix hammer line according to the position of the mark.

6 Conclusions

In Yaotian project, we have tried and used a variety of advanced equipment and methods, including innovative using the methods such as static crushing and wire saw in pier overall slotted process, using advanced UWB NDC pouring underwater in bottom sill project. With these advanced equipment and innovative construction methods, the project of adding the new maintenance gate slots in Yaotian Station can be carried out smoothly under the conditions of underwater construction.

Underwater construction has the following several advantages:

(1) Not limited by the operating conditions of the reservoir, no diversion, no abandoned water construction, does not affect the normal operation of the reservoirs and power plants, saving water and power resources;

(2) Normally flood discharge in the flood season, does not affect the safety of the normal release flood water;

(3) Compared with the cofferdam construction, the preparatory work of underwater construction is less than cofferdam construction, it can maximize the construction schedule;

(4) Working underwater directly, without setting a large cofferdam to create dry land construction conditions, greatly reducing the project cost.

Underwater diving construction technology has developed very mature in China currently. But this is the first case in China of direct building new structures such as gate slots underwater. In the next period of time, with the development of economic construction, China's hydraulic and hydro-power engineering will also be at the peak period. The project of adding the new maintenance gate slots in Yaotian Station, provides new technology and new method which is technical feasible, safe and reliable, cost-effective for the similar projects in ports, hydraulic, bridges, etc, and also provides more choice and method for new construction underwater.

References

[1] Enhancement expanding capacity, reconstruction project and tender announcement about underwater construction project of adding maintenance gate of sluice gate in YaoTian hydropower Station in Hunan Leiyang. [EB/OL]. China procurement and tendering network, 2014.

[2] WangZhan. The composition of the wire saw[DB/OL]. China Network, 2009.

[3] WangZhan. The purpose and characteristics of the wire saw [DB/OL]. China Network, 2009.

[4] UWB-II underwater non-dispersible concrete type flocculent, anti-washout flocculating agents [J/OL]. Oil-Gasfield Surface Engineering.

Study on Adsorption and Release Characteristics of Phosphorus in the Bottom Sediment of Yellow River Xiaolangdi Reservoir

Deng Congxiang[1], Zhao Qing[2], Wu Guangqing[1], Li Xiaobing[3]

(1. Xiaolangdi Project Management Center, Ministry of Water Resources. Zhengzhou, 450002, China;
2. Zhao Qing, The Yellow River Henan Bureau Zhengzhou, 450002, China;
3. Li xiaobing ali Water Conservancy Bureau, Zhengzhou, 450002, China)

Abstract: Phosphorus is one of the key factors affecting the eutrophication of reservoir. Based on the unique particularity and representative of the Xiaolangdi Reservoir, this paper details the research methods and contents of the experimental study on phosphorus adsorption and release characteristics in bottom sediment from Xiaolangdi Reservoir. At the same time, it provides the results of the experiment, and strives to provide references for other similar problems through the introduction.

Key words: Reservoir bottom sediment, Phosphorus adsorption, Phosphorus release, Characteristics

1 Profile of Xiaolangdi Reservoir Project

Xiaolangdi Water Control Project is on the main stream of the Yellow River about 40 km northwest of Luoyang city Henan Province. Its dam site is 130 km upward from Sanmenxia Dam Project, 128 km downward from Huayuankou in Zhengzhou. It is the last canyon – reservoir with larger capacity in the middle and low reaches of the Yellow River, which controls the catchment area of 694,000 km^2 accounting for 92.2% of the total area of the Yellow River Basin.

Xiaolangdi Water Control Project is high dam and one – stage development engineering. The hub project is a first – calss engineering, whose main buildings are Grade I building with one – thousand – year frequency design flood and ten – thousand – year frequency checking flood. The reservoir presents a long – narrow shape, with surface area 272 km^2, the highest water level 275 m and the maximum dam height 154 m. Its original total capacity is 12.65 billion m^3, which contains the capacity of main stream 8.58 billion m^3, the capacity of tributaries 4.07 billion m3 and the late effective storage capacity 5.1 billion m^3.

2 Nitrogen and phosphorus in Xiaolangdi Reservoir

The ratio of nitrogen and phosphorus in Xiaolangdi Reservoir is between 104 ~ 472, which is far greater than that of required by the algae growth in theory. Therefore, it can be considered that the limiting nutrient element of algae growth in the reservoir is phosphorus.

3 The bottom quality of the Xiaolangdi Reservoir

The inflow and sediment of Xiaolangdi Reservoir are much higher with the mean annual average runoff 43.32 billion m3 and average sediment discharge 1.351 billion t. The reservoir bottom sediment is formed by the depositing of the suspended load and sediment from coming water, whose surface has a strong affinity and plays the role of phosphate buffer, so that a dynamic phosphorus cycle is formed between various media in the reservoir area.

Monitoring data shows that the sediment particles of the Xiaolangdi Reservoir are much fine, the particles less than 0.02 mm accounting for more than 80%, in which the organic matter content is 1% to 2%, moisture content is 2% to 3%, total nitrogen content is in 1,000 mg/kg – order – magnitude, total phosphorus content is about 600 mg/kg, and pH value is relatively higher than water, between 8.4 ~ 8.7.

4 Test methods

According to the water body and the existing form and concentration of phosphorus in bottom sediment of Xiaolangdi Reservoir, by simulating the natural environment in laboratory, the test data was obtained which result in a series of achievements through calculation, conversion and drawing.

5 Test plan and results

5.1 Test methods and results of phosphorus adsorption in reservoir bottom sediment

5.1.1 The dynamic test of phosphorus adsorption in reservoir bottom sediment

The test was under the static condition by selecting a series of 200 mL test solution, in which the sediment concentration is 2.0 g/L, the phosphorus concentration is 2.0 g/L, and placing in a constant temperature (20 ℃) oscillating box. The results are shown in Fig. 1.

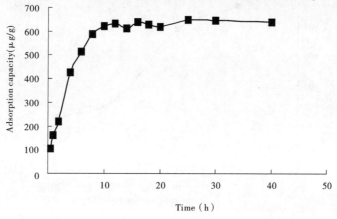

Fig. 1 Dynamic curve of phosphorus adsorption in reservoir bottom sediment

It can be seen from the Fig. 1 that the equilibrium adsorption time of phosphorus in reservoir bottom sediment is 10 h.

5.1.2 The isotherm adsorption test of phosphorus in reservoir bottom sediment

The test was in a constant temperature oscillating box by selecting nine water sample each 200 mL with initial concentration of phosphorus respectively 0.4 mg/L, 0.6 mg/L, 0.8 mg/L, 1.0 mg/L, 1.2 mg/L, 1.4 mg/L, 1.6 mg/L, 1.8 mg/L and 2.0 mg/L. The results are shown in Fig 2.

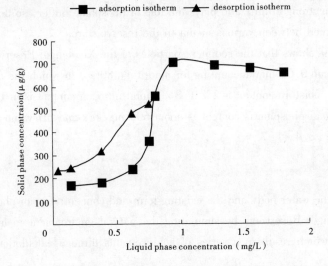

Fig. 2 The isotherm of phosphorus adsorption and release in reservoir bottom sediment

5.1.3 The influence of environmental condition on equilibrium adsorption capacity of phosphorus in reservoir bottom sediment

5.1.3.1 The influence of environmental temperature on phosphorus adsorption in bottom sediment

The adsorption test was under the condition of 5 ℃, 10 ℃, 15 ℃, 20 ℃. The results are shown in Fig. 3.

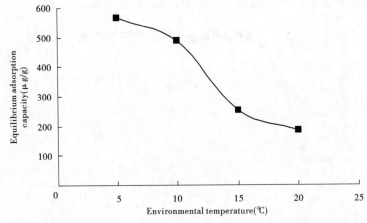

Fig. 3 The influence of environmental temperature on phosphorus adsorption in bottom sediment

5.1.3.2 The influence of sediment concentration on phosphorus adsorption in bottom sediment

The adsorption test was under the condition of 20 ℃ and different sediment concentration of 0.1 g/L, 0.2 g/L, 0.5 g/L, 1.0 g/L, 2.0 g/L, 3.0 g/L. The results are shown in Fig. 4.

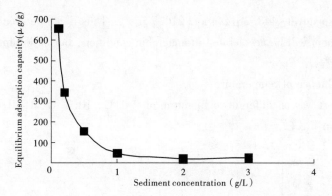

Fig. 4 The influence of different sediment concentration on phosphorus adsorption in bottom sediment

5.1.3.3 The influence of water disturbance on phosphorus adsorption in bottom sediment

The adsorption test were under the condition of 20 ℃ with different disturbing velocity of 0.5 r/s, 1.0 r/s, 1.5 r/s and 2.0 r/s. The results are shown in Fig. 5.

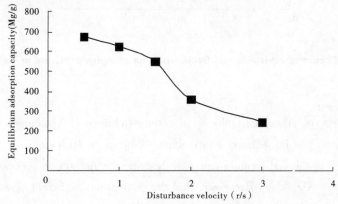

Fig. 5 The influence of water disturbance on phosphorus adsorption in bottom sediment

5.2 Test methods and results of phosphorus release in reservoir bottom sediment

5.2.1 The isotherm release test of phosphorus in reservoir bottom sediment

After separating the phosphorus from the sediment which had adsorbed phosphorus in isotherm environment in a high - speed centrifuge, then adding 200 mL distilled water, the phosphorus was released. Ten hours later, the phosphorus released from the liquid phase was tested, by calculation, the phosphorus concentration in sediment was obtained, so the release isotherm curve of phosphorus in sediment was drawn.

For comparison, the test results were plotted in the same coordinate system with adsorption isotherm curve. See Fig 2.

5.2.2 The influence of environment condition on phosphorus release in reservoir bottom sediment

The release test was under the conditions of anaerobic, aerobic and different temperature, by preparing a series of test solution, whose sediment concentration is 2.0 g/L, phosphorus concentration is 2.0 mg/L.

5.2.2.1 The influence of dissolved oxygen

According the test and calculation, the conditions to achieve equilibrium release in anaerobic and aerobic are respectively 448.2 μg/g and 216.7 μg/g. Thus it can be seen that the phosphorus release in sediment will be accelerated in anaerobic condition, but be inhibited in aerobic condition.

5.2.2.2 The influence of temperature

The release test was on different temperature of 5.0 ℃, 10.0 ℃, 15.0 ℃, 20.0 ℃. The results are shown in Fig. 6.

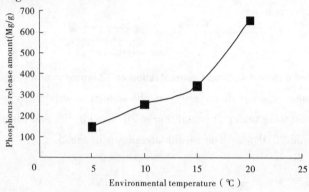

Fig. 6 The influence of environmental temperature on phosphorus release in bottom sediment

6 Conclusions

(1) The phosphorus adsorption velocity in bottom sediment of Xiaolangdi Reservoir is fast, and phosphorus adsorption in sediment can reach equilibrium in 10 hours.

(2) Freundlich isothermal formula can well describe the desorption processes of phosphorus in bottom sediment of Xiaolangdi Reservoir, and the correlation coefficient of mathematical fitting can reach 0.989. The phosphorus desorption from sediment showed desorption delay and limiting residual, and the limiting residual amount of the phosphorus in bottom sediment of Xiaolangdi Reservoir is 218.6 μg/g.

(3) Temperature has an effect on phosphorus adsorption and release in bottom sediment of reservoir area, and the results are in the opposite direction.

(4) Under the same conditions of the initial concentration of liquid phase phosphorus and other experimental conditions, the equilibrium adsorption capacity of unit mass sediment decreases with the increase of sediment concentration.

(5) The phosphorus equilibrium adsorption capacity in reservoir sediment decreases with the increase of disturbing velocity.

References

[1] Ma Jingan, Li Hongqing. Introduction to the Domestic and Foreign Rivers and Reservoir Eutrophication Status [J]. Resources and Environment in Yangtze River Basin, 2002,11(6): 575-578.

[2] Zheng Airong, Sheng Haiwei, Li Wenquan. Study on the Existing Forms of Phosphorus in Sediment and its Bioavailability [J]. Marine Journal, 2004, 26(4): 49-57.

[3] Xu Yishi, Xiong Huixin, Zhao Xiulan. Progress in Phosphorus Adsorption and Release of Sediment Study

[J]. Chongqing Environmental Science,2003,25(11): 147-149.
[4] Wang Ying, Guo Shihua. Random Dynamic Model of Phosphorus Release in Sediment of Reservoir[J]. Journal of Hydraulic Engineering, 2003,(11):71-77,84.
[5] Cheng Jingsheng, Zhang Yu, Yu Tao,et al.. Study on Dissolution and Degradation of Organic Matter in Yellow River Sediment[J]. Journal of Environmental Science, 2001,24(1): 1-5.
[6] Lin Ronggen, Wu Jingyang. Adsorption and Release of Phosphate by Yellow River Sediment[J]. Marine Journal, 1994,16(4):82-90.

Study on the Relationship between the Mass Concrete Temperature and Cooling Water Parameters

Tan Kaiyan[1], Duan Shaohui[2], Yu Yi[3], Hu Shuhong[2], Yan Qiao, Zhang Zhikui[1]

(1. Gezhouba Group Testing Co., Ltd., Yichang, 443002, China;
2. Jinping Construction Bureau, Ertan Hydropower Development Co., Ltd., Chengdu, 610021, China;
3. College of Hydraulic and Environmental Engineering, China Three Gorges University, Yichang, 443002, China)

Abstract: Water cooling measures is a very effective temperature control measures, which was used in mass concrete of Water Resources and Hydropower Engineering. However, water cooling is also a double - edged sword, which could cause damage in concrete in the case of improper control. In a project, usually, subject to numerical calculation, engineering experience, concrete internal temperature and cooling water's temperature, the cooling water flow rate is adjusted. Currently, due to lack of a simple and practical program, temperature control parameters are often out of limits in temperature control process, which affects the quality of concrete. Therefore, it is particularly important to study the relationship between concrete internal temperature and cooling water flow rate for ensuring the quality of the project. In this paper, based on the heat conservation law and combined engineering test data analysis, a set of relatively simple, practical and efficient formula were summarized. It can guide the preparation of water cooling program, and play an active role to ensure the quality of the project.

Key words: cooling water, concrete, temperature control

1 Summary

Mass concrete is extensively used in hydropower projects, especially in concrete dams. During the construction process of massive concrete, temperature control is very important and complex[1]. Since 1930s piped water cooling technology was used in the Hoover dam for the first time, it is still in use today, which plays a good role in preventing temperature stress cracks. In order to carry out better engineering application, piped water cooling was studied at home and abroad. Calculation method of the cooling for phase II was studied by US Bureau of Reclamation, and with separation of variables the strict solution for plane problems and approximate solution for space problems without pyrogen have been obtained[2]; Zhu Bofang academician studied a calculation method of cooling for phase I, and with the integral transformation obtained the strict solution for plane problems and approximate solution for space problems with pyrogen, and proposed finite element analysis method of piped water cooling, non - metallic pipe cooling calculation method and equivalent heat conduction equation considering piped water cooling[3]. He also proposed a complex algorithms considering the cooling effect of the piped water: cooling effect of piped water cooling for the upper part of the dam of phase I is calculated with an accurate calculation method, and that of phase II and phase III in the wide range for the lower part of the dam is done in conventional sparse grid with two equivalent heat conduction equation and three water – cooled computing[4]. Currently there are fewer studies on the relationship between concrete temperature and water cooling parameters. Relations between concrete temperature, piped water temperature and water flow rate are very complicated, and they are related to concrete mix proportion, age, temperature control process, water cooling process and water cooling parameters. In this paper, study on the relationship between concrete internal temperature and piped cooling water flow rate is summarized, and is illustrated through reference data from a project.

2 General information for piped water cooling

In the current concrete dam construction, high – density polyethylene HDPE pipes are widely used for cooling water pipes, with specifications of external diameter of 32 mm and internal diameter of 28 mm. Compared with steel pipes, their advantage is light – weight, easy transport, less joints, convenient construction and mature application technology, and their disadvantage is relatively high pipe thermal resistance. The pipes are usually arranged in layers with vertical equal – spacing of 1.5 m, and horizontally the pipe in a layer is laid in a serpentine way with horizontal spacing of 1.0 m or 1.5 m and fixed with U – shaped fasteners.

Concrete internal temperature is measured with thermometers installed in the position with vertical distance ranging from 0.7 m to 0.8 m to the pipes (intermediate position between two pipe layers), where the measured values can basically reflect the cooling effect of corresponding cooling pipes.

3 Formula derivation

3.1 Condition assumption

3.1.1 Constraint condition assumption

In general, the date of watering cooling of phase II is 2 to 4 months away from the pouring date, when adjacent monoliths have been poured and their free surface has been covered with insulation board, and when internal temperature is less different in a cooling area after mid – term water cooling, and when a single dam monolith is less influenced from adjacent monoliths and environmental conditions. Therefore, this paper assumes that a single dam monolith is free from climatic conditions and adjacent monoliths.

In addition, due to watering cooling being a long term and cumulative process, the thermal resistance of cooling water pipes will not considered. Considering water flow and temperature sensors installed near to pouring position behind the dam, the heat loss and the heat exchange along the pipes is negligible.

3.1.2 Cooling pipe assumption

A dam monolith with thickness of 1.5 m is selected in this paper. Due to actual controlled concrete being upper portion of the old concrete and under portion of the fresh concrete for the phase II of water cooling, cooling water pipes are laid along the surface of the old concrete. Considering the adjacent dam monoliths being the same basic conditions, it is assumed that water cooling pipes only work in a single dam monolith in order to facilitate the calculation. Cooing water pipe equivalent figure is shown in Fig. 1.

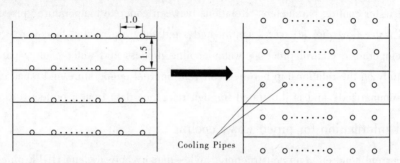

Fig. 1 Cooling Pipe Equivalent Figure　(unit:m)

3.2 Formula derivation

According to the heat balance equation of $Q_1 = Q_2$, heat loss of concrete is equal to the increased heat of cooling water.

3.2.1 Formula derivation of $T \sim \sum q_i t_i \Delta T_i$

(1) Concrete internal heat loss:

$$Q_1 = C_1 m_1 \Delta T_1 = C_1 \rho_1 V_1 \Delta T_1$$

Where, C_1——specific heat capacity or concrete, taken as 0.931×10^3 J/(kg · ℃) for concrete C40;

ρ_1——density of water, taken as 2.4×10^3 kg/m^3;

V_1——volume of the dam monolith, m^3;

ΔT_1——temperature difference of concrete before and after water cooling, ℃.

(2) Increased heat of cooling water:

$$Q_2 = C_2 m_2 \Delta T_2 = C_2 \sum m_i \Delta T_i = C_2 \rho_2 \sum V_i \Delta T_i = C_2 \rho_2 \sum q_i t_i \Delta T_i$$

Where, C_2——specific heat capacity of water, taken as 4.183×10^3 J/(kg · ℃);

ρ_2——density of water, taken as 1×10^3 kg/m^3;

q_i——flow rate, m^3/h;

t_i——water cooling time corresponding to the flow rate q, h;

ΔT_i——water temperature difference between intake and outlet corresponding to water cooling time t_i, ℃;

$\sum q_i t_i$——amount of cooling water, m^3.

(3) According to the equation of $Q_1 = Q_2$, the following equation is determined:

$$C_1 \rho_1 V_1 \Delta T = C_2 \rho_2 \sum q_i t_i \Delta T_i$$

or:

$$\frac{\Delta T_1}{\sum q_i t_i \Delta T_i} = \frac{C_2 \rho_2}{C_1 \rho_1 V_1}$$

Setting $\dfrac{C_2 \rho_2}{C_1 \rho_1 V_1} = k$, relationship between concrete internal temperature and cooling water parameters is obtained as follows.

$$T = -k \sum q_i t_i \Delta T_i + T_0 \tag{1}$$

Where, $\sum q_i t_i \Delta T_i = \sum V_i \Delta T_i$;

$\sum V_i$ ——amount of cooling water, m^3;

T_0——initial temperature of cooling water, ℃.

3.2.2 Formula derivation for the relationship of $T \sim \sum q_i t_i$

The relationship of $T \sim \sum q_i t_i$ refers to the relationship between concrete internal temperature and amount of cooling water. After studying some measured data the author found that water temperature difference ΔT_i between intake and outlet was not much different during the water cooling of phase II. Average temperature difference between intake and outlet of cooling pipes for a dam monolith within a certain elevation range is shown in Fig. 2.

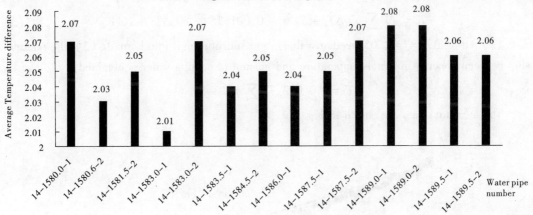

Fig. 2 Average Temperature Difference Figure

The above Fig. 2 indicates that water temperature difference between intake and outlet is about 2.05 ℃. And after studying some measured data, it is the average temperature difference ranges from 2.0 ℃ to 2.1 ℃. Therefore, when calculating can be taken as a constant l, and substituting it into the Formula (1) the relationship between concrete internal temperature and cooling water parameters is obtained as follows during water cooling of stage II.

$$T = -kl \sum q_i t_i + T_0 \tag{2}$$

It should be noted that, for different projects, due to the impact of regional differences, climate and other conditions, values of l may fluctuate, and in order to ensure the accuracy of the calculation, the values shall be corrected according to analysis of measured data.

3.3 Example of calculation

For example of one concrete arch dam, the basic information for a dam monolith is as follows:

　　Pouring time: 15:35 Dec. 8 ,2009 ~ 05:20 Dec. 9 ,2009

　　Elevation: El1,589 ~ El1,590.5 m

　　Poured Quantity: 2,782.8 m^3

　　Two cooling pipes of Pipe14 - 1589.0 - 1 in the front half dam monolith and Pipe 14 - 1589.0 - 2 in the back half, with the same water cooling conditions. They are considered to control half of the dam monolith with quantity of $2,782.8 \times \frac{1}{2} = 1,391.4 (m^3)$.

　　Specific heat capacity of water: 0.931×10^3 J/(kg·℃);

　　Density of concrete: 2.4×10^3 kg/m^3;

　　Initial temperature of cooling water: $T_0 = 16.8$ ℃;

3.3.1　Calculation

　　(1) Value k can be obtained:

$$k = \frac{C_2 \rho_2}{C_1 \rho_1 V_1} = \frac{4.2 \times 10^3 \times 1 \times 10^3}{0.931 \times 10^3 \times 2.4 \times 10^3 \times 1 = 1,391.4} = 0.001,35$$

Substituting it into Formula(1), the following formula can be obtained:

$$T = -k \sum q_i t_i \Delta T_i + T_0 = -0.001,35 \sum q_i t_i \Delta T_i + 16.8 \quad (3)$$

　　(2) Setting $\Delta T_i = l = 2.05$ (median value), and substituting it into Formula (3), the relationship between concrete internal temperature and amount of cooling water is obtained:

$$T = -0.002,77 \sum q_i t_i + 16.8$$

The measured data are shown in Fig. 3.

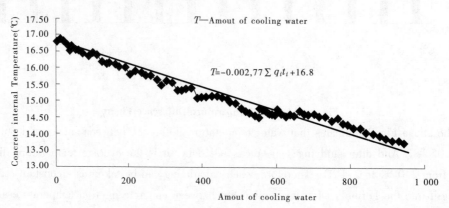

Note: The straight line in the figure refers to theoretical calculation results.

Fig. 3　Relationship between Concrete Internal Temperature and Amount of Cooling Water of Dam Monolith No. ①

　　As per the above method, four additional dam monoliths are selected for analysis, and poured concrete quantity and relationship between concrete internal temperature and amount of cooling water are listed in Table1.

Table 1　Relationship of T and Amount of Cooling Water

No.	Poured concrete quantity(m^3)	Relationship
②	2 398.2 × 1/2 = 1 199.1	$T = -0.003\ 21 \sum q_i t_i + 16.35$
③	2 356.8 × 1/2 = 1 178.4	$T = -0.003\ 3 \sum q_i t_i + 17$
④	4 295.5 × 1/2 × 1/2 = 1 073.9	$T = -0.003\ 6 \sum q_i t_i + 15.95$
⑤	1 489.5 × 1/2 = 744.8	$T = -0.005 \sum q_i t_i + 17.2$

Measured data and theoretical calculation are shown in Fig. 4.

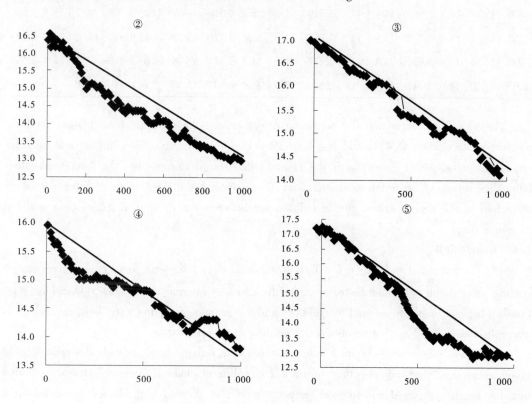

Note: The straight line in the figure refers to theoretical calculation results; lateral axis refers to amount of cooling water(m^3) and vertical axis refers to concrete internal temperature(℃)

Fig. 4　Relationship between Concrete Internal Temperature and Amount of Cooling Water of Dam Monolith No. ②③④⑤

The above figure indicates that the relationship between the measured data of internal temperature and the amount of cooling water $\sum q_i t_i$ is linear, and the measured data match well with the theoretical calculations.

3.3.2　Error analysis

Setting $\sum q_i t_i = 0, 150, 300, 450, 600, 750, 900, 1\ 050$ respectively, theoretical calculations, measured data and errors are listed in the Table 2.

Table 2 Error Analysis Table

Amount of cooling water	①			②			③			④			⑤		
	Theo. temp.	Actual Temp.	Error	Theo. temp.	Actual Temp.	Error	Theo. temp.	Actual Temp.	Error	Theo. temp.	Actual Temp.	Error	Theo. temp.	Actual Temp.	Error
0	16.8	16.8	0.00%	16.3	16.3	0.00%	17	17	0.00%	15.95	15.95	0.00%	17.2	17.2	0.00%
150	16.38	16.15	1.45%	15.8	15.7	1.08%	16.5	16.6	0.57%	15.61	15.2	2.66%	16.45	16.75	1.79%
300	15.97	15.6	2.37%	15.3	14.8	3.63%	16	16.2	1.17%	15.26	15	1.73%	15.7	15.65	0.32%
450	15.55	15.15	2.66%	14.9	14.5	2.45%	15.5	15.8	1.80%	14.92	14.8	0.78%	14.95	14.9	0.34%
600	15.14	14.7	2.98%	14.4	14.1	2.31%	15	15.3	1.51%	14.57	14.5	0.48%	14.2	13.8	2.90%
750	14.72	14.4	2.24%	13.9	13.5	2.91%	14.5	15	3.17%	14.23	14.15	0.53%	13.45	13.2	1.89%
900	14.31	13.9	2.93%	13.4	13.1	2.37%	14	14.5	2.91%	13.88	14.05	1.21%	12.7	12.9	1.55%
1,050	13.89	13.6	2.14%	12.9	13	0.54%	13.5	13.7	0.84%	13.54	13.6	0.48%	11.95	—	—

The table 2 indicates that the maximum error between the theoretical calculations and the measured data is about 0.4 ℃ and less than 3%. Further, the errors rise initially and then decrease gradually, which shows that in the case of long-term water cooling, the theoretical calculations are getting closer to the measured data. That is to say, the closer concrete internal temperature fells to the design value, the less difference between the theoretical calculations and the measured data.

3.4 Conclusion

(1) The formula $T = -kl\sum q_i t_i + T_0$ indicates that after long term water cooling, amount of cooling water is the only main factor to affect the concrete internal temperature. Therefore, relationship between concrete internal temperature and the amount of cooling water is linear, which is also reflected in $T \sim \sum q_i t_i$ curve obtained according to measured data.

(2) Linear relationship obtained from measured data reflects from one side the rationality of given assumption, and indicates that during water cooling of phase II concrete internal hydration heat has nearly completed and internal temperature is less affected from climatic conditions and adjacent monoliths.

(3) The relationship of $T \sim \sum q_i t_i$ also indicates that concrete internal temperature falls evenly and slowly with the increase of amount of cooling water, which ensure that cooling rate meet design requirement. The water cooling method of long term and small temperature difference can effectively reduce the large temperature gradient around the pipes, cooling rate of concrete and increase rate in stress, which can take full advantage of concrete creep effect and reduce the risk of concrete cracking[5].

4 Conclusions

Temperature control of concrete is a complex and systematic problems. With the continuous study on temperature control of mass concrete, in modern water conservancy construction of concrete temperature control is gradually developing to paperless direction and computer automation technology is applied increasingly. By simplifying boundary conditions and from the perspective of heat conservation, in this paper the relationship between concrete internal temperature and amount of cooling water is studied and analyzed and some conclusions are reached. Mr. Zhu Bofang proposed that early and slow cooling with small temperature difference is the development direction of pipe water cooling for concrete dams[6], which serve as a guideline to some extent to cooling rate and amplitude control in order to reduce temperature stress of concrete and to improve strength of concrete. In this paper, a relatively simple calculation method is provided, which can provide reference for temperature control automation.

References

[1] Yuan Guangyu, Hu Zhigen, et al. Construction of Hydro Project [M]. Beijing: China Waterpower Press, 2005.
[2] Zhu Bofang. Temperature Stress and Temperature Control of Mass Concrete [M]. Beijing: China Electric Power Press, 1999.
[3] Zhu Bofang. Equivalent Equation of Heat Conduction of Concrete Considering Water Cooling Effect [J]. Journal of Hydraulic Engineering, 1991(3)28 - 34.
[4] Zhu Bofang. Equivalent Equation of Heat Conduction of Concrete Considering Effect from External Temperature [J]. Journal of Hydraulic Engineering 2003(3),49-54.
[5] Liujun, Huangwei, Zhouwei, et al.. Long Term water Cooling with Small Temperature Difference of Mass Concrete [J]. Journal of Wuhan University, 2011(5):549 - 553.
[6] Zhu Bofang. Early and Slow Cooling with Small Temperature Difference is the Development Direction of Pipe Water Cooling for Concrete Dams [J]. Hydroelectric Technology, 2009, 40(1):44 - 50.

Analysis of the No. 6 Main Transformer's Temperature Rise in XLD Hydropower Station

Cui Peilei, Lu Feng, Zhang Yang, Chen Meng, Ben Xupeng

(Yellow River Water and Hydropower Development Corporation,
Jiyuan, 459017, China)

Abstract: According to the phenomena of No. 6 main transformer temperature rising during the operation compared to the others in the Xiaolangdi hydropower plant, a series of experiment and research are carried out. Certain effective measures are taken to solve the problem after comprehensive analysis in detail. Then six main transformer oil coolers are cleaned and repaired thoroughly by taking use of the opportunity of generator units maintenance to assure good cooling effect, and then to ensure the safe and stable operation of the main transformers.

Key words: Xiaolangdi, main transformer cooling system, temperature rising, generator units maintenance

1 Overview

The Xiaolangdi water control project consists of two hydropower plants, the Xiaolangdi hydropower station and the Xixiayuan against regulation station, a total of ten hydroelectric generating units, including the Xiaolangdi 6 × 300 MW mixed – flow hydro – turbine governing system and the reversing power plant Xixiayuan 4 × 35 MW axial flow impeller turbine units. From the relationship of administrative subordination these two hydropower stations are under control of the Henan province and are very important in the Henan power grid during the frequency and load regulation to accomplish a certain emergency task. Every main transformer of the six Xiaolangdi station with the capacity of 360 MVA adopts the form of oil – immersed transformer type, using one generator and transform connection mode, increasing the voltage from 18 kV to 220 kV bus voltage generator, sending to the outdoor substation of the Yellow River by 220 kV dry – type cable. The main transformer uses the forced oil circulation water cooling way and the oil temperature is not higher than 90 ℃ when running. The main transformer concludes three groups of cooler, which are usually a group of main use, a set of backup, a set of auxiliary and regular switch operation.

2 Background

The 6 main transformers of the Xiaolangdi water control project were put into commercial operation on January 9, 2000. In June 2013 the water – sediment regulation during large flow drainage and large power generation, the main transformers operated normally and the oil temperatures did not appear abnormal compared with previous operation. In September 2013, during the time of the Xiaolangdi large flow drainage and the big power generation, the oil temperature of the NO. 6 main transformer was to be found to be 10 ℃ higher than that of the other main transformers. To identify the reasons of the NO. 6 main transformer temperature rising, a series of investigation and experiment were conducted.

3 Analyzation and inspection of the No. 6 main transformer temperature rising

According to the structure and operation condition of the main transformer, there are mainly two possible reasons causing the main transformer temperature rise. Firstly, the transformer internal faults lead to the problem, which are mainly the main electrical circuit fault, insulation fault and core multipoint ground connection fault of ontology, etc. Secondly, transformer external malfunctions are also the reason, which are insufficient transformer oil cooling water flow, small valve opening, cooling system and cooling system pipe blockage.

3.1 The No. 6 main transformer chromatographic data analysis

In 2005, the insulation oil chromatogram online monitoring device, which was installed on the NO. 6 main transformer, can real-time monitoring the data of oil chromatographic and there will be an alarm signal when it is abnormal. Meanwhile, in accordance with *Guide to the analysis and the diagnosis of gases dissolved in transformer oil* guideline in DL/T 722—2014, the main transformer insulation oil chromatographic tests are conducted a laboratory chromatography test every six months, and compared to the on-line monitoring data. From July to October, 2013, the main transformer insulation oil chromatographic tests are shown in Table 1.

Table 1 From July to October the main transformer insulation oil chromatographic test data (units: ul/L)

Num	Time	H_2	CO	CH_4	C_2H_4	C_2H_6	C_2H_2	CO_2	Total hydrocarbon
1	October 2013	8.9	774.2	62.8	16.1	11.5	0.4	3,986.9	90.8
2	September 2013	7.9	771	63.2	16.2	11.9	0.2	3,706	91.5
3	August 2013	7.9	770.8	63.6	16.2	11.6	0.2	3,596.8	91.6
4	July 2013	8.3	768.1	63.1	16.3	11.2	0.3	3,585.9	90.6

According to the guidelines (of Guide to the anduysis and the dingnosis of gases dissolved in transformer oil DL/T 722—722, during the operation of the transformer insulation oil hydrogen content is not more than 150 μL/L, total hydrocarbon (methane, ethane, ethylene, acetylene) content is not more than 150 μL/L, acetylene content is not more than 5 μL/L, the ratio of carbon dioxide and carbon monoxide < 7. When the values are beyond the above, three ratio method is adopted in a comprehensive analysis and judgment. Currently the NO. 6 main transformer maximum hydrogen content is 8.9 μL/L, total hydrocarbon (methane, ethane, ethylene, acetylene) content in a maximum of 91.6 μL/L, a maximum of 0.4 μL/L content of acetylene, the maximum ratio of carbon dioxide and carbon monoxide 5.14, and all of the test results meet the national standards. Compared to the test data, after test instrument error and test personnel testing error removing, the experimental data remain unchanged basically. All the test data show that internal partial discharge and damp, local overheating of the high and low voltage winding and iron core, thermal decomposition of solid insulating materials don't exist in the NO. 6 main transformer.

3.2 measurement and analysis of the NO. 6 main transformer core earthing current

If transformer core appears two or more grounding connections, the core will form current circulation internally and the main transformer temperature will rise. When the NO. 6 main transformer core earthing current was measured, compared to that of the NO. 4 main transformer, the two main transformers core earth current values both were 1 mA, and the NO. 4 main transformer temperature is not high. So it can be judged that the NO. 6 main transformer iron – core earthing fault didn't exist.

3.3 Inspection and analysis of the No. 6 main transformer submersible pump

Xiao Langdi six Main transformers' cooling system adopts the forced oil circulation by water cooling. Based on the three submersible pumps start – stop tests of the No. 6 and No. 4 main transformer, the submersible pumps outlet oil flow values were measured and both No. 6 and No. 4 main transformer submersible pump outlet oil flow values were the same, which can determine that the No. 6 main transformer submersible pumps operated normally.

3.4 Inspection and analysis of the No. 6 main transformer cooling water system

By comparing the No. 6 and No. 4 main transformer cooling water system, both of the water flow rates were the same, which was about 300 L/h. When each valve was tested to be switched on and off, water flow remained unchanged. And after the sandbox of the No. 6 main transformer cooling water examined, the inlet and outlet pressure difference was 0.02 MPa, no more than 0.06 MPa, which is the design allowable value. After through inspection, comparison and analysis the No. 6 main transformer cooling water system valves and pipeline were not blocked, which can be judged that the No. 6 main transformer cooling water system operated normally.

3.5 Inspection and analysis of the No. 6 main transformer cooler heat exchanging

By measuring the inlet and outlet of oil pipeline temperatures of both the No. 6 and No. 4 main transformer cooler heat exchanger, the No. 6 main transformer cooler inlet and outlet oil pipeline temperatures difference was about 2 ℃, while that of the No. 4 main transformer cooler was about 4℃. It can be determined preliminarily that the No. 6 main transformer temperature was higher than that of the other main transformers due to the No. 6 main transformer cooler lower heat exchange efficiency and the worse cooling effect.

4 Solutions to the No. 6 main transformer temperature rise

Every main transformer oil cooler has three groups. One group is mainly used, another is utilized in spare condition, and the last is in auxiliary operation. By decomposing and examining thoroughly the first oil cooler of the No. 6 main transformer. There was not found out sundry jam in the No. 6 main transformer cooler tubes but very thin yellow mud attached to the inner wall of the copper tube. After the inner wall of the copper tube was cleaned and the cooler was operated normally again, the temperatures of oil pipeline heat inlet and outlet of the cooler were measured and the temperature difference was about 5 ℃. Then the second and third cooler copper tube were also cleaned up. After three groups of cooler cleaning up, the No. 6 main transformer temperature dropped significantly(see table 2), about up to 10 ℃. The problem of the No. 6 main transformer rise was resolved.

Table 2 The No. 6 main transformer cooler after treatment compared with the No. 4 main transformer cooler

Num	date	No. 4 main transformer			No. 6 main transformer		
		Cooling operation mode	Power (MW)	Oil temperature (℃)	Cooling operation mode	Power (MW)	Oil temperature (℃)
1	10.30	One main use, one auxiliary and one spare	296	47	One main use, one auxiliary and one spare	289	44
2	10.31	One main use, one one auxiliary and one spare	320	40	One main use, one auxiliary and one spare	296	43
3	11.01	One main use, one auxiliary and one spare	230	45	One main use, one auxiliary and one spare	230	39
4	11.02	One main use, one auxiliary and one spare	193	47	One main use, one auxiliary and one spare	200	41

5 Supplementary measures

(1) After the No. 6 main transformer cooler cleaned up, the main transformer temperatures decreased clearly. The next step is to handle with the other coolers arranged by the Xiaolangdi power station.

(2) Strengthen the management of equipment operation and revise the maintenance procedures. Clean up the the main transformer coolers on a regular basis and add it into the maintenance procedures.

(3) Add temperature measuring devices on the oil pipeline inlet and outlet of the coolers.

(4) Develop technical transformation and study the online cleaning device of the main transformer oil coolers to carry out online cleaning.

6 Conclusion

The six main transformers of the Xiaolangdi hydropower station has been operating continuously more than 10 years. But thorough overhaul has not still implemented. Because of the Yellow

River silt, pipeline corrosion and wall scaling factors, the heat exchange effect of main transformer oil coolers will worsen, which is a big potential security problem. To guarantee the safe and reliable operation of the main transformers, the six main transformers oil coolers are cleaned and repaired thoroughly by taking advantage of the chance of A level maintenance of every generator unit to achieve a better cooling effect.

Practical Application of Rapid Dam Construction Techniques in Construction of Tingzikou Hydroproject Dam

Kong Xikang, Liu Hui

(SiChuan Ertan International Engineering Consulting Co., Ltd., Chengdu, 611130, China)

Abstract: Rapid construction is the characteristic and advantage of roller compacted concrete construction. Interlayer bonding quality and temperature control and crack prevention are the key technologies of roller compacted concrete construction. To give full play to the characteristics of rapid construction of roller compacted concrete, after analyzing the construction period, equipment system configuration, and construction intensity, the roller compacted concrete placement methods are optimized, researched and verified according to the arrangement characteristics of Tingzikou Hydroproject structures as well as arrangement of construction roads. The methods like direct placing with dump truck, large – span belt conveyor + material transfer by dump truck, and slow descending articulated chute + material transfer with dump truck are adopted, effectively solving the problem of rapid concrete placing. By taking such measures as dynamic control of VC value, human assistance in construction, and application of antiseepage coatings, rapid and continuous construction of roller compacted concrete is realized and a powerful guarantee is provided for interlayer bonding quality control as well as for temperature control and crack prevention.

This article mainly explains the planning of road for concrete placing, placement method design, interlayer bonding quality control as well as temperature control and crack prevention technologies for roller compacted concrete of the Tingzikou Hydroproject. It also introduces the actual application status and implementation effects of the rapid damming technology with roller compacted concrete applied in roller compacted concrete construction of the Tingzikou Hydroproject.

Key words: Tingzikou Hydroproject, mix proportion, optimization of placement method, interlayer bonding control, temperature control and crack prevention

1 Project overview

1.1 General

The Tingzikou Hydroproject, located in Cangxi County, Guangyuan, Sichuan Province, is the only control project in the development of the Jialing River. As a comprehensive project, it is used mainly for flood control, irrigation, rural water supply and power generation and concurrently for navigation and sediment detention. The normal water level of the reservoir is 458 m, the dead water level is 438 m, the design flood level is 461.3 m, the check flood level is 463.07 m, and the total capacity of the reservoir is 4.067×10^9 m^3. The reserved flood control capacity of the reservoir is 1.06×10^9 m3 (1.44×10^9 m^3 for abnormal operation). The irrigation area is 2 921 400 mu, the installed capacity of the hydropower station is 1 100 MW, and the navigation structure is a vertical ship lift with a capacity of 2×500 t. The Project is a large – scale (Grade I) hydro – project.

The total length of dam axis is 995.4 m, with the elevation of dam crest being 465 m, the maximum dam height being 116 m and the maximum dam width in the direction of water flow being 107.18 m.

Eight crest orifices, 5 bottom orifices and an energy dissipation structure are arranged at the river bed dam monolith; a power house of hydropower station at dam toe is arranged on the left side of the river bed; a vertical ship lift is arranged on the right side, and non – overflow dam monoliths are arranged on the left and right banks

The total volume of concrete for the dam is about 4.50×10^6 m^3, among which RCC is about 2.53×10^6 m^3, mainly distributed in the non – overflow dam monoliths on the left and right banks, and at the positions below the elevation of 435 m of the crest orifice dam monolith, below the elevation of 370 m of the bottom orifice and bottom orifice gate dam monoliths and below the elevation of 410 m of the power house dam monolith.

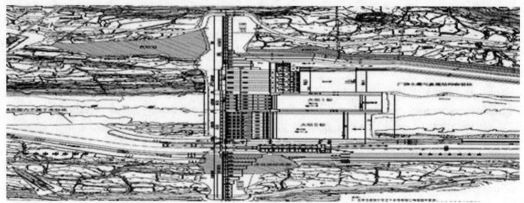

Fig. 1 Layout of Tingzikou Hydroproject

1.2 Meteorological condition

The mean annual precipitation of the dam site is 995.8 mm; the historical maximum monthly and daily precipitations are 477.8 mm and 204.3 mm respectively. The mean precipitation in each month from May to September is over 100 mm and the precipitation from May to September accounts for 78.7% of the annual precipitation. The mean annual temperature is 16.6 ℃, with the highest mean monthly temperature of 26.3 ℃ appearing in July and August, and the lowest of 5.8 ℃ appearing in January. The mean annual wind speed is 1.9 m/s, the dominated wind direction is NNW, and the mean annual maximum wind speed is 13.2 m/s. The mean annual water surface evaporation is 1 318.6 mm. The mean annual sunshine percentage is 35%. The mean annual relative humidity is 73%. And the mean annual ground temperature is 19.2 ℃.

1.3 Configuration of aggregate production and concrete mixing system

A natural sand and gravel processing system with a processing capacity of 2 000 t/h and an aggregate production capacity of 1 600 t/h is arranged on the left bank and can provide enough finished aggregates for the project at the peak concrete placing intensity of 2.5×10^5 m^3/month.

A concrete production system is respectively arranged on the left and right banks, with a straight line distance of about 1.5 km from dam axis. The system is provided with 4 forced concrete mixing plants in total, 2 with a capacity of 2×4.5 m^3 (the design production capacity is 320 m^3/h), 1 with a capacity of 2×6.0 m^3/h (the design production capacity is 320 m^3/h) and 1 with a capacity of 2×3.0 m^3/h (the design production capacity is 240 m^3/h). The production capacity for normal temperature concrete is about 960 m^3/h while that for pre-cooling concrete is about 750 m^3/h.

2 Construction characteristics

(1) In terms of structural arrangement, the design of power house at dam toe is adopted for Tingzikou Hydroproject. Four diversion penstocks are arranged and 5 bottom orifices and 8 crest orifices are arranged at the river bed dam monolith. Four layers of galleries are arranged in the RCC construction area and 4 drainage galleries are arranged transversely at the dam foundation, with a minimum distance of 4 m from the upstream face. The structures are so close to each other, and RCC rises at the same pace with normal concrete at some dam monoliths. Therefore, the above-mentioned structural arrangement is not conducive to rapid RCC construction, which is the key and difficult part of the construction of RCC.

(2) According to the construction schedule, the construction period for RCC is from October 2010 to the end of 2012, including two high temperature seasons and three low temperature seasons, during which the construction of RCC will be carried out without stopping. The technique for "interlayer bonding, temperature control and crack prevention" is the key and difficult part of quality control, as well as one of the factors that will affect the rapid RCC construction.

(3) The aggregates for RCC of Tingzikou Hydroproject are all made of natural sands and gravels from the river bed of the Jialing River at the downstream of the dam site. The quality variation of raw materials from the natural yard is relatively large and will affect the stability of finished aggregates. The exceeding and inferior diameter ratio and gradation continuity of finished aggregates, the fineness modulus of natural sands and the crusher dust content will directly affect the performance of RCC. Finished coarse aggregates are mostly flaky pebbles with smooth surface, and have smaller specific surface and poor mortar coating ratio. The small fineness modulus of natural sand and high argillaceous content may easily lead to quick dehydration on the surface of the aggregates during construction. How to ensure the performance and compactability of raw materials and concrete is the key and difficult part of construction.

3 Mix proportion

The alkali-reactive aggregates with low crusher dust content for Tingzikou Hydroproject are all made of natural sands and gravels from the river bed of the Jialing River. To prevent concrete from being damaged, the alkali content in concrete shall not be more than 2.5 kg/m^3 according to

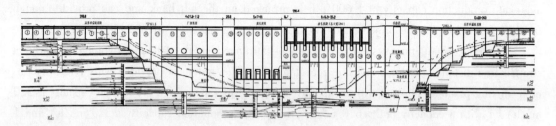

Fig. 2 Upstream Elevation View of Tingzikou Hydroproject Dam

the design requirements. The total alkali content in concrete is kept within 0.6 ~ 0.8 kg/m^3 by selecting Grade I flyash as admixture and keeping alkali content in cement under 0.6%. The maximum mixing amount of Grade I flyash is up to 60%, and the mixing of flyash reduces water consumption in the production of concrete, improves workability, lowers hydration heat temperature rise, and improves the strength and durability of concrete. Meanwhile, crusher dust is also mixed in to replace sand and the mixing amount of crusher dust is within 16% ~ 18%.

Tab. 1 Mix proportion of RCC for Tingzikou Hydroproject dam

Design strength	Grade	Mixing amount of flyash (%)	Water - binder ratio	Material consumption (kg/m^3)					Chemical admixture (%)	
				Water	Cement	Flyash	Sand	Stone	Water - reducing agent Jm - 2	Air entraining agent Jm - 2000
R_{90}150	III	60	0.5	75	60	90	661	1 584	0.9	0.09
R_{90}200	III	55	0.5	90	99	81	728	1 452	1.08	0.108

Note: the water - reducing agent is naphthalene based set retarding water - reducing agent; the air entraining agent is modified rosin acid salts nonionic resin surfactant; and the flyash is Grade I high - quality flyash.

4 Optimization of placement method

For rapid RCC construction, the transportation method for concrete placing is one of the key factors. According to a lot of experience in construction, placing by dump truck, which can significantly reduce intermediate procedures, speed up construction and meanwhile reduce the concrete's absorbing the heat due to temperature rise, is the most effective method for rapid RCC construction.

Optimization and improvement have been conducted to the original placement scheme at the beginning of the Tingzikou Hydroproject Dam Project. Subject to the actual condition, the placement scheme of direct placing by dump truck together with transporting by negative pressure articulated chute and belt conveyor is selected. Direct placing by dump truck, transporting materials by large - span belt conveyer + material transfer by dump truck, and transporting materials by articulated chute + material transfer by dump truck, and other placement methods are adopted, effectively solving the problem of concrete placing and providing a strong guarantee for rapid RCC construction, quality control of interlayer bonding, and temperature control and crack prevention.

4.1 Non-overflow dam monoliths on the left and right banks

Original construction scheme: For non-overflow dam monoliths on the left bank, placing by dump truck + spreader is adopted for $1^{\#} \sim 11^{\#}$ dam monoliths; transporting by articulated chute + dump truck is adopted for $12^{\#} \sim 16^{\#}$ dam monoliths below the elevation of 444.5 m and placing by dump truck + spreader is adopted for $12^{\#} \sim 16^{\#}$ dam monoliths above the elevation of 444.5 m. For non-overflow dam monoliths on the right bank, direct placing by dump truck via the filling construction roads at the upstream or downstream is adopted for $39^{\#} \sim 43^{\#}$ dam monoliths; direct placing by dump truck + spreader is adopted for $44^{\#} \sim 46^{\#}$ dam monoliths and direct placing by dump truck is adopted for $47^{\#} - 50^{\#}$ dam monoliths. Problems of the scheme: the largest planned placing area for the non-overflow dam monoliths on the left bank is 1 600 m². One spreader (TB105) with a conveying capacity of 80 ~ 100 m³/h is planned to be provided. However, subject to the dam type, there are no conditions for the dump truck to transport while placing at the non-overflow dam monoliths, therefore, the spreader must be moved when the placement exceeds the placing scope of the spreader when continuous placement method is employed. Considering the impact of the dumping of materials by dump truck and the position change of the spreader, the actual placing intensity is about 50 m³/h and the placing area is limited to be below 1 000 m², which will affect the construction progress; and if the spreader fails, there will be no spare equipment for replacement, which will directly affect the normal concrete construction. The receiving platforms of the cable cranes are arranged at the downstream of $1^{\#} \sim 10^{\#}$ non-overflow dam monoliths on the left bank and $47^{\#} \sim 50^{\#}$ non-overflow dam monoliths on the right bank. The arrangement is carried out as soon as possible for the placing of normal concrete at the river bed dam monolith. However, placing by spreader is slow. The open terrain at the dam toe of $1^{\#} \sim 11^{\#}$ non-overflow dam monoliths on the left bank provides conditions for direct placing by dump truck via the filling construction roads at dam tow, and the lifting of $39^{\#} \sim 43^{\#}$ dam monoliths after suspending the construction of $44^{\#} \sim 46^{\#}$ dam monoliths provides conditions for continuous placing and direct placing by dump truck for $44^{\#} \sim 46^{\#}$ dam monoliths. The fact that the placing intensity of placing by dump truck is only subject to the layout and reserved number of placing entrances, and the placing intensity of placement from one single placing entrance can be controlled at about 200 m³/h, the requirements for placing area planning and construction progress can be fully met.

The above scheme is adjusted as below: for $1^{\#} \sim 11^{\#}$ non-overflow dam monoliths on the left bank and $39^{\#} \sim 50^{\#}$ non-overflow dam monoliths on the right bank, direct placing by dump truck is adopted. For $12^{\#} \sim 16^{\#}$ dam monoliths which are chute dam monolith, subject to the terrain, the construction method of transporting by box type articulated chute + truck is adopted to easily ensure placing intensity and construction quality since placing through chutes is a relatively mature process. This part is constructed from bottom to top through continuous placing, with a maximum placing area of 3 000 m². The placing intensity is required to be not less than 200 m³/h and the calculated section dimension of the chute not less than 486 mm × 486 mm. The actual section dimension of the adopted box type chute is 700 mm × 700 mm, fully meeting the requirements for placing intensity. Construction practices have proved that the construction of the non-overflow dam monoliths on the left and right banks are completed on time or ahead of schedule and the placing intensity, construction progress and quality are all ensured by adjusting the scheme.

4.2 Power house dam monolith

Original construction scheme: for the part below the elevation of 415.00 m, RCC is all directly placed by dump truck via the temporary filling placement roads for Phase II rock – earth cofferdam at the upstream. For the part above the elevation of 415.00 m, normal concrete is placed with cable crane. In this scheme, the placement method of RCC is relatively simple, however, subject to the actual terrain, when the height difference is too large, the amount of filling works for the placement roads will be so great as to affect the construction safety and construction progress. When the construction proceeds to the large corbel above the elevation of 397.00 m at the upstream, it is very difficult to conduct direct placing by dump truck due to the restriction of the corbel reinforcement. The safety and quality of construction also cannot be ensured. Affected by the layout of cable cranes and the type selection of dump truck, the placing intensity of a single cable crane is about 60 m^3/h which is used not only for the concrete placing for the power house dam monolith and the bottom orifice dam monolith, but also for the erection of the metal structures of the bottom orifice dam monolith, so that it can not fully guarantee to meet the demands of normal concrete placing.

Therefore, during the implementation phase, the construction scheme is adjusted as below: ①With in the elevation of 350 ~ 397 m, RCC is all directly placed by dump truck crossing the anti – seepage area at the upstream through the steel viaduct, via roads from lower foundation pit roads for cofferdam at the upstream to filling placement roads in the dam body. ②For the part below the elevation of 415 m, the concrete is transported into the placing surface by belt conveyor arranged in front of dam and is transported by dump truck in the placing position. ③One more M900 stationary tower crane is provided at downstream side between 18# ~ 19# dam monoliths, to assist the cable crane in concrete encasing around the penstock of the power house dam monolith and normal concrete placing above the elevation of 415 m.

4.3 Bottom orifice dam monoliths

Original construction scheme: for the part below the elevation of 357 m, RCC is directly placed by dump truck via the filling placement roads at the downstream side. Within the elevation of 357 ~ 371 m, RCC is directly placed by dump truck via roads from lower foundation pit roads for cofferdam at the upstream to filling placement roads on the dam surface. Problems of the scheme: placing from the downstream must pass the filling placement roads for the stilling basin of the bottom orifice, which will affect the normal construction of the stilling basin of the bottom orifice.

Therefore, during the implementation phase, the RCC construction scheme is adjusted as below: for the part below the elevation of 371 m, RCC is all directly placed by dump truck crossing the anti – seepage area at the upstream through the steel viaduct from the filling construction roads for cofferdam at the upstream. The adjusted scheme will not affect the construction of the stilling basin of the bottom orifice, and is conducive to the unified construction planning of the construction working face of bottom orifice dam monolith and power house dam monolith and the continuous placing. In the adjusted scheme, the placing area is divided according to the mixing capacity for employing sequence placement method, and the amount of filling works for the main construction roads is reduced to improve the rapid RCC concrete construction.

4.4 Crest orifice dam monolith

Original construction scheme: ①For the part below the elevation of 379 m, the concrete is placed by dump truck via the filling placement roads for the stilling basin of the side orifice at the downstream. ②Within the elevation of 379 ~ 423 m, the concrete is placed with 1# belt conveyor + articulated chute, among which the 1# belt conveyor with a total length of about 140 m is placed on the platform at the elevation of 402 m of the right guide wall of the open diversion channel. ③Within the elevation of 423 ~ 434.5 m, the concrete is placed with 2# belt conveyor crossing the open channel + articulated chute, among which the receiving opening of the belt conveyor is arranged on the platform at the elevation of 434 m at the upstream side of 44# dam monolith on the right bank while the feed opening is arranged at the elevation of 439 m of 36# dam monolith. The bottom of the belt conveyor is supported by concrete buttress (409 ~ 439 m) and the total length of the belt conveyor is 170 m. ④ For the part below the elevation of 410 m, the concrete for the overflow face is placed by 1 C7050 tower crane and above the elevation of 410 m, the concrete for the overflow face and the pier is placed with 2 pumps. Problems of the scheme: for the part below the elevation of 379 m, the filling placement roads for the stilling basin will occupy the construction working face of the stilling basin of the crest orifice and the road filling volume will be relatively large; theoretically, the conveying capability of the provided belt conveyor is about 240 m^3/h, however, considering the impact of the unloading of dump truck, slope, length and other factors, the actual conveying capability is less than 200 m^3/h. The maximum placing area of continuous placing according to the planning is 5 000 m^2, so the conveying capability can not meet the demands of construction. And there are no backup plans. If the equipment fails, the normal construction will be affected. The receiving opening of 1# belt conveyor with a total length of 140 m is arranged on the platform at the elevation of 402 m on the right wall of the open diversion channel at the downstream of the dam body. The chute is 50 m long. 2# belt conveyor with a total length of 170 mm crosses the open channel via the tower type chain bridge. The cost is high but the safety factor is low. Meanwhile, the belt conveyors will affect the normal construction of concrete above the elevation of 409 m of 36# dam monolith. When concrete pumps are adopted to place concrete for the overflow face and the pier, the maximum length of the pump pipe is up to 200 m. The pipes are easy to be plugged but difficult to be cleaned. And the hydration heat generated during the pumping of concrete may easily lead to quality problems such as concrete cracking.

The adjusted scheme: ①Stage I, the concrete is directly placed by dump truck crossing the anti-seepage area through the steel viaduct via roads from lower foundation pit roads for rock-earth cofferdam at the upstream to filling placement roads on the dam surface and constructed to the elevation of 409 m from the foundation. ②Stage II, the concrete is directly placed by dump truck crossing the transverse joint via the filling construction roads on the right guide wall and crest orifice gate chamber dam monolith until the elevation of 420 m. ③Stage III, for the part above the elevation of 420 m, the construction method of transporting materials by belt conveyor crossing the river + dump truck is adopted since the dump truck can't place concrete directly. The belt conveyor crossing the river is 78.9 m long and crosses the open diversion channel. The receiving point of the belt conveyor is arranged at 39# non – overflow dam monolith while the discharge opening arranged at 36# dam monolith. The RCC construction ends when it reaches the elevation of 434.5 m. ④Two more D1100 tower cranes will be provided, with one arranged at 28# dam monolith and one at 33# dam monolith, for the normal concrete placing for the pier of crest orifice dam monolith and the overflow face above the elevation of 410 m. The adjustment of scheme effectively solves the problem of occupying the working face of the stilling basin of the crest orifice. Placing by dump truck ensures the placing intensity of RCC and meanwhile guarantees the construction progress and construction quality of the crest orifice dam monolith.

4.5 Construction scheme for upstream lock head

Original construction scheme: ①Stage I: below the elevation of 387 m, the concrete is placed by dump truck via the filling roads in the open channel after Phase III Closure; within the elevation of 387 ~ 410 m, RCC is directly placed by dump truck via the berm at the elevation of 387 m and the construction roads with a gradient not greater than 10% ②Stage II: for the part above the elevation of 410 m, the construction method of transporting materials by belt conveyor + dump truck is adopted since the dump truck can't place concrete. The receiving opening of the 140m – long belt conveyor is arranged on the 2 – 2# road at the elevation of 435 m. The dam body is constructed from the elevation of 410 m to the elevation of 427 m. For the part above the elevation of 427 m, the normal concrete is placed with cable crane. Problems of the scheme: ①If the filling road is constructed from the berm at the elevation of 387m to the elevation of 410 m, the filling volume will be too much; and due to the width of the berm at the elevation of 387m, the road will definitely occupy the working face of the ship chamber of the navigation project Due to the road slope, the safety factor is low and the potential safety hazards are great. ② For the part above the elevation of 410m, the concrete is placed with belt conveyor, however, the belt conveyor is so long that safety is not ensured and the failure rate is relatively high, which will affect the construction progress and construction quality and can't guarantee to meet the demands of high – intensity continuous RCC placing for upstream Lock Head. ③January to May 2013 is the peak period for the installation of crest orifice pier and radial gate. The utilization rate is low since the cable crane can not fully meet the demands of normal concrete placing above the elevation of 427 m.

Therefore, the construction scheme is adjusted as below: ①For upstream lock head and 38# dam monolith within the elevation of 370 ~ 380 m, the concrete is directly placed by dump truck via the filling construction roads in the open channel at the dam toe. ②Within the elevation of 380 ~ 392 m, the RCC or the pre – cast concrete is directly placed by dump truck via the steel viaduct erected on the berm over the open channel at the elevation of 387 m and the reserved placing slope roads. Within the elevation of 392 ~ 427 m, the concrete is transported by vertical vacuum chute + dump truck. The diameter of the vacuum chute is 800 mm/600 mm and the placing intensity is 500~800 m³/h. The adjustment reduces the road filling volume, further improves the safety factor, and does not affect the construction of the ship chamber, and ensures the concrete placing intensity, creating favorable conditions for the continuous placing for and the lifting of the upstream lock head. The concrete placing height of the upstream lock head continuously increases by 47 m within two months. For the part above the elevation of 427 m, normal concrete is placed with cable crane + D1100 tower crane + HBT80C concrete pump. And a series of safeguard measures have also been worked out to ensure the rapid concrete placing for upstream lock head, which is successfully constructed to the required elevation of 454 m in June 2013, meeting the requirements for lowering gate for water storage of the project.

5 Interlayer bonding control

Roller compacted concrete is under thin – base – layer continuous construction, and interlayer bonding quality is of vital importance to roller compacted concrete construction, since the quality is in relation to permeation resistance, anti – sliding stability and overall performance of dam concrete, and is the key to rapid dam construction technology. The interlayer bonding quality of roller compacted concrete in roller compacted concrete construction for Tingzikou Hydroproject is improved mainly through adequate construction preparation, interlayer interval control, VC dynamic control, human assistance and other measures, with main methods as below.

5.1 Adequate construction preparation

Adequate construction preparation is the premise of ensuring rapid construction of roller compacted concrete. Before roller compacted concrete construction, detailed concrete placement surface design and construction plans are prepared. Layer boundary lines and concrete section lines are clearly marked with red paint in each section. Work tasks and work areas of each construction team and equipment operator are identified; possible problems during construction are fully estimated, and plans against the problems have been developed in advance; concrete production and transportation system, construction equipment after concrete placement (vibrating roller, leveling machine, joint sawing machine, concrete vibrator, nuclear density meter, VC measuring instrument, etc.) and spraying facilities in high – temperature period are kept in sound operating conditions. Construction tools and appliances, thermal insulation, moisturizing and rain – proof materials are all ready and under inspection one by one to ensure completeness of tools and appliances, rate of equipment usability and dependability. For pre – cooling concrete production, pre – cooling system is started 4 h in advance to provide full cooling for aggregates. Inspection is conducted to check whether aggregates and mixing water in primary and secondary air cooling silos have dropped to preset temperature.

5.2 Timely placing, spreading and rolling and interlayer interval control

The placing interval between upper and lower layers of roller compacted concrete is in inverse proportion to interlayer bonding quality and interlayer cohesion of continuous placing layers. Initial setting time for the roller compacted concrete of Tingzikou Dam generally ranges from 6 h to

8 h, and is properly extended to 8 to 10 h in high-temperature season through additive formula and concrete mix proportion adjustment. Interlayer interval is controlled to be 1 to 2 h shorter than initial setting time to ensure bonding effect and interlayer cohesion of upper and lower layers of roller compacted concrete.

Roller compacted concrete of the dam is under large-placement surface thin-layer continuous or intermittent paving and under construction by leveling course method, paving thickness is controlled at 35 cm for uncompacted layer, and base layer thickness upon compaction is about 30 cm. According to placing position, strip paving sequence, influence to cooling water pipe, etc. reversing placement method or bank-off advancing placement method is adopted for superimposed dumping in turn in quincunx shape. Placing, spreading and rolling are conducted in time after dumping to ensure that lower concrete layer can be covered within allowed interlayer interval. Since natural aggregates are adopted for Tingzikou Hydroproject, the aggregates are smooth in appearance and small in specific surface area, and are easy to show white color in high-temperature or windy seasons. As the strip which does not meet conditions for immediate rolling upon placing and spreading, the whole strip is under static rolling for two times for water retention at first to prevent aggregate whitening and quick water loss, and to realize the thermal insulation effect of mitigating temperature rebound.

Cooling water pipes are in staggered layout: on the basis of production capacity of mixing system, roller compacted concrete of the dam is generally under multi-dam monolith integrated continuous construction. Staggered layout is adopted in construction to avoid the influence of cooling water pipe laying on concrete placing, spreading and rolling; each dam monolith is taken as a cooling water pipe access unit, and priority is given to paving for the cooling water pipes in lower layer of earlier rolled dam monolith according to specific interlayer placing interval. Feeding, placing, spreading and rolling will be conducted after pipe laying to avoid influence of cooling water pipe laying to the greatest extent.

5.3 Interlayer construction treatment

(1) Surface roughening is conducted for concrete construction joint, with control standard of exposed coarse aggregate and slightly exposed small aggregate; joint surface washing and cleaning are conducted before placing so as to eliminate water, slag and other sundries from section surface; a layer of 1.5 cm thick mortar layer with higher strength is paved uniformly, and new concrete is covered immediately after placing to prevent influence on interlayer bonding quality due to weak interlayer caused by water loss of mortar.

(2) It is strictly prohibited to spray or pour water during normal rolling. It is allowed to conduct proper spraying for moisturizing when concrete is dry due to water loss, thus avoiding poor compactability due to too fast VC loss and further mitigate influence on interlayer bonding quality. Rolling for roller compacted concrete is required to be completed within 2 h from feeding; hot lift (the layer without initial set in interlayer interval) is generally unnecessary for treatment; in case of influence of high temperature and rain, cementing material, neat paste or mortar is determined to be paved on the layer as the case may be. Initial set is generally not allowed, in case of small

area of initial set, mortar, cementing material or neat paste is paved for treatment.

(3) During roller compacted concrete construction, driving route and speed of transportation trucks are planned according to actual placing conditions to avoid contamination to construction layer and harm to interlayer bonding quality due to peeling by repeated vehicle wheel pressing, etc.

5.4 VC value dynamic control

Under determined construction process and environmental conditions, VC value of roller compacted concrete mixture is an important factor that affects construction efficiency and quality, and refers to an important indicator reflecting performance of roller compacted concrete mixture and the key to assurance of the compactability and interlayer bonding quality of roller compacted concrete. An optimal VC value can improve pore structure of concrete mixture, reduce friction among aggregates, facilitate exciting force transfer, and limit aggregate separation and honeycomb to a small range; and roller compacted concrete will feature sound compactability and will be easy to meet designed volume weight, thus improving overall shear strength of concrete, realizing good layer bleeding effects and facilitate interlayer bonding. Therefore, VC value greatly affects performance of roller compacted concrete. VC value dynamic control is a necessary and important condition for compaction for roller compacted concrete, and is the key to assurance of compactability and interlayer bonding of roller compacted concrete. In the roller compacted concrete construction for Tingzikou Dam, VC value is adjusted at any time according to different construction seasons, periods, temperature, humidity, wind forces and directions and change of other surrounding environment, and it is better to range from 3 to 5 s in general period. During high – temperature and windy period, VC value is adjusted to 1 to 3 s and is minimized to the greatest extent with the criterion of no rolling collapse on construction site. In addition, additive dosage is increased on the premise of unchanged mix proportion parameters to lower VC value under water reducing and set retarding effects of additive and further postpone initial setting time, thus to ensure placing compactability and bleeding effects.

It has been proved through construction practice that dynamic control of VC of roller compacted concrete on the basis of measured air temperature, relative humidity, rolling suspension time, precipitation conditions, wind force, wind direction and relevant data effectively improves compactability, liquefied bleeding and interlayer bonding quality of roller compacted concrete.

5.5 Human assistance construction

During dumping, spreading and leveling, coarse aggregates will always be separated and gathered around stock pile and the edge of placing strip and so cause aggregate honeycomb at interlayer bonding position and affect aggregate filling and packing effects by cementing material. It is one of the major factors causing seepage and interlayer bonding quality defects of roller compacted concrete. Concrete in rolling layer is under control by specified rolling parameters and under timely test, and whether to increase rolling times is determined according to concrete surface bleeding conditions and test results of nuclear density meter; in regard of poor bleeding, bug hole, honeycomb and other conditions in rolling layer, human assistance is adopted to timely eliminate the defects and add mortar and perform repair rolling, thus to ensure concrete interlayer bonding quality.

To ensure liquefied bleeding effects of roller compacted concrete, a worker is allocated for each vibrating roller during rolling by vibrating roller; aggregate distribution or fine aggregate filling is conducted for the positions with bug hole or poor bleeding effects, and proper cementing materials and net paste are sprayed on the position with bug hole before repair rolling, and repair rolling will be conducted till surface bleeding, realization of certain elasticity and the standard for completing concrete rolling is met. After rolling of each layer, compaction measurement is conducted with nuclear density meter by 1 point/100 m^2; repair rolling is conducted for unqualified positions according to test results, and construction for the next base layer shall not be turned to before all the positions are up to standard through test.

6 Temperature control and crack prevention

(1) Lowering hydration heat of concrete by optimizing concrete mix proportion: main methods include adopting high – dosage flyash, improving water – reducing rate of additive, lowering water consumption and substituting sand by crusher dust; flyash dosage in roller compacted concrete is improved to 60%, water – reducing rate of additive is improved by 25% to 30%, water consumption for Gra. 3 and Gra. 4 roller compacted concrete is respectively lowered to 75 kg/m^3 and 90 kg/m^3, and 16% ~ 18% crusher dust is additionally mixed to substitute sand; in this way, temperature rise by hydration heat is effectively lowered, and concrete performance is improved.

(2) Placing temperature and pouring temperature of roller compacted concrete are important factors that affect internal temperature of dam body; the higher placing temperature of roller compacted concrete is, the higher pouring temperature will be, and the earlier time for internal temperature of dam monolith to reach maximum temperature will be, the poorer effect of lowering peak value by water cooling will be. ① Key point lies in controlling mixer outlet temperature; mixer outlet temperature is controlled within 12°C after primary and secondary air cooling for aggregates, mixing by adding cold water and ice. Special attention is paid to inspection on internal temperature of coarse aggregate to prevent too fast temperature rebound of roller compacted concrete due to failure of complete cooling for coarse aggregates; ② The key lies in temperature rebound control during transportation, placing and spreading, the number of dump truck is in reasonable allocation to reduce transportation and waiting time, thus to avoid failure of concrete dumping for a long time; sun shade is set during dump truck transportation to prevent exposure under sunshine; ③ thermal insulation materials are pasted around dump body of dump truck, and truck body is watered to lower temperature in high – temperature period, thus mitigating increase of roller compacted concrete temperature due to too high temperature of dump body of dump truck;④ Spraying is conducted for placing surface to create microclimate for cooling and moisturizing, microclimate at placing surface is improved by water spraying with spraying machine and spray nozzle in dry and high – temperature period in the daytime, and temperature at placing sur-

face can be generally lowered by 4 to 6 ℃, and cooling effect is better by spraying cold water when necessary; ⑤ Thermal insulation material is covered in time to mitigate temperature rebound, after spreading, leveling and rolling of roller compacted concrete, concrete is covered by PVC coiled material to lower concrete temperature by 1 to 2 ℃.

(3) Water cooling is one of the important means to lower peak internal temperature and prevent thermal crack of dam body concrete. Cooling water pipe of Tingzikou Hydroproject is made of I. D. 25 mm HDPE tubing, with laying distance of 1.5 m × 1.5 m or 1.5 m × 2.0 m, and with total length of single pipeline not more than 250 m. Roller compacted concrete construction elevation is 3 to 6 m, and single - section placing time is 2 to 4 d. To effectively lower peak internal temperature of dam body, water cooling is applied during placing. In cooling water pipe laying layer, water cooling is conducted immediately after spreading, leveling and rolling for roller compacted concrete to mitigate adverse influence of placing temperature, delay temperature rise rate and lower peak temperature. According to design requirements, after 8 ~ 10 ℃ water cooling for each dam monolith for 15 d, water cooling shall be continued with water at ordinary temperature for 7 to 10 d to ensure that internal and external temperature of roller compacted concrete of dam body meets requirements and to effectively control the probability of thermal crack at dam body.

(4) Key point in concrete crack control lies in prevention. Most of the cracks at roller compacted concrete dam refer to surface crack. Generally, surface anti - crack steel bar is not set for roller compacted concrete dam, thus surface crack is easy to develop into deep crack in certain conditions, which affects permeation resistance indicators and overall performance of concrete dam and is hard to handle. Therefore, it is of vital importance to strengthen surface protection for concrete dam. The issue is attached importance to during construction for Tingzikou Hydroproject, and the following protective actions are taken: ① After form stripping in low - temperature season, foam thermal insulation coiled material is paved on surface of concrete within age for surface thermal insulation, and surface watering or moist curing is conducted in high - temperature season to prevent crack due to too high internal and external temperature difference of concrete; ② clay is filled for powerhouse, bottom outlet and the place below upstream elevation 352 m for protection and improving permeation resistance of roller compacted concrete at foundation; ③ cement - based permeable crystalline material is brushed for the place below upstream dead water level 438 m of the dam to form impervious coating, thus preventing permeation, maintain moisture, prevent crack, and improve concrete durability; ④ Polystyrene board is set for bottom outlet, powerhouse, crest orifice and other key positions of river bed and dam monoliths for permanent thermal insulation, thus preventing concrete crack in late period.

7 Conclusion

(1) Roller compacted concrete construction can be carried out in the way of flow operation and large - area continuous placing, to increase the concrete construction intensity, so as to maximize the service capability of supporting equipment systems and improve the equipment utilization rate and the construction mechanization degree. Rapid construction is the most prominent charac-

teristic and advantage. The road for concrete placing and placement method should be rationally planned. The method of direct placing with dump truck should be adopted as possible to reduce transfer impact, which is one of the important means to ensure rapid construction. In roller compacted concrete construction of the Tingzikou Hydroproject, the construction road arrangement and placing scheme are optimized, with the method of placing with dump truck as the main scheme and with the method of placing with belt conveyor as the supplement, giving full play to the characteristics and advantages of rapid construction of roller compacted concrete. Interlayer bonding quality control as well as temperature control and crack prevention technologies are the key technologies in roller compacted concrete construction and are the important conditions for ensuring continuous rapid construction. By taking means of mixture proportion optimization design, interlayer interval time control, and dynamic control of VC value, the problem of working performance of roller compacted concrete is effectively solved, guaranteeing the compactability of concrete, liquefied concrete bleeding effect and interlayer bonding quality and achieving the construction progress goal and construction quality goal of roller compacted concrete.

(2) Practices have proved that the rapid construction technology got obvious application effects in roller compacted concrete construction for the dam of the Tingzikou hydroproject, fully embodying the characteristics of rapid construction of roller compacted concrete. The maximum construction intensity of roller compacted concrete per day per berth was 15 840 m^3, the maximum placing intensity was 924 m^3/h, and the maximum continuous construction layer rising height was 27 m. The monthly output of above 20×10^4 m^3 roller compacted concrete was completed in 5 successive months. In 2011, the roller compacted concrete construction output reached the peak, completing 135.3 $\times 10^4$ m^3 roller compacted concrete construction in the whole year. The overall efficiency indicator of the project is spectacular. After completion of construction, the concrete quality was checked by coring. Complete long core samples of Grade III and Grade II natural - aggregate roller compacted concrete, with the diameter of ϕ190 mm and length of 15.88 m, 18.88 mand19.95 m respectively were obtained in monoliths $20^{\#}$, $35^{\#}$ and $43^{\#}$ respectively. The Quality Inspection Center for Water Conservancy Projects of Sichuan Province authenticated that the surface of core samples was smooth and compact, the aggregates were uniformly distributed, no evident layering was found, and the completeness and continuity were good, which renewed the long core sample record of the natural - aggregate roller compacted concrete in China. This also indicated that interlayer bonding quality was well controlled in roller compacted concrete construction for the dam of the Tingzikou Hydroproject.

(3) A steel trestle is employed to span from the dam upstream face over the anti - seepage area for construction. How to solve the impact on the quality of the anti - seepage area is critical. When spanning over the anti - seepage area, the concrete ramp placed beforehand should not occupy the anti - seepage area. The width of the anti - seepage area reserved in Tingzikou hydroproject construction is about 6 m in general. The lower part of the steel trestle is placed with metamorphic concrete or normal concrete. The concrete ramp should be provided with key and dowel, and the surface should be subject to surface roughening. In this case, the problem of the impact

of placing with dump truck from the upstream face on the anti-seepage area quality is solved, and the anti-seepage effect and construction quality of the anti-seepage area are guaranteed.

(4) Rapid construction of roller compacted concrete still has much worthy of exploration and research. In addition to selection of an appropriate placing road and placement method to increase placing intensity, interlayer bonding quality control as well as temperature control and crack prevention technologies, which guarantees the construction continuity and construction quality, adapting hydroproject layout and structure design to the construction characteristics of roller compacted concrete, increasing the layer thickness of roller compacted concrete, and reducing the impact of the concrete temperature control link and turnover formwork design on rapid construction also worth every colleague's concern.

Discussion on Technological Development Direction of Grout Enriched Vibrated RCC

Zhang Zhenyu

(Gezhouba Group Testing company, Yichang, 443002, China)

Abstract: In this paper, development process of Grout Enriched Vibrated RCC(GEVR) is summarized, and current construction situation is analyzed, and current problems are mentioned. Finally four proposals are put forward, which are improving the performance of RCC, improving grout performance, developing integrated equipments for grout adding and concrete vibrating, and improve methods and standards of construction.

Key words: Grout Enriched Vibrated RCC (GEVR), construction technology, current situation, problem, proposal

1 Introduction

Grout enriched Vibrated roller compacted concrete(GEVR) refers toa type of concrete with characteristics of conventional concrete by adding an appropriate amount of cement grout (generally between 4% to 7% of the total GEVR) in a RCC mixtureand then compacting with plug-in vibrators, which is mainly used in face area of a dam. GEVR was first used in Yantan Hydropower Station in Guangxi provincein 1980s and after nearly 30 years of continuous development, construction technology matures gradually. In China this technology has completely replaced the conventional impervious structure for RCC dam, in which conventional concrete is placed concurrently with RCC. GEVR provides a solution of poor cementation and poor compaction between two different concretes, and ensures compaction and smoothness of concrete surface, and simplifies field management, and speeds up the construction progress, and gives full play to the speedy construction process of thin continuous and thin-lift placement for RCC.

2 Current construction technology of GEVR

2.1 Grout mix design

GEVR is formed by adding grout with good flow in field spread RCC material, which has the properties such as vibrability, adhesion, high strength, impermeability, crack-resistance etc., therefore groutis one of the most crucial factors to guarantee concrete quality. Currentstandards only require that raw materials for grout mixtures should be the same as that for RCC, and water-cement ratio for grout be not greater than that for RCC. Therefore, mix design for grout mixtures for different projects is different from each other, and little research on performance of grout mixtures has been conducted.

2.2 Grout adding method and grout dosage

At present, there are three kinds of grout adding methods including surface sprinkling, trench adding and hole adding, among which hole adding is most used. Due to different adding methods and mix proportions, grout dosage is also different, with general fluctuation ranging between 4% and 6%.

2.3 Construction method

Currently grout for GEVR is usually mixed in mixing station outside the field, and then is delivered to site through pipeline, and finally is added into RCC with help of small loaders or other containers. GEVR is usually compacted by manual vibrators. Some manufactures produced some hole – making and grout – metering device to control grout adding, however due to low work efficiency the overall effect is not satisfied.

3 Existing problems for construction technology of GEVR

3.1 Performance requirements for grout mixtures is not clear

Raw material of grout mixtures is one of the most crucial factors to guarantee concrete quality. However currently fewer specifications are available for grout material and mix design, which cause no unified specification to guide the mix design and poor performance of grout and GEVR. Meanwhile, recognition of impervious ability and other abilities of GEVR is also affected, thereby restricting the promotion of GEVR. Grout mix proportions of GEVR for some Chinese projects are shown in Table 1.

Table 1 Statistics Table of Grout Mix Proportions of GEVR for Some Chinese Projects

No.	Project	Design Requirement for RCC	Parameters of Grout Mix Design				Grout Density (kg/m^3)
			Water–binder Ratio	Fly Ash (%)	Water Reducing Agent (%)	Air Entraining Agent (%)	
1	Mianhuatan	R180200	0.50	55	0.6	—	1 567
2	Longshou	C9020F300W8	0.45	40	0.7	—	1 598
3	Linhekou	R90200D50S8	0.51	50	0.7	0.007	1 755
4	Baise	R90200D50S10	0.50	58	0.6	0.03	1 656
5	ZHaolaihe	C9020F150W8	0.48	50	0.6	—	1 613
6	Auxiliary Dam of Yixing	C9020F100W8	0.45	50	0.6	—	1 634
7	Longtan	C9025F150W12	0.40	50(Grade I)	0.4	—	1 744

Table 1

No.	Project	Design Requirement for RCC	Parameters of Grout Mix Design				Grout Density (kg/m³)
			Water-binder Ratio	Fly Ash (%)	Water Reducing Agent (%)	Air Entraining Agent (%)	
8	Guangzhao	C9025F150W12	0.45	50	0.7	—	1 693
9		C9020F100W10	0.50	55	0.7	—	1 730
10	Jinanqiao	C9020F100W10	0.52	50	0.5	—	1 683
11	Gelantan	C9020F100W8	0.43	40(Slag)	0.8	—	1 805
12	Gongguoqiao	C18020F100W10	0.46	40	0.7	—	1 780
13	Guandi	C9025F100W6	0.45	50	0.7	—	1 618
14		C9020F50W6	0.48	50	0.7	—	1 549
15	Lianhuatai	C18020F200W6	0.47	60	0.6	—	1 572
16	Xiangjiaba	C18025F150W10	0.42	50(Grade I)	0.4	—	1 685

3.2 Poor performance of GEVR

Currently, only cement and fly ash are used as cementitious materials for RCC, and since internal temperature rise needs to be controlled, strength grade of RCC is low, generally not greater than C20. Poor performance of RCC and grout results in limited room for improvement of performance of GEVR, and in order to ensure impermeability of the dam, it is necessary to increase the strength grade by one level. In the case of high dam, thickness of GEVR has to be increased to ensure impermeability of the dam, which results in more workload and higher risk of cracking. In some cases, despite strength grade of GEVR reaching C25, dosage of fly ash was basically below 50%, and during construction cooling measures such as water cooling pipes had to be taken to reduce temperature rise of concrete, which deviated from the original intention of adopting RCC and GEVR.

3.3 Manual operation can not guarantee accuracy and uniformity of grout-adding

Current manual placement can not guarantee accuracy and uniformity of grout-adding, and vibration quality, which results in poor quality of GEVR and seepage of the dam. The promotion of GEVR is restricted seriously. In order to ensure the uniformity, batching plants are used for mix the GEVR in some projects, and then GEVR is transported to site and compacted manually. However it can not solve the problems of poor cementation and poor compaction between two different concretes.

4 Discussion on technological development direction of GEVR

4.1 Improving the performance of RCC and reducing the types of grout mixtures

Improve the performance of RCC by using new raw materials and design method and avoid simply relying on GEVR to guarantee impermeability and durability of the dam. Existing research shows that compound use of different kinds or different fineness of mineral admixturecan make use of their different characteristics. And compound use also can generate superimposed effect, and greatly increase the dosage of mineral admixture and reduce costs. Therefore, the pattern of single cementitious material system of cement with fly ash should be changed, and compound cementitious material system, which can improve the performance of RCC effectively, should be developed. Super low heat RCC with good impermeability and binding properties should be developed, and current dam structures of three graded RCC + two graded RCC + GEVR should gradually be changed to structures of three graded RCC + GEVR or two graded RCC + GEVR, so as toreduce interference in the construction process of different graded concretes, and reduce the types of grout mixtures, and greatly improve construction efficiency and construction quality of RCC.

4.2 Improving the performance of grout and reducing the thickness of GEVR

In the case of high flow, conventional cement grout is less stabilized and prone to bleeding and settlement, which is more with increasing of water cement ratio. Therefore, it is necessary to conduct modified research of cement grout so as to reduce bleeding rate, to extend the time of water loss and the time of bleeding stability, to minimize segregation and bleeding of suspension static grout, to obtain grout with high flow and good stability, and to make it easier to conduct placing, permeating and vibrating. Clear requirements should be put forward for the performance of grout mixtures, and at the same time mix design method should be standardized so as to improve the performance of GEVR, thereby reducing the thickness of GEVR and decreasing the quantity of GEVR and cracks.

4.3 Develop integrated construction equipment and improving the construction speed and quality of GEVR

Currently, grout adding and vibration of GEVR are conducted manually, and due to lack of effective pre – job training and supervision, it is hard to guarantee accuracy and uniformity of grout, and vibration quality of GEVR, thereby resulting in seepage. Also due to the poor efficiency of manual operation, GEVR construction often influence and restrict speedy placement of RCC, thereby seriously restricting the promotion of RCC and GEVR. Therefore, it is urgent to develop integrated construction machinery of GEVR, which can measure the dosage of grout accurately, and guarantee uniformity of grout adding and uniform vibration. At present, some manufacturers have produced some equipment, however for some reasons they has not been widely used.

4.4 Improving relevant standards and specifications and ensure the quality of all aspects under control

Currently fewer specifications are available for grout material and mix design, thereby causing no unified specification to guide the mix design. Therefore detailed mix design method of grout mixtures should be developed so as to specify performance requirements of grout, and standardize indoor mixing method of GEVR, and supplement and improve existing quality standards and construction method, and ensure the quality of all aspects under control.

5 Conclusions

Since GEVR was first used in Yantan Hydropower Station in Guangxi provincein 1980s, after nearly 30 years of continuous development, construction technology of GEVR has made considerable progress. However due to the complexity of construction technology and continuous development of people's understanding on things, many technical problems still need to be optimized and improved, and some aspects such as material selection, mix design, construction equipment development and standards establishment still need to be improved.

Development and Application of Digital Huangdeng Dam's Construction Management Information System

Xiang Hong[1], Yang Mei[1], Zheng Aiwu[2], Gong Yongsheng[2], Deng Yongjun[2]

(1. Power China Kunming Engineering Corporation Limited, Kunming, 650051, China;
2. Huangdeng Lancang River Hydropower Co. Ltd. Kunming, 650214, China)

Abstract: In order to effectively solve the problems like dynamic quality monitoring, intelligent temperature control, dynamic adjustment and control of the construction schedule, comprehensive integration and efficient management of the construction information, remote, mobile, real-time, convenient construction management and control in the course of the Huangdeng dam construction, "Digital Huangdeng Dam's Construction Management Information System" was put forward which was charged by owner, under the technical consulting service of the designing institution and domestic research units. And the technologies like computer, wireless network, handheld data-capture, data sensor (internet of things) and database were used synthetically to develop a intelligent control and management information system for dam construction quality based on the WINDOWS platform, then comprehensive quality monitoring of dam concrete from raw materials, production, transportation, pouring as well as operation can be realized. Finally, the system was applied in engineering practice after the system development, field tests and trials links.

Key words: information, digitization, intelligentialize, quality control, Huangdeng Hydropower Station

1 Background and significance of system development

Huangdeng Hydropower Station is the sixth step in eight steps plan from Quzika to Miaowei on the upper reaches of the Lancang River, which is a large scale water conservancy and hydropower project has comprehensive benefits, mainly for power generation, and also for flood control, irrigation, water supply, soil and water conservation, tourism and so on. The hub is mainly composed of such buildings like RCC gravity dam, dam flood discharge and drain buildings, underground water diversion and power generation system, and the dam with maximum height of 203 m is the highest RCC gravity dam under construction in China.

The construction management, construction quality and schedule control of the Huangdeng hydropower station are fairly difficult because of the large construction scale, tight schedule and complicated construction conditions. The key technical issues to achieve high quality and high strength safety construction are as follows: How to effectively carry out the dynamic construction quality control? How to efficiently integrate and analyze the construction information in the course of the construction? How to realize the remote, mobile, real-time, convenient construction management and control?

In order to effectively solve the problems like dynamic quality monitoring, intelligent temperature control, dynamic adjustment and control of the construction schedule, comprehensive integration and efficient management of the construction information, remote, mobile, real – time, convenient construction management and control in the course of the Huangdeng dam construction, it is necessary to develop a dam construction quality automatic monitoring system which has the characteristics of real – time, continuity, automation and high accuracy, to monitor the roller – compacted and temperature control of the dam concrete effectively, establish the quality and schedule dynamic real – time control and early warning mechanism under the conditions of ensuring the inspection items specified in the specification and using technical potential, realize timely adjustment and optimization of construction plan and measures, achieve the comprehensive quality monitoring in full life – cycle of dam concrete from raw materials, production, transportation, pouring to operation, and make the dam construction quality and progress of the Huangdeng Station are always under control. In the meantime, a digital dam comprehensive information integrated system of Huangdeng Station should be set up to carry out dynamic acquisition and digital processing of information like progress and construction quality during the process of engineering design, construction and operation, build integrated digital information platform and 3D virtual model, realize various engineering information integration and data sharing, and realize dynamic update and maintenance of comprehensive information during the whole life cycle of the project, therefore to provide information application and support platform for engineering decision and management, dam safe operation and health diagnosis.

In this, "Digital Huangdeng Dam Construction Management Information System" (here after referred to as "Digital Huangdeng") has been put forward under the trend of the development of digital management and monitoring for hydropower projects. The significances of this system are to develop a unified, normative and standard scientific management system to change the extensive and traditional management mode, to achieve normalized and standardized management, to improve enterprise management, to standardize enterprise production and management activities, to achieve the safe, efficient and harmonious of construction projects by scientific and economic management mode.

2 Overall design of the system

"Digital Huangdeng" is a system for RCC gravity dam construction quality intelligent control and management informatization based on the WINDOWS platform, which can realize comprehensive quality monitoring of dam concrete from raw materials, production, transportation, pouring, as well as operation. Modularization was used in the system development, according to the needs of different functions, the system was divided into two relative independent and interrelated subsystems, by uses functional: integrated management platform and engineering information management system; intelligent monitoring and quality evaluation system for construction process. Each system is further refined into a number of subsystems, and interfaces were reserved in order to add other subsystems according to project requirements. The general structure of the system is shown in Fig. 1.

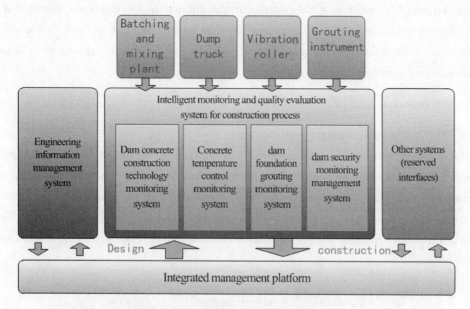

Fig. 1 The general structure of the system

2.1 Integrated management platform

The subsystems are established for different business processes, and the integrated management platform is the medium of business application systems interaction, so that these business application systems can coordinate each other to form a whole information system.

2.2 Engineering information management system

Engineering information management system can realize informatization of the information collection and management for owners, designer, supervisor, research institutions and contractors of all parties involved in the construction, provide assistance to the construction acceptance of the project and the transfer of operations by fully inheriting design achievements, managing construction process and forming complete digital archives of the project.

2.3 Intelligent monitoring and quality evaluation system of construction process

To establish a real-time intelligent monitoring system on the indicators like quality control of dam construction, the process of temperature control during the construction period, the quality control of foundation grouting, the control of stress and strain, and the early warning can be emitted to the relevant person in real time when the control index exceeds the warning index to discover the problems existing in construction site, put forward specific solution, ensure the quality of dam concrete construction. The system includes four subsystems:

(1) Dam concrete construction technology monitoring system;
(2) Dam concrete temperature control monitoring system;
(3) Dam foundation grouting monitoring system;
(4) Dam security monitoring management system.

2.4 Other systems (reserved interfaces)

Other subsystems can be added into the system according to the project requirement based on the reserved interfaces function of the system, such as runtime post evaluation system which is established based on BIM model generated after the dam construction period, and its functions will include running the monitoring equipment model in BIM, such as effective management of monitoring information, visual query and analysis, and so on.

3 System function and implementation

3.1 System integrated management platform

The system integrated management platform is the core of the system structure, which is responsible for data exchange and coordination among various subsystems, adopts unified entrances, takes navigation function as the core design of (as shown in Fig. 2), and after login, users can enter the specified subsystem to operate business function through tiled navigation. The data exchange bus can make the data sharing and business collaboration of the independent and heterogeneous subsystems.

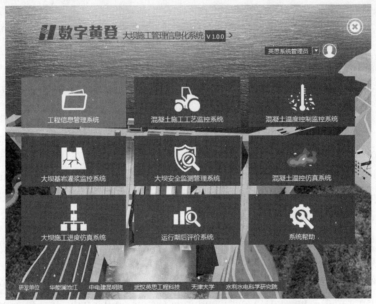

Fig. 2 The entrance of the system

3.2 Engineering information management system

The functional framework of the engineering information management system is shown in Fig. 3, including: design result management, construction schedule management, construction design management, preparation and open of deck management, test inspection management, quality assessment management, construction data management, automatic, remote, mobile, convenient management and control, and it can provide a comprehensive, fast and accurate information service and decision support for management of the dam design, construction, and operation.

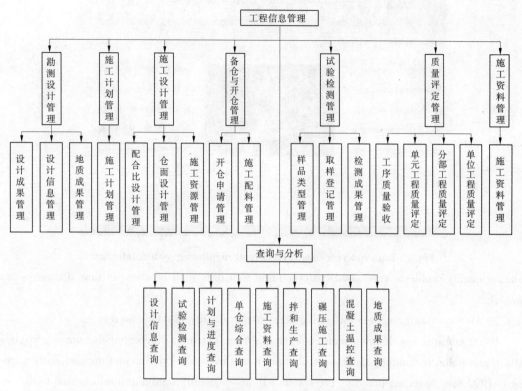

Fig. 3　Functional framework of the system

3.3　Dam concrete construction craft monitoring system

The dam concrete construction craft monitoring system can conduct effective monitoring of the links like mixing, pouring, rolling, mortar adding and so on, build dynamic real-time control and early warning mechanism of construction quality, made dam construction quality and progress is always in a controlled state; and integrate the dam roller compacted quality, the quality of mortar adding, heat lift, deck construction, dam surface detection, dam construction progress and other information to realize the visualization of construction information. Mainly including:

(1) Concrete mixing plants production data collection and analysis;

(2) GPS monitoring of dam concrete rolling construction quality;

(3) RCC heat lift monitoring;

(4) Real-time monitoring of the impervious layer mortar adding job;

(5) Construction site PDA acquisition system;

(6) Real-time simulation and control of RCC construction;

(7) Real-time acquisition and analysis of rainfall information. System interface is shown in Fig. 4.

3.3.1　Concrete mixing plants production data collection and analysis system

Using wireless data transmission technology and database technology to realize the data collection, transmission and analysis of concrete mixing plant production data, then the information like ingredients list and actual components of concrete can be query in real time when concrete

Fig. 4 Dam concrete construction craft monitoring system interface

pouring quality problems arise, and corresponding measures shall be taken in time if there is any problem.

3.3.2 GPS monitoring system for rolling construction quality of dam concrete

Using satellite – positioning technique, real – time dynamic differential technique, wireless data transmission technique, database technique to realize real – time monitoring and early warning of RCC dam in rolling process: the system will automatically send alarm information to the vehicle driver, on – site supervision and construction personnel when the rolling machine running speed and vibration state is not reached the standard, automatically and notably prompt details and spatial location of the substandard when the number of rolling passes and the thickness of the compaction are not up to standard, and put the alarm and prompts information in exception database for reference at the same time.

3.3.3 RCC heat lift monitoring system

Based on the data collected from the GPS monitoring system for dam concrete construction quality, RCC heat lift flat monitoring system can determine whether the layer of concrete was poured in the specified heat lift time through the monitoring of concrete paving and rolling process and automatic monitoring analysis of its duration; for the placement of close to the prescribed heat lift time, send prompt to the site supervision and construction management personnel in time to urge the site operator to intensify construction, which can ensure the layer of concrete is poured in the heat lift time; for the placement of exceeded heat lift time, send prompt also to remind the site operator to deal with the layer of concrete as warm lift or cold lift.

3.3.4 Real – time monitoring system for the impervious layer mortar craft

Real – time monitoring system for the upstream impervious layer mortar adding job can monitor operation of mortar adding equipment through the monitoring and analysis of metamorphic concrete mortar adding processes in upstream impervious layer, and control the construction technology and construction quality in real time by feedback mechanisms, so as to ensure the mortar adding quality of impervious layer metamorphic concrete, and to provide effective management control platform for field construction and supervision.

3.3.5 Construction site PDA acquisition system

Using PDA technology, database technology and hierarchical warning technology to automatically gather and dynamically analyze the engineering dynamic information that difficult to automatically collect, mainly including detection information of mixer outlet concrete, deck concrete, nucleon density and so on. The system automatically judges the input data, and establishes the hierarchical warning system.

3.3.6 Real – time simulation and control system of RCC construction

The system emulation technology, 3D modeling technology, database technique, cybernetics are used to achieve real – time monitoring and feedback control of the dam construction schedule by decomposition and coordination system analysis of dam construction, so as to provide technical support and analysis platform for the progress control and decision making in the process of project construction.

3.3.7 Real – time acquisition and analysis system of rainfall information

Through establishing a number of rainfall information monitoring stations in the dam area, and using wireless data transmission technique, sensor technology and database technique to collect and analyze rainfall information in real time, and to alarm rainfall value influence to the dam construction.

3.4 Dam concrete temperature control intelligent monitoring system

Function of dam concrete temperature control intelligent monitoring system: the first one is the whole process real – time automatic collection and monitoring of the temperature control elements like mixer outlet temperature, placing temperature, pouring temperature are achieved to ensure data real – time and accurate; the second one is the intelligent water cooling "without manual intervention" for the whole dam is realized to improve the quality of water cooling construction; the third one is automatic early warning based on real – time monitoring information and decision support for the intervention measures are achieved; the last one is the data sharing of temperature control construction, monitoring, assessment, early warning to provide the foundation for on – site construction management.

3.4.1 Temperature control information monitoring and collection system

To achieve the automatic and semi – automatic real – time acquisition of the temperature control elements like aggregate temperature, mixer outlet temperature, placing temperature, pouring temperature, deck microclimate, internal temperature process, temperature gradient, inlet and outlet water temperatures, and flow discharge of water cooling, air temperature and so on, special equipment (shown in Fig. 5) and software were developed, which can realize the automatic acquisition of temperature control information most possibly to ensure data real – time and accurate.

Fig. 5　Acquisition instrument of mixer outlet, placing and pouring temperature

3.4.2 Intelligent water cooling system

Various factors such as adiabatic temperature rise, temperature gradient, cooling rate and concrete thermodynamic parameters are considered to develop a parameter predicts model of intelligent concrete water cooling, which can automatically suggest flow parameters according to the requirements of cooling water, achieve "no artificial intervention" intelligent water cooling, improve the quality of water cooling construction and reduce construction error rate simultaneously. The logic principle of intelligent water cooling system is shown in Fig. 6.

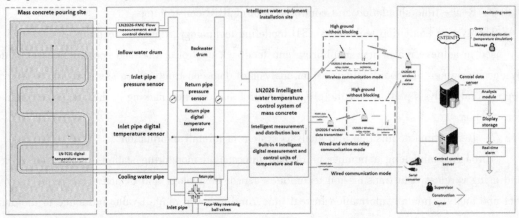

Fig. 6 The logic principle of intelligent water cooling system

3.5 Monitoring system of dam foundation grouting

Monitoring system of dam foundation grouting consists of four parts: grouting design management, construction process management, grouting results management, geophysical detection management, whose functional framework is shown in Fig. 7. The system can manage the design, scheduling, construction, acceptance and other stages of bedrock grouting construction, provide data management and the analysis of the whole process of the unit, the hole, the section, the filling time, the time history from definition of unit hole section and the construction process data, collection of quality inspection data to final acceptance of the single hole and result analysis, ensure the integrity of design, construction, inspection and other grouting data, and significantly improve the efficiency of the grouting results data collation.

3.6 Management system of dam safety monitoring

Management system of dam safety monitoring mainly manages the dam safety monitoring apparatus and data in construction period and operation period, to provide a reliable data base for the simulation calculation and inversion analysis of temperature, stress, displacement and so on by conducting the reorganization, statistics, analysis and early warning of the monitoring data, which includes 5 parts: basic definition management, monitoring data management, data reorganization management, early warinig management, query and analysis. The function framework is shown in Fig. 8.

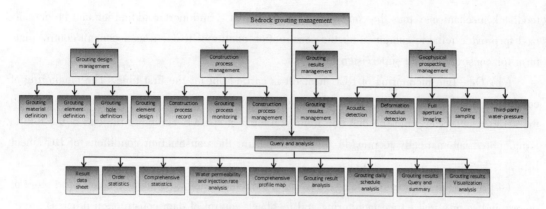

Fig. 7 The functional framework of dam foundation grouting monitoring

Fig. 8 The functional framework of dam safety monitoring management

4 Conclusion

(1) "Digital Huangdeng" put forward the idea of the whole life cycle safety management for hydropower projects, which based on dam unified BIM model, centered on structural risk assessment, and take the GIS as a collaborative platform, established 3D platform of Huangdeng RCC dam, carried out the dynamic acquisition and digital processing of the engineering progress information and construction quality information during the construction and operation of the dam, based on the "Digital Dam" integrated information dynamic management system of the B/S mode, build a comprehensive digital information platform and 3D virtual model for dam construction of Huangdeng hydropower station, realize the integrated and visual management of various engineering information under the virtual "Digital Dam" environment, and realize dynamic update and maintenance of comprehensive information during the whole life cycle of the project to provide information application and support platform for engineering decision and management, dam safe operation and health diagnosis.

(2) The on-line and real-time monitoring of the RCC rolling process and the mortar adding job of the metamorphic concrete in upstream impervious layer are realized using satellite-positioning technique, real-time and dynamic differential technique, wireless data transmission technique, database technique, real-time feedback control technique, graphical analysis technique. And the technology and quality of construction can be controlled in real time through the

feedback mechanisms, thus the construction process of RCC and mortar adding job can be normalized to provide reliable guarantee for the construction quality of RCC, a management control platform for construction and supervision in the site.

(3) On – line monitoring of RCC heat lift is carried out for the first time. The monitoring of concrete spreading and rolling process is carried out by means of the test equipment installed on the scraper and roller, and the duration from beginning of spread, scrape to the end of the rolling is monitored automatically to provide support to ensure the construction conditions of RCC heat lift.

(4) Using simulation technology, 3D modeling technology, database technology, control theory and so on, real – time monitoring and feedback control of dam construction progress are achieved through the decomposition coordination system analysis of the dam construction. And the dynamic update and maintenance of progress information are realized in the whole project construction life cycle to provide technical support and analysis platform for the progress control and decision making in the construction process of the Huangdeng Hydropower Station.

(5) Automatic collection and whole process real – time monitoring of the temperature control elements like aggregate temperature, mixer outlet temperature, placing temperature, pouring temperature, deck microclimate, internal temperature process, temperature gradient, inlet and outlet water temperature, and flow of water cooling are realized to ensure real – time and veracity of data, which can provide a direct basis for the real and comprehensive assessment of the temperature control in construction and a strong technical support for the completion inspection and acceptance of the dam.

(6) The real – time early warning and intervention of temperature control in construction are realized by the implementation of the intelligent monitoring system of temperature control, especially the implementation of intelligent water cooling can realize the personalized and intelligent water cooling without manual intervention.

(7) A real – time dynamic information monitoring system is established to monitor the implementations of dam foundation grouting. Grouting information is collected in real – time by using the grouting auto – recorder with the function of wireless data transmission, and alarm signals will be emitted under the suboptimal situation to take appropriate measures in time. Through information management and data summary of the collected grouting information, the analysis results such as grouting construction hydrograph, grouting quantity histogram and grouting schedule display drawings can be get as the acceptance materials of the foundation grouting.

(8) The dam concrete of Huangdeng Hydropower Station has be poured from March 2015, and the each subsystem of "Digital Huangdeng" has been running on the line and in normal operation, which established a dynamic data acquisition, comprehensive analysis, real – time feedback and decision support platform by comprehensive succession and management design results for the construction progress and quality of the dam, realized the comprehensive monitoring and joint scheduling of all links such as raw materials, production, transportation, pouring, curing, quality inspection and so on, carried out the whole process, all – day, real – time and online mo-

nitoring on dam construction, overcome the disadvantages of conventional quality control methods like large interference of human factors, extensive management and so on, effectively ensured and improved the quality control level and efficiency of the construction process, made the dam construction quality always be in a real controlled state, provided a new way for the quality control of concrete dam construction, and this is also the important development direction of the future promotion of hydropower construction technology and management level.

I Index and Application in Risk Monitor of Water Conservancy and Hydropower Engineering

Zhang Xiangyu

(Dongfeng Power Plant of Guizhou Wujiang Hydropower Development Co., Ltd. Qingzhen, 551408, China)

Abstract: The paper first gives a brief introduction to the risk characteristics of water conservancy and Hydropower Engineering, and the risk index recommended in Bulletin 41 of ICOLD and the deficiency analysis in itself, In order to promote the risk management of water conservancy and Hydropower Engineering, and then further introduces the *I* Index method to propose by the author based on water conservancy and hydropower engineering properties and with over 20 years' frontline field experience of the author in risk monitoring, analysis, control and research in water conservancy and hydropower engineering construction and operation. Finally, Combined with the application by the method in recent years in the "Technical specification for concrete dam safety monitoring" (DL/T 5178—5178), the "Technical specification for concrete dam safety monitoring" (SL 601—2013), and other hydropower and water conservancy industry dam safety monitoring technology specification perfect to explore. The application of *I* index method in the risk monitoring in water conservancy and hydropower engineering are discussed and summarized in this paper.

Key words: *I* Index, water conservancy, Hydropower Engineering, risk monitoring, application

1 Introduction

Currently, the risk control level varies greatly among nations. Most developing countries, whose reservoir and dam projects are still under construction, lack of systematic and in-depth research and practice in risk control of dam; whereas, most developed countries with intensive research experience are in the leading position; relevant risk control develops especially rapidly in developed counties and regions as America, Canada, Australia and Western Europe. The 81st Annual Meeting of ICOLD was held in Seattle, USA in August 2013, including an international symposium with the theme of "Changes of Times——Development and Management of Infrastructure". The author attended the meeting and heard the speeches from experts, had an access to the source of wisdom, especially the review and evaluation on main risk assessment practices in dam safety control from Mr. Zielinski, Chairman of Professional Committee of Dam Safety of ICOLD from Canada. In order to promote the development of water conservancy and hydropower engineering, the author innovates, supplements and completes the existing problems of *I* Index based on over 20 years' frontline field experience of the author in risk monitoring, analysis, control and research in water conservancy and hydropower engineering construction and operation, according to the reference country, industry and local current regulations, the risk characteristics of Water Conservancy and Hydropower projects, referenced in relevant information and literatures from the 2012 Academic Annual Conference of CHINCOLD, the 2013 Academic Annual Conference of ICOLD and the 2013 Academic Annual Conference of CHINCOLD & the 3rd International Symposium on Rockfill Dams. Proposed I index method; part of the contents of the method has been in the year June as the annual meeting of the International Commission on Large Dams 82 communication materials[1], this paper further explored.

2 Risk characteristics of water conservancy and hydropower engineering

Firstly, risks may exist in project planning, investigation, design, construction, first impound period and operation period with different characteristics. In addition, each dam differs in structure, geological conditions, operation conditions and surrounding environment, the failure probability and post - failure loss are significantly different for dams even with same engineering grade or for the same dam in different periods. Therefore, engineering risks are characterized by dynamic changes. Secondly, the reservoir - dam risks are related with traditional and non - traditional factors; the former includes equipment and facilities in respect of design, construction and installation, structure and gate control, climate, geological conditions, high slope, operation, project investment, social and economic benefit, post - failure social and economic impact, rules and regulations on operation management and personal quality, population distribution and quality in the upstream and downstream risk areas, emergency plans and disaster relief measures; the latter includes the upstream reservoir - dam failure, war and terrorist attacks. Lastly, the safety risks strictly observe the probability distribution and functions of load, structure, of geological conditions, and the risk analysis method for failure losses is thereby considered. The process is rather complicated and many losses are difficult to be quantified. Therefore, the risk analysis method based on the structure system failure and its losses is impractical.

In addition, China has built all kinds of total reservoir dam has reached over eighty - seven thousand seats, although has promulgated Other laws and regulations, such us People's public of China Production Safety Law, but the actual needs of the economy and booming construction and reservoir dam risk management compared to the risk management of the reservoir dam face there is a big gap, the need to constantly strive to improve.

3 Risk index and its shortcomings

3.1 Risk index

To Risk monitor the dams, the Risk Index method is recommended in Bulletin 41 of ICOLD. See Tab. 1 for related risk factors. In the method, the dam risks are evaluated in respect of environment, design of reservoir capacity and dam, construction, operation and downstream situations, basically considering the potential factors and losses of reservoir - dam failures.

Tab. 1 Assessment on hazardous conditions suggested in estimation table for risk index of dam

Sub-index a_i	External or environmental conditions (Index E)					State/reliability of dam (Index F)				Potential hazard to residents/economy (Index R)	
	Seismic intensity V(cm/s)	Risk of reservoir bank collapse	Risk of overdesign flood	Reservoir functions (water storage type and management)	Corrosive environment effects (climate and water)	Structural configuration	Foundation	Flood discharging facilities	Maintenance state	Reservoir capacity (10 000 m^3)	Downstream facilities
	(1)	(2)	(3)	(4)	(5)	(6)	(7)	(8)	(9)	(10)	(11)
1	$V<4$	Minimum	Small (concrete dam)	Multi-year regulating, annual regulating	Weak	Proper	Good	Reliable	Good	$W<10$	No economic value or residents
...	...	...	...	...	...	...	...	...	...	...	...
6	Large	landslide	Probable			Improper	Poor or bad	Insufficient or useless	Undesirable		

The table shows that a_i is 1 – 6 according to different degrees of the above factors. The risk indices of E, F and R are respectively calculated according to Formulas (1) – (3). And then the total risk index of structure is calculated according to Formula (4).

$$E = 1/5 \times (\sum a_i) \quad (i = 1 \sim 5) \tag{1}$$

$$F = 1/4 \times (\sum a_i) \quad (i = 6 \sim 9) \tag{2}$$

$$R = 1/2 \times (\sum a_i) \quad (i = 10 \sim 11) \tag{3}$$

$$\text{Total index } a_g = EFR \tag{4}$$

The overall risk of dam can be captured from the calculated risk index a_g from Formula (4). Graded evaluation is thereby carried out on dam risk. Then monitoring items and number of measurements are selected according to risk grades.

3.2 Shortcomings

Firstly, in the above-mentioned Risk Index method, the arithmetic average method is adopted to calculate E, F and R, which cannot correctly reflect the characteristic that the more dangerous sources and potential hazards exist, the greater risks will be. Furthermore, the interaction and function of uncertainty of each risk factor in each stage are not taken into account. Secondly, as mentioned above, engineering risks are characterized by dynamic change. the Risk Index method fails to fully reflect the inherent characteristic of project risks through identification of risk factors (E and F in Tab. 1) and potential hazards. Finally, the risk factors (E and F in Tab. 1) and potential hazards cannot be comprehensively identified via the Risk Index method, and the effect on risks in reservoir, dam and hydropower station due to risk monitoring, construction and control of emergency rescue system has not been considered. Meanwhile, the six-degree classification for sub – index a_i is still not scientific. Considering the inadequacy in calculation and identification of risk factors and potential hazards, the method shall be further improved.

4 *I* Index Method

As noted above, the Risk Index method shall be further improved in respect of calculation and identification of dangerous source (including indices *E* and *F*) and potential hazards (such as power grid accident). Therefore, the method shall be innovated, supplemented and complemented based on relevant risk characteristics and named *I* index method. See Tab. 2 for risk factors considered in *I* Index method. In addition to original factors, the engineering risks are evaluated in terms of planning, environment reconstruction, risk monitoring and power grid accidents, From the water, land and air – round, scientifically considering the potential factors and losses of reservoir – dam failure, Shipping, bamboo and log rafting and other water conservancy and hydropower engineering wreck.

a_i is $1-q$ according to different degrees of the factors in Tab. 2. The risk indices E, S, H, M and R are respectively calculated according to Formulas (5) – (9). And then the total risk index of structures safety is calculated according to Formula (10).

$$E = K_e a_i^{max} \times \left(\prod (1 + a_i / \sum a_i) \right) \ (i = 1 \sim m, \ a_i \text{ is sorted in descending order}) \tag{5}$$

$$S = K_s a_i^{max} \times \left(\prod (1 + a_i / \sum a_i) \right) \ (i = m+1 \sim n, \ a_i \text{ is sorted in descending order}) \tag{6}$$

$$H = K_h a_i^{max} \times \left(\prod (1 + a_i / \sum a_i) \right) \ (i = n+1 \sim o, \ a_i \text{ is sorted in descending order}) \tag{7}$$

$$M = K_m a_i^{max} \times \left(\prod (1 + a_i / \sum a_i) \right) \ (i = o+1 \sim p, \ a_i \text{ is sorted in descending order}) \tag{8}$$

$$R = K_r a_i^{max} \times \left(\prod (1 + a_i / \sum a_i) \right) \ (i = p+1 \sim q, \ a_i \text{ is sorted in descending order}) \tag{9}$$

$$\text{Total } I \text{ index } a_g = ESHMR \tag{10}$$

Where, K_e, K_s, K_h, K_m and K_r are corresponding influence coefficients of E, S, H, M and R determined according to the influence and effect of each factor on the sub – index; the overall risk of engineering can be captured from the calculated *I* index a_g from Formula (10). Graded evaluation is thereby carried out on engineering risk. Then Scheme safety monitoring items and number of measurements are selected based on risk grades. Limited by space, no more details will be presented here. See Discuss on Refernces[7].

4.1 Risk monitoring

4.1.1 Number of measurements

In term of time, dam failure mainly occurs in first impound period and aging period featuring higher structural risk. For the former, considering the first operation of the new project and people's poor understanding of security situation of actual structure, the risk of structures and reservoir bank will be increased; for the latter, the risks will change with material aging, operation environment and conditions and changes of population, resources and environment in risk areas. For details, see Discuss on Current Safety Monitoring Standard and Risk Control Based on Sustainable Development (to be published).

4.1.2 Reliability of layout of measuring point

Current general specifications provide little about layout of measuring point, except "spacing" and "important part". Therefore, related risk analysis theory shall be used to prevent omission of important safety information and unnecessary repetition. To obtain the safety information of the whole structure through limited measuring points, the measuring points shall be characterized by typicalness, exhibition of maximum risk and sensitivity. In addition, there is no need to evenly arrange the measuring points, instead, it shall be denser for monitored physical quantities with large gradient and high risk positions and sparse for monitored physical quantities with small spatial gradient and low risk positions; for typical and sensitive positions, especially the position with maximum risks, redundancy and cross - check shall be considered for layout of measuring points so as to improve the reliability of risk monitoring system. Moreover, the functionality and compatibility shall also be considered.

4.1.3 Design and implementation of automation of monitoring system

As mentioned above, the dam failure mainly occurs in first impound period and aging period; in term of implementation time, the automation falls into automation for new dam and automation renovation for old dam. For the former, more automatic monitoring items and points shall be arranged for high - risk new dams. For immature items and points (mostly due to unreliable monitoring instrument), the automation may be delayed; for the latter, the arrangement of monitoring items and layout of monitoring points can be optimized through structure risk assessment by analyzing artificially observed data, etc.. The automatic monitoring items and points can be reduced for low - risk structure, and increased for high - risk dam in addition to necessary reinforcement measures. The automatic telemetering station shall be located in the place without shake or vibration from dam and volute, to prevent failure of timely collecting corresponding data during shake or vibration.

5 Conclusion

(1) In this paper, the International Commission on Large Dams Section 41 Proceedings of the shortcomings of the risk index pushed discussed, and in accordance with the principles of electrical engineering reliability, combined the risk characteristics of Water Conservancy and Hydropower projects, puts forward the I index method.

(2) Due to space limitations, this article only briefly introduced the I index, are not discussed in detail and did not give examples of the use of this method, the authors will be further described in subsequent related publications.

Tab. 2 Assessment on Hazardous Conditions Suggested in Estimation Table for Risk Index of Dam

Sub-index a_i	External or environmental conditions (Index E)								State/reliability of reservoir, dam and hydropower station (Index S)						Potential hazards (Index H)				Risk monitoring (Index M)		Emergency rescue (Index R)	
	Waters Risk	Land risk	Airspace Risk	Leaking water risks	Seepage Control System mation	Key layout	Design conditions for survey and planning	Reservoir and dam break	Inlet and outlet cable and power grid structure	Configuration and quality of structure	Foundation	Flood discharging facilities	Maintenance state	Prevention against vandalism	Relative capacity, black-start	Reservoir capacity, W	Risk areas	Power grid accident	External conditions	Reservoir dam and hydropower station	Potential hazards	System construction	Plan, drill
1	1	2	3	4	5	6	7	8	m	$m+1$	$m+2$	$m+3$	$m+4$	$m+5$	n	$n+1$	$n+2$	o	$o+1$	$o+2$	p	$p+1$	q
…	…	…	…	…	…	…	…	…	…	…	…	…	…	…	…	…	…	…	…	…	…	…	…
	Measures reliable, minimal risk	Measures reliable, minimal risk	Measures reliable, minimal risk	minimal risk	Very weak	No hidden dangers	Reliable and unchanged	Low risk	Reasonable and reliable, with Proper conditions, no need of black-start		Good	Reliable	Good	Strong	<1%, strong black-start capability	$W <$ 100,000 m³	No risk in economic value/person	Not splitting	Complete and reliable	Complete and reliable	Complete and reliable		Reliable and practical plan
…	…	…	…	…	…	…	…	…	…	…	…	…	…	…	…	…	…	…	…	…	…	…	…
9	Unreliable, the great risk	the great risk	the great risk	the great risk	Strong	Change of main structure	Worse with high risk	High risk	Black-start will lead to over pressure of generator terminal	Improper or abnormal condition	Bad or big reversible deformation	Useless	Big reversible deformation of block	Poor	<3.5%, without black-start capability	$W >$ 4,000,000 m³	Urban cascade reservoir group	Transnational blackouts	Reliable personnel and equipment	Unreliable personnel and equipment	Unreliable personnel and equipment	The system is not constructed or becomes a mere formality	No plan or poor feasibility without drilling

Note: the values for risk control and emergency rescue in the table are probable values and the actual calculated values will be 1/9 of the probable values; $\alpha 8$ is 0 when there are no reservoirs and dams upstream.

(3) I index in addition to be used in Water Conservancy and Hydropower Engineering, industrial and civil architecture engineering civil engineering risk research, evaluation and scheme optimization, study and evaluate the risk can also beused in other fields; I index method not only can be used forrisk assessment has been building dam operation, also can be used in theoptimization of water conservancy and hydropower engineering construction research, assessment and program; in addition, the current risk design in related fields in domestic has not yet commenced, this paper is a new attempt.

References

[1] Zhang Xiangyu - I index innovation and application in risk control of water conservancy and hydropower engineering[C]. International symposium on dams ina global environment al challenges. Indonesia. 2014.

[2] GB／T28001—2001 Occupational Health and Safety Management System Specification [S], Beijing: China Standard Press, 2001.

[3] Zhang Xiangyu, Ying Futian. The concept of to improve current specification and risk precontrol be based on the sustainable development [J]. Guizhou Electric Power Technology, 2014.

[4] Zhang Xiangyu. Hagen risk index and its application in the study of theinnovation scheme in Hydropower Station Dam treatment[A]. Chinese Dam Association Series——Technical Progress rockfill dam construction and hydropower development [C]. Zhengzhou: Yellow River Conservancy Press, 2013. 526 – 535.

[5] Jia Jinsheng, Zhang Jiyao, Ma Qihong, etc. . The 2006 International Symposium on hydropower sets [C]. Beijing: China Water Conservancy and Hydropower Press, 2006. 326 – 328.

Quality Management by the Owner in Hydropower Project

Zhang Jinshui, Chen Meng, Wu Guangqing

(Xiaolangdi multi-purpose dam project management center of
the Ministry of water resources, Zhengzhou, 45000, China)

Abstract: The large hydropower stations belong to the major infrastructure of the country. In the course of development and construction management, the series of management system has been widely implemented including the owner's responsibility system, the bidding system, the construction supervision system and the project contract management system. In this situation, the owner faces a new topic on the quality management of the whole project, which will be discussed in this article. Before the project begins to construct, the owner should do a good job about quality management plan including to determine the quality of management objectives, to develop quality policy, to establish quality management agencies, etc. After the project begins, the contents of the quality plan should be implemented well. The author thinks that the quality management by the owner is mainly to form a set of mechanism and system, that is, all stuff participation in quality management, all stuff quality awareness, scientific management through the whole process, the concept of creating high quality engineering, forming the system of the continuous improvement of quality.

Key words: Hydropower project, project, owner, quality management

1 Necessity and importance to strengthen quality management by the owner

1.1 Importance of engineering quality

Investments of the large hydropower stations are huge, the projects are various, the construction processes are complex. The quality problems will cause incalculable damages to the country and the society. As the sponsor of the power station construction management, to ensure the engineering standard to production is the ultimate goal of power station during the construction management, and the quality objective is one of the most important factors. Therefore, the quality management in the construction and management by the owner, become an important content of project management.

1.2 The change of quality management system

Under the planned economy system, the financial allocation is implemented in capital during the construction of large hydropower stations. Construction and implementation of project construction headquarters set up by the state, the headquarters will be entirely responsible for the hydropower station design, construction and funds. In the quality management, the headquarters specially set up the quality management department who directly exercises their quality management responsibilities. Since the construction of innovation system of Lubuge Hydropower Station, large – scale infrastructure construction, including the construction of hydropower stations in the country has generally implemented responsibility system by the owner, bidding system, construction supervision and project contract management system. Owners overall responsibility for the project, quality control mainly rely on supervision unit management in the implementation of specific projects, by the supervision unit of the project quality management. The quality management of the project is carried out through the supervision unit. The changes of the system of engineering construction require that the owner quality management mode will be changed and adjusted to meet the needs of quality management in the new situation.

1.3 Quality management situation

In recent years, the power industry began to flourish. Hydropower development has also entered a golden period. Currently hydropower resources across the country, especially in southwest China's hydropower resources, began large – scale development and construction, which makes domestic hydropower construction scale to an all – time high. A large number of sources began to development and construction, which brings enormous business opportunities to the hydropower construction enterprise. At the same time, it also brings management challenges, behind winning the bid with amounts and complete production, enterprise management and human resources failed to keep up with. And the quality of the project is to dry out by the construction companies, the emergence of these problems of construction enterprises has brought the engineering construction quality management more difficult.

2 The main content of quality management by the owner

According to the characteristics of the large hydropower station owner of quality management, quality management by the owner should mainly do the following aspects of the work.

2.1 Quality management plan before commencement of construction

After establishment of the power station construction management, we must carry out quality planning. The contents mainly include:

1) Determination of the quality of management objectives

The owner should base on hydropower station's own characteristics and difficulties of design and construction technology, to establish a clear power plant quality management objectives. Quality management is divided into specific targets and macroeconomic objectives, specific targets such as excellent rate, pass rate, acceptance rate for only 1 time. Macro objectives such as up to standard production, create quality engineering, a Luban Award or other local awards. The overall quality goals of the power station should be accord with the actual and realizability, to avoid the formulation of the high and the quality target that can not be achieved. Conforming to the actual quality management objectives would have practical guidance, and incentives for all units involved in the whole process of quality management.

2) Establishing the quality policy

The quality policy is always the quality purpose and direction of the organization by the organization's top management official release, is the guiding ideology and commitment of managers on the quality, is a concrete manifestation of the quality culture, it reflects the general quality management ideas and management philosophy of managers. It will be throughout the life span hydropower station construction, to guide all crew involved in the construction of hydropower station for daily quality management and quality construction. Quality consciousness of top management, often determines the level of quality awareness of the enterprise. it is reflected by the quality policy. Quality policy of the standard for the higher, there will be more high quality level of consciousness.

3) Explicit quality management system

Quality planning should be based on the characteristics and size of power plants, selected quality standard system for the power station. Quality standards are in the process of power plant construction, suitable for the unified standard of various professional and all contractors, which is carried out on the basis of various quality management activities, it is also an important factor controlling the construction cost. Each station construction, especially large – scale hydropower station construction has its own unique characteristics, and under the current hydropower development boom, the national various records and even world record had been broken, so in the selection of quality standards, the corresponding national standards or specifications should be chosen. But for large projects or special projects, if necessary, it should develop specific and suitable quality standard system. For the record – breaking large hydropower station, if necessary, there should also have their own quality standards on individual projects. But the quality of each power station itself should not be lower than the corresponding national standards. If the construction quality standards set too high, it will increase the difficulty of construction, and increase the construction cost and construction period, if the quality standards set too low, it will be adverse to the long running after the project construction.

4) Establishment of quality management agencies

Quality management planning also programme the quality management organization. According to the overall management of hydropower planning and management ideas, the owners should determine quality management organizations, define quality management responsibility, arrange the source of financial support for quality management. At present, in the domestic large and medium – sized hydropower station construction, the owner units generally set up the quality management committee, the members of the committee composed of the participation of the unit personnel. The director of the committee is generally the deputy general manager of the production of the owner units, it is the highest decision – making body of the owner of quality management. The Committee set up under the Quality Management Office, usually affiliated in the engineering department, responsible for the daily affairs of quality management committee, on behalf of the owners to perform quality management committee of the relevant decisions. To reflect the importance of the quality and enhance the quality of management, in some owner unit, quality management office will be set up independently, and have a certain staffing.

5) Financial security of quality management

It needs funds to carry out daily quality management activities. In the current system of national budget, if there is no quality management fee, the owner unit can show in the contract and be listed in the management fees. Quality management planning requires a clear source of funding and the direction and focus of the use of funds.

In addition to the above, quality management plan also be based on the engineering practice, the essential point of quality management and quality control measures, etc.

2.2 Contents of the quality plan after commencement of the project

For the quality management activities during the construction period, the most important thing is according to the requirements of the quality management plan and implementation of the contents of the planning of the quality management. Through the beforehand control, process control, quality problems processing, the goal of quality management in the planning of quality management is realized. Quality management activities are mainly as follows:

(1) Hydropower is the complex with more professionals and more projects. During quality management, according to various professions and various characteristics of the project, through the sensitivity analysis of project quality, according to the influence degree of the quality control points on the whole power station and the project itself, quality grading management system of the whole plant and various purposes will be made. For different levels of quality control points, classification management and hierarchical management, the quality grading management system should run through the whole process of quality control in advance, quality process control, quality problem disposal, quality evaluation, etc.

(2) The key point of the quality control is pre – control. On account of the determined the impact on the quality of the classification and the quality of power plant or project, determining the quality control points. By the analysis of various factors affecting the quality, efforts will be taken to strengthen the pre – control work of the quality pre – control issues. Factors affecting quality are factors of people, equipment, the environment, the use of materials, construction materials and methods. Analysis of factors might influence the quality from several aspects, the examination should be done in the project plan and the sub – parts of the above – mentioned various aspects, to do programming and standardization.

(3) Giving full play to the role of supervision. In the process mainly to strengthen control and management of several links, to strictly perform the procedure of sign recognition system, on – site supervision system will be carried out. , by ensuring the quality of each process to ensure the quality of projects and programs of each product. Supervision unit is fully authorized in quality management by the owner units, to establish supervision's authority of the quality management in the field, to create good conditions for the quality control management in the process of supervision.

(4) Correctly handling the quality issues. For the quality problems that have appeared, first of all, the reason is analyzed. According to the classification management system, the quality treatment scheme is determined. General quality issues must be confirmed by the supervision, the

quality problem of the large deal must be confirmed by the quality management committee, the quality of the major issues must be approved by the local government and the competent department for confirmation. After the deal confirmed, the implementation of the treatment scheme should also be incorporated into the scope of quality management. After processing of quality problem is completed and test qualified, the quality problem report is formed. In addition to the statement of the report on the causes of quality problems, quality issues deal with program and processing results. The most important is to sum up the thinking and experience of this kind of problem, and analyze the significance and guidance of the next quality control and management.

(5) Using the new technology, improving the level of quality management. Along with the social enterprise is paying more and more attention to the quality of products and the development of information technology, quality management technology has been developing rapidly. A lot of practical quality management technology and quality management tools, such as quality control charts, pareto chart, sampling inspection, etc is born at the right moment. New concept of the total quality management also gets promotion and application in the construction enterprises. In the process of quality management, it is dynamic management from making full use of the quality management tools and strengthening scientific quality management.

3 Establishment and perfection of quality management system by the owner

Quality management is a subsystem in the hydropower construction management system, compared with other subsystems, it has its own characteristics and relative independence. So, quality management can form a relatively perfect quality management system.

3.1 Three – grade – quality management system

Hydropower station owner unit of quality management shall establish a three grade quality management system including the owner, the supervision unit and the project contractor. Hydropower station owner units of the quality management system is based on the macro level of the entire project management, the main contents include execution of the supervisor's quality objective requirements, management of the quality department for the supervision unit and project contractor, the macro quality management of the hydropower stations. The quality management unit will be subject to contract tenders or to the professional division. The quality management system of supervision units is based on the quality of the supervision of the project management, the main contents are carried out by the hydropower station owner unit which include quality management objectives and quality management decisions, daily quality management for the supervision of the project, etc. The quality management unit includes two levels: a. the quality management which is divided by the project or the working surface; total quality management based on unit project or process. Project contractor's quality management system is the basis of the fundamental system, mainly includes quality objectives and quality management decision implemented by project owner units and supervision units, all levels within the scope of the project quality management, etc. The management unit is from each face, each specialized to each process, each link of all levels.

3.2 Organizational security and investment security of the quality management

Organizational security mainly refers to in all levels of management systems, the full – time quality management system and quality management personnel which adapts to strength of management should be set up. Quality management agencies also have some independence, in terms of quality management can achieve the quality of a veto. For example, the project owner unit set up quality committee and office, the supervision unit set up quality director, project contractors set up quality management, etc. Investment protection refers to all levels of the system should be quality management of capital investment, and quality management capital planning should be done well, used for the daily management activities and quality of incentive.

3.3 The quality's assessment and the incentive mechanism

A quality assessment and incentive mechanism should be established in all levels of the quality management system. The project owner considers each section as a unit, set quality amount of the fund according to the size of the project. And the owner unit set up quality management as a whole fund. If a unit of quality is not good, this part of the money will be rewarded to other well done units, strengthening incentives substances in larger dimensions. A similar mechanism of work area or construction group should be established for the unit all items in the contractor's quality management system. The use of the fund monthly will be recorded by owner units, and the internal quality management is strengthened from external supervision, to ensure the quality of investment in management.

4 Concrete measures for the owner's quality management

Hydropower construction projects from the decision – making to the acceptance of delivery, the owner will take the corresponding quality of means and management measures in view of the different emphasis in different periods. The quality control of construction project was finally completed and the expected quality target is reached. As the owner unit quality management, management measures are mainly in the following areas:

4.1 Quality control for design

To provide timely and accurate design drawings, to process engineering major technical problems, design work is the basis of engineering construction, is of great significance to ensure the engineering quality. The owner's quality objective requirements are: It should take a "unified planning, rational distribution, local conditions, comprehensive development, supporting the construction" approach, to be applicable, reasonable, economic, disaster prevention and safety. To achieve this goal, the following measures should be taken to control the design quality.

(1) According to the project requirements for approval, the information, the design outline or plan competition file is developed. Design bidding and scheme competition is organized, to evaluate design scheme.

(2) To carry out investigation, design and the qualification examination of research units, survey, design, research institutes, the investigation, design and scientific research units are preferred. Sign the contract and in accordance with the contract implementation, and to strengthen

the quality control of contract implementation

(3) Review of plan. Control design quality, design review to ensure that the project is designed to meet the requirements of the design brief, in line with national guidelines and policies relating to the construction, in line with current design specifications, standards, in line with national conditions and reasonable process, advanced technology, give full play to social projects, economic and environmental benefits.

(4) Review of design drawings. Design drawings are the results of the design work, but also the construction of the direct basis. So design stage quality control, eventually want to reflect on the review of design drawings.

Preliminary design drawings for review. The preliminary design is to determine what technical solutions. The focus of the review is: the technical proposal adopted by the conform to the requirements of the overall plan, whether to achieve the quality standards of project decision – making stage. Technical design drawings for review. Technical design is the concrete of the preliminary design technology plan. The focus of the review is: whether the professional design meets the predetermined quality standards and requirements. Construction drawings design for review. The construction drawing is the detailed drawings and descriptions of the engineering objects such as equipment, facilities, buildings, pipelines and so on. It includes the dimensions, layout, selection, structure, relationship, construction and installation quality of the engineering object, it is the direct basis for the construction, which is the design phase of a focus on quality control. The focus of the review is: the use of functional quality objectives and levels.

(5) Construction cooperation and completion acceptance. Owners organize the design unit to cooperate the construction, the task has two aspects: firstly, it is a problem occurred in the process of construction design that to solve the quality problem of the construction unit and the owners; Secondly, the design changes and processing budget revisions. Final acceptance is not only the construction quality of the final assessment, but also the final approval of the quality of design. There is a quality problem elimination period, when the design or construction quality problem is found during the acceptance period, it limits the period of design and construction units to eliminate quality problems within a time limit.

4.2 Quality control for supervision

One of the main contents of the construction supervision is the quality control of the construction project. In order to ensure the correct implementation of the project, as well as supervision units can effectively control the quality of the implementation process, commissioned supervision should be specific by the owners, and the supervisor should be given the corresponding powers. In terms of quality control, the following contents should be entrusted:

(1) Supervising and urging the construction unit to establish and improve the construction management system and quality assurance system, and supervise the implementation.

(2) Inspect the quality of the raw materials, semi – finished products, finished products, construction fittings and equipments, and carry out the necessary testing and monitoring.

(3) Supervising the construction unit in strict with the technical standards and construction

design documents, control project quality. Urge the construction unit to implement preventive measures at important project.

(4) Checking construction quality, to carry out inspection of the hidden projects and to participate in the analysis and disposal of the engineering quality accident.

(5) Organizing the pre-acceptance at engineering phase of acceptance and final acceptance, and the engineering construction quality assessment is put forward.

To ensure that the supervision unit has the ability to perform the supervision contract, the owner must review the qualification of the supervision units, optimizing the supervision units. On this basis, the owner based on the supervision and control of the above content specific implementation process, to make the supervision engineer better to perform the duties stipulated in the supervision contract, to achieve the purpose of controlling the quality of the project construction.

Supervision and control to supervision unit by the owners, mainly through the monthly supervision report and on-site supervision. One of the supervision monthly report contents main is controlling the engineering construction quality. It includes: the unit acceptance of work; current unit project once acceptance rate statistics; control chart of unit works excellent and good rate; acceptance situation of division of engineering; test situation of the construction; the quality accidents; suspension of construction instruction; quality analysis of the present phase (including the reasons of engineering quality problems in current engineering quality analysis and countermeasures of quality list). On-site supervision is that the owner appoints the stationed on-site management personnel. According to monthly supervision report, through on-site investigation, appropriate measures on the quality issue are taken to urge the supervisor and construction units, to jointly improve the quality control to achieve the expected goal of quality control.

4.3 Quality control for construction

Before commencement of construction, it should open tender for the selection of the contractor who is adapted to the tasks and signing the project contract. The owner's quality control is mainly commissioned supervision. According to the contract between the owner and the contractor, the supervisor will make the comprehensive supervision, so the project quality activities are completely in control of the contractor's supervision among the effective implementation of quality control. However, in practice, some of the problems cannot be solved by the supervisor and the constructor themselves, the owner also needs to do a lot of work. Quality control in the construction phase by the owners is mainly the following aspects:

(1) Determining the active control of factors affecting the quality during the project quality control process (including personnel, materials, equipment, equipment, construction order and methods, etc). After engineering quality control is explicitly, quality supervision organization will be further improved. Such as the owner can establish the quality management department, directly responsible for the supervisor and the constructor, and coordinate the relationship between them.

(2) Doing the quality inspection well and implementation of test methods. Quality inspection method comprises the following steps: operator self-inspection, mutual inspection within the

team, the handover inspection during the process, the inspector's inspection, and the quality supervision inspection by the owner, supervisor, the designer and government department.

(3) Acceptance of the division of engineering and the concealed engineering. For different types of division of the project, it should be checked by relevant departments. Because different types of branch engineering project have different content, quality inspection and assessment is different. It should be strictly in accordance with national standards, ministerial standards and industry standards.

(4) Reviewing quality problems (accident) report, to participate in the field of quality supervision meetings. When quality problems (accidents) occur in construction, it should be timely paid attention to prevent the induced major quality accident. By organizing special to take investigation and analysis of causes and characteristics, the report of engineering problem (accident) and treatment plan should be censored, which is filled by supervision and construction units.

5 Conclusions

Quality management of the large hydropower stations are an important part of management during the construction of power plants. In the present situation of large hydropower stations are built, it is a new subject that the project owner how to strengthen the management of construction quality. This paper argues that the owner's quality management is mainly to form a set of mechanism and system. The leadership attaches great importance to engineering quality, full participation in quality management, the whole process of strict control and scientific management are formed. The continuous improvement of quality and the improvement of the system are formed and good running.

The Impact of Primary Frequency Regulation to Henan Grid by Xiaolangdi Power Station

Li Yindang, Zheng Wei, Li Yiding, Cheng Chao, Zhou Ming

(Xiaolangdi Hydraulic Power Plant, Jiyuan, 459017, China)

Abstract: Frequency is one of the most important factor in power system, and primary frequency regulation is one function of Hydroelectric generating set by adjusting the frequency both statically and dynamically. Hydroelectric Generating Set in Xiaolangdi makes a great impact in Henan Grid because of its multiple advantages. However, there are still some restrictions when realizing all its functions, such as the dispatching regulation of determining the electricity by hydropower, governor's own factors, based on which, following suggestions have been proposed.

Key words: primary frequency regulation, governor, frequency response, restrictions, Henan Grid

1 Overview of Xiaolangdi Hydraulic Power Station

Xiaolangdi Hydraulic Power Station is a comprehensive large-scale hydro-junction project which locates in main yellow river, 40 km north away from Luoyang, Henan Province. There are six main functions: flood control, ice prevention, deposition reduction, irrigation, water supplement and power generation. It is equipped with six 300 000 kW - Francis Hydroturbine Generator Set, with a rated head of 112 m, and compiled into Henan provincial electric main grid. Its main task is to peak regulation and frequency regulation.

The installed capacity of Xiaolangdi Hydraulic Power Station occupies 4.2% of total installed capacity of Henan Grid, in which it occupies 49% of total installed capacity of hydropower. And it also occupies 70% of available installed capacity of hydropower in Henan Grid. Thus, Xiaolangdi Hydraulic Power Station makes a great impact in implementing the peak regulation and frequency regulation in Henan Grid.

2 The effect of primary frequency regulation in grid and the technical requirement for Hydroelectric Generating Set from grid

The primary frequency regulation adjust the governor to control frequency and keeps the balance between unit output and system load. Primary frequency regulation make quick response towards changes in system load and frequency and make sure that the frequency within a proper range. Hydroelectric Generating Set's main function in power system is to adjust frequency regulation and peak regulation. Compared with thermal power unit, it has the following advantages: the process of regulation is simple, the speed is high in load regulation as well as the high regulation range. Primary frequency regulation could make instant power support when a great load change happens in power system, which enhance the reliability of power system. And the regulation towards the load fluctuation in a short period could decrease the usage of secondary frequency regulation and optimize the regulation of system. In conclusion, the ability of generating set's primary frequency regulation has a direct influence towards the power system's proper operation.

Primary frequency regulation of Hydroelectric Generating Set is realized by the governor as well as its ability and function. The requirement for primary frequency regulation of Hydroelectric Generating Set from power dispatching center of Central China Power Grid has following requires: The Man-made Dead Band of primary frequency regulation should be constrained into ±0.05 Hz; permanent slip ratio≤4%; the restriction amplitude of load change should be constrained into ±10% of the rated load; the speed dead-band of governor should be constrained <0.04%; Hydroelectric Generating Set with a rated head more than 50 m should have a lag time of response from primary frequency regulation no more than 4 s; all the load adjustment of set's primary frequency regulation should reaches the 90% of the theoretical value no later than 15 s. In the first 45 s when the frequency change of power grid surpass the set's dead-band of primary frequency regulation, the average value between set's actual output and the response target bias should be within ±5% of the theoretical value.

When frequency bias equals 50Hz, the following formula is used to adjust the output:

$$\Delta P = \frac{\Delta f^*}{b_p} P_{max} \qquad (1)$$

In the above formula: Δf^* is the frequency bias and should be calculated by percentage, $\Delta f^* = \Delta f/f_0$; b_p is the factor of permanent slip; P_{max} is the maximum power of set converted by the limitation of opening.

Its response speed is determined by the transient slip of governor and the speed of servomotor. Generally speaking, start and shut down period of the hydroelectric generating set's guide vane is within 20 s, however, the setting of factor of primary frequency regulation such as full stop of guide vane, permanent slip ratio and temporary slip ratio are restricted by the adjustment calculation and can not be chosen arbitrarily; meanwhile, the dead-band of speed must be set up and within a small range of frequency fluctuation, the set shouldn't make any output so as to avoid the frequent operation of servomotor. When the change of frequency exceed a certain set value, governor should adjust the output of the set according to a previously set slope.

Parameters of Xiaolangdi Hydraulic Power Station: man-made dead-band is ±0.02 Hz; permanent slip ratio is 4%; the restriction amplitude of load change is 5% of the rated load; the speed dead-band of governor is lower than 0.04%; the lag time of load response is 0.15 s; period between the start and shut down of guide vane is 17 s. Those parameters above is much more higher than the technical requirement of power grid regulation.

3 The realization of primary frequency regulation

In Xiaolangdi Hydraulic Power Station, type VGCR211 digital governor is designed with a comprehensive frequency regulation strategy which could be used in various complex load working condition of power system to improve the steady operation of power system.

The speed adjustment system in Xiaolangdi take VGCR211 Duplex Microcomputer Hydroturbine Governor as the main support equipment of hydraulic turbine, the whole system is imported from American Company VOITH and employs traditional combination of hydraulic servo and regulator. The whole system is consisted of test unit, Duplex Microcomputer Hydroturbine Governor, VCA1 card, electrical position feedback, electrohydraulic servo valve and hydraulic servo. Servomotor and the position of main distributing value employ the electrical feedback circuit, the feedback of power is derived from the LCU power sensor. And it employs two independent PLC Duplex Microcomputer Hydroturbine Governors which are mutual relational. Both PLC have the same hardware and based on the software with the same algorithm. If PLC1 encounters a problem, the system will be switched to PLC2 by the WATCHDOG. Each PLC is equipped with two CPU and used on speed controlling, power and opening controlling (opening restriction). It can be switched easily among three modes to work in parallel operation and isolated operation. The collection of power grid's frequency could be switched between two modes: ether or not influenced by the frequency. Three modes are implemented individually in different stage of parallel running of sets. In the starting period, the function of opening controlling comes into use. While the speed of sets reaches the rated speed, the function of speed controlling comes into use and will make the non differential regulation according to the system's frequency. In the stage of paralleling running, the governor will be switched into power controlling, and follows the given value of power.

Under each of the three modes above, it must follow minimization principle, that is to say, the restriction value is determined by the minimum restriction value determined by any of the three modes. The sets could implement speed control under both load and non – load condition. Then speed controlling employs parallel PID regulation principle. The permanent slip ratio factor is derived from the factors after integration so as to eliminate the error caused by integration. The speed controlling has sets of PID parameters, and are used individually under non – load mode, isolated running mode and parallel running mode. Either when the system frequency exceeds 2% n_r (rated power), or when the sets are broken from the system and the governor is affected by the staff, the governor could always be switched from paralleling running mode into isolated running mode. If the system frequency are within the proper range again, the governor can be switched from isolated running mode into parallel running mode only under the condition of staff affection.

4 The duration of performance of primary frequency regulation

In normal condition, system frequency will be within the range between 98% and 102% of the rated power, thus, the speed controlling will always use the parallel running mode. When the system load increases dramatically or the circuit breakers from other stations shut down, the system power will decrease sharply, and under such condition, the governors of Xiaolangdi will automatically begin primary frequency regulation, increase the generating sets' output to increase the system frequency. If the gap between the system frequency and the rated frequency is still out of range, the station staff can increase the guide vane's opening degree by increasing the previous set speed, and finally enhance the set's output until it reaches the frequency requirement, while this issue is actually belongs to the secondary frequency regulation.

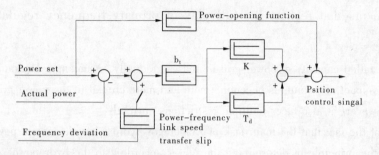

Fig. 1 Xiaolangdi Hydropower Station VGCR211 governor power control mode schematic diagram

Under regular condition, when the sets are in the mode of paralleling running, the governor will implement the mode of power control so as to maintain output always meet the demand of the given power. The VGCR211 governor in Xiaolangdi can import the system's frequency signal when the sets are using power control (Fig. 1). System frequency could have an impact on the set's output. The function of Frequency Influence enables sets to be involved in adjustment of system frequency even in the mode of power controlling. In Fig. 1, it can be told that the frequency – power close – loop is the stage of Frequency Influence. It is conclued that frequency bias could make an impact on the opening degree of guide vane, moreover, have an influence on set's output. Rs is the speed adjustment ratio of Frequency Influence, and this value is different from the value of permanent slip ratio bp when in mode of speed controlling. When the Frequency Influence comes into use, it follows the Formula(2):

$$P_{Totolset} = P_{Set} + \frac{(n_r - n_{Actul})}{n_r R_s} \times P \tag{2}$$

In Formula(2): $P_{Totolset}$ is the total given power, P_{Set} is the man – made given power, P is the set's rated output, n_r is the rated speed, n_{Actul} is the actual speed. According to the Formula(2), under the power controlling mode, when the system frequency affects the function, the generating set's output will adjust automatically according to the system frequency.

Frequency Influence function in Xiaolangdi could increase or decrease the set's output according to the power – frequency characters. In the mode of power controlling, the slip ratio equals to 6%. In this case, when the actual frequency of system is lower or higher than 0.1% of the rated power, the sets will increase or decrease 75 MW output. When it $(\frac{n_r - n_{Actul}}{n_r})$ is below 0.1%, or in other words, a man – made malfunction range with a value of 0.1%, that means to guarantee the steady operation of governor system and sets when the system frequency's fluctuation is very low. When the system frequency bias is beyond the dead – band of frequency adjustment, Frequency Influence function will make adjustment towards the output according to the change of system frequency in spite of the affection from the staff. Thus, the frequency regulation under the mode of frequency controlling is still an issue of primary frequency regulation while this function could be controlled manually by the button of "Artificial Band" in host computer and the "Freuency Influence" button on the governor's control panel.

5 The elements that restrict the full play of primary frequency regulation in Xiaolangdi

As the installed capacity of power grid increases gradually, Xiaolangdi Hydraulic Power Station's positive impact and status is becoming more and more crucial in the power grid's frequency regulation. However, it also faces multiple restrictions currently.

Because of the fact that the main task of Xiaolangdi hydro – junction is not power generating and it follows the principle of determining the power according to the hydropower, which means that the power amount generated in one single day is determined by the discharge volume, the start – up mode is now facing severe restrictions. Meanwhile, due to the fact that the installed capacity of thermal power unit in Henan Power Grid is becoming much more larger, the demand of hydropower is now becoming more and more obvious, which leads to a severe conflict between water regulation and power regulation. Such situation restricts the full performance of Xiaolangdi's frequency regulation usage in a way.

Primary frequency regulation depends on governors, thus, the ability and performance of governor affect the result of a frequency regulation hugely. Since Xiaolangdi's governor has been put into operation, Generator was forced to stop because of governor twitch which due to the quality of oil quite often.

Meanwhile, the settings of several parameters of the governor such as parameters of VCA1 card, active power signal collection of the sets also affect the performance of the primary frequency regulation negatively.

6 Some suggestions about solving the restrictions towards full performance of primary frequency regulation in Xiaolangdi

From the power grid's perspective, the main task is to enhance the stability of power grid, furthermore, to enhance the relationship between regional power grid. Meanwhile, establishing pumped storage power station with proper capacity, fully expressing its ability in regulating the system is also vital to the stability of load and frequency of the power grid. We should also enhance the research of improving the ability and performance of primary frequency regulation of thermal power units, and develop larger capacity, higher performance thermal power units .

From our station's perspective, we should adjust the restriction towards the discharge volume properly when it is needed by the power grid, fully express the excellent ability and performance of Hydroelectric Generating Set, make full use of anti regulation function of Xixiayuan reservoir and further the study about combined optimal operation between the two reservoirs. At the same time, the maintenance for the equipment must be carried out. That is, check the oil quality in the governor system and filer them regularly so as to avoid the negative affection towards the sets caused by the oil problem. We should also choose the optimized operation parameters of VCA1 card and make the performance of governor at its best condition. Finally, we should enhance the stability and accuracy of the active signal's collection and use multiple ways to do that as a back-up.

7 Conclusion

In fact, Xiaolangdi Hydraulic Power Station is not the main frequency regulation power plants in Henan Power Grid. But based on the fact that the deficiency of hydroelectric resource of Henan Power Grid and the fact that the frequency regulation ability of Xiaolangdi Hydraulic Power Station is extremely Superior, the switch from back-up to work of the sets only takes 3 min, and within 10 s after it is put to work, it could gives a 300 MW output and the up-down load rate could reaches 30MW/s compared with those 600 MW thermal power units with better performance in the provincial power grid, which only could reaches 15 MW/min at maximum, it is concluded that Xiaolangdi Hydraulic Power Station is both qualified and suitable for burdening the main task of peak regulation and frequency regulation in Henan Power Grid. In the period of ninth water and sediment regulation in Yellow River which happened in 2010, in order to deeply implement and ensure the national principle of energy-saving and power generation, make full use of Yellow River's hydropwer resources, the power dispatching center of Henan Province arrange Xiaolangdi Hydraulic Power Station to make full operation all day. Although 37 thermal power units with AGC functions and a total capacity of 1 520 MW were put into work by Henan Power Grid, they still cannot reach the same results of peak regulation, frequency regulation and speed which was achieved by a single set of Xiaolangdi Hydraulic Power Station. According to the actual operation condition in recent years, the positive influence that Xiaolangdi Hydraulic Power Station made to Henan Power Grid is becoming more and more vital, especially that it handled a series of abnormal situation such as the '7.1'power grid shock properly. In conclusion, increasing the operation stability of Xiaolangdi Hydraulic Power Station and improving its frequency regulation performance is much more crucial to the operation of Henan Power Grid.

On – line Monitoring of High – speed Flow Sediment Concentration in the Sediment Tunnels of the Xiaolangdi Project on the Yellow River by Vibration Method

Song Shuke, Zhang Jinshui, Ma Zhihua, Xin Xingzhao

(Xiaolangdi project construction & management center, Zhengzhou, 45000, China)

Abstract: The sediment concentration of the high – speed flow through the Sediment Tunnels of the Xiaolangdi Project on the Yellow River is close to 1 000 kg/m^3 at utmost; the sand grain size distribution is relative thin; the water flow is three – phase by water, sand and air with flow speed up to 40 m/s. All of these bring huge challenges on the on – line monitoring of the sediment concentration.

In this paper, the vibration method is used to realize on – line monitoring of the sediment concentration contained in the hyper – concentration flow. The sampling obtained from the device in the high – speed flow is diverted into the fixed steel tube, through monitoring the changes of the steel tube natural frequency to realize the online measurement of high sediment concentration in high speed flow, which provides the very important monitoring information for the water – sediment regulation of the Yellow River, sedimentation reduction, and regulation of the reservoir.

Key words: High speed flow, sediment concentration, vibration method, on – line monitoring

Xiaolangdi Project is located in the last valley in the middle reaches of the Yellow River, the only controlling project obtained larger capacity downstream of the Sanmenxia Project, controlling the Yellow River basin area of 92.3% and nearly 100% of sediment load. A large amount of sediment deposition in the reservoir area is discharged to the downstream river channel mainly during the flood season of every year during the water – sediment regulation and through three sediment tunnels. Therefore, the sediment concentration monitoring through sediment tunnels of the Xiaolangdi Project provide the basic data for the water – sediment regulation, the reservoir sedimentation reduction, and the reservoir regulation, which has great significance on carrying out the water – sediment regulation and flood control.

1 On – line monitoring method on sediment concentration

1.1 Water flow characteristics of Sediment Tunnles of the Xiaolangdi Project

There are 3 sediment tunnels in Xiaolangdi Project. The radial gates control the muddy water flow. It is pressure flow before the radial gate, and free flow behind the gate. The free flow is discharged through a channel with 35 m long. Pressure flow is two phase flow by water and sand, but free flow is three-phase flow by water, sand and air, with speed up to 40 m/s, the air is form by air added artificially and surface permeability air to prevent cavitation. The flow sediment concentration through the sediment tunnels of the Xiaolangdi Project is high. According to the actual measurement, the highest value close to 1 000 kg/m^3, sand grain size distribution is relative thin. These features bring enormous challenges on online monitoring sediment concentration.

1.2 Monitoring methods on sediment concentration

According to the different measuring principle, the sediment concentration measurement methods can be divided into direct method and indirect method. Direct measurement methods include drying method and specific gravity method, characterized by sampling measurement, high accuracy, but can not be real-time online measurement. Indirect methods include infrared ray method, spectro-photometry, capacitance method, the vibration method, ultrasonic method, laser method and γ ray method, etc. γ ray method and vibration method in the indirect measurement methods are quick, easy, accurate and reliable measurement as the field applications, which can better realize monitoring on the sediment concentration. The other indirect measurement methods are only suitable for measurement on low sediment concentration.

The basis of measuring sediment concentration by ray method is that the transmission intensity of γ ray will reduce after refraction, scattering and absorption of sediment particle in flow. γ ray method is a quick, easy, accurate way to measure the sediment concentration, but there are some limitations in dynamic measurement, the grain size distribution change and requirements by environmental protection will restrict the method in the application of the Xiaolangdi Project.

Vibration method, using the vibration theory, based on vibration period of resonance rod in different sediment concentration to derive different concentration in the mud. In a certain composition of the sediment proportion and the particle size, and the same sediment particle velocity, the resonance rod vibration period T (s) and sediment concentration ρ approximate linearly proportional relationship is: $\rho = aT + b$. Among them: a, b are constant, can be decided in advance rate through the experiment. The test for a resonance rod with certain material, shows that rod density is proportional to the square of the vibration period. In the actual measurement, movement of the rod is greatly influenced by water depth, flow velocity. Therefore, the metal pipe is, instead of resonance rod, the commonly used measuring equipment. Then the materials and volume of the pipe is fixed, the density of the measuring tube is completely determined by the density of fluid in pipe. If the sediment concentration changes, that means that the density in the whole pipe has changed, the tube body vibration period also will change. At this time, measuring the density of mud, from the relationship between the vibration period and density of the pipe body, sediment concentration can be calculated. Yunnan University and Yunnan Hydrological Station cooperatively developed ZN-1 type vibrating sediment concentration measuring instrument, its vibration pipe is installed in the lead stomach, and make the flow freely through the vibration tube. It is mainly used for low velocity measurement environment, when the sediment concentration is 10 ~ 830 kg/m^3, the relative error is less than 5% of the cumulative frequency 85% of measuring points. The High-tech engineering research and development center, Institute of Water Resources and Hydropower Research of the Yellow River Conservancy Commission based on real-time monitoring on sediment content in Sanmenxia Project for a long time, developed a type MDS51 on-line monitoring device. The device adopts the sampling measurement with measurement range $0 \sim 1\,000$ kg/m^3, a resolution of 0.1 kg/m^3, and it is more suitable for the online measurement of high-speed, high sediment concentration flow.

2 Realization of online monitoring on sediment concentration

2.1 Composition of the online monitoring system

MDS51 type on-line monitoring device is composed by three parts, i.e. sediment concentration meter, water drawing equipment, water air separation and measurement device, as shown in figure 1, the key technology is reasonable water sampling.

The flow to be measured the sediment concentration through the water device is transferred to the device of water air separation and measuring by pipeline, the sediment concentration meter installed in the monitoring center, is connected through the cable with sediment concentration measuring device, installed in south side wall of the Sediment Tunnels exit vertical to the dam. The sensor output frequency signal into the measuring instrument after the photoelectric isolation, is sent to the computer, to accurately detect frequency and to process the signal, and calculate the sediment concentration. The sediment concentration signals on the one hand are transferred to the D/A converter, 4-20 mA current signal output (or 0 to 5 V voltage signal) monitoring; on the other hand to the digital display, display sediment concentration values. The measuring instrument is equipped with digital switch, used for measurement and calibration, sediment concentration of the measuring range of $0 \sim 1\,000$ kg/m^3, precision can reach $+/-1\%$ FS.

2.2 Water drawing device and water and air separation device

Water drawing device installed at the exit of the Sediment Tunnels, is composed of elastic turbulence and funnel. The high speed flow contained sands at the exit of the Sediment Tunnels became mainstream, also form a dispersed flow. To stabilize the flow discharge, a disturbance device (Figure 2), changing the direction and reducing the velocity of the high speed flow to become a low-speed flow into the designed funnel to collect the muddy water to the pipeline, through the nozzle to water and air separation device. Due to the velocity reduced, rising air overflow, water and air separation is realized. After the separated air is removed by air hole, the water flow contained sands access to the device for measuring and flow out. When the flow is larger, the redundant muddy water will flow out from the overflow hole.

The sensing device is composed of measuring container and sensors shown in Figure 3. The measuring vessel has a larger top diameter and a small bottom one, the upper top mounted a vent. The sensor device is installed in the lower part. When the drawing muddy water into the sensor device, with the same flow discharge, the larger diameter, the velocity is further reduced, which is more advantageous to bubble rising out. When the flow discharge is greater than the pipe drainage discharge, the vent discharges water and air, the sensor installed in the lower part will meas-

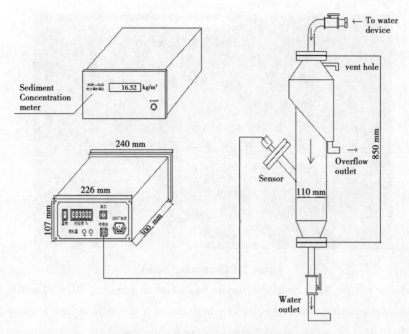

Figure 1 Sediment concentration in Sediment Tunnels on‑line measurement system schematic diagram

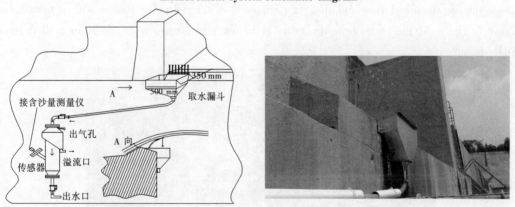

Figure 2 Installation plan diagram and the water device photo

ure muddy water sediment concentration. Through the twice water and air separations by funnel and sensing devices, excluding the muddy waters without air, thus realize online monitoring of sediment concentration.

2.3 Test and improvement of water drawing device

The Siphon method is used to taking the water detained in the hole to the water funnel, and the sensing device by a 10 m long led tube and then into the drain discharge, the sediment concentration meter display normal and stable, the vent shows the moisture separation normal. Muddy of about 700 kg/m^3 sediment concentration is poured into the funnel, and it is observe if any deposition in the 10 m led tube. The data displayed on the measuring instrument and the calculated value from sampling (displacement method) for the sediment concentration value are almost the same, with about 20 s delay from the funnel to the sensing device production.

Figure 3 The sensing device

The drainage test for No. 3 Sediment Tunnel was made on June 26, 2014. The site actual situation shows that the elastic turbulence device can draw water normally at both beginning end stages of flow discharge. However, at discharge into the middle stage, the velocity increases, effect is poor due to the elastic turbulence, very little water out. Through the laboratory simulation, using a tubular disturbed flow and then to compose the water drawing device with a funnel, as shown in Figure 4, the vortex tube diameters of 50 mm and 80 mm respectively in a 30 degrees Angle layout.

Figure 4 Water funnel tubular disturbed flow device (before discharge)

During June 30 ~ July 4 the discharge water is clean. By observation the water device composed by the tubular disturbed flow and funnel is operated normally. The discharge flow from the sensing device is normal, but the transferred data appear zero value sometimes. After analysis showed that when the drain valve is fully open, the drawing flow and the drainage flow at the end part is in critical condition, when the drawing flow discharge is insufficient, bubbles will be produced, the measuring instrument remote transmission value will be 0. Through improvement and perfection, it will meet the needs of sediment concentration in the Sediment Tunnels' flow on –

line monitoring, can be put into formal operation.

On July 9, after discharging flow is over, on-site inspection found the four tubular disturbed fluid roots installed remaining only three, as shown in Figure 5. The one washed away, through inspection, was not caused by loose screw and nut, but at the design phase, in order to increase the effect of elastic turbulence when planting bar set aside a 40 cm long screw. The screw moved frequently under high speed flow for a long time, causing fatigue, screw root fracture damage. While other screws are less than 10 cm long, were not damaged after the high speed flow impact.

Figure 5 Water funnel tubular disturbed flow device (after discharge)

3.3 Monitoring effect

In 2014 the water-sediment regulation of the Yellow River last 12 days from June 28 to July 9. By installing a tubular disturbed flow and funnel combination, water can be taken out stably for long-term from high speed flow. The sensor is stable, the meter works well, and the data can be transferred for long distance. From 23 o'clock on the night of July 5 to July 9, 8 times man-made samples were compared, the results are shown in Table 1. From Table 1 it can be seen that the result is consistent with the artificial replacement method sampling monitoring results. The vibration method for on-line monitoring sediment concentration of high speed of sands contained stream flow is successful.

Table 1 The sediment concentration of No. 3 Sediment Tunnel 3 (unit: kg/m^3)

Time	Online Value	Man-made sampling Value	Deviation
2014-7-5 10:02	4.9	4.7	0.2
2014-7-5 12:20	11.9	7.9	4
2014-7-5 15:00	31.4	31.8	-0.4
2014-7-5 16:10	64.4	68.1	-3.7
2014-7-5 17:03	99.2	103.4	-4.2
2014-7-6 10:35	78.5	74.2	4.3
2014-7-7 08:50	75.3	73.8	1.5
2014-7-9 09:10	29.5	30.7	-1.2

Conclusion

To lead the high – speed flow real – time sampling into the fixed steel tube through monitoring the change of the steel tube natural frequency, the online measurement of sediment concentration for high sediment concentration and high speed can be realized. Through practice, changing the elastic turbulence mode into tubular disturbed flow mode the key sampling problem under the condition of high speed flow was solved, water and air separation function of the funnel and sensing devices is good, meet the needs of measuring. It has proved that the data by the sediment concentration meter measured are accurate and reliable, and can provide the important on – line monitoring information for the water – sediment regulation of the Yellow River, the reservoir sedimentation reduction, and scheduling.